Advanced Physical Oceanographic Numerical Modelling

NATO ASI Series

Advanced Science Institutes Series

A series presenting the results of activities sponsored by the NATO Science Committee, which aims at the dissemination of advanced scientific and technological knowledge, with a view to strengthening links between scientific communities.

The series is published by an international board of publishers in conjunction with the NATO Scientific Affairs Division

A Life Sciences B Physics	Plenum Publishing Corporation London and New York
C Mathematical and Physical Sciences	D. Reidel Publishing Company Dordrecht, Boston, Lancaster and Tokyo
D Behavioural and Social Sciences E Engineering and Materials Sciences	Martinus Nijhoff Publishers The Hague, Boston and Lancaster
F Computer and Systems Sciences G Ecological Sciences	Springer-Verlag Berlin, Heidelberg, New York and Tokyo

Series C: Mathematical and Physical Sciences Vol. 186

Advanced Physical Oceanographic Numerical Modelling

edited by

James J. O'Brien

Secretary of the Navy Professor, Meteorology and Oceanography,
The Florida State University, Tallahassee, Florida, U.S.A.

D. Reidel Publishing Company

Dordrecht / Boston / Lancaster / Tokyo

Published in cooperation with NATO Scientific Affairs Division

Proceedings of the NATO Advanced Study Institute on
Advanced Physical Oceanographic Numerical Modelling
Banyuls-sur-mer, France
June 2-15, 1985

Library of Congress Cataloging in Publication Data

NATO Advanced Study Institute on Advanced Physical Oceanographic Numerical Modelling (1985 : Banyuls-sur-Mer, France)
Advanced physical oceanographic numerical modelling.

(NATO ASI series. Series C, Mathematical and physical sciences; vol. 186)
"Published in cooperation with NATO Scientific Affairs Division."
"Proceedings of the NATO Advanced Study Institute on Advanced Physical Oceanographic Numerical Modelling, Banyuls-sur-Mer, France, June 2–15, 1985"—T.p. verso.
Includes index.
1. Oceanography—Mathematical models—Congresses. 2. Oceanography—Data processing—Congresses. I. O'Brien, James J. II. North Atlantic Treaty Organization. Scientific Affairs Division. III. Title. IV. Series; NATO ASI series. Series C, Mathematical and physical sciences; vol. 186.
GC150.2.N38 1985 546'.00724 86–20203
ISBN 90–277–2329–X

Published by D. Reidel Publishing Company
P.O. Box 17, 3300 AA Dordrecht, Holland

Sold and distributed in the U.S.A. and Canada
by Kluwer Academic Publishers,
101 Philip Drive, Assinippi Park, Norwell, MA 02061, U.S.A.

In all other countries, sold and distributed
by Kluwer Academic Publishers Group,
P.O. Box 322, 3300 AH Dordrecht, Holland

D. Reidel Publishing Company is a member of the Kluwer Academic Publishers Group

---

Printed in The Netherlands

Dedication

To Bob
    who taught me to teach myself

To Sheila
    who encouraged me for a lifetime

To Ruth
    who supplied the support to complete this book

TABLE OF CONTENTS

CONTENTS (continued)

Chapter 14

Chapter 15

Chapter 16

PREFACE

This book is a direct result of the NATO Advanced Study Institute held in Banyuls-sur-mer, France, June 1985. The Institute had the same title as this book. It was held at Laboratoire Arago. Eighty lecturers and students from almost all NATO countries attended.

The purpose was to review the state of the art of physical oceanographic numerical modelling including the parameterization of physical processes. This book represents a cross-section of the lectures presented at the ASI. It covers elementary mathematical aspects through large scale practical aspects of ocean circulation calculations. It does not encompass every facet of the science of oceanographic modelling. We have, however, captured most of the essence of mesoscale and large-scale ocean modelling for blue water and shallow seas. There have been considerable advances in modelling coastal circulation which are not included. The methods section does not include important material on phase and group velocity errors, selection of grid structures, advanced methods to conservation in highly nonlinear systems, inverse methods and other important ideas for modern ocean modelling. Hopefully, this book will provide a foundation of knowledge to support the growth of this emergent field of science.

The NATO Advanced Study Institute was supported by many organizations. The seed money, of course, was received from the NATO Science Committee. Many national organizations provided travel money for participants. In France, CNES, IFREMER, and CNRS provided funds to support the French participants. In the U.S., NSF (Oceanography Section) and ONR provided support for several U.S. participants. Partial support was received by some participants from Belgium, Canada, Norway, Germany and the United Kingdom.

For my own research support, I am particularly grateful to the Office of Naval Research and the National Science Foundation which have generously supported my research group over the past 15 years. The continuous support by ONR and NSF has permitted me to train 15 Ph.D. students and make many discoveries in oceanic and atmospheric science.

The ASI and this book could not have been completed without the dedicated help and hard work of many people. The Co-director of the ASI was Dr. Michel Crepon, France. He was very instrumental in arranging many aspects of the Institute, especially the ambience and organization. I could never have done it without his help.

Dr. Bernard Saint-Guily was the local host in Banyuls-sur-mer. Everyone appreciates his great personal attention to the details of the Institute. Mrs. Ruth Pryor assisted in numerous ways from the emergence of the plan to conduct the ASI to the final preparation of the book. Her expert organizational and management skills were responsible for a successful meeting. In Banyuls, Madame Clara was superb in arranging the accommodations and social events. Nicole and her husband were extremely friendly and helpful to all the international visitors. We all applaud them.

I would like to thank all the speakers who cooperated and made the ASI a wonderful historical event in oceanography. In addition I thank the wonderful group of students who demonstrated an indepth love for learning. Here I thank them again for the excellent fishing rod with which I have captured many fine fish.

During the preparation of the book, Rita Kuÿper performed outstanding technical typing and other secretarial duties. I award special thanks to Mia Shargel and Judy Weatherly for the technical editing of all the manuscripts. Mrs. Shargel in particular worked many long hours in organizing the technical details of the manuscript. It could not have been done without her. I thank Mark Luther for his assistance with figures and technical assistance.

Over the past two years numerous other persons besides those cited above helped me with the ASI and the book. I thank each of you and apologize to those whom I have not mentioned personally.

I sincerely hope that this book will be used to teach the next generation of ocean modellers the mathematical and physical aspects of this subject. We have tried to provide the basic concepts and current ideas in many areas for using modern high speed computers in understanding the ocean's physical structure.

James J. O'Brien
The Florida State University
Tallahassee, FL 32306, USA
May, 1986

INTRODUCTION

J. J. O'Brien
Mesoscale Air-Sea Interaction Group
The Florida State University
Tallahassee, FL 32306-3041
USA

1. BACKGROUND

Modelling the ocean using large mainframe computers is a young science. After WWII, John von Neumann encouraged the meteorological community to use the first general purpose computers for numerical weather prediction. For the past 35 years, atmospheric scientists have had early access to each generation of faster computers. In meteorology, the majority of graduate students have used computers to test models or analyze large data bases collected by the world wide meteorological community or special field programs. In oceanography the story is quite different. Scientists at oceanographic institutions did not encourage their students to use the computer for models. However in meteorogical laboratories, such as GFDL, NCAR, NPG, UCLA, etc., several meteorologists were enlisted to build ocean circulation models. They anticipated the requirement to build coupled ocean-atmospheric models to study the earth's climate. After 1970, a small number of university professors in the U.S. and Europe recognized that fast computers are powerful for understanding ocean circulation using numerical solutions of partial differential equations.

By 1980, considerable progress had been made and the international oceanographic community began to support the investment of resources for ocean modelling and fast computers. The promise of a new age in understanding ocean variability is at hand. Since 1975, there has been a doubling of trained ocean modellers every 5 years. Now there is a need for monographs to teach and explain the science of ocean computer modelling. This book attempts to fill such a need.

There are several important physical differences between large scale atmospheric flow and ocean circulation. The atmosphere has larger space scales and smaller time scales than the ocean. A meteorologist will be content to use 100-200 km grid boxes and integrate for a few days to a month; the oceanographer should use grid boxes of 2-10 km. The meteorologist will be able to estimate good initial conditions; the ocean modeller will rarely have good initial conditions from data. This places severe constraints on the

J. J. O'Brien (ed.), Advanced Physical Oceanographic Numerical Modelling, 1–3.

problems which are attempted in ocean modelling. Most ocean models are not global but regional, and most ocean models have poor vertical resolution due to limitations in available computer resources.

Another problem arises. The domain may be not an enclosed basin but have open boundaries. Special attention to boundary conditions is a requirement in ocean modelling.

In order to model coastal regions even more horizontal resolution is required. It is not unusual to use grid sizes of 1-2 km (or less) in complex coastal circulation problems or estaurine circulation studies.

The ocean modeller has one advantage. Most ocean flows are wind driven. A good knowledge of the wind field and bottom topography allows the modeller to calculate very realistic ocean current patterns. One does not have to rely on esoteric formulations of turbulent parameterizations. This will probably change in the future as ocean models are developed that resolve the details of observed ocean data.

There are many scales of ocean motion which may be modelled using computers. This book tries to cover several, however, there is an emphasis on large and meso-scale ocean modelling. The remainder of this chapter contains two additional introductory articles on ocean modelling by M. Cane and B. Semtner. The latter traces the history of the development of large world ocean circulation models.

The rest of the book is divided into three sections. Chapters 2-6 are introductory chapters on the mathematics and numerical analysis needed for understanding ocean modelling. Chapters 7-12 discuss special types of ocean models and problems.

The remainder of the book covers special aspects of ocean modelling. Chapter 2 is a review of some mathematics used in subsequent chapters. Chapter 3 covers finite difference methods for elliptic partial differential equations. The first part reviews iterative methods. The more difficult and special direct methods are outlined by M. Luther. The application of multi-grid methods and the island problem are discussed by T. Jensen. Proehl, et al. give an application to equatorial oceanography.

Chapter 4 uses parabolic partial differential equations to introduce the motion of linear stability analysis. A special application section by Gargett discusses mixing in the ocean.

Chapter 5 reviews various time integration schemes and introduces artificial viscosity. Chapter 6 reviews finite difference schemes for hyperbolic partial differential equations such as the advection equation and the shallow water wave equations.

Chapter 7 begins the second part of this book. Semtner explains the formulation of world ocean models. Holland reviews quasigeostrophic models. B. L. Hua applies these for important GFD turbulence problems.

Chapter 9 is a special set of papers on using vector computers for ocean modelling. Ronday, Luther and Klinck explain the special programming techinques for using vector computers efficiently.

Chapter 10 is on world tidal modelling. Chapter 11 is on shallow sea modelling of tides and currents. Chapter 12 is a discussion of the important subject of open boundary conditions.

The remainder of the book is a series of papers on various aspects of ocean modelling. This monograph does not cover all aspects of ocean modelling; the reader should consult the literature for additional information.

Should this book be used as a text? I suggest that second year graduate students may find this a useful book to learn the essentials of ocean modelling. If the students have some background in physical oceanography, partial differential equations and numerical analysis, a good one semester course can be fashioned from this monograph. It is suggested that Chapters 2-6 plus 9 and 12 be used for the course. The instructor may then choose from the remainder of the book those topics of particular interest to the class. There are many homework problems inserted in Chapters 3-6 for the student to solve. It is also useful if the students know Fortran and have access to sufficient computer time to test those methods used in ocean modelling and outlined in this book.

INTRODUCTION TO OCEAN MODELING

Mark A. Cane
Lamont-Doherty Geological Observatory
of Columbia University
Palisades, NY 10964

1. OVERVIEW

The fundamental physical laws used in numerical models of the ocean circulation are known: the foundation is the Navier-Stokes equations. However, every ocean model is only an approximation of the complete physics. Sometimes simplifications are deliberately introduced in order to isolate a subset of the physics believed to be essential to the phenomena under study, but in all cases computational feasibility demands approximation.

As a discipline, numerical ocean modeling is largely concerned with how to choose approximations and how to analyze their consequences. In choosing, one must consider both those approximations that alter the physical system and the approximations involved in converting the continuous equations that describe that system into a discrete set of equations which can be integrated numerically. This chapter is concerned with the former and chapters 3-6 with the latter, but, as discussed below, the choices in either step have implications for the other.

Models are used, broadly speaking, for two purposes: understanding; and simulation or prediction. Of course, the two uses are not totally distinct. If one is using a model to understand the ocean it is comforting to know that it is capable of realistic simulations. Yet, it is difficult to achieve realistic simulations - especially of new conditions - without understanding what physics is essential and how to model it. In principle, however, one could achieve understanding with a carefully simplified idealized model or use a trustworthy simulation model as a "black box" to study such things as CO_2 warming or fisheries.

Whatever its use, running a model amounts to conducting a set of experiments much like laboratory experiments. Occasionally a properly designed set based on an *a priori* hypothesis yields an immediate result, but more often, especially with simulation experiments, the model runs *per se* give little or no insight. Interpretation is required. It may be that making sense of numerical model results has become more of a methodological challenge than the

J. J. O'Brien (ed.), Advanced Physical Oceanographic Numerical Modelling, 5–21.

making of models. In contrast with a field experiment, one may be daunted by the recognition that the source of the numbers is, after all, only a model.

Yet, there are advantages in doing experiments in a model. First, all the data you could possibly want is available; second, you can vary parameters, thereby doing controlled experiments rather than observing only those few cases nature selects for you. Both meteorological experience and common sense indicate that the most powerful use of models will be in conjuction with data (see the chapter on data assimilation for a discussion of how this might be done).

Since numerical modeling is relatively new to oceanography, it is worthwhile to look for lessons in numerical weather prediction (NWP), an area which has a longer history and is probably the most highly developed example of simulation/prediction modeling in geophysical fluid dynamics (GFD). There are many ways to predict the weather -- many sorts of "models": "feeling it in your bones" or other intuitive methods; persistence (assuming that tomorrow's weather will be like today's); statistical methods, such as linear regression on today's temperatures; analogues (find a past weather map like today's and assume that tomorrow's weather will follow as in the past case); and forecasts of the evolution of the atmosphere from an observed initial state on the basis of physical laws, that is, by using the hydrodynamical equations that govern the atmosphere.

The forecasting method, by far the most effective, is the one we will be concerned with. The governing equations are nonlinear and cannot, as a rule, be solved by analytic means but require approximate numerical methods. This is true of many other GFD problems, including those in oceanography. The idea of predicting the weather on the basis of physical principles - using the dynamical equations for all they are worth -- was apparently first stated explicitly by V. Bjerknes in 1904. In 1922, Lewis Richardson, in what is surely the most remarkable meteorology book ever written, presented the details of a scheme for "Weather Prediction by Numerical Process."

Richardson thought of almost everything; one could construct a general circulation model (GCM) on the basis of this book that would not be very different from modern GCMs. The summary chapter could serve for a modern NWP course; his point of view is quite modern. In his introduction, he states the fundamental idea of NWP:

> "The fundamental idea is that atmospheric pressures, velocities, etc. should be expressed as numbers, and should be tabulated at certain latitudes, longitudes and heights, so as to give a general account of the state of the atmosphere at any instant, over an extended region, up to a height of say 20 kilometres. The numbers in this table are supposed to be given, at a certain initial instant, by means of observations.
>
> It is shown that there is an arithmetical method of operating upon these tabulated numbers, so as to obtain a new table representing approximately the subsequent state

of the atmosphere after a brief interval of time, δt say. The process can be repeated so as to yield the state of the atmosphere after $2\delta t$, $3\delta t$, and so on."

His introductory example is the shallow water equations, still the most heavily used model problem in the field. He considered almost every issue, though he did not always make the choice now regarded as correct. For example, he considered using pressure as a vertical coordinate but decided to retain height.

Richardson attempted one real forecast as an example of his method. Actually, he only tried to compute tendencies at a single point. The labor was enormous and the result was disappointing: he predicted a rate of pressure change of 145 mb/6 hrs, whereas the observed change was about 1 mb/6 hrs. He understood that this result was due to a spurious value for the convergence of the winds due to the poor quality of the observations, but he did not understand the dynamics thoroughly enough to see how to solve the problem. At the time the observations were totally inadequate to his needs, but the same problem would arise even now if we did not know how to treat it.

There have been four advances since Richardson's day that make modern NWP possible. Two are technological and two are conceptual. They are the development of electronic computers, the enhancement of the observing network, improvements in theoretical understanding of atmospheric motions and advances in numerical analysis.

1.1 Development of Electronic Computers

Richardson envisioned the mechanics of the calculation as follows [Richardson, 1922, p. 219]:

"If the time-step were 3 hours, then 32 individuals could just compute two points so as to keep pace with the weather, if we allow nothing for the very great gain in speed which is invariably noticed when a complicated operation is divided up into simpler parts, upon which individuals specialize. If the co-ordinate chequer were 200 km square in plan, there would be 3200 columns on the complete map of the globe. In the tropics the weather is often foreknown, so that we may say 2000 active columns. So that 32 x 2000=64,000 computers would be needed to race the weather for the whole globe. That is a staggering figure. Perhaps in some years' time it may be possible to report a simplification of the process. But in any case, the organization indicated is a central forecast-factory for the whole globe, or for portions extending to boundaries where the weather is steady, with individual computers specializing on the separate equations. Let us hope for their sakes that they are moved on from time to time to new operations.

After so much hard reasoning, may one play with fantasy? Imagine a large hall like a theatre, except that the circles and galleries go right round through the space usually occupied by the stage. The walls of this chamber are painted to form a map of the globe. The ceiling represents the north polar regions, England is in the gallery, the tropics in the upper circle, Australia on the dress circle and the antarctic in the pit. A myriad computers are at work upon the weather of the part of the map where each sits, but each computer attends only to one equation or part of an equation. The work of each region is coordinated by an official of higher rank. Numerous little "night signs" display the instantaneous values so that neighbouring computers can read them. Each number is thus displayed in three adjacent zones so as to maintain communication to the North and South on the map. From the floor of the pit a tall pillar rises to half the height of the hall. It carries a large pulpit on its top. In this sits the man in charge of the whole theatre; he is surrounded by several assistants and messengers. One of his duties is to maintain a uniform speed of progress in all parts of the globe. In this respect he is like the conductor of an orchestra in which the instruments are slide-rules and calculating machines. But instead of waving a baton he turns a beam of rosy light upon any region that is running ahead of the rest, and a beam of blue light upon those who are behindhand.

Four senior clerks in the central pulpit are collecting the future weather as fast as it is being computed, and despatching it by pneumatic carrier to a quiet room. There it will be coded and telephoned to the radio transmitting station.

Messengers carry piles of used computing forms down to a storehouse in the cellar.

In a neighboring building there is a research department, where they invent improvements. But there is much experimenting on a small scale before any change is made in the complex routine of the computing theatre. In a basement an enthusiast is observing eddies in the liquid lining of a huge spinning bowl, but so far the arithmetic proves the better way. In another building are all the usual financial, correspondence and administrative office. Outside are playing fields, houses, mountains and lakes, for it was thought that those who compute the weather should breathe of it freely."

This passage indicates the greatest limitation on Richardson's progress. Had he had a modern computer I suspect that he would have largely solved the other problems. (Note that Richardson's computational organization amounts to Massive Parallel Processing, a computer architecture only coming into use now, some 60 years later).

1.2 The Observing Network

It is only since World War II that the observing network has approached adequacy for providing upper air observations. Even with satellites, data are still sparse from over oceans and in the southern hemisphere. For oceanography the problems are far more severe. There are almost no subsurface data to initialize with, and with the exception of a few special experiments, little prospect of significant improvement. Furthermore, the ocean circulation, unlike the weather, is forced first and foremost. To calculate it one needs to know the surface forcing. The surface wind stress, which is the primary forcing, is not very well known over the ocean as a whole. (The equatorial Atlantic and Northern hemisphere midlatitudes may be adequately covered; the tropical Pacific is certainly not). Surface fluxes of heat and salt are very poorly known at present. In the end, these quantitites may be best obtained from atmospheric models and SST observations rather than from direct measurements. Perhaps surface salinity and precipitation can be obtained from satellites.

1.3 Improved Theoretical Understanding of Atmospheric Motions.

Whereas Richardson's formulation of the equations was complete and quite complicated, our understanding has increased through analysis of simple idealized models. Only by applying the knowledge thus gained has it become possible to make successful forecasts with a model as complicated as Richardson's. This is true not only for physical reasons, but because this knowledge is crucial for designing numerical methods for solving the equations of geophysical fluid dynamics.

We mention here only a few historical milestones [see Phillips, 1970]. In 1939, Rossby demonstrated that a model which treats the atmosphere as a homogenous, nondivergent fluid is surprisingly realistic. Charney [1948] employed the technique of scale analysis to derive an approximate, but consistent set of equations, the quasi-geostrophic equations, that would eliminate Richardson's failure. Briefly, Richardson's equations are too general. They allow motions of little meteorological importance which were artificially excited by his initial data. Charney's equations describe only synoptic scale motions in middle and high latitudes.

Charney [1947] and Eady [1949] identified cyclones with theoretically calculated hydrodynamically unstable waves, suggesting that successful forecasts could be made without considering complicated nonadiabatic processes.

In 1949, the first numerical predictions were made by Charney, Fjortoft and von Neumann [1951] on the ENIAC computer. They were quasi-two-dimensional and based largely on Rossby's nondivergent model. It took almost another twenty years before forecasts were made with models which could rival Richardson's in complexity.

1.4 Advances in Numerical Analysis

Though many types of equations can now be solved reliably with canned routines that require the user to understand little, GFD problems (with some exceptions) are not yet among them. Simulating geophysical flows is still something of an art that demands an understanding of the nature of the physical system as well as of numerical techniques. The simulation is never perfect; different methods preserve different features of the original continuous equations. One must understand what is important for the problem at hand as well as the characteristics of numerical schemes.

While numerical solution of GFD problems has not yet advanced to the cookbook stage, there has been considerable progress since Richardson's day. In 1928, Courant, Fredrichs, and Lewy published an analysis of linear computational stability and derived the celebrated CFL condition. This question of whether the numerical solution would converge to the true solution is one of the few that Richardson overlooked entirely. This concept, central in numerical mathematics, is discussed in Chapters 4-6. For now we note that a particular first order scheme for solving the wave equation $u_t + cu_x = 0$ is stable if,

$$-1 < \frac{c\Delta t}{\Delta x} < 0.$$

Richardson violated the CFL condition; his timestep was too large. The condition depends on Δt and Δx, which we control, and on the wave speed, c, which comes from the equation. We must understand the equations to know what c is. Note that the bigger c is, the smaller the timestep must be. This is why we want to filter the fast waves that are unimportant in our problem while keeping those that are relevant. For large scale oceanography we filter sound waves and perhaps gravity waves. For other problems, gravity waves are the signal (e.g., surface wave modeling).

Another kind of computational instability, called nonlinear instability, was first pointed out by Phillips [1959]. It can occur even when no linear stability criteria are violated. It is both harder to understand and harder to treat. This problem and its solution are tied up with the difference between the way energy is transferred to small scales and finally dissipated in real geophysical fluids and the way this happens in numerical models. The continuous fluid contains all scales of motion down to the small scales directly dissipated by molecular friction; the numerical version has no scales shorter than the size of a grid box, typically tens or hundreds of kilometers. Again, numerical difficulties lead us back to the need to understand the properties of the equations thoroughly.

Before proceeding more formally, let me state some useful rules of thumb:

(1) Both physically and numerically, the more accurate your model is, the more sensitive it tends to be. More accurate numerical methods usually have more stringent stability criteria. Crude physical parameterizations will often hold up under all sorts of conditions whereas "good" ones can give disastrous results when their forcing conditions are unrealistic.

(2) When a model runs into conditions it can't handle, ask yourself what nature does under similar circumstances.

(3) When your scheme doesn't work, it is usually, but not always, a programming bug.

2. FILTERING APPROXIMATIONS

Observation to an Oceanographer:

> What you do is really boring. You work on the same problem for years and the best you can do is an approximate answer. You are not even sure it's right.
>
> Jacob Cane, age 11
> to his father

2.1. Governing Equations.

The starting point for our approximations are the Navier-Stokes equations on a rotating earth. They will not be derived here but will simply be stated. Derivations may be found in any number of basic books and articles. (For oceanographers, Veronis [1973] is most to the point.) The momentum equations are,

$$d_3 \underset{\sim}{V}/dt = \rho^{-1}\nabla_3 p - 2\underset{\sim}{\Omega}x\underset{\sim}{V} - \nabla_3\phi + \underset{\sim}{F}_3 \tag{1}$$

where $\underset{\sim}{V}$ = three-dimensional velocity relative to the rotating earth; ρ = density, p = pressure; $\underset{\sim}{\Omega}$ = earth's rotation; $\underset{\sim}{F}$ represents frictional forces; d_3, ∇_3 = three-dimesional material derivative and divergence, respectively. ϕ is the "geopotential", including both the effects of terrestrial gravity and centrifugal forces due to the earth's rotation.

We immediately begin to approximate. First, take z = 0 at mean sea level and ϕ = gz with g constant. This is a very good approximation for a few km below sea level, although it ignores tidal forces. We also assume the surfaces ϕ = constant are spherical (though in fact they are elliptical; cf Veronis [1973] for a thorough discussion). Let us write r = a + z where r is the radial coordinate and a is the (mean) radius of the earth. Whenever r appears undifferentiated we replace it by the constant value a since the ocean is a shallow fluid (i.e., z<<a).

Let θ = latitude, y = aθ = distance northward, and v = d_3y/d_3t; λ = longitude, x = a cosθ λ = distance eastward, and u = d_3x/d_3t; z = height above sea level and w = d_3z/d_3t, and rewrite (1) in spherical coordinates.

$$\frac{d_3\underset{\sim}{u}}{dt} + (f + \frac{u\ \tan\theta}{a})\underset{\sim}{k}\ x\ \underset{\sim}{u} + \frac{w\underset{\sim}{u}}{a} + w2\Omega\cos\theta\underset{\sim}{i} = -\rho^{-1}\nabla p + \underset{\sim}{F} \tag{2}$$

where $\underset{\sim}{i}$, $\underset{\sim}{j}$ and $\underset{\sim}{k}$ are unit vector eastward, northward, and upward; $\underset{\sim}{u}$ = (u,v);

$$\frac{d_3}{dt} \equiv \frac{d}{dt} + w\frac{\partial}{\partial z}; \ \frac{d}{dt} \equiv \frac{\partial}{\partial t} + u\frac{\partial}{\partial x} + v\frac{\partial}{\partial y} \equiv \frac{\partial}{\partial t} + \underset{\sim}{u}\cdot\nabla; \ \nabla \equiv$$

$$\equiv \underset{\sim}{i}\frac{\partial}{\partial x} + \underset{\sim}{j}\frac{\partial}{\partial y};$$

$f = 2\Omega\sin\theta$ is the vertical component of the rotation vector $\underset{\sim}{\Omega}$. The terms proportional 1/a are curvature terms that arise because of the curvature of the earth; analysis is often done on planes (f-plane, beta plane), where curvature terms may be neglected. Note that they are all nonlinear.

In deriving model systems it is important to maintain the conservation principles of the physical system [cf Lorenz, 1960]: e.g., energy, angular momentum, potential vorticity. Quadratic invariants like energy are often considered explicitly in constructing numerical schemes. Spurious sources or sinks must be avoided over long integrations because a systematic $O(\varepsilon)$ error at each timestep accumulates over $O(1/\varepsilon)$ timesteps to become an $O(1)$ error.

The equations above do not possess an angular momentum principle, i.e., a statement of the form

$$\frac{d_3}{dt} r\cos\theta(u + \Omega r\cos\theta) = r\cos\theta F_x - \frac{1}{\rho}\frac{\partial}{\partial\lambda}p$$

The simplification based on $z \ll a$ has already introduced a spurious source of angular momentum. The problem is in the horizontal component of the Coriolis term; i.e. $2\Omega w\cos\theta + uw/a$. It is a bit tricky to get this rigorously right [cf Veronis, 1973; Phillips, 1966]. Note that to keep an energy principle, $2\Omega w\cos\theta$ in the du/dt equation and $2\Omega u\cos\theta$ in the dw/dt equation must go together.

Neglecting the horizontal component of the Coriolis acceleration makes the local vertical -- the direction determined by gravity -- special. (For a homogenous fluid the axis of rotation might be singled out). Stratification is being used as a constraint on the flow. This is true for a homogenous model which retains only the vertical component of the Coriolis acceleration. GFD is defined by the specialization to rotating, stratified fluids.

In light of the difficulties, why simplify at all? The answers are as before: to increase understanding by focusing on the relevant physics; to be able to solve. It is not feasible to integrate the equations on all scales down to the molecular dissipation scale. Filtering out sound waves and buoyancy oscillations allows a much longer timestep for large scale simulations.

2.2. The Hydrostatic Approximation

We know that the hydrostatic balance $p_z = -\rho g$ holds approximately on the large scale. In fact, one cannot hope to compute (or observe) the imbalance between p_z and $-\rho g$ well enough to estimate the vertical acceleration. An error in ρ of 1 part in 10^6 implies an acceleration of 1m/s per day whereas vertical velocities in the ocean rarely exceed 1m/s.

One uses the technique of scale analysis to obtain a systematic and consistent approximation, with a precise statement of the conditions under which the approximation holds, a statement which can be checked a postiori. Scale analysis is now common in GFD, and expositions can be found in a number of textbooks. (An early and still illuminating example is Charney's [1948] derivation of the quasi-geostrophic equations).

The hydrostatic approximation holds if [Phillips, 1973]

$$H/L \ll 1 \text{ and } H/C_s \ll \tau$$

where H, L and τ are characteristic vertical, horizontal and time scales for the motions, and C_s is the sound speed. In other words, the motions are shallow and slow (compared to C_s). Buoyancy oscillations have been ruled out so static instability is no longer handled by the equations "automatically". Modelers must intervene, usually with some form of convective adjustment. In any case, for large scale ocean motions, dw/dt is too small to compute and hydrostatic balance is forced on us. This does not mean that dw/dt = 0, only that it is negligible compared to other terms in the force balance.

To summarize, with $H/L \ll 1$ and $\tau \gg C_s/H$ the momentum equations (2), become:

$$\frac{d}{dt}\underset{\sim}{u} + [f + \frac{u\tan\theta}{a}]\underset{\sim}{k}\times\underset{\sim}{u} + \frac{1}{\rho}\nabla p = \underset{\sim}{F}, \tag{3}$$

$$\frac{\partial p}{\partial z} = -\rho g; \tag{4}$$

and the continuity equation is,

$$-\frac{1}{\rho}\frac{d\rho}{dt} = \nabla\cdot\underset{\sim}{u} + \frac{\partial w}{\partial z}, \tag{5}$$

where

$$\frac{d}{dt} \equiv \frac{\partial}{\partial t} + \frac{u}{a\cos\theta}\frac{\partial}{\partial\lambda} + \frac{v}{a}\frac{\partial}{\partial\theta} + \frac{w\partial}{\partial z} \tag{6}$$

To complete the system one also needs

(i) an equation of state $F(\rho,T,p,S) = 0$ (7)

(ii) a thermodynamic equation: $dT/dt = Q$ (8)

(iii) a salinity equation: $dS/dt = E-P$ (9)

Also, boundary conditions. For example,

at the bottom: $w = \underset{\sim}{u} \cdot \nabla D(x,y)$;
at the top: $w = \underset{\sim}{0}$ (rigid lid) or $w = dz_{top}/dt$
$p = P_{ATM}$
at the sides: $\underset{\sim}{u} \cdot \underset{\sim}{n} = 0$ or $\underset{\sim}{u} = 0$; etc.

These are the so-called "primitive equations," rife with approximations but as primitive a set as anyone integrates for the large scale. Imposing hydrostatic balance eliminates a prognostic equation for w and replaces it with a diagnostic one, the hydrostatic equation. The system of equations is no longer completely hyperbolic, making it mathematicaly less elegant and, from some points of view, more difficult to analyze. Associated with these changes is a change in the meaning of kinetic energy from $1/2\rho\ (u^2 + v^2 + w^2)$ to $1/2\rho(u^2 + v^2)$.

Almost all the sound waves have been eliminated (the Lamb wave being the exception). What were formerly compressional adjustments proceeding at speed C_s are now instantaneous. The diagnostic relation for w amounts to the statement that the vertical velocity does what is needed to ensure hydrostatic balance. There is an analogue with the quasi-geostrophic system. In the complete (i.e., primitive equation) system, geostrophic balance is brought about by inertia-gravity waves. (This is Rossby's geostrophic adjustment problem). In the geostrophic or quasi-geostrophic system the inertia-gravity waves have been filtered out, and geostrophic balance is achieved instantly (or more precisely, rapidly compared to the timescales on which geostrophic motions evolve).

In oceanography, it is usual to make the Boussinesq approximation as well. This amounts to treating density as constant everywhere except where it is coupled to gravity (i.e. in (4)). In practice, it means setting the left hand side of (5) to zero and taking ρ to be constant in (3). By itself, the Boussinesq approximation filters sound waves but not buoyancy oscillations.

The example given above for the shallow fluid approximation [eq. (2)] illustrates the fact that scaling does not automatically guarantee conservation principles. This is disappointing, but not implausible. Two different derivations of O(1) equations must agree to O(1) if the scaling is carried through correctly, but differences at $O(\varepsilon)$ are not ruled out.

In summary, filtering approximations

(i) Eliminate motions we do not wish to follow explicitly (this is very important for the numerical implementation because it allows a much larger timestep),

(ii) Eliminate imbalances too small to be calculated accurately,

(iii) Result in diagnostic relations in place of prognostic ones,

(iv) Simplify the system of equations, thereby highlighting the important physics (though the mathematical structure may become less straightforward).

2.3. Grid Scale vs. Sub-grid Scale

While sound waves are not important to the ocean circulation, convection is. Equations that don't describe convection can be used by including a parameterization. We make a parameterizability assumption: the effect of such processes on the large scale can be expressed entirely in terms of large scale variables. This is done for all processes that are too fast or too small in spatial scale; e.g., convective overturning in the ocean, dissipation of momentum, mixing (diffusion) of heat. Ocean models have spatial resolutions of 0 (20 km) to 0(500 km) and timesteps of hours or days. Dissipation, etc. of these scales by smaller ones is parameterized.

This means the equations are a model for the grid scale motions and larger. There is not a simple truncation at these scales. Truncation implies that the model will approach the exact equations as the grid spacing Δx approaches zero. When we include the effect of subgrid scale processes this implies that physical parameterizations change as Δx approaches zero. In the usual numerical procedures one makes Δx small enough to obtain convergence. Not so here. This fact makes much of the usual numerical analysis philosophy inapplicable without modification.

Unfortunately the separation of scales is not clean; at the scales of our grids there is often no "spectral gap," no clear division between those processes at the scales that are resolved by the grid and those which are not. A related issue is the applicability of the parameterizability assumption. A prominent example of an open question is whether or not it is possible to model the large scale ocean circulation without explicitly calculating the mesoscale eddies. Can the effects of the eddies on the large scale be parameterized?

The sub-grid scale parameterizations enter the equations via the terms for friction and heating by non-adiabatic processes (F and Q in eqs (3) and (7)). The Reynolds stresses associated with sub-grid scale motions are a classic example. Many of the most important sub-grid scales processes are small scale in the vertical. The frictional layers at the bottom and top (ocean mixed layer) are especially important. In modeling these processes there is a strong interplay between the physical parameterizations and the vertical structure of the model. A "correct" physical parameterization will perform poorly if the variables it needs are not carried along by the model. A Richardson number dependent mixing will perform quite differently depending on the vertical resolution. Vertical representation is one of the more important and far reaching modeling choices.

2.4. Vertical Representations: Modes, Levels, Layers, and others.

Vertical representation is a greater issue when the number of degrees of freedom in the vertical is low. Specialization based on an understanding of the physics can be the most cost-effective approach. For example, if one is interested only in the upper ocean it is often appropriate to assume the ocean is motionless at depth (i.e. below the main thermocline).

2.4.1. Modes. One form of vertical representation is spectral, in terms of vertical basis functions, e.g.,

$$u = \sum_{n=0}^{\infty} u_n(x,y,t)\, A_n(z),$$

$$w = \Sigma w_n B_n(z).$$

There are many possible choices for the basis functions, e.g.,

$$A_n = \cos n\pi z/D \quad B_n = \sin n\pi z/D.$$

A valuable special choice is the vertical structure function (normal mode), e.g. B_n such that

$$\partial_{zz} B_n + \frac{N(z)^2}{gh_n} B_n = 0$$

where h_n is an eigenvalue, the equivalent depth and the B_n satisfies the appropriate boundary conditions at top and bottom. The choice captures the structure of the ocean's (linear) response. For each mode u_n, v_n, p_n satisfy the shallow water equations. That is why the shallow water equations comprise such a powerful model. Using these natural modes is not advantageous if many modes are retained and the interaction among modes is computed (i.e. if the flow is significantly nonlinear). For more than 0(10) modes the labor of computing the interaction coefficients becomes too large; for highly nonlinear or viscous flows these modes which characterize the linear inviscid dynamics so well are not necessarily meaningful. They are an inefficient way to represent the ocean mixed layer, and are especially awkward if surface heating is included.

2.4.2 Generalized Vertical Coordinate [Sundqvist, 1979; Kasahara, 1974]. The usual vertical coordinate z is replaced by a generalized vertical coordinate $s=s(x,y,z,t)$. We assume a single valued monotonic relation between s and z so that we also have $z = z(x,y,s,t)$.
Define

$$h(x, y, s, t) \equiv \partial z/\partial s \tag{10}$$

so

$$\frac{\partial}{\partial z} = \frac{\partial s}{\partial z}\frac{\partial}{\partial s} = \frac{1}{h}\frac{\partial}{\partial s}; \tag{11a}$$

and, with ξ =x, y, or t,

$$\left(\frac{\partial}{\partial \xi}\right)_z = \left(\frac{\partial}{\partial \xi}\right)_s - \frac{\partial s}{\partial z}\left(\frac{\partial z}{\partial \xi}\right)_s \frac{\partial}{\partial s} \tag{11b}$$

Now

$$\frac{d_3}{dt} = (\frac{\partial}{\partial t} + \underset{\sim}{u}\cdot\nabla)_s + \frac{d_3 s}{dt}\frac{\partial}{\partial s} \tag{12}$$

But working in the z system and then using (11a,b),

$$\frac{d_3}{dt} = (\frac{\partial}{\partial t} + \underset{\sim}{u}\cdot\nabla)_s + W_e \frac{1}{h}\frac{\partial}{\partial s} \tag{13}$$

where

$$W_e \equiv [w-(\frac{\partial z}{\partial t} + \underset{\sim}{u}\cdot\nabla z)_s] = h\frac{d_3 s}{dt} \tag{14}$$

is the volume flux per unit area across constant s surfaces.

If s = z, then W_e = w, and we have a level model. If s = constant is a material surface then W_e = 0: a layer model. Here h is a measure of layer thickness.

Physically, the choice of s and W_e are tied together. Usually, choosing a meaning for one determines the other. It is more typical to choose s but a choice of "entrainment velocity" could be allowed to determine s.

In terms of the s coordinate the hydrostatic relation is

$$\frac{\partial p}{\partial s} = h\frac{\partial p}{\partial z} = -\rho g h.$$

Define ϕ = gz (geopotential). Then ignoring horizontal metric terms, which don't change with the coordinate changes, use of (11) converts the primitive equations to the form

$$\frac{d_3\underset{\sim}{u}}{dt} + f\underset{\sim}{k}x\underset{\sim}{u} = -\rho^{-1}\nabla_s p - \nabla_s\phi + \underset{\sim}{F}; \tag{15a}$$

$$\frac{\partial}{\partial t}(\rho h) + \nabla_s\cdot(\rho h\underset{\sim}{u}) + \frac{\partial}{\partial s}(\rho W_e) = 0; \tag{15b}$$

$$\frac{d_3 T}{dt} = Q \tag{15c}$$

(The first of these is the momentum equation, the second the continuity equation, the last the thermodynamic equation).

2.4.3 Choices for s

(A) The familiar level model: s = z; h ≡ 1

(B) Pressure coordinates: s = p; h = 1/ρg; define

$$\omega = \frac{dp}{dt} = -\frac{1}{g\rho}W_e$$

Now the hydrostatic relation is

$$\partial\phi/\partial p = -1/\rho$$

and in the momentum equation the pressure gradient force (PGF) is just $-\nabla\phi$, a linear term. In the continuity equation $\partial/\partial t\ (\rho h) = 0$ so

$$\nabla\cdot\underset{\sim}{u} + \partial\omega/\partial p = 0$$

and the system looks incompressible. Thus the p coordinate makes it look Boussinesq: p is a mass coordinate; in a Boussinesq system so is z.

Pressure coordinates are very popular in meteorology, largely for the properties pointed out above. They are rarely used in complex numerical models, however, because the lower boundary conditions are awkward.

(C) Isentropic coordinates: $s = \rho$ (or ρ_θ, potential density); $h = \partial z/\partial\rho$.

Note that h is like a layer thickness; defining

$$h_j = \int_{\rho_j-\Delta\rho}^{\rho_j+\Delta\rho} h\,d\rho$$

is a formal way of deriving a layer model. The PGF is just ∇M, where $M = p/\rho + \phi gz$ is the Montgomery streamfunction. The hydrostatic relation may be written in the form $\partial M/\partial\rho = -p/\rho^2$ and the momentum equation is now

$$\frac{d_3\underset{\sim}{u}}{dt} - f\underset{\sim}{k}\times\underset{\sim}{u} = -\nabla M + \underset{\sim}{F} \tag{16a}$$

The continuity equation is

$$\frac{\partial}{\partial t}h + \nabla\cdot(h\underset{\sim}{u}) = -\frac{1}{\rho}\frac{\partial}{\partial\rho}(\rho W_e); \tag{16b}$$

$W_e = 0$ if there is no cross isopycnal flow. This is what we expect in most of the ocean interior, in which case (16a,b) become almost two-dimensional; the system of equations is coupled vertically only via M. This two-dimensionality, the embodiment of the idea that flow is along isopycnals, accounts for the appeal of isopycnal coordinates. However, there are technical problems related to layers that "disappear" before a boundary is reached, and, more generally, questions about how to treat boundaries. We expect a lot of mixing at the ocean surface so the blessings of isopycnal coordinates are mitigated there.

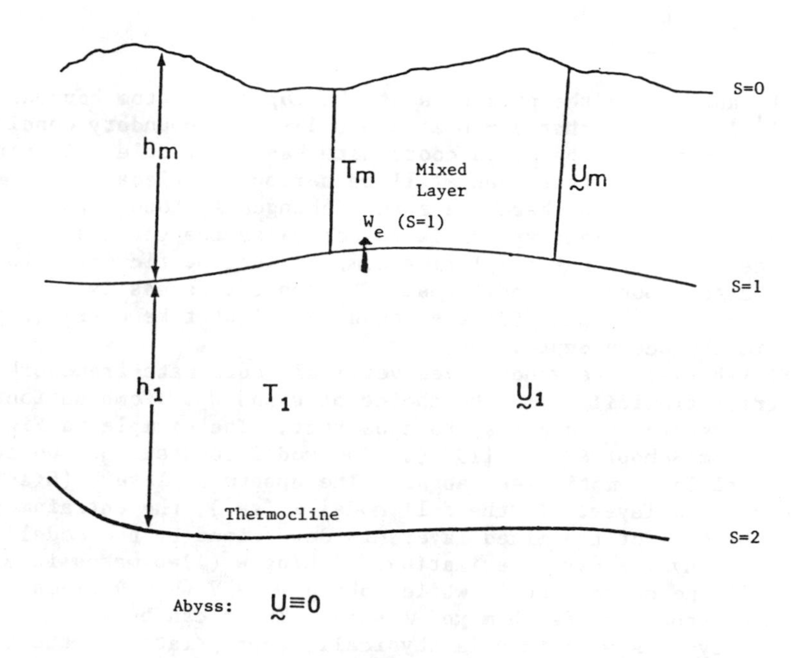

Figure 1. Hybrid model (After Schopf & Cane, 1983)

(D) Sigma coordinates: The meteorological original takes the form

$$\sigma = \frac{p - p_T}{p_B - p_T}$$

where P_T and P_B are the pressures at the top and bottom boundaries respectively. Notice that $\dot{\sigma} = 0$ at $\sigma = 0,1$ so the boundary conditions are straightforward; the sigma coordinate has no trouble following complex terrain, but there can still be serious problems with truncation errors when there are abrupt changes in topography.

The idea of a sigma system is to normalize the vertical coordinate in a way that simplifies computation, particularly for the top and bottom boundary conditions. The concept is easily generalized; for example, for the ocean one might take $\sigma = z/D(x,y)$ where D is the ocean depth.

(E) Hybrids: The generalized vertical coordinate framework allows great flexibility in the choice of s and W_e. Combinations of schemes -- hybrids -- are easy to construct. The example in Figure 1 is taken from Schopf & Cane [1983]. The model consists of two active layers overlying a motionless abyss. The uppermost layer $(0<s<1)$ is the ocean mixed layer. In the full model $W_e(s=1)$, the entrainment rate at the base of the mixed layer, is determined by the model's mixed layer physics parameterization. Taking $W_e(1)=0$ makes it a layer model (no entrainment), while taking $W_e = \nabla \cdot (h_1 \underset{\sim}{u}_1)$ gives a level model (constant depth mixed layer). There can be a discontinuity at $s = 1$; this is physically appropriate for the base of the mixed layer, because we expect a sharp change in properties there. If the region below, $1<s<2$, were split up into many layers (e.g., in the spirit of sigma coordinates) the usual finite difference assumptions about continuity would presumably apply, and discontinuities would be inappropriate.

It is widely recognized that the ocean surface layer is special. Heat and momentum are input there, mixing is enhanced, and momentum is concentrated, possibly making nonlinearities significant. The representation shown in Figure 1 is well matched to the physics. I suspect that the best model with only a few degrees of freedom in the vertical would consist of a mixed layer and a few normal modes. A precise formulation is left as an exercise for the reader.

Acknowledgements. The preparation of this chapter was supported by grant OCE 84-44718 from the National Science Foundation. The assistance of Karen Streech is deeply appreciated. Special thanks to Jim O'Brien and Ruth Pryor for making it all happen so happily.

REFERENCES

Charney, J.G., 'On the scale of atmospheric motions,' Geofys. Publik., 17., 17 pp., 1948.

Courant, R., K.O. Friedrichs, and H. Lewy, 'Uber die partiellen differenzengleichungen der mathematischen physik.' Math. Annalen, 100, 32-74, 1928.

Eady, E., 'Long waves and cyclone waves,' Tellus, 1(3), 33-52, 1949.

Kasahara, A., 'Various vertical coordinate systems used for numerical weather prediction,' Mon. Wea. Rev., 102, 504-522, 1974.

Lorenz, E.N., 'Energy and numerical weather prediction,' Tellus., 12, 364-373, 1960.

Mesinger, F. and A. Arakawa, Numerical methods used in atmospheric models, Vol. I. WMO/ICSU Joint Organizing Committee, GARP Publication Series No. 17, 64 pp.; Vol II, 499 pp., 1976.

Phillips, N.A., 'An example of non-linear computational instability,' in The Atmosphere and the Sea in Motion, Rossby Memorial Volume, Rockefeller Institute Press, 501-04, 509 pp., 1959.

Phillips, N.A., 'The equations of motion for a shallow rotating atmosphere and the traditional approximation,' J. Atmos. Sci., 23, 626, 1966.

Phillips, N.A., 'Models for weather prediction,' Annual Review of Fluid Mechanics, Vol. 2., 251-289, 1970.

Phillips, N.A., 'Principles of large scale numerical weather prediction,' in Dynamic Meteorology, edited by P. Morel, D. Reidel, Boston, 1973.

Richardson, L.F., Weather Prediction by Numerical Process, 236 pp, Cambridge University Press, London, 1922; reprint, Dover Publications, New York, 1965.

Rossby, C.G., 'Relation between variations in the intensity of the zonal circulation of the atmosphere and the displacements of the semi-permanent centers of action.,' J. Marine Res., 2, 38-55, 1939.

Schopf, P. and M.A. Cane., 'On equatorial dynamics, mixed layer physics, and sea surface temperature,' J. Phys. Oceanogr., 13, 917-935, 1983.

Sundqvist, H., 'Vertical coordinates and related discretization,' in Numerical Methods used in Atmosphere Models, Vol. II, general editor A. Kasahara, GARP Publication Series No. 17, 499pp. (p. 3-50), September, 1979.

Veronis, G., 'Large scale ocean circulation,' Advances in Applied Mechanics, 13, 1973.

HISTORY AND METHODOLOGY OF MODELLING THE CIRCULATION OF THE WORLD OCEAN

A. J. Semtner, Jr.
National Center for Atmospheric Research
Boulder, Colorado 80307-3000
USA

ABSTRACT. This paper recounts the historical development of numerical models of the world ocean circulation and describes the methodology of constructing a global ocean model by means of judicious numerical choices. The choices deal with a variety of considerations such as numerical stability and accuracy, computational efficiency, treatment of irregular geometry, and values of physical parameters.

1. INTRODUCTION

Almost every scientific field has its own "grand problem," requiring the synthesis of many different techniques to understand complex natural phenomena in a comprehensive and unified way. In physical oceanography, the grand problem must certainly be to understand the circulation of the world ocean. This requires not only the correct knowledge of many physical processes but also the ability to predict their evolution and interaction in a very irregular domain. Because of the nonlinearity of the governing equations and the complex geometry, numerical solution techniques are the only practical means of proceeding.

Numerical models of the world ocean have been in use over the past fifteen years. However, progress in simulating ocean circulation has been slow because of limitations both in observational data and computer power. Now the problem is becoming tractable because of recent advances in observational techniques, both in-situ and remote (especially satellite), and in computer technology, with machines capable of 10 gigaflops promised by the year 1987 [S. Cray, recent announcement]. It is therefore an opportune time to review the status of global modelling and to discuss the techniques which will allow improved modelling in the future.

Section 2 reviews the early beginnings of ocean modelling, which culminated in the early 1970's with the construction of two global models. Section 3 discusses developments up to the present. Section 4 describes the proper choice of numerical techniques needed to construct a world ocean model. The detailed formulation of one such model is given in a companion paper in this volume [Semtner, 1986]. For additional information on how existing models of global circulation can be improved by physical and numerical upgrades, the reader is referred to Semtner [1984].

J. J. O'Brien (ed.), Advanced Physical Oceanographic Numerical Modelling, 23–32.

2. EARLY BEGINNINGS OF WORLD OCEAN MODELLING

2.1 Developments in the United States

The author finds it convenient to begin with developments in the United States, by virtue of personal contact with many of these activities while at the Geophysical Fluid Dynamics Laboratory (GFDL) in Princeton, New Jersey and at the Department of Meteorology of the University of California in Los Angeles (UCLA).

Kirk Bryan began ocean modelling at GFDL by applying techniques of numerical weather prediction to the solution of the barotropic vorticity equation in a rectangular domain [Bryan, 1963]. Somewhat independently and and slightly later, William Holland developed a similar model at the Scripps Institution of Oceanography as a numerical extension of an analytical study of western boundary currents [Holland, 1967]. Since Bryan's goal was to develop baroclinic ocean models suitable for coupling to atmospheric climate models, he moved rapidly to three-dimensional box models and carried out a number of multi-level primitive equation studies in collaboration with Michael Cox [cf. Bryan and Cox, 1968]. This effort was followed by the development of a more general model suitable for cases with irregular coastline, variable bottom topography, and multiple connectedness [Bryan, 1969].

The 1969 model of Bryan was the first model capable of being used for global simulations. However, consideration of computer power (or lack thereof) led to regional applications as an interim step. Gill and Bryan [1971], Holland [1971], Holland and Hirschman [1972], and Cox [1970] applied the improved model in turn to the Southern Ocean, oceanic tracers, the North Atlantic, and the Indian Ocean.

In the meantime, parallel development of multilevel primitive equation models was occurring at UCLA, where Haney [1971] constructed a three-dimensional box model under the direction of his advisor, Akio Arakawa, and a visitor, Kenzo Takano. This model departed from that of Bryan [1969] in that different choices were made in certain numerical features, such as the location of the boundary, the vertical placement of gridpoints, and the treatment of the vertically integrated flow.

The early US effort to develop global ocean models culminated in 1973 with the construction by Cox [1975] of such a model based on Bryan's formulation, using three overlapping coordinate systems for the two polar caps and a low latitude belt. The gridsize was ambitiously chosen to be 200 km, and as a consequence, the model could only be run for limited integrations. However, the use of observed data for initial conditions together with a limited integration with observed wind stress made possible the first quasi-diagnostic simulation of global ocean circulation.

2.2 Developments outside the United States

The first-ever applications of numerical techniques to large scale ocean circulation were carried out by Artem Sarkisyan [1955]. Subsequently, efforts by Sarkisyan and his colleagues focused on diagnostic studies of regional extent, such as for the North Atlantic. Elsewhere in Europe, Hans Friedrich constructed a multi-level model in West Germany [Friedrich, 1967].

The most ambitious effort outside the United States to model the global-scale ocean circulation was carried out by Takano, who began his work in Japan and carried out some additional computations at UCLA in 1973. His model is described in

Takano [1974]. In a fashion similar to that of Haney [1971], coastlines were located on temperature gridpoints and temperature gridboxes of partial size (1/2, 1/4, and 3/4 of normal size) complicated the calculation, unlike the case of Bryan's [1969] model, where coastlines were located on velocity gridpoints and momentum gridboxes of partial size could be ignored. Takano's approach was novel in regard to the treatment of multiple connectedness, where a method of "hole relaxation" allowed a simpler handling of islands than in Bryan's [1969] model. Takano was able to reduce his model's computational requirement by neglecting bottom topography, bottom friction, and momentum advection. This allowed the calculation of the vertically averaged flow to be done only once, followed by a multi-year integration of the three dimensional equations for the baroclinic velocity fields and the density field. A simulation for the world ocean circulation was carried out using 400 km gridsize, observed wind forcing, and idealized (zonally constant) atmospheric temperatures for computing surface heat flux. The resulting fields of predicted surface temperature, near surface vertical velocity, and surface heat flux showed many realistic features of the known ocean circulation [Takano, 1975].

3. PROGRESS SINCE 1973

3.1 Development of New Codes

A number of new codes for multi-level primitive equation modelling have been constructed since 1973. Semtner's [1974] code is mainly based on Bryan's formulation, but it includes certain of Takano's methods generalized to the case of variable bottom topography. It also was designed to run efficiently on vector computers. Subsequently Semtner constructed a box model [Semtner and Mintz, 1977] that allowed one-dimensional bottom topography and was highly conserving of momentum (as with Haney's model) and of mean square vorticity [following Arakawa, 1972]. Two attempts were made in the late 1970s to construct multi-level models based on a different horizontal arrangement of model gridpoints (the so-called C grid), along the lines of the UCLA atmospheric general circulation model of Arakawa and Mintz [1974]. These models, by Jeong-Woo Kim at Oregon State University [Kim, 1979] and by Michael Cox at GFDL [personal communication], appeared vulnerable to computational noise in high vertical wavenumbers, and they are no longer in use. Two relatively recent codes have appeared, one by Young-June Han [1984] at Oregon State University and the other by Michael Cox [1984] at GFDL. Both of the models are based mainly on Bryan's formulation, and the Cox model has a preprocessor that facilitates various applications, including running on Cyber-type machines.

3.2 Additional Coarse-grid Simulations

Following the pioneering but computationally constrained calculations of Cox and of Takano, some further coarser-grid simulations of global ocean circulation have been carried out, as for example, by Bryan and Lewis [1979], Washington et al. [1980], Meehl et al. [1982], and Han [1984]. However, progress has been slow in global modelling for several reasons: (i) attention to mesoscale eddy problems using filtered models; (ii) attention to regional models; and (iii) insufficient computer power to increase resolution significantly. Nevertheless, the time is now ripe for further progress in global modelling in view of what has been learned from both limited-domain and coarse-grid models and in view of anticipated increases in computer power and in the amount of observed data.

4. DESCRIPTION OF A BASIC MODEL FOR GLOBAL APPLICATIONS

The construction of any numerical model involves making a large number of choices between competing possibilities. In this section, the rationale is given for the various choices made in constructing the model described in Semtner [1986]. Most of these choices were made implicitly by Bryan [1969], although some of the theoretical justifications have only emerged since 1969. A general reference for some of the material presented in this section is that of Mesinger and Arakawa [1976].

4.1 Order of Finite Differencing

Finite differencing is not the only technique available for modelling fluid dynamics. However, spectral methods, which are often employed in global atmospheric models, are impractical for the complicated geometry of the world ocean because of the lack of suitable basis functions. Finite-element methods are sometimes used in coastal ocean models, but their utility has not been fully demonstrated for three-dimensional problems such as global ocean modelling. Although they have the advantage of allowing resolution to be increased in certain regions, they can be computationally demanding, especially if complexities of the mesh cause much of the code to be nonvectorizable. For these reasons, finite differencing is the method of choice.

Fourth-order finite differencing is sometimes used in atmospheric models because of the increased accuracy of the method. However, the geometry of the world ocean makes it difficult to establish boundary conditions for the difference equations. Thus second-order differencing is chosen here.

Sometimes nonuniform grid spacing is used in regional models to give enhanced resolution of western boundary processes or of the equatorial undercurrent. The global ocean's irregular boundaries make this technique rather ineffective for resolving boundary currents. Enhanced resolution near the equator may have merit for equatorial models, but may not be justifiable in terms of the additional calculations required if the entire global domain is considered. We therefore opt for uniform grid spacing in both latitude and longitude.

4.2 Vertical Grid Structure

Ocean models almost always employ a stretched grid in the vertical to allow better resolution of upper-ocean processes. However, two different selections have been made with regard to placing the gridpoints within a vertical column of gridboxes. Bryan's [1969] model places the gridpoints at the centers of the gridboxes, whereas Takano's [1974] model places them such that adjacent gridpoints are equidistant from the box interface between them. The former method is more consistent with our thinking about finite differencing, while the latter is more consistent with a requirement for energetic consistency that simple averaging be used in computing vertical advection of heat and salt. The former method is chosen here, although simple modifications in the computed depths of the gridpoints will implement the other scheme. To the author's knowledge, there is no theoretical guidance as to which scheme might actually be better in terms of accuracy. Presumably, the distinction between the two becomes less important as vertical resolution is increased.

4.3 Treatment of Topography

Topography is usually treated in atmospheric models by the use of sigma coordinates, which follow the terrain. The method breaks down when the variations in topography are comparable to the depth of the fluid itself (hence the problems in modelling atmospheric circulation near the Himalayas). The method chosen by Bryan was to use a variable number of gridboxes in the vertical, stacked downward until the bottom is reached at each of the horizontal gridpoints. It is possible for some applications to vary the thickness of the lowest box in each column [as in Semtner and Mintz, 1977]; however, increased vertical resolution of future global models will probably make this step unnecessary and help to keep the numerical scheme simple and vectorizable. Therefore Bryan's method is chosen.

4.4 Horizontal Arrangement of Gridpoints

The original Bryan model employs a staggered arrangement of gridpoints in which temperature points can be thought to be at the centers of gridboxes and u and v velocity components reside together at the centers of the vertical edges of the boxes (the so-called B scheme). This method allows a relatively straightforward advection calculation (without a lot of averaging operations to define values of variables where they are needed). It also allows a semi-implicit treatment of the Coriolis term, without which the timestep is severely limited in coarse-grid models. However, the placement of u points at the centers of the meridional faces of gridboxes and of v points at the centers of the latitudinal faces (the so-called C scheme) is even simpler (less averaging) and might be selected if resolution was fine enough not to require semi-implicit treatment of the Coriolis terms. Which method should be used?

Theoretical guidance as to which scheme is better was first provided by Arakawa [cf. Mesinger and Arakawa, 1976]. Analysis of the dispersion relations for various finite-difference versions of the shallow water equations shows that the C scheme does a better job than the B scheme in reproducing the dispersion relation for the differential equations, provided that the gridsize is smaller than the radius of deformation. However, when the gridsize is larger than the radius of deformation, the situation is reversed. A recent study by Wajsowicz [1986] extends the theory to the case of planetary waves and obtains somewhat similar conclusions. This suggests that the choice of scheme depends on grid resolution; and indeed it is not surprising that most coarse-grid model studies have used the B scheme and most eddy resolving studies have used the C scheme. The exception to the latter statement is that multi-level models based on the B scheme have been used in some eddy resolving studies [e.g. Semtner and Mintz, 1977].

Cox [personal communication] constructed a multi-level model based on the C scheme, as suggested by the work of Arakawa. However, noise in high vertical wavenumbers made the model less useful for eddy-resolving studies than B-scheme models. Evidently, poor resolution of high baroclinic modes was more deleterious with the C scheme than poor phase speeds for low baroclinic modes in the B scheme.

Therefore caution is urged regarding use of the C scheme in models having high vertical resolution, and for the time being, we choose to continue with the B scheme.

4.5 Location of Lateral Boundaries

The Bryan formulation is based on temperature gridboxes being the basic building blocks of the model. The boundary of the domain consists of a rectilinear curve passing through u,v points whose values remain identically zero throughout the calculation. As is shown in Section 4 of Semtner [1986], this approach does not conserve momentum at the boundary, but it does conserve total energy as long as the no-slip condition is used. Another approach is to make momentum gridboxes the basic building blocks of the model and to locate the boundary on temperature points (as in the models of Takano and Haney). This requires the use of partial temperature boxes if one wants to conserve heat. However, momentum is conserved and either free-slip or no-slip boundary conditions may be used. The approach with momentum gridboxes is probably superior to the one with temperature boxes, but the complexity of implementing calculations involving partial gridboxes is prohibitive (and probably nonvectorizable) in complex geometries. Thus we opt for the more straightforward approach of Bryan for purposes of global modelling.

4.6 Surface Boundary Condition on Vertical Velocity

The boundary condition on vertical velocity that is most physically correct is the kinematic boundary condition. However, this restricts the length of model timestep drastically because of the presence of high-speed (but innocuous) external gravity waves. To eliminate these and allow a longer timestep without seriously distorting the low frequency waves of primary interest, the rigid-lid condition is used, whereby the vertical velocity is taken to be zero at the top of the ocean.

4.7 Treatment of Momentum Advection

The original treatment of momentum advection by Bryan is conserving of kinetic energy but not of mean square vorticity. Arakawa [1966] has shown the desirability of conserving both quantities in order to prevent a false computational cascade of energy to high wavenumbers. Arakawa and Mintz [1974] have demonstrated that this problem becomes worse with increasing resolution. There is a formulation of momentum advection for the C scheme which is exactly equivalent to the Arakawa [1966] vorticity formulation, and there is an approximately equivalent version for momentum advection in the B scheme [Arakawa, 1972], which was used in the study of Semtner and Mintz [1977]. These schemes involve momentum exchanges between a momentum gridpoint and the eight closest points in the horizontal, whereas the Bryan method only involves four points. It may be the case that Arakawa's methods will have to be used as grid resolution of global models increases, but in the meantime, the complexity of setting boundary conditions for the methods with eight fluxes precludes its routine use. Thus the advection formulation of Bryan is chosen here.

4.8 Methods of Timestepping

It is well known [cf. Mesinger and Arakawa, 1976] that forward timestepping works well for diffusive processes, whereas leapfrog timestepping works well for advective processes. Accordingly, the appropriate time level of variable is employed for each set of forcing terms within a basic leapfrog structure. To remedy the time splitting that occurs slowly with the leapfrog scheme, an occasional forward timestep or Euler-backward timestep is needed. The former timestep is unstable in principal but only excites a small amount of high frequency noise when used in this fashion. The latter timestep has the advantages of conditional stability with built in high frequency damping, but requires more time levels than the basic leapfrog scheme. If sufficient memory is available, the Euler-backward scheme should be used.

4.9 Computation of Pressure

It is well known that pressures are very high in the deep ocean but that the gradients of presssure are very small. It is possible to run into problems of truncation error under these circumstances, especially on computers with word lengths as short as 32 bits. The trouble manifests itself in spurious large amplitude gravity waves, as seen in the vertical velocity field. The problem can be eliminated by using perturbation density rather than full density in all calculations. It suffices simply to subtract a constant density of 1.0 gm/cm**3 from all density values.

4.10 Energetic Consistency

It is important to conserve total energy in any numerical model. This is relatively easy to achieve by using centered differencing and by applying equally weighted averaging operators whenever values of variables are needed other than at their natural gridpoints. A proof of this following Bryan's [1969] argument is given elsewhere in this volume [Semtner, 1986].

4.11 Solution of Elliptic Problems

The rigid-lid condition complicates the solution of the governing equations and creates an elliptic problem to be solved at every timestep for the mass transport streamfunction. Further information is given in the companion paper on the model.

4.12 Treatment of Islands

Bryan's method to obtain the value of the mass transport steamfunction on islands requires the storage of a two-dimensional Green's function for each of the islands. Takano's method saves space and is more elegant in that streamfunction values on islands are obtained by line integration. Therefore it is preferable to generalize Takano's method to the case of variable bottom topography, the details of which are given in the companion paper

4.13 Values of Diffusion Constants

Choices of diffusion constants are often dictated by considerations of the available computer time, whereas they should in fact be dictated by physical considerations. If eddies are to be resolved in a calculation, the diffusion constants should be small enough to eliminate poorly resolved scales of motion without interfering with eddy dynamics except as is physically justified. If eddies are not resolved, then parameterizations of their effects in terms of larger scale predicted quantities is required.

Unfortunately, only gross estimates of eddy effects are known, but we do know that their order of magnitude effect is that of a diffusivity of 10**3 m**2/sec. In a recent comparison of ocean basin simulations with and without resolved eddies, Cox [1986] indicates that even the use of the above constant as a parameterization of eddy heat diffusion gives some skill in reproducing time-averaged fields.

4.14 Horizontal Grid Spacing

Horizontal grid spacing must be adequately small to allow the chosen parameterizations to be used without excessive computational noise. This is problem dependent, but in the case of a world ocean model with parameterized eddies, the gridsize should probably be smaller than 100 km. With the improvements anticipated in computer power over the next few years, world ocean simulations with this gridsize should be computationally possible.

4.15 Vertical Grid Spacing

Vertical grid spacing should be adequate to resolve the vertical structure of known oceanic phenomena and to resolve the bottom topography. It is suggested that with a horizontal gridsize of 100 km, the maximum number of vertical levels should be chosen between 10 and 20.

4.16 Length of Timesteps

The final consideration is the choice of timestep. The limitation on stability is typically about the same number of minutes as the gridsize in kilometers. Longer timesteps are possible on temperature and salinity, but some caution is in order unless a steady state solution is sought (cf. the companion paper on the model).

Acknowledgments. The National Center for Atmospheric Research is sponsored by the National Science Foundation.

REFERENCES

Arakawa, A., 1966: Computational design for long-term numerical integration of the equations of fluid motion. *J. Comput. Phys.*, **1**, 119–143.

Arakawa, A., 1972: Design of the UCLA general circulation model. *Numerical Simulation of Weather and Climate*, Tech. Rept. No. 7, Department of Meteorology, University of California, Los Angeles, 116 pp.

Arakawa, A. and Y. Mintz, 1974: *The UCLA Atmospheric General Circulation Model*, Department of Meteorology, Univ. California, Los Angeles, 300 pp.

Bryan, K., 1963: A numerical investigation of a nonlinear model of a wind-driven ocean. *J. Atmos. Sci.*, **20**, 594–606.

Bryan, K., 1969: A numerical method for the study of the circulation of the world ocean. *J. Comput. Phys.*, **4**, 347–376.

Bryan, K., and M.D. Cox, 1968: A nonlinear model of an ocean driven by wind and differential heating: Parts I and II. *J. Atmos. Sci.*, **25**, 945–978.

Bryan, K., and L.J. Lewis, 1979: A water mass model of the world ocean. *J. Geophys. Res.*, **84**, 2503–2517.

Cox, M.D., 1970: A mathematical model of the Indian Ocean. *Deep-Sea Res.*, **17**, 47–75.

Cox, M.D., 1975: A baroclinic numerical model of the world ocean: preliminary results. *Numerical Models of Ocean Circulation*, Nat. Acad. of Sciences, Washington, D.C., 107–120.

Cox, M.D., 1984: A primitive equation three-dimensional model of the ocean. *GFDL Ocean Group Tech. Rept. No. 1*, GFDL/NOAA, Princeton University, Princeton, 250 pp.

Cox, M.D., 1986: An eddy resolving numerical model of the ventilated thermocline. Submitted to *J. Phys. Oceanogr.*.

Friedrich, H., 1967: Numerical computations of the wind-induced mass transport in a stratified ocean. *Proc. Symp. Math.-Hydrodyn. Invest. Phys. Processes in the Sea, Moskau 1966.* Mitt. Inst. Meersk. Univ. Hamburg, Nr. 10, 134.

Gill, A.E., and K. Bryan, 1971: Effects of geometry on the circulation of a three-dimensional southern hemisphere ocean. *Deep-Sea Res.*, **18**, 685–721.

Han, Y.-J., 1984: A numerical world ocean general circulation model. *Dyn. Atmos. Oceans*, **8**, 141–172.

Haney, R.L., 1971: *A numerical study of the large-scale response of an ocean circulation to surface heat and momentum flux.* Ph.D. Thesis, Department of Meteorology, Univ. of California, Los Angeles, 91 pp.

Holland, W.R., 1967: On the wind-driven circulation in an ocean with bottom topography. *Tellus*, **19**, 582–600.

Holland, W.R., 1971: Ocean tracer distributions. *Tellus*, **23**, 371–392.

Holland, W.R., and A.D. Hirschman, 1972: A numerical calculation of the circulation in the North Atlantic Ocean. *J. Phys. Oceanogr.*, **2**, 336–354.

Kim, J.-W., 1979: *Design and preliminary performance of the OSU four-level oceanic general circulation model.* Oregon State Univ. Climatic Research Inst., Corvallis, Oregon, 40 pp.

Meehl, G.A., W.M. Washington, and A.J. Semtner, 1982: Experiments with a global ocean model driven by observed atmospheric forcing. *J. Phys. Oceanogr.*, **12**, 301–312.

Mesinger, F., and A. Arakawa, 1976: Numerical Methods Used in Atmospheric Models. *GARP Publication Series*, No. 17, World Meteorol. Organization, Geneva, 64 pp.

Sarkisyan, A.S., 1955: (title unknown.) *Izv. Akad. Nauk SSSR*, Ser. Geofiz. **6**, 554.

Semtner, A.J., 1974: An oceanic general circulation model with bottom topography. *Numerical Simulation of Weather and Climate*, Tech. Rept. No. 9, Department of Meteorology, University of California, Los Angeles, 99 pp.

Semtner, A.J., 1984: Modeling the ocean in climate studies. *Annals of Glaciology*, **5**, 133–140.

Semtner, A.J., 1986: Finite-difference formulation of a world ocean model. *Proceedings of the NATO Advanced Study Institute on Advanced Physical Oceanographic Numerical Modelling*, D. Reidel Publishing Co., Dordrecht.

Semtner, A.J., and Y. Mintz, 1977: Numerical simulation of the Gulf Stream and mid-ocean eddies. *J. Phys. Oceanogr.*, **7**, 208–230.

Takano, K., 1974: A general circulation model for the world ocean. *Numerical Simulation of Weather and Climate*, Tech. Rept. No. 8, Department of Meteorology, University of California, Los Angeles, 47 pp.

Takano, K., 1975: A numerical simulation of the world ocean circulation: preliminary results. *Numerical Models of Ocean Circulation*, Nat. Acad. of Sciences, Washington, D.C., 121–129.

Wajsowicz, R.C., 1986: Free planetary waves in finite-difference numerical models. Submitted to *J. Phys. Oceanogr.*

Washington, W.M., A.J. Semtner, G.A. Meehl, D.J. Knight, and T.A. Mayer, 1980: A general circulation experiment with a coupled atmosphere, ocean and sea ice model. *J. Phys. Oceanogr.*, **10**, 1887–1908.

SOME USEFUL MATHEMATICAL PRELIMINARIES

J. J. O'Brien
Mesoscale Air-Sea Interaction Group
The Florida State University
435 Oceanography/Statistic Building
Tallahassee, FL 32306-3041

1. INTRODUCTION

The reader is expected to have a good background in calculus and applied mathematics through partial differential equations. Several topics related to the finite difference calculus and elementary numerical methods are reviewed in this chapter. The reader is encouraged to seek an alternative source for an in-depth discussion of each topic.

2. GAUSS ELIMINATION

A set of linear equations, $Ax = b$, where A is a square matrix of order n and x and b are column vectors, can be solved by a variety of techniques. Gauss elimination is a two-stage process. The first stage involves using elementary arithmetic operations to reduce A to an upper triangular matrix. The second stage is called *back-substitution*. Consider the case when A is order 3 and upper triangular.

$$a_{11}x_1 + a_{12}x_2 + a_{13}x_3 = b_1$$

$$a_{22}x_2 + a_{23}x_3 = b_2$$

$$a_{33}x_3 = b_3$$

The solution x is given simply by

$$x_3 = b_3/a_{33}$$

$$x_2 = (b_2 - a_{23}x_3)/a_{22}$$

$$x_1 = (b_1 - a_{13}x_3 - a_{12}x_2)/a_{11}$$

Now suppose we have a general 3 x 3 system. If we can reduce it to

J. J. O'Brien (ed.), Advanced Physical Oceanographic Numerical Modelling, 33–49.

upper-triangular form, we can complete the solution by back-substitution

$$a_{11}x_1 + a_{12}x_2 + a_{13}x_3 = b_1 \tag{1}$$

$$a_{21}x_1 + a_{22}x_2 + a_{23}x_3 = b_2 \tag{2}$$

$$a_{31}x_1 + a_{32}x_2 + a_{33}x_3 = b_3 \tag{3}$$

We want to operate on the equations to eliminate the a_{21} and a_{31} elements.

Let us multiply (1) by $m_{21} = a_{21}/a_{11}$ and subtract the result from (2). Similarly, multiply (1) by $m_{31} = a_{31}/a_{11}$ and subtract from (3). The result is

$$a_{11}x_1 + a_{12}x_2 + a_{13}x_3 = b_1$$

$$(a_{22} - a_{12}m_{21})x_2 + (a_{23} - a_{13}m_{21})x_3 = b_2 - b_1m_{21}$$

$$(a_{32} - a_{12}m_{31})x_2 + (a_{33} - a_{13}m_{31})x_3 = b_3 - b_1m_{31}$$

In matrix form we have

$$\begin{bmatrix} a_{11} & a_{12} & a_{13} \\ 0 & a_{22}^{\{1\}} & a_{23}^{\{1\}} \\ 0 & a_{32}^{\{1\}} & a_{33}^{\{1\}} \end{bmatrix} \begin{bmatrix} x_1 \\ x_2 \\ x_3 \end{bmatrix} = \begin{bmatrix} b_1 \\ b_2^{\{1\}} \\ b_3^{\{1\}} \end{bmatrix}$$

Define $m_{32} = a_{32}^{\{1\}}/a_{22}^{\{1\}}$ and multiply second equation by m_{32} and subtract from third. This leaves the third in the form

$$(a_{33}^{\{1\}} - a_{23}^{\{1\}}m_{32})x_3 = b_3^{\{1\}} - b_2^{\{1\}}m_{32}$$

or

$$a_{33}^{\{2\}}x_3 = b_3^{\{2\}}$$

This simple system is now in upper-triangular form and is easily solved by back-substitution. Of course the diagonal elements a_{11}, $a_{22}^{\{1\}}$ and $a_{33}^{\{2\}}$ cannot be zero.

Basically, the process is to replace all elements below a_{11} by zero; next, replace all elements below a_{22} by zero, and so forth until the system is upper-triangular. Then we complete the solution by back-substitution.

In general, we wish all m_{ij} multipliers to be bounded above by unity. This usually requires an interchange of rows as the solution progresses. For most Gauss elimination procedures arising in the solution of differential equations this is not necessary, since the matrices are well-behaved. Bad behavior is usually associated with a poorly designed physical problem.

Gauss Elimination can be used to solve several systems, $Ax = y$, where A remains unchanged. Once we have m_{ij} stored in the zeroed positions (we only have to compute them once!), we can easily solve for each x, given y.

We can also use this method to find the inverse, A^{-1}, of A. Let $y_1^T = [1, 0, \ldots 0]$. Let $y_2^T = [0, 1, \ldots, 0]$ and so on until $y_n^T = [0, 0, \ldots 0, 1]$. Now if we solve the system n times for these y_i vectors it is equivalent to solving the system

$$AX = I$$

with solution

$$X = A^{-1}$$

where, of course, X is order n.

3. TRIDIAGONAL SYSTEMS OF LINEAR EQUATIONS

In several methods to solve partial differential equations by finite difference methods we need to solve a tridiagonal system of linear equations many times. There is a special variant of Gauss elimination which is called the tridiagonal algorithm, or the "up-down" algorithm. We will illustrate the method using a second-order ordinary differential equation with endpoint boundary conditions.

Consider the solution to the second order ordinary differential equation

$$y'' + 2\,\alpha(x)y' + \beta(x)y = g(x) \tag{4}$$

with boundary conditions

$$y(0) = q$$

$$y(1) = p$$

Divide the interval [0,1] into n + 1 sections of width Δx and replace (1) by the finite difference equations for all the interior points.

$$\frac{y_{j+1} + y_{j-1} - 2y_j}{\Delta x^2} + 2\alpha_j \frac{(y_{j+1} - y_{j-1})}{2\Delta x} + \beta_j y_j = g_j \tag{5}$$

where

$$y_j = y(x_j) \;\; \alpha_j = \alpha(x_j) \;\; \beta_j = (x_j) \;\; g_j = g(x_j), \text{ and } x_j = j\Delta x.$$

Now (5) is a set of linear equations

$$'a_j y_{j-1} + b_j y_j + {'c_j} y_{j+1} = {'g_j} \;;\quad j = 1(1)n \tag{6}$$

where

$$a_j = (1 - \alpha_j \Delta x)/\Delta x^2$$

$$b_j = (\beta_j \Delta x^2 - 2)/\Delta x^2$$

$$c_j = (1 + \alpha_j \Delta x)/\Delta x^2$$

$$'a_j = a_j \quad j = 2(1)n\ , \quad 'a_1 = 0$$

$$'c_j = c_j \quad j = 1(1)n-1\ , \ 'c_n = 0$$

$$'g_j = g_j \quad j = 2(1)n-1$$

$$'g_1 = g_1 - a_1 q; \quad 'g_n = g_n - c_n p$$

The reader should recognize that the boundary conditions are now associated with the forcing function, $g(x)$, and the algebraic problem (6) is similar to a forced problem with homogeneous boundary conditions.

Equation (6) is a tridiagonal set of linear equations in the unknowns $y^T = [y_1, y_2, \ldots, y_n]$ and can be written

$$My = g$$

where

$$g^T = ['g_1, 'g_2, \ldots, 'g_n]$$

The matrix M is tridiagonal with principal diagonal elements $m_{jj} = b_j$; the element to the left of the diagonal is $'a_j$ and to the right is $'c_j$. All the other elements are zero.

$$m_{jj} = b_j$$

$$m_{j,j-1} = 'a_j$$

$$m_{j,j+1} = 'c_j$$

This system is easily solved by a "forward sweep" and a backward substitution. The algorithm is simply Gaussian elimination. The important difference is that we only work with vectors, a, b, c, d, g, w, and y and not with the large sparse matrix, M.

Forward Sweep

$$d_1 = c_1/b_1$$

$$d_i = c_i/(b_i - d_{i-1}'a_i), \quad i = 2(1)n$$

$$w_1 = 'g_1/b_1 \tag{7}$$

$$w_i = ('g_i - 'a_i\, w_{i-1})/(b_i - d_{i-1}\, a_i); \quad i = 2(1)n$$

Backward Substitution

$$y_n = w_n$$
$$y_j = w_j - d_j\, y_{j+1} \qquad j = n-1(-1)1 \tag{8}$$

The forward sweep reduces M to an upper-triangular matrix with principal diagonal element unity and $m_{j,j+1} = d_j$. The backward substitution solves for y.

The matrix M is a banded matrix. Other banded matrices arise with 5, 7 and higher number of non-zero elements clustered around the principal diagonal. There are well-documented subroutines for solving these systems also. In this book, the most common banded matrix which arises is the tridiagonal one.

The need for the tridiagonal algorithm will arise frequently in the methods outlined in the book. In many cases, the vectors a, b, c are constants, but w is a function of another independent variable, usually time, and occasionally another spatial coordinate. In these cases, we perform the forward sweep to determine d only once.

In order for this algorithm to be useful, it is necessary that the d_j are not zero.

Theorem: If the tridiagonal matrix, M, satisfies the weak row-sum criterion:

$$|a_j| + |c_j| \leqslant |b_j| \,, \qquad |c_j| < |b_j| \,, \qquad j = 1(1)n \tag{9}$$

then each d_j is not zero [Stommel and Haines, 1980, p. 107]. This is satisfied in the applications to partial differential equations that appear in this book. However, the algorithm does not provide a general solution to all problems.

Exercise:

The Yoshida Jet problem is a simple oceanographic application for this method. Consider a "reduced-gravity" oceanic layer on an infinite equatorial ocean which is at rest and an easterly wind which starts to blow at t = 0

$$\frac{\partial u}{\partial t} = \beta y v + \tau^x/\rho H_1$$
$$\frac{\partial v}{\partial t} = -\beta y u - g\frac{\Delta\rho}{\rho}\frac{\partial h}{\partial y} \tag{10}$$
$$\left(\frac{1}{H_1} + \frac{1}{H_2}\right)\frac{\partial h}{\partial t} = -\frac{\partial v}{\partial y}$$

(u,v) are eastward and northward flow; h is change of thickness of upper layer from H_1. Negative h means upwelling.

Eliminate inertial oscillations from the solution by neglecting,

$\frac{\partial v}{\partial t}$. If τ^x is a constant, we can derive the ordinary differential equation (ODE)

$$L^4 \frac{\partial^2 v}{\partial y^2} - y^2 v = aLy \tag{11}$$

subject to boundary conditions, $v(0) = 0$, $v(y \to \pm\infty) = 0$. See O'Brien & Hurlburt [1974] for details on the physics and applications.

$$L = (c/\beta)^{1/2}$$

$$c = (g\frac{\Delta\rho}{\rho} H_1 H_2/(H_1 + H_2))^{1/2}$$

$$a = \tau^x/\beta L H_1 \rho$$

Use the tridiagonal algorithm to solve the ODE (11) with boundary conditions, $v(0) = 0$, $v(\text{large } y) = -aL/y$
Let $\tau^x/\rho = 10^{-1}$ m^2 s^{-2}, $\beta = 2 \times 10^{-11}$ m^{-1} s^{-1}, L = 275 km, H_1 = 100 m, ρ = 1 Kg m^{-3}. Construct a plot of v and u at time = $(\beta L)^{-1}$. We need solve only for positive y. Let $\Delta y = 0.1$ L and y_{max} = 8L. Figure 2.1 shows the solution.

4. VECTOR AND MATRIX NORMS

It is useful to have simple measures of size for vectors and matrices. These are extremely helpful in understanding convergence and stability of numerical schemes.

Definition: A vector norm is a mapping which associates a non-negative real number with each vector.

It is denoted by $||x||$. It must satisfy the conditions

1) $||x|| > 0$ and $||x|| = 0$ iff $x = 0$

2) For any scalar α, $||\alpha x|| = |\alpha| \; ||x||$

3) For <u>all</u> vectors x and y $||x + y|| \leq ||x|| + ||y||$

This latter is called the triangle inequality.

Definition: Euclidean or L_2 norm is the length of a vector

$$||x||_2 = |x| = \sqrt{\sum_{j=1}^{n} |x_j|^2}$$

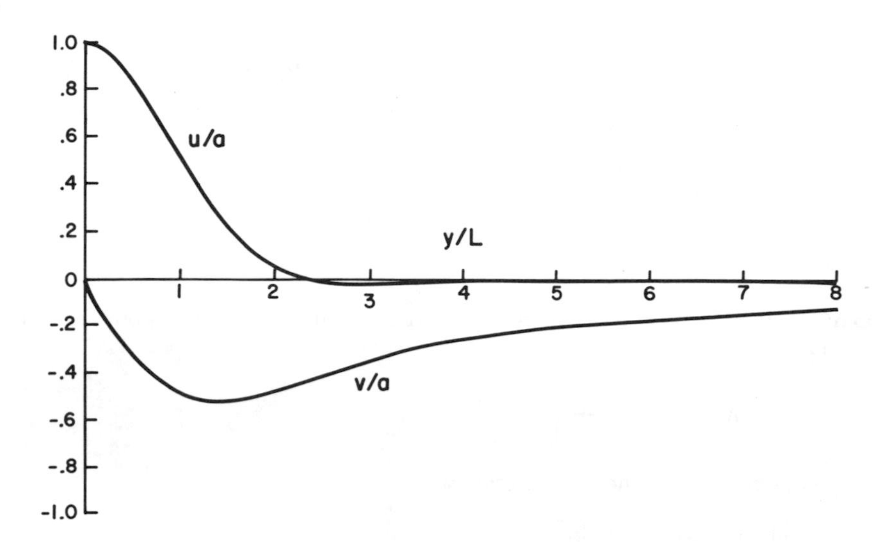

Figure 1. The solution for the Yoshida Jet Problem. The north-south coordinate is normalized by the equatorial radius of deformation, L. The north-south flow v/a is anti-asymmetric about the equator; at $y/L > 2.5$, the solution is the steady Ekman flow, $v = -aL/y$. The east-west flow, u/a, is a function of time, the curve is drawn at $t = (\beta L)^{-1}$.

Definition: The L_p norm (p is a positive integer) is defined

$$||x||_p = \left(\sum_{j=1}^{n} |x_j|^p \right)^{1/p}$$

Definition: The maximum (or infinity norm) denoted L_∞ is defined

$$||x||_\infty = \max_j |x_j|$$

The most commonly used norms are L_2 and L_∞. The Matrix norms considered here are those induced by vector norms [Isaacson and Keller, 1966].

Definition: If $||x||$ is a vector norm, then the corresponding matrix norm is defined

$$||A|| = \sup_{x \neq 0} \frac{||Ax||}{||x||}$$

Properties of the matrix norm

(1) $||A|| = \max_{||x||=1} ||Ax||$

(2) $||A|| > 0$ and $||A|| = 0$ iff $A = 0$

(3) $||\alpha A|| = |\alpha| \, ||A||$

(4) $||A + B|| \leqslant ||A|| + ||B||$

(5) $||AB|| \leqslant ||A|| \, ||B||$

(6) $||A||_2 = \sigma(A^T A)^{1/2}$

(7) $||A||_1 = \max_k \sum_{j=1}^{n} |a_{jk}|$ (maximum column sum)

(8) $||A||_\infty = \max_j \sum_{k=1}^{n} |a_{jk}|$ (maximum row sum)

(9) $\sigma(A) \leqslant ||A||$ for any matrix norm.

The notation in Properties (6) and (9), $\sigma(A)$, is the spectral radius, i.e., the absolute value or modulus or modules of the maximum eigenvalue. Property (9) is very important since $||A||_1$ and $||A||_\infty$ are quite easy to calculate and our basic theorem for iteration processes is based on $\sigma(A)$.

We can use these properties to prove some useful theorems. The proof of property (9) is as follows:

Theorem: For any matrix A, $||A|| \geqslant \sigma(A)$ where $||A||$ is any norm.

Proof:

$$\lambda x = Ax \quad \text{by def. of eigenvalues and eigenvectors}$$

$$||\lambda x|| = |\lambda| \; ||x|| = ||Ax|| \leqslant ||A|| \; ||x||$$

by virtue of definition of $||A||$

$$|\lambda| \leqslant ||A|| \quad \text{for any eigenvalue of A}$$

therefore

$$\sigma(A) \leqslant ||A||$$

Theorem [Isaacson and Keller, 1966, p. 14]: The following three statements are equivalent

(a) A is convergent (i.e., $\lim_{m\to\infty} A^m = 0$)

(b) $\lim_{m\to\infty} ||A^m|| = 0$ for any matrix norm

(c) $\sigma(A) < 1$

Theorem: $\lim_{r\to\infty} A^r = 0$ if $||A|| < 1$ for any norm

Proof: $||A^r|| = ||AA^{r-1}|| \leqslant ||A|| \; ||A^{r-1}|| \leqslant ||A||^2 \; ||A^{r-2}|| \ldots$

$$\leqslant ||A||^r$$

The reader can complete the proof.

Corollary: The series

$$I + A + A^2 + \ldots + \ldots$$

converges iff A is convergent and the sum is $(I-A)^{-1}$.

If A is convergent, $\sigma(A) < 1$. The eigenvalues of I-A are $1-\lambda(A)$, therefore I-A is non-singular. Inspect the following identity.

$$(I - A)(I + A + A^2 + \ldots + A^m) = I - A^{m+1}$$

Premultiply by $(I - A)^{-1}$ and take the limit as $m\to\infty$

$$I + A + A^2 + \dots = (I - A)^{-1}$$

This is an important result for iterative processes to solve elliptic PDEs.

5. BIG OH

It is important to understand the rate at which a function approaches zero. We need this to estimate the dependence of the error upon the mesh spacing in a difference approximation. Let the error be $\varepsilon(k)$ where $k \equiv \Delta x$, the minimum distance between grid points. Note that k is a positive number which is less than unity, since Δx should be normalized by an appropriate physical length scale.

Definition: $\varepsilon(k)$ is order p at $k = 0$ if there are constants M and $k_o > 0$ such that

$$|\varepsilon(k)| \leqslant Mk^p \quad \text{if} \quad |k| \leqslant k_o$$

A term of order p is written

$$\varepsilon(k) = O(k^p)$$

If $p = 2$ and $k=\Delta x$, then reducing Δx by 2 reduces the error by a factor of 4.

6. CONSISTENCY, STABILITY AND CONVERGENCE

Suppose $u(x,t)$ is the exact solution to the initial value problem

$$\frac{\partial u}{\partial t} = L(x,t) \tag{12}$$

and $u(n\Delta t, j\Delta x) = U_j^n$ is the solution to a finite difference approximation to (12). The approximation must be <u>consistent</u>, <u>stable</u>, and must <u>converge</u> in order to be useful in physical problems.

Definition: Consistency - A finite difference approximation is consistent with a differential equation if the difference equation converges to the correct differential equation as the space and time grid spacing tends to zero.

Definition: Stable - Let U_j^n be the numerical solution to the difference equation and u_j^n be the exact solution . The difference approximation is stable if $Z_j^n = U_j^n - u_j^n$ remains bounded as n tends to infinity for fixed Δt.

In Chapter 4, we will introduce the terms neutral or dissipative scheme to refer to the behavior of a particular scheme to conserve or reduce the variance of U_j^n.

Definition: Convergence - If the difference between the theoretical solutions of the differential and difference equations at a fixed point (x,t) tends to zero uniformly as Δt, $\Delta x \to 0$ and n, $j \to \infty$, then the finite difference approximation converges to the continuous equation.

Theorem [Isaacson and Keller, 1966, p. 541]: Lax Equivalence theorem: Given a properly posed linear initial value problem and a finite difference approximation to it that satisfies the consistency condition, stability (as Δx and $\Delta t \to 0$) is the necessary and sufficient condition for convergence.

Definition: Truncation error - The local difference between the finite difference approximation and the Taylor Series representation of the continuous problem at a fixed point is the truncation error.

Example:

Consider the one-dimensional advection equation with constant speed, c.

$$\frac{\partial u}{\partial t} + c\frac{\partial u}{\partial x} = 0 \tag{13}$$

The Taylor series expansions for second order derivatives are

$$U_j^{n+1} - U_j^{n-1} = 2\Delta t\left(\frac{\partial u}{\partial t}\right)_j^n + \frac{\Delta t^3}{3}\left(\frac{\partial^3 u}{\partial t^3}\right)_j^n + \ldots$$

$$U_{j+1}^n - U_{j-1}^n = 2\Delta x\left(\frac{\partial u}{\partial x}\right)_j^n + \frac{\Delta x^3}{3}\left(\frac{\partial^3 u}{\partial x^3}\right)_j^n + \ldots$$

Combining we obtain

$$\frac{(U_j^{n+1} - U_j^{n-1})}{2\Delta t} = \frac{-c}{2\Delta x}[U_{j+1}^n - U_{j-1}^n]$$

$$+ \left(\frac{\partial u}{\partial t} + c\frac{\partial u}{\partial x}\right)_j^n$$

$$+ \frac{\Delta t^2}{6}\left(\frac{\partial^3 u}{\partial t^3}\right)_j^n + \frac{c\Delta x^2}{6}\left(\frac{\partial^3 u}{\partial x^3}\right)_j^n + \ldots$$

The first line is the finite difference approximation; the second line is zero because of (13); the third line is the truncation error. The difference approximation is consistent if the truncation error, which is $O(\Delta t^2 + \Delta x^2)$, goes to zero as Δt, $\Delta x \to 0$.

7. APPENDICES

7.1. The Fundamentals of Matrix Algebra

Definition: A matrix A of order m by n is a rectangular array of mn elements having m rows and n columns denoted [a] or A. The elements a_{ij} of [a] may be real, complex, or functions. We denote the order of A as (m,n). The element a_{rs} lies in the rth row and the sth column.

Definition: If n = m, then A is a square matrix of order n. The determinant formed from the elements of A is denoted det A or $|A|$. The diagonal of the square matrix A that contains a_{11}, a_{22}, ..., a_{nn} is called the principal diagonal.

Definition: If det A vanishes, A is said to be a singular matrix.

Definition: If det A $\neq$ 0, A is nonsingular. The inverse matrix, A^{-1} exists only if A is nonsingular.

Definition: The maximal order of any non-singular minor of A is called the rank of A.

Definition: A square matrix A whose elements $a_{ij} = 0$ for $i > j$ is called upper triangular.

Definition: A square matrix A whose elements $a_{ij} = 0$ for $i < j$ is called lower triangular.

Definition: The trace, Tr A, of a square matrix A is the sum of the elements of the principal diagonal.

Definition: The transpose of a square matrix A, denoted A^T or A', is obtained by a complete interchange of the rows and columns of A; i.e., the ith row of A is identical with the ith column of A'.

Definition: A column vector is a matrix of order (m,1).

Definition: A row vector is a matrix of order (1,m).

Definition: A scalar is a matrix of order (1,1).

Definition: A diagonal matrix, denoted D, is a square matrix with $d_{ij} = 0$, if $i \neq j$. In other words, only elements of the principal diagonal may be non-zero.

Definition: The identity or unit matrix denoted I is an order n diagonal matrix with all principal diagonal elements equal to unity.

Definition: The null or zero matrix is a matrix with all elements equal to zero and denoted $\mathcal{O}$.

Definition: Equality: A = B if they both are of the same order and $a_{ij} = b_{ij}$ for all i and j.

Definition: Addition: C = A + B, if A and B are of the same order and $c_{ij} = a_{ij} + b_{ij}$ for all i and j.

Definition: A symmetric matrix is a square matrix whose elements are symmetric about the principal diagonal; in other words

$$A = A^T \quad \text{or} \quad a_{ij} = a_{ji}$$

Definition: A skew-symmetric matrix has the property

$$A = -A^T \quad \text{or} \quad a_{ij} = -a_{ji}$$

Definition: Scalar Multiplication: B = kA where k is a scalar if $b_{ij} = ka_{ij}$ for all i and j.

Definition: Matrix Multiplication: If A is order (m,n) and B is order (p,q), the product AB is not defined unless n = p. If n = p, A and B are said to be conformable.

If C = AB then A and B are conformable and

$$c_{ij} = \sum_{k=1}^{p} a_{ik} b_{kj}$$

and C is order (m,q).

If A and B are square, then

$$|C| = |A| \; |B|$$

In general matrix multiplication is not commutative, i.e.,

$$AB \neq BA$$

Hence we must introduce the terms premultiplication and post-multiplication. In C = AB, B is premultiplied by A or A is post-multiplied by B. The identity matrix is the premultiplication and postmultiplication neuter, i.e., IA = AI = A for any matrix A.
Matrix multiplication obeys the associative law of continued products, i.e., (AB)C = A(BC).

The notation A^n indicates the product of a square matrix A multiplied by itself n times; e.g. $A^3 = AAA$.
If AB = BA, A and B are said to commute; this is a very special property.

Definition: Matrix Division or the Inverse Matrix: If A is a square matrix such that

$$AX = I$$

then we denote $X = A^{-1}$ and call it A-inverse or the inverse of A.

Define C_{ij} as the co-factors of the elements a_{ij} of A. $C_{ij} = (-1)^{i+j} |M_{ji}|$ where M_{ji} is the minor without the jth row and i^{th} column. Define the adjoint matrix of A, denoted adj A, as the square matrix with elements C_{ij}.

The inverse of A is given by

$$A^{-1} = \frac{\text{adj } A}{\det A}$$

(In practice, this formula is never used for matrices larger than order 5).

Theorem: The inverse of a matrix is unique. (Prove it by assuming there exists another inverse.)

Properties of the inverse

1) $|A^{-1}| = |A|^{-1}$

2) $(A^{-1})^{-1} = A$

3) $(ABC)^{-1} = C^{-1} B^{-1} A^{-1}$ [Prove it!]

Special matrices arising in physical problems

Let A have complex elements $a_{rs} = p_{rs} + i\, q_{rs}$ where $i^2 = -1$.

(a) The conjugate of A, denoted A*, has elements $a^*_{rs} = p_{rs} - i\, q_{rs}$.

(b) The associate of A is the transposed conjugate of A.

(c) If $A = A^{-1}$, A is involutory.

(d) If $A = A^*$, A is real.

(e) If $A = (A^T)^{-1}$, A is orthogonal. (Note: $(A^T)^{-1} = (A^{-1})^T$) (The inverse is the transposed matrix.)

(f) If $A = A^{T*}$, A is Hermitian. (The matrix is identical to its own associate matrix.)

(g) If $A = (A^{*T})^{-1}$, A is unitary. (The inverse is the associate matrix.)

7.2. Solution of N Linear Equations in N-unknowns

If A is a square matrix of order (n,n) and X and Y are column vectors of order n, then

$$Ax = y$$

is a set of n linear equations in the unknowns $(x_1, x_2, \ldots, x_n)$.

If A is nonsingular, then the solution is

$$x = A^{-1}y$$

If $y = 0$, then the equations are homogeneous.

(a) If $|A| \neq 0$, then the trivial solution $x = 0$ is the only solution.

(b) If $|A| = 0$, then there are infinitely many solutions.

If $y \neq 0$, then the equations are not homogeneous. If A is non-singular, the solution x is unique. A central problem arising in the solution of finite-difference analogues of partial differential equations is the solution of large-sets of linear algebraic equations such as $Ax = y$.

7.3. Eigenvalues and Eigenvectors

Let M be a square matrix of order n and x be a column vector of nth order. The product Mx generates a new column vector y

$$Mx = y.$$

The new vector y can be conceived as a transformation of the original vector x. (We might say M maps x into y.) If the new vector y is in the same direction as x then

$$y = \lambda x$$

where λ is a scalar, i.e.,

$$Mx = \lambda x$$

which can be written as the homogeneous set of equations

$$(M - \lambda I)\, x = 0$$

There is no solution x unless det $(M - \lambda I)$ vanishes.

Definition: $K = \lambda I - M$ is called the characteristic matrix of M.

Definition: $p(\lambda) = \det K$ is called the characteristic function of M, or, since $p(\lambda)$ is polynomial of nth degree in λ, it is also called the characteristic polynomial of M.

Definition: The equation $p(\lambda) = 0$ is called the characteristic equation of M. From the fundamental theorem of algebra, p has n roots λ_j, $j = 1(1)n$. In general, the roots are complex and may be multiple.

Definition: The roots, λ_j, of $p(\lambda) = 0$ are called the eigenvalues of M. Other names found in the literature include characteristic values, secular values, proper values, latent roots.

Theorem: The sum of the eigenvalues of any matrix A is TrA.

Definition: Two matrices A and B are similar if there exists a non-singular matrix R such that

$$B = R^{-1} A R$$

Definition: The spectral radius of a matrix A, denoted $\sigma(A)$, is defined as the magnitude of the eigenvalue λ_j of A with the largest magnitude

$$\sigma(A) = \max_j |\lambda_j|$$

7.3.1. Four basic theorems concerning eigenvalues and eigenvectors of a matrix.

Theorem I: If λ_i, $i = 1(1)n$ are the eigenvalues of A, then the eigenvalues of A^k are $(\lambda_i)^k$. More generally, if p(x) is a polynomial, the eigenvalues of p(A) are $p(\lambda_i)$.

Theorem II: If A is real and symmetric, all eigenvalues are real. (Also if A is Hermitian, all eigenvalues are real.)

Theorem III: Any similarity transformation RAR^{-1} applied to A leaves the eigenvalue of the matrix unchanged.

Proof: Let λ be an eigenvalue of A and x the associated eigenvector, then by definition

$$Ax = \lambda x$$

and $$RAx = \lambda Rx$$

Let $y = Rx$ so that $R^{-1}y = x$

then

$$RAR^{-1}y = \lambda y$$

thus λ is an eigenvalue of RAR^{-1} and y is the associated eigenvector.

Theorem IV: Cayley-Hamilton Theorem. Every square matrix satisfies its own characteristic equation in a matrix sense. If det $K(\lambda) = 0$ is the characteristic equation for a matrix M with eigenvalues λ, then $K(M) = 0$.

8. REFERENCE

Isaacson, E., and H. B. Keller, Analysis of Numerical Methods, 541, John Wiley and Sons, Inc., New York, 1966.

THE STEADY PROBLEM

J. J. O'Brien
Mesoscale Air-Sea Interaction Group
The Florida State University
435 Oceanography/Statistics Building
Tallahassee, Florida 32306-3041

1. INTRODUCTION

We introduce the solution of partial differential equations (PDEs) by considering elliptic partial differential equations which arise in oceanography and meteorology. Most of the elliptic equations encountered are well-posed linear problems for which it is easily shown that a unique solution exists. Therefore we concentrate on finding the solution rapidly and accurately. Later, when time-dependent problems are considered, we need to concentrate on stability and accuracy, and speed becomes a lesser priority. Thus numerical solutions for elliptic problems provide the framework for introducing finite differences for PDEs. These approximations lead to large sets of algebraic equations which require solution.

In physical systems, elliptic problems arise naturally in steady state problems. The Stommel problem (7.1) is an example of a modified Poisson equation. The Munk problem (7.2) leads to a biharmonic elliptic equation.

Numerical methods for solving elliptic equations are either iterative or direct. In iterative methods we guess the solution and improve it with a recursive algorithm which reduces the error. A direct method finds the solution without iteration. Direct methods are usually fast but limited to very special physical problems.

There are many common elliptic equations.

Laplace's Equation:

$$\frac{\partial^2 u}{\partial x^2} + \frac{\partial^2 u}{\partial y^2} + \frac{\partial^2 u}{\partial z^2} = 0 \tag{1}$$

Poisson's Equation:

$$\frac{\partial^2 u}{\partial x^2} + \frac{\partial^2 u}{\partial y^2} + \frac{\partial^2 u}{\partial z^2} = -g(x, y, z) \tag{2}$$

J. J. O'Brien (ed.), Advanced Physical Oceanographic Numerical Modelling, 51–72.

or

$$\nabla^2 u = -g$$

or, with non-constant coefficients

$$\nabla \cdot [f(x, y, z)\, \nabla u] = -g$$

Helmholtz Equation:

$$\nabla^2 u + f(x, y, z)u = -g(x, y, z)$$

or

$$\frac{\partial^2 u}{\partial x^2} + \frac{\partial^2 u}{\partial y^2} + \frac{\partial^2 u}{\partial z^2} + fu = -g(x, y, z) \qquad (3)$$

or, with non-constant coefficients

$$\nabla \cdot [h(x, y, z)\, \nabla u] + fu = -g$$

These require boundary conditions on all boundaries. The boundary conditions may be Dirichlet, Neumann, mixed or periodic. Any combination of these lead to a unique solution (sometimes to within an additive constant) to the linear problems.

There are several ways elliptic equations can arise: e.g., a useful model of a real fluid is one which assumes the fluid is incompressible. This filters out the sound waves from the solutions. In an incompressible fluid, the equation of continuity is

$$\nabla \cdot \underline{v} = 0$$

In numerical calculations, it is essential that this condition be satisfied at every grid point. This is accomplished by forming the "Divergence Equation" from the equations of motion.

$$\frac{\partial \underline{v}}{\partial t} + \underline{v} \cdot \nabla \underline{v} = -\frac{1}{\rho} \nabla p - g\underline{k} + \underline{F} \qquad (4)$$

If we take the divergence of the above and set it to zero, then

$$\frac{1}{\rho} \nabla^2 p = \frac{1}{\rho^2} \nabla \rho \cdot \nabla p + \nabla \cdot \underline{F} - \nabla \cdot (\underline{v} \cdot \nabla \underline{v}) \qquad (5)$$

This is Poisson's equation for p, which must be solved at every time step that we solve the equations of motion. Equations (4) and (5) form a convenient set for the velocity and pressure. Additional equations such as the first law of thermodynamics and an equation of state need to be introduced if density, ρ, is not a constant. As we shall see in this chapter, solving elliptic equations is time consuming. A fast solution for (5), needed every time step, is a major requirement for realistic models of incompressible flow. In practice, we do not entirely neglect the time rate of change of the divergence. Instead we add to the right hand side of (5)

$[\nabla \cdot \underline{v}/\Delta t]$ evaluated at the old time.

A similar quantity at the new time is assumed to be zero. If the finite difference solution for the velocity divergence is zero, we are adding nothing. In practice the velocity divergence is not zero, and this correction keeps this physical error small.

There is another situation in which elliptic equations frequently occur. Any time we have a physical problem with two-dimensional, nondivergent flow or transport, we need only calculate the rotational part of the flow. If $\nabla \cdot \underline{v} = 0$, then we can define a stream function and solve

$$\nabla^2\psi = \underline{k} \cdot \nabla x \underline{v}$$

to find the stream function. This is an elliptic equation for ψ if we know the curl of the vector field, $\underline{v}$. We use this in the next section for calculating the Sverdrup flow.

2. STATIONARY WIND-DRIVEN MASS TRANSPORT

The numerical methods for elliptic PDEs can be understood more easily if we study a model problem carefully. The solution of the Dirichlet problem in a rectangle is a convenient choice since it is linear and contains only second-order derivatives.

In a classical paper, Sverdrup [1947], it was shown that the time-independent wind stress curl is related to the northward mass transport in the ocean. Thus if we have an estimate of the wind field, we can calculate the steady mass transport field. Since this vector field is nondivergent, the scalar mass transport stream function is a convenient graphical representation of the Sverdrup solution.

The Sverdrup model problem

$$\beta v = \underline{k} \cdot \text{curl } \underline{\tau}/\rho \tag{6}$$

$$\frac{\partial u}{\partial x} + \frac{\partial v}{\partial y} = 0 \tag{7}$$

where u, v are vertically averaged mass transport. Equation (7) allows us to define a mass transport stream function

$$\frac{\partial \psi}{\partial y} = -u \tag{8}$$

$$\frac{\partial \psi}{\partial x} = v$$

For simplicity, assume a rectangular basin $x \, \varepsilon[0, a]$, $y \, \varepsilon[0, b]$. From prescribed $\underline{\tau}$, we can calculate v everywhere; then u can be calculated from

$$u(x, y) = \int_x^a \frac{\partial v}{\partial y} dx \tag{9}$$

where the boundary condition is $u(a, y) = 0$ along the "eastern" boundary. From (8) we can construct the Poisson's equation

$$\frac{\partial^2 \psi}{\partial x^2} + \frac{\partial^2 \psi}{\partial y^2} = \frac{\partial v}{\partial x} - \frac{\partial u}{\partial y} = \zeta \tag{10}$$

To solve (10) we need boundary conditions on all boundaries. For this simple example, we choose the northern and southern boundary to be on a zero in curl $\underline{\tau}/\rho$; thus $\psi = 0$ is appropriate. On the western boundary, we have to approximate the boundary condition, by solving

$$\frac{\partial^2 \psi}{\partial y^2} = - \frac{\partial u}{\partial y} \quad , \quad \psi(0, 0) = 0 \; , \quad \psi(0, b) = 0$$

This can easily be solved by the tridiagonal algorithm (see chapter 2). On the eastern boundary, $u = 0$; therefore $\psi = 0$.

We approximate the domain by a uniform finite difference grid, $\Delta x = \Delta y = \Delta$ with

$$x_j = (j-1)\Delta \quad , \quad j = 1(1)J \quad , \quad J - 1 = \frac{a}{\Delta}$$

$$y_k = (k-1)\Delta \quad , \quad k = 1(1)K \quad , \quad K - 1 = \frac{b}{\Delta}$$

At $j = 1$ and J for all k, and at $k = 1$ and K for all j, we know $\psi(x, y)$. At each grid point we approximate

$$\psi(j\Delta, k\Delta) = Q_{j,k}$$

and we know the curl of the mass transport, ζ_{jk} at each point.

Using standard second-order finite differences, we approximate (10) with

$$\frac{\partial^2 \psi}{\partial x^2} = \frac{Q_{j+1,k} + Q_{j-1,k} - 2Q_{j,k}}{\Delta^2} + 0(\Delta^2)$$

$$\frac{\partial^2 \psi}{\partial y^2} = \frac{Q_{j,k+1} + Q_{j,k-1} - 2Q_{j,k}}{\Delta^2} + 0(\Delta^2)$$

This leads to the set of linear equations

$$Q_{j+1,k} + Q_{j-1,k} + Q_{j,k+1} + Q_{j,k-1} - 4Q_{j,k} = \Delta^2\zeta_{j,k} = L_{j,k} \qquad (11)$$

$$j = 2(1)J - 1$$

$$k = 2(1)K - 1$$

In practice this is a very large set of equations; the product, JK, may be on the order 10^3-10^4. It is too large for standard Gauss elimination. The equations for J = 5, K = 6 is given in detail in Figure 1.

$$\begin{bmatrix} -4 & 1 & 0 & 1 & 0 & 0 & 0 & 0 & 0 & 0 & 0 & 0 \\ 1 & -4 & 1 & 0 & 1 & 0 & 0 & 0 & 0 & 0 & 0 & 0 \\ 0 & 1 & -4 & 0 & 0 & 1 & 0 & 0 & 0 & 0 & 0 & 0 \\ 1 & 0 & 0 & -4 & 1 & 0 & 1 & 0 & 0 & 0 & 0 & 0 \\ 0 & 1 & 0 & 1 & -4 & 1 & 0 & 1 & 0 & 0 & 0 & 0 \\ 0 & 0 & 1 & 0 & 1 & -4 & 0 & 0 & 1 & 0 & 0 & 0 \\ 0 & 0 & 0 & 1 & 0 & 0 & -4 & 1 & 0 & 1 & 0 & 0 \\ 0 & 0 & 0 & 0 & 1 & 0 & 1 & -4 & 1 & 0 & 1 & 0 \\ 0 & 0 & 0 & 0 & 0 & 1 & 0 & 1 & -4 & 0 & 0 & 1 \\ 0 & 0 & 0 & 0 & 0 & 0 & 1 & 0 & 0 & -4 & 1 & 0 \\ 0 & 0 & 0 & 0 & 0 & 0 & 0 & 1 & 0 & 1 & -4 & 1 \\ 0 & 0 & 0 & 0 & 0 & 0 & 0 & 0 & 1 & 0 & 1 & -4 \end{bmatrix} \begin{bmatrix} Q_{2,2} \\ Q_{3,2} \\ Q_{4,2} \\ Q_{2,3} \\ Q_{3,3} \\ Q_{4,3} \\ Q_{2,4} \\ Q_{3,4} \\ Q_{4,4} \\ Q_{2,5} \\ Q_{3,5} \\ Q_{4,5} \end{bmatrix} = \begin{bmatrix} L_{2,2} - Q_{1,2} - Q_{2,1} \\ L_{3,2} - Q_{3,1} \\ L_{4,2} - Q_{4,1} - Q_{5,2} \\ L_{2,3} - Q_{1,3} \\ L_{3,3} \\ L_{4,3} - Q_{5,3} \\ L_{2,4} - Q_{1,4} \\ L_{3,4} \\ L_{4,4} - Q_{5,4} \\ L_{2,5} - Q_{1,5} - Q_{2,6} \\ L_{3,5} - Q_{3,6} \\ L_{4,5} - Q_{4,6} - Q_{5,5} \end{bmatrix}$$

Figure 1. The matrix of coefficients for (11) when J is 5 and K is 6.

The set of equations (11) is block tridiagonal.

$$\begin{bmatrix} B & I & \phi & \phi \\ I & B & I & \phi \\ \phi & I & B & I \\ \phi & \phi & I & B \end{bmatrix} \begin{bmatrix} S_1 \\ S_2 \\ S_3 \\ S \end{bmatrix} = \begin{bmatrix} z_1 \\ z_2 \\ z_3 \\ z \end{bmatrix}$$

$S_i^T = [Q_{2,i}, Q_{3,i}, Q_{4,i}]$, and

$$B = \begin{bmatrix} -4 & 1 & 0 \\ 1 & -4 & 1 \\ 0 & 1 & -4 \end{bmatrix}$$

This block tridiagonal structure can be used to advantage in many schemes to solve elliptic PDEs.

Exercise:

Show that the general second-order PDE can be reduced to a block tridiagonal set of equations using second-order finite difference approximations.

$$a_1 P_{xx} + a_2 P_{xy} + a_3 P_{yy} + a_4 P_x + a_5 P_y + a_6 P = a_7$$

where $a_i = a_i(x, y)$, $i = 1(1)7$.

3. THE ITERATIVE SOLUTION OF LINEAR EQUATIONS

We will consider finite difference approximations to elliptic partial differential equations such as Poisson's equation. These methods all require the solution of a linear system of equations. The order of this system may be very large -- in some cases over 100,000. However, these linear systems frequently have special properties which permit us to construct particularly efficient iterative methods for their solution.

3.1. General Remark on the Convergence of Iterative Methods

We will consider iterative methods for the solution of the matrix equation $Ax = b$, where x and b are n-dimensional vectors, and A is a matrix of order n. We are trying to find the zero of the vector function $f(x) = Ax - b$. We can convert this to a fixed point problem by defining the function g(x) by $g(x) = x - f(x) = (I - A)x + b$. Then we are looking for fixed points; that is, vectors x such that $x = g(x)$. The easiest iterative method for this problem is to choose an initial guess $x^{(0)}$ for the vector and then define $x^{(\nu+1)} = g(x^{(\nu)})$ for $\nu \geqslant 0$. Under the proper conditions the vectors $x^{(\nu)}$ will converge to the solution x. An introductory course in numerical analysis will usually consider conditions under which this iterative process will converge for scalar functions g(x) of a single unknown x. We need to consider this question for matrix equations in the form

$$x = Mx + b$$

Here M is a matrix of order n, x and b vectors of dimension n. The iterative process is

$$x^{(\nu+1)} = Mx^{(\nu)} + b, \qquad x^{(0)} \text{ given.}$$

If we denote the error $x - x^{(\nu)}$ by $e^{(\nu)}$, then by subtraction of the above equations we obtain

$$e^{(\nu+1)} = Me^{(\nu)}$$

From this equation we have $e^{(1)} = Me^{(0)}$, $e^{(2)} = Me^{(1)} = M^2e^{(0)}$, and by induction we can prove

$$e^{(\nu)} = M^{\nu}e^{(0)}.$$

This convergence is dependent on the νth powers of M. We need to find conditions under which M^{ν} will approach zero. If M is convergent, then the process will converge to the desired answer; i.e., if $\sigma(M)<1$, then the error will eventually approach zero.

3.2. Iterative Methods for the Solution of Elliptic PDEs

Express any square matrix as the sum

$$A = D - E - F$$

where D is diagonal and E and F are strictly lower and upper triangular nth order square matrices, respectively.

We wish to solve the linear problem $Ax = b$ which has solution $x = A^{-1}b$. Using the above partition, we have

$$Dx = (E + F)\,x + b$$

We consider three simple methods which are related. For most applications, Method 3 or SOR is the fastest and, therefore, the most desirable. However, the theory for SOR depends on Methods 1 and 2. We also outline some more complicated iterative methods 4 and 5 that are faster but more difficult to implement. The reader is cautioned to be careful to separate the theory which proves that the algorithm works from the implementation of the method on the computer.

Method 1

This technique is the basic one on which we will build other techniques. It should not be used computationally. (This is true for scalar computers such as the CYBER 760; it may not be true for vector computers such as the CYBER 205.) This technique has been called

(a) Ordinary iteration

(b) Gauss-Jacobi iteration

(c) Point Jacobi

(d) Point Total Step Iterative Method

(e) Unaccelerated Richardson's Method

(f) Method of Simultaneous Displacements.

The names (b) and (f) are most common in the modern American literature.

Define the iteration scheme

$$x^{(m+1)} = D^{-1} (E + F) x^{(m)} + D^{-1}b, \qquad m \geq 0 \qquad (12)$$

in component form, (12) can be written

$$x_i^{(m+1)} = - \sum_{\substack{j=1 \\ j \neq i}}^{n} (a_{ij}/a_{ii}) x_j^{(m)} + b_i/a_{ii} \qquad i = 1(1)n \qquad (13)$$

where m is an integer scan counter. This equation (13) is the form used on the computer. Since D is a diagonal matrix, multiplication by D^{-1} is equivalent to dividing each component equation by the coefficient of the diagonal term.

Define: $M_1 = D^{-1} (E + F) = L + U$

where $L = D^{-1}E$ is strictly lower triangular and $U = D^{-1}F$ is strictly upper triangular.

Our basic theorem for iterative processes states that Method 1 converges if $\sigma(M_1) < 1$, i.e., M_1 is convergent.

On the computer we make any guess $x^{(0)}$ and compute successive approximations $x^{(1)}$, $x^{(2)}$, etc. Observe that we need to carry two arrays for $x^{(k)}$ and $x^{(k+1)}$.

Method 2

This second method is used extensively to solve many problems. It has been called

(a) Gauss-Seidel Iteration

(b) Point Gauss-Seidel

(c) Point Single-Step Iterative Method

(d) Liebmann or Unextrapolated Liebmann Method

(e) Method of Successive Displacements.

The names (a) and (e) are most common.

In matrix form, Method 2 can be written

$$(D - E)\, x^{(m+1)} = F\, x^{(m)} + b$$

In component form (for programming) we can write

$$a_{ii}\, x^{(m+1)} = - \sum_{j<i} a_{ij}\, x_j^{(m+1)} - \sum_{j>i} a_{ij}\, x_j^{(m)} + b_i \tag{14}$$

Define: an iteration matrix $M_2 = (D-E)^{-1} F$, then Method 2 converges if $\sigma(M_2) < 1$, i.e., M_2 is convergent.

We observe that Method 2 requires the storage of only one vector, x, since we actually replace components of $x^{(k)}$ by components of $x^{(k+1)}$ as quickly as we calculate them. Method 2 is intuitively faster than Method 1.

Method 3

This technique has been called

(a) SOR -Successive Over-Relaxation

(b) Extrapolated Liebmann

(c) Relaxation Method of Successive Displacements

(d) Gauss-Seidel Relaxation

(e) Gauss-Southwell Relaxation.

SOR is the most common. Recall that $A = D-E-F$ and $D^{-1}A = I-L-U$.

Define: A residual vector, R, which is precisely the error vector at any stage in the calculation is defined

$$R = D^{-1}b - x^{(m)} + Lx^{(m+1)} + Ux^{(m)} \tag{15}$$

It is very important to observe that the coefficient of component x_j is -1 in the second term on right. In problems with non-constant coefficients, we should normalize R such that the coefficient of the component x_j is minus one.

Define: an interative process

$$x^{(m+1)} = x^{(m)} + \alpha R^{(m)} \tag{16}$$

where α is called a relaxation parameter (or coefficient). In general, we can show that $0 < \alpha < 2$ is the condition for the method to work. There is another way to define this scheme. Let $\hat{x}^{(m+1)}$ be the Gauss-Seidel value; then Method 3 can be written

$$\hat{x}^{(m+1)} = Lx^{(m+1)} + Ux^{(m)} + D^{-1}b$$
$$x^{(m+1)} = x^{(m)} + \alpha(\hat{x}^{(m+1)} - x^{(m)}), \tag{17}$$

This set (17) clearly shows that SOR is an error correction method. If $\alpha > 1$, we are over correcting the error. It remains to be shown that this is faster than the Gauss-Seidel scheme in which we don't overcorrect the error at any stage in the calculation.

Note if $\alpha = 1$, then Method 3 $\equiv$ Method 2; $1 < \alpha < 2$, is called over-relaxation; $0 < \alpha < 1$, is called under-relaxation. We need to derive the iteration matrix for SOR. Using (15) and (16), we obtain

$$x^{(m+1)} = x^{(m)} - \alpha x^{(m)} + \alpha L x^{(m+1)} + \alpha U x^{(m)} + \alpha D^{-1}b$$

Solving for $x^{(m+1)}$

$$(I - \alpha L)x^{(m+1)} = (I - \alpha I + \alpha U)x^{(m)} + \alpha D^{-1}b$$

$$x^{(m+1)} = (I - \alpha L)^{-1} (I - \alpha I + \alpha U)x^{(m)} + (I - \alpha L)^{-1} \alpha D^{-1}b$$

$$x^{(m+1)} = M_3\, x^{(m)} + (I - \alpha L)^{-1} \alpha D^{-1}b$$

where M_3, the iteration matrix for Method 3 is

$$M_3 = (I - \alpha L)^{-1} (I - \alpha I + \alpha U)$$

Method 3 converges if $\sigma(M_3) < 1$, but $M_3(\alpha)$, hence we want to learn how to find the optimum α for which $||M_3||$ is a minimum.

In outlining SOR, we have yet to consider A as a matrix from an elliptic PDE. There is a general theorem which is useful.

Theorem: (Kahan) Let M_3 be the iteration matrix for SOR for an arbitrary matrix A, where $A = I - L - U$, and

$$M_3(\omega) = (I - \omega L)^{-1} (\omega U + (1 - \omega)I)$$

then the spectral radius of M_3 satisfies

$$\sigma(M_3) \geqslant |\omega - 1|$$

Proof: Define $\phi(\lambda) = \det(\lambda I - M_3)$ then inserting the definition of M_3

$$\phi(\lambda) = \det[\lambda I - (I - \omega L)^{-1} (\omega U + (1 - \omega)I]$$

and,

$$\phi(\lambda) = \det[\lambda(I - \omega L) - (\omega U + (1 - \omega)I)] \tag{18}$$

since det [AB] = det A det B and

$$\det (I - \omega L) = 1$$

Let $\lambda_i(\omega)$ be the roots of $\phi(\lambda)$ and also the eigenvalues of M_3, then from the theory of polynomials

$$\phi(0) = (-1)^n \prod_1^n \lambda_i(\omega)$$

where Π means the product, but, from (18)

$$\phi(0) = (\omega - 1)^n$$

hence, the spectral radius must satisfy

$$\sigma(M_3) \geqslant |\omega - 1|$$

This is an important result which tells us that for all problems, $Ax = b$, SOR may converge if $0 < \omega < 2$ but will not for other values.

Exercise:

Reconsider the Kahan theorem when we have not normalized the matrix A above to have the form $A = I-L-U$ but, instead $A = \gamma I-L-U$, where γ is a constant. Show that the spectral radius must satisfy $\sigma(M_3) \geqslant |\omega\gamma-1|$.

Before we study the application of SOR to elliptic PDEs, it is important to understand how to program this method. Since SOR is an iterative method, we must program the computer to stop either if the method isn't working or if the method has worked. Therefore, we must specify a maximum number of scans (m), ν_{max}, and an upper bound, ε, on the error vector if $\nu_{max} < m$, the method isn't working fast enough; if $\varepsilon \geqslant ||R||$, we have found the solution to the accuracy desired. Frequently, $||R||_\infty$ is used but $||R||1$, may also be employed conveniently. In a program, we compute

$$x_j^{(m+1)} = x_j^{(m)} + \alpha R_j \quad , \text{ for each } j$$

and $||R||_\infty$ for each cycle, m, where R_j is a component of R. If, after the calculation of a new vector, $x^{(m+1)}$, $\varepsilon > ||R||_\infty$, we have found the solution. This simplicity should not be confused with the theory in the next section.

The details of the following analysis are beyond the scope of this book; the reader is referred to Issacson and Keller [1966]. A useful theorem is:

Theorem: (Keller) Let α and β be non-zero scalars and let γ_1, γ_2, γ_3, and β, be arbitrary scalars, then

$$\det\,[\gamma_1 I - (\beta\gamma_2 L + \beta^{-1}\gamma_3 U)]$$

is independent of β.

The characteristic equation for M_3 is

$$\phi(\eta) = \det[\eta I - (I - \alpha L)^{-1}((1 - \alpha)I + \alpha U)]$$

where α is the relaxation parameter. Since $|I - \alpha L| = 1$ we can rewrite this as

$$\phi(\eta) = \det[\eta(I - \alpha L) + (\alpha - 1)I - \alpha U]$$

$$= \det[(\eta - 1 + \alpha)I - (\eta\alpha L + \alpha U)]$$

$$= \det[(\eta - 1 + \alpha)I - \eta^{1/2}(\eta^{1/2}\alpha L + \eta^{-1/2}\alpha U)]$$

Use of Keller's Theorem yields

$$\phi(\eta) = \eta^{1/2}\det[\frac{\eta + \alpha - 1}{\alpha\eta^{1/2}} I - (L + U)]$$

If $(\eta + \alpha - 1)/(\alpha\eta^{1/2}) = \lambda$, then the det $[\lambda I - (L + U)]$ can be identified as the characteristic equation for Method I, i.e., we have derived a nonlinear relationship between the eigenvalues of Methods I and III.

If we solve for η, we obtain

$$\eta = [\frac{\alpha\lambda}{2} \pm \left((\frac{\alpha\lambda}{2})^2 + 1 - \alpha\right)^{1/2}]^2$$

The complete analysis of this equation is beyond the scope of this book. However, we can state several properties for the model problem (11),

a) $\lambda_{j,k} = 1 - 2\sin^2(\frac{\pi j}{2(J - 1)}) - 2\sin^2(\frac{\pi k}{2(K - 1)})$

b) Since each λ is real, $|\eta| \geq 1$, if $\alpha \leq 0$

c) For $\alpha > 0$, η is monotonically decreasing until

$$|\eta| = |1 - \alpha| ,$$

then η is complex

d) If $\alpha > 2$, $|\eta| \geq 1$

e) $|\eta|$ is a minimum when $\alpha = \alpha_{opt}$ (α-optimum) and

$$\alpha_{opt} = \frac{2}{1 + (1 - \lambda_{1,1}^2)^{1/2}} \quad ; \; 1 < \alpha_{opt} < 2$$

f) When $\alpha > \omega_{opt}$, $\sigma(M_3) = |\eta_{min}| = \alpha - 1$

$$\text{where } \lambda_{1,1} = 1 - 2\sin^2\left(\frac{\pi}{2}\frac{1}{J - 1}\right) - 2\sin^2\left(\frac{\pi}{2}\frac{1}{K - 1}\right)$$

The relationship between $\sigma(M_3)$ and α is given in Figure 3.1

Exercise:

Derive the formula for $\lambda_{1,1}$ and α_{opt} for the case $\Delta x \neq \Delta y$.

4. Elliptic Equations with Neumann or Mixed Boundary Conditions

If we need to solve a model with Neumann boundary conditions; i.e., specifications of normal first derivatives, we need to recall that the solution is only unique to within a constant. The extra constant can usually be determined in physical problems by considering some integral property of the solution. If we apply SOR to the solution of this problem, the method usually will not converge unless special care is taken. The trick is to specify the answer at one grid point to be the value unity and to apply SOR and skip the special point. After the solution converges, we calculate some global property to the solution and rescale the answer at every grid point such that the global constraint is satisfied.

Experience shows that the same problem occurs with mixed boundary conditions; e.g. Luther [1982], and Romea [private communication].

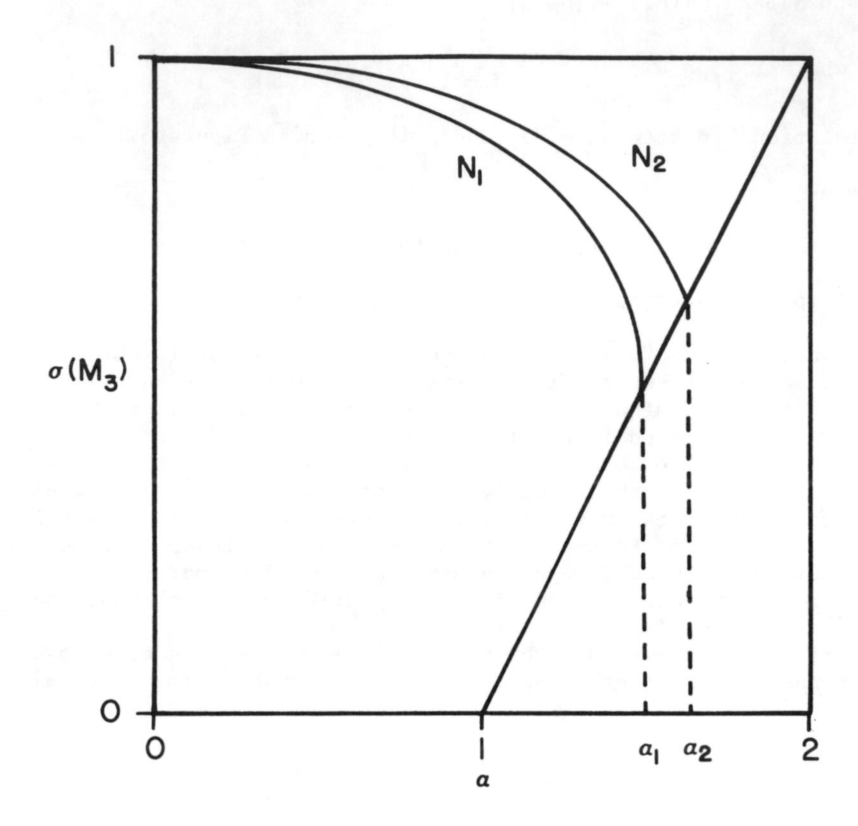

Figure 1. If N_1 = JxK and $N_2 > N_1$, we observe that the optimum relaxation parameter, α_{opt}, is larger for more grid points. If $\alpha<0$ or $\alpha>2$, $\sigma(M_3)>1$. For the model problem $1<\alpha_{opt}<2$, recall that $\alpha = 1$ is the Gauss-Seidel Method II. As the total number of grid points increases, α_{opt} shifts toward 2. It is better to overestimate α_{opt} due to the shape of the curve near α_{opt}.

5. OTHER ITERATIVE METHODS

Method 4: Line Relaxation

In Method 3, SOR, we correct each unknown in order. It is possible to accelerate the process by correcting an entire row at one time. For all the unknowns in a specific row, we can write

$$Q_{i-1} + BQ_i + Q_{i+1} = Z_i$$

where Q_i are the unknowns in row i and Z_i are the knowns and required boundary conditions associated with row i. In line relaxation, we sweep out the error for the row i using the tridiagonal algorithm using

$$- B\hat{Q}_i^{(m+1)} = [Q_{i-1}^{(m+1)} + Q_{i+1}^{(m)} - Z_i] \tag{19}$$

where $\hat{Q}_i$ is a Gauss-Seidel type value for all the unknowns in row i. We complete the iteration using

$$Q_i^{(m+1)} = Q_i^{(m)} + \alpha(\hat{Q}_i^{(m+1)} - Q_i)$$

In (19) it is important to recognize that, at any stage in the calculation, we know everything on the right hand side of (19). Since B is tridiagonal for any Poisson or Helmoholtz problem, Equation (19) can be solved for $\hat{Q}$ using the tridiagonal algorithm. In effect we are assuming that we know the answer on the rows below and above the row i and calculating the exact answer if $\alpha = 1$. Line relaxation is faster than SOR by $\sqrt{2}$, and the extra programming is relatively easy.

Method 5: Alternating-Direction SOR

If the reader understands Method 4, the following idea is straightforward. In essence, we use line relaxation by rows for one scan or sweep and then switch to line relaxation by columns. There are several hundred articles written about this method. It is very popular for time-dependent parabolic problems.

In Methods 3 and 4 we only use one relaxation parameter; in Method 5 we can use a sequence of parameters.

6. EQUIVALANCE OF ITERATIVE SCHEMES TO SOLUTION OF TIME-DEPENDENT PROBLEMS

The steady state problem $\nabla^2 u = -g$ can be written as the matrix problem

$$\nabla^2 u_{ij} = +\delta^2 g_{ij}$$

where $\delta^2 = \dfrac{\Delta x^2 \ \Delta y^2}{2(\Delta x^2 + \Delta y^2)}$.

and where the coefficient of the centerpoint in the operator ∇^2 is +1. The usual SOR scheme is

$$u_{ij}^{(\nu+1)} = u_{ij}^{(\nu)} + \alpha[\delta^2 g_{ij} - \nabla^2 u_{ij}] \tag{20}$$

Suppose we wish to solve the time-dependent problem

$$\frac{\partial u}{\partial t} = K \nabla^2 u + h(x,y)$$

which is approximated by the finite difference scheme

$$u_{ij}^{n+1} = u_{ij}^{n} + \left[\frac{K\Delta t}{-\delta^2} \nabla^2 u_{ij} + \Delta t \ h_{ij}\right] \tag{21}$$

Note that (20) and (21) are equivalent if

$$\alpha = \frac{K\Delta t}{\delta^2}$$

Thus Method 3, SOR, is equivalent to solving the time-dependent problem for the steady state solution! In fact, it is easily shown that any iterative solution to a steady problem is equivalent to finding the steady-state solution to a time-dependent problem. In the next chapter we will show that the condition $\alpha \leq 2$ is equivalent to the linear stability requirement for solving (21).

7. SOME OCEANOGRAPHICAL EXAMPLES OF ELLIPTIC EQUATIONS

7.1. The Stommel Problem

Stommel [1948], in a classical paper, explains why the strong boundary currents such as the Gulf Stream and Kuroshio are found on the west side of oceans rather that on the east. His model is a good elementary example of an elliptic problem.

Consider a rectangular basin $x \in [0, a]$; $y \in [0, b]$ on a beta plane, $f = f_0 + \beta(y - y_0)$, where β is the variation of the Coriolis parameter f with latitude. Stommel extended the Sverdrup balance [(6) and (7)] to include bottom friction

$$\beta v = \underline{\bar{k}} \cdot \text{curl } \underline{\tau}/\rho - r\zeta \tag{22}$$

where r is the linear bottom friction coefficient with units, s^{-1}, and ζ is the vertical component of vorticity. Introducing the streamfunction, ψ, as in 3.2, we can rewrite (22) as

$$r\nabla^2\psi + \beta\frac{\partial\psi}{\partial x} = k \cdot \text{curl } \underline{\tau}/\rho \tag{23}$$

In our rectangular basin we regard the east and west boundaries as walls on which the normal velocity vanishes. We will choose the north and south boundaries where the curl of the wind stress vanishes. Hence a natural boundary condition is $\psi = 0$ on all boundaries.

Exercise:

Solve the Stommel Problem (23) in a square basin using SOR. Let $x \in [0, a]$, $y \in [0, b]$, $\Delta x = a/41$, $\Delta y = b/21$, $a = 2000$ km, $b = 2000$ km, $\beta = 2.25 \times 10^{-11}\ m^{-1}\ s^{-1}$. Choose r such that $2\Delta x < r/\beta$.

$$\psi(j\Delta x, k\Delta y) = Q_{j,k}$$

$$\underline{k} \cdot \text{curl } \underline{\tau}/\rho = -\alpha \sin(\pi y/b)$$

This wind stress simulates strong eastward winds near $y = b$ and strong westward winds near $y = 0$. Choose α such that Q_{max} exceeds $10^7\ m^3\ s^{-1}$.

Figure 2 shows the solution of this classical problem which can be solved analytically. We can see a packing of the isolines of the streamfunction on the western side; this represents the western boundary current. Over most of the basin the physical balance is the Sverdrup balance. In the western boundary region the physical balance is

$$r\psi_{xx} + \beta\psi_x = 0$$

A simple scale analysis indicates that the boundary layer width is

$$L_s = r/\beta$$

we can have as narrow or as wide boundary current as we wish by adjusting the bottom friction! We wouldn't want to adjust β unless we were interested in an ocean on a planet other than Earth.

In the above exercise, we required the smallest scale, $2\Delta x$, to be less than the width of the boundary current, r/β. If we use SOR or another iterative method we are really finding the steady solution to the time-dependent Stommel problem. If we don't resolve the boundary layer by selecting Δx to be small enough, the solution will not converge since Rossby waves will not be able to develop the western boundary current.

The lesson here is straightforward. We need to understand the linear physics of any numerical calculation in order to select the appropriate parameters for the problem. In the Stommel problem, the linear term, $\beta\psi_x$, representing advection of planetary vorticity, makes the numerical solution impossible unless we resolve the boundary layer. In using the computer to model the ocean many similar situations arise.

We have used SOR to solve the exercise for several values of the relaxation coefficient to demonstate the usefulness of SOR over Gauss-Seidel. Define the error, $\varepsilon(\nu)$,

$$\varepsilon(\nu) = \sum_{j,k} (Q_{jk} - \psi(j\Delta x, k\Delta y)^2 / \sum_{j,k} \psi^2(j\Delta x, k\Delta y)$$

where ψ is the analytical solution at the grid points.

For the model problem, $\beta = 0$, the optimum relaxation parameter is $\omega_{opt} = 1.83$. In the Table 3.1, we present the number of scans, ν, where ε is reduced to various values

Table 1. Convergence of the Stommel Problem

ε / α	10^{-1}	10^{-2}	10^{-3}	10^{-4}	10^{-5}
1.0	84	147	202	252	294
1.76	18	29	35	39	49
1.83	15	24	40	48	58
1.90	13	48	57	66	74

There are two important facts to glean from Table 3.1. If we use Method 2, Gauss-Seidel (α=1), it requires 5 times more iteration than Method 3. Also the presence of the beta term redues slightly the optimum relation parameter. However, even a poor guess on α_{opt} allows much faster convergence than Gauss-Seidel.

Table 2. Number of scans necessary for error of model to analytical solution, in the Stommel problem. $\omega_{opt} = 1.83$

α \ ε	10^{-1}	10^{-2}	10^{-3}	10^{-4}	10^{-5}	10^{-6}
1.76	18	29	35	39	45	
1.83	15	24	40	48	58	
1.90	13	48	57	66	74	78
0.9	98	172	237			
1.0	84	147	202	252	294	323
1.1	70	122	167	208	242	226

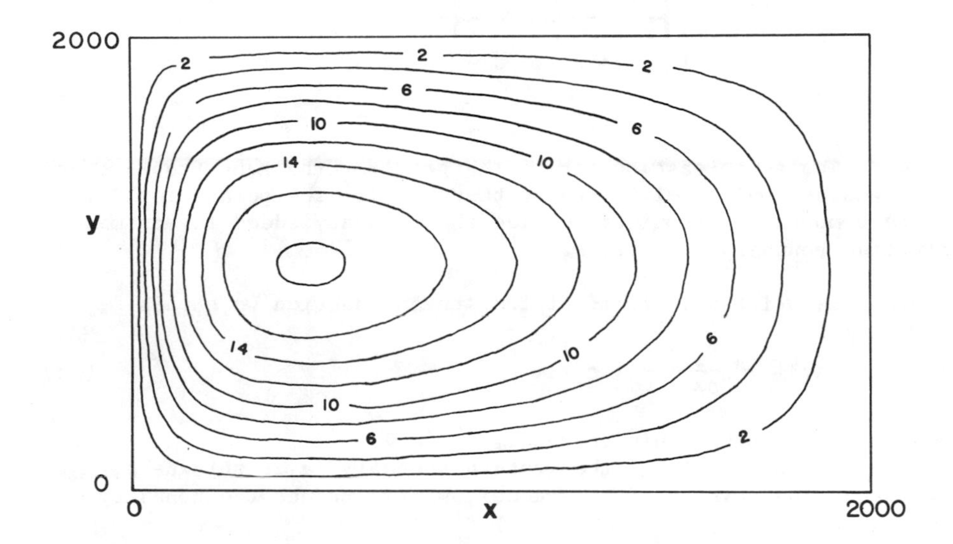

Figure 2. The solution to the Stommel Problem with $r = \beta\Delta x$ and $\alpha = 1.0 \times 10^{-9}$ m s^{-2}. The contours are in units of 10^6 m^3 s^{-1}.

7.2. The Munk Problem

Munk [1950] revisited the western boundary problem and choose to use Laplacian horizontal friction instead of bottom friction. He used the vorticity balance

$$\beta v = \underline{k} \cdot \text{curl}\underline{\tau}/\rho + A\nabla^2\zeta \tag{24}$$

where A is the constant eddy viscosity.

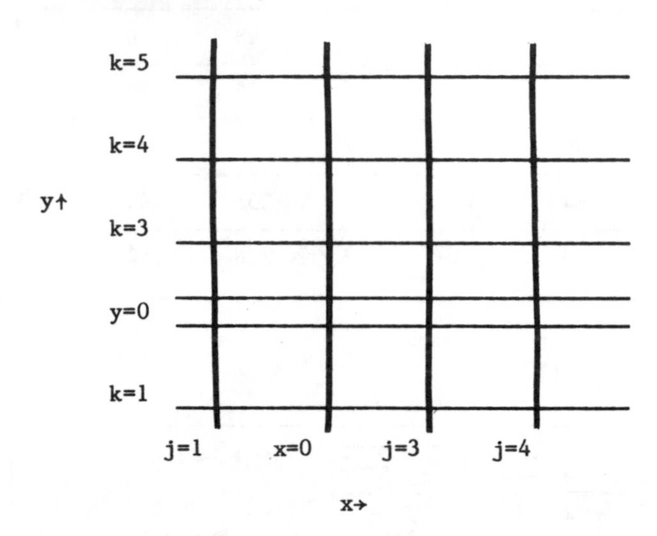

Figure 3. The south-east corner of the grid for the Munk Problem; j=2 is the western wall, k=1 is the southern wall. The points for j=1 or k=1 are computational points outside the boundary added to accommodate derivative boundary conditions.

When (24) is written in terms of the stream function we obtain

$$-A\nabla^4\psi + \beta\frac{\partial\psi}{\partial x} = \underline{k} \cdot \text{curl}\underline{\tau}/\rho \tag{25}$$

where ∇^4 is the biharmonic operator.

First we need to find the western boundary layer thickness, L_m. If we study the physics of this model, we can deduce the boundary layer balance

$$-A\psi_{xxxx} + \beta\psi_x = 0$$

which leads to the scale

$$L_m = (A/\beta)^{1/3}$$

If we decide to solve (25), we must choose A large enough to resolve the western boundary current or the solution will not converge. The oceanographic literature is filled with values of A ranging from 10^2 -10^4 m^2 s^{-1}. The particular choice is usually based on numerical reasons rather than physical reasons.

To solve (25) using SOR, it is suggested that the methods of Smith (1968) or Ehrlich (1971) be adopted. In this method we rewrite (25) as a coupled set of equations by defining

$$\nabla^2\psi = \chi \tag{26}$$

such that

$$A\nabla^2\chi - \beta\psi_x = -\underline{k} \cdot \text{curl}\,\underline{\tau}/\rho \tag{27}$$

Before we can attempt to solve this coupled set of elliptic equations we need to explore the boundary conditions for the problem. The easiest way is to recall the physics of the problem. Introducing horizontal friction means that at any solid boundary both the normal and tangential velocity components should be zero. In the Stommel problem, we set $\psi = 0$ on the boundaries. This satisfies the condition of no normal velocity. In order to satisfy the other condition, we require

$$\frac{\partial\psi}{\partial n} = 0 \;, \; n \text{ is normal direction}$$

This is accomplished by placing an extra grid point outside the boundary.

Consider Figure 3. The boundary condition is $\psi_x = 0$. We approximate this by adding the equation

$$\psi_{1,k} - \psi_{3,k} = 0 \tag{28}$$

where $\psi_{j,k}$ is the streamfunction. When we solve (26) we only solve for points inside the basin, thus only the boundary points are involved. However when we solve for (27), we need boundary conditions for χ. Along the north-south boundary, $\psi = 0$, therefore $\psi_{yy} = 0$. Hence $\nabla^2\psi = \psi_{xx}$. An approximation for $\psi_{xx} = \chi$ is

$$\psi_{1,k} + \psi_{3,k} - 2\psi_{2,k} = (\Delta x)^2\chi_{2,k}$$

which using (28) reduces to

$$\chi_{2,k} = 2(\psi_{3,k} - \psi_{2,k})/\Delta x^2$$

Similar boundary conditions may be derived on any rigid boundary. If the boundary is not a rigid wall, then we have to return to the physics to derive an approximation.

Suppose the northern boundary is at a position where curl $\underline{\tau}/\rho$ = 0. We know that v = 0 but not the east-west flow, u. However it will be a maximum or minimum. The condition $u_y = 0$ applies. This implies that

$$\psi_{yy} = 0$$

is an appropriate boundary condition and the boundary condition for χ is

$$\chi = 0 .$$

We have derived boundary conditions for the Munk problem.

Exercise:

Solve (26) and (27) subject to appropriate boundary conditions and the wind stress distribution given for the Stommel problem. The solution is shown in Figure 4.

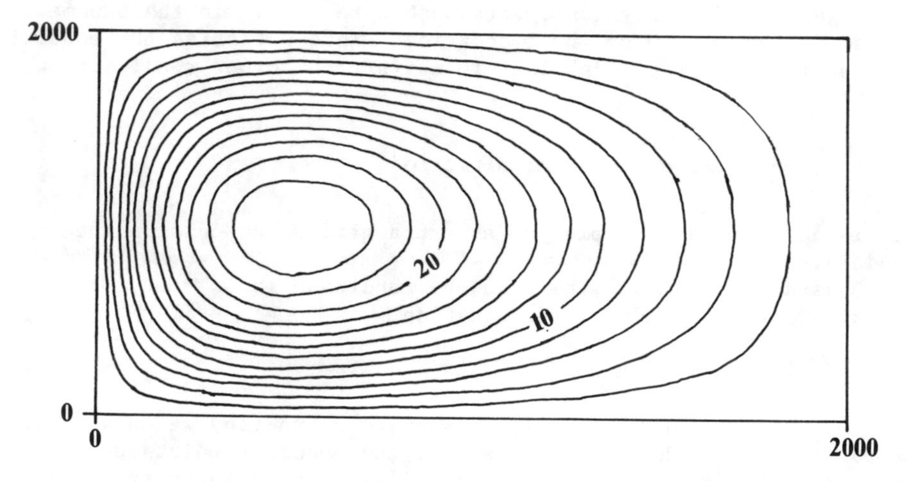

Figure 4. The solution to the Munk problem with $A = 1.8 \times 10^3$ m^2 s^{-1} and $d = 10^{-10}$ m s^{-2}; d is the amplitude of forcing.

ADVANCED METHODS FOR STEADY PROBLEMS--DIRECT ELLIPTIC SOLVERS

Mark E. Luther
Mesoscale Air-Sea Interaction Group
The Florida State University
Tallahassee, Florida 32306-3041
USA

1. INTRODUCTION

The iterative methods for elliptic equations described in the preceding section have many attractive qualities. They require a minimal amount of computer storage, they almost always converge to the true solution to the partial differential equation and they are simple to implement in most situations. They have the disadvantage, however, that they can be very expensive, since they usually require a large number of iterations. There is another class of solvers, the direct methods, that treat the finite difference equations as a large linear system and employ some tools from linear algebra to operate on the matrix of the finite difference coefficients. These methods take advantage of the sparseness and regular structure of the coefficient matrices to minimize storage requirements and operation counts. Although they usually require more storage than the iterative methods, they require fewer operations. We describe four direct methods that can be generally divided into two categories. The first three methods seek to reduce the order of the system of equations by appropriate transformations, while the fourth method performs a lower-upper decomposition on the entire system of equations.

2. DIRECT METHODS FOR SOLVING ELLIPTIC PDEs

Consider Poisson's Equation on a rectangle:

$$\nabla^2\phi = g, \qquad \text{with boundary conditions}$$

$$\phi(x,o) = \alpha(x), \qquad \phi(x,b) = \beta(x) \quad \text{(Dirichlet in y)}$$

$$\phi(o,y) = \phi(a,y) \qquad \text{(Periodic in x)}$$

In finite difference form, this gives rise to a five-point stencil on ϕ_{jk} as shown in Figure 1 with

J. J. O'Brien (ed.), Advanced Physical Oceanographic Numerical Modelling, 73–86.

$$\phi_{j,o} = \alpha_j \ , \quad \phi_{j,m} = \beta_j \ , \quad \phi_{o,k} = \phi_{n,k}$$

In matrix form, the finite difference equations are:

$$M\phi = g \tag{1}$$

Assume we have an n x m (x,y) solution mesh and $\Delta x = \Delta y = 1$.

The matrix M can be thought of as an m-1 by m-1 tridiagonal matrix of n by n sub-matrices as in Figure 2. The diagonal elements of M are general matrices A, and the off-diagonal elements are the identity matrix I. The vectors ϕ and g are vectors of m-1 subvectors, each of length n, i.e., each subvector ϕ_k is the vector of solution values along the kth row of the mesh. The 1's in the corners of A come from the periodic boundary conditions in x. The matrix M is very sparse and has a very regular block structure. All the direct methods take advantage of this structure. Irregularly shaped boundaries destroy this structure and so prohibit the use of direct methods. Direct methods are thus restricted to use with very simple geometries.

Advanced Method 1: Eigenvalue-Eigenvector Expansion

(Generalized form of Hockney's Method or Fourier Transform method)

Define the matrix Q such that

$$Q^{-1}AQ = \Lambda = \text{diag}\ (\lambda_1, \lambda_2, \dots \lambda_n)$$

Q is thus the matrix whose columns are the eigenvectors of A, and the λ_ν's are the corresponding eigenvalues. We also define the transforms

$$\Phi_k = Q^{-1}\ \phi_k \qquad \text{and } G_k = Q^{-1}g_k \tag{2}$$

For the kth row of (1) we have

$$I\phi_{k-1} + A\phi_k + I\phi_{k+1} = g_k. \tag{3}$$

Substituting (2) into (3) and multiplying by Q^{-1} we get

$$I\Phi_{k-1} + Q^{-1}AQ\Phi_k + I\Phi_{k+1} = G_k \tag{4}$$

For each value of the eigenvalues λ_ν we get a tridiagonal system of m-1 equations:

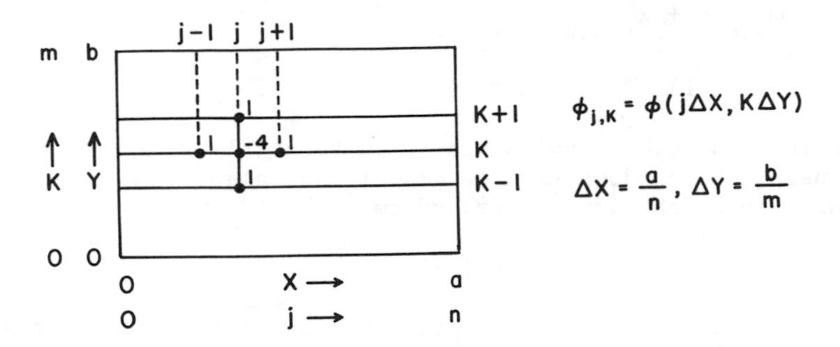

Figure 1. Schematic of finite difference mesh for Poisson's equation

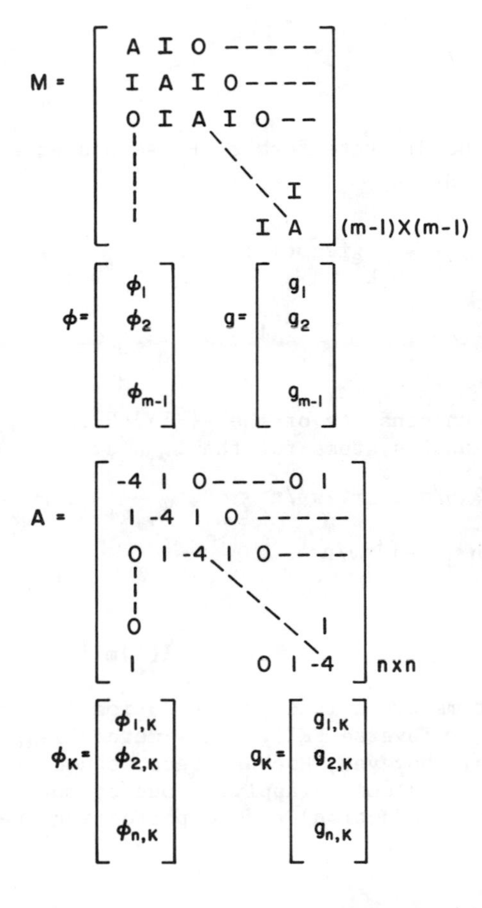

Figure 2. Diagram of matrices in equation 1

$$\Phi_{\nu,k-1} + \lambda_\nu \Phi_{\nu,k} + \Phi_{\nu,k+1} = G_{\nu,k} \tag{5}$$

$$\nu = 1(1)n\ , \qquad k = 1(1)m-1$$

Each of these n tridiagonal systems can be easily solved for the $\Phi_{\nu,k}$'s using the up-down (L-U decomposition) algorithm. The solution is then given by the inverse transform

$$\phi_k = Q\Phi_k\ , \qquad k = 1(1)m-1 \tag{6}$$

In practice, we compute the transform (2) using the Fast Fourier Transform (FFT) and (6) using the inverse FFT. There are several very efficient FFT packages available.

If we write the Fourier Transforms of (2) as:

$$\phi_{jk} = \sum_\nu \Phi_{\nu,k}\, e^{i2\pi\nu j/n}$$

$$g_{jk} = \sum_\nu G_{\nu k}\, e^{i2\pi\nu j/n} \tag{7}$$

and substitute these into the discrete form of Poisson's equation, then for the point (j,k) we get

$$\sum_\nu \Phi_{\nu,k-1} e^{i2\pi\nu j/n} + \sum_\nu \Phi_{\nu,k} e^{i2\pi\nu(j-1)/n} - 4\sum_\nu \Phi_{\nu,k} e^{i2\pi\nu j/n}$$

$$+ \sum_\nu \Phi_{\nu,k} e^{i2\pi\nu(j+1)/n} + \sum_\nu \Phi_{\nu,k+1} e^{i2\pi\nu j/n} = \sum_\nu G_{\nu,k} e^{i2\pi\nu j/n}$$

for $\nu = 1(1)n$. From the orthogonality of the $e^{i2\pi\nu j/n}$ terms, this gives n independent tridiagonal systems for the $\Phi_{\nu,k}$'s:

$$\Phi_{\nu,k-1} + (-4 + e^{i2\pi\nu/n} + e^{-i2\pi\nu/n})\Phi_{\nu k} + \Phi_{\nu,k+1} = G_{\nu k}$$

so that $\lambda_\nu = -4 + e^{i2\pi\nu/n} + e^{-i2\pi\nu/n}$

$$= -4 + 2\cos(2\pi\nu/n)$$

for $k = 1(1)m-1$.

Once these tridiagonal systems are solved, the solution is recovered using (7) (back-transform, or inverse FFT). This method is generally attributed to Hockney [1965]; however, Hockney recommended reducing the order of the tridiagonal systems by applying one or more passes of cyclic reduction (or odd-even reduction) before performing the Fourier Transforms.

Advanced Method 2: Cyclic Reduction (also called Odd–Even reduction or Recursive Cyclic Reduction)

For the kth row (k even) of (1) we have:

$$\begin{aligned} I\phi_{k-2} + A\phi_{k-1} + I\phi_k &= g_{k-1} \\ I\phi_{k-1} + A\phi_k + I\phi_{k+1} &= g_k \\ I\phi_k + A\phi_{k+1} + I\phi_{k+2} &= g_{k+1} \end{aligned} \tag{8}$$

Multiply the first equation by I, the second by -A and the third by I and add to get

$$I\phi_{k-2} + (2I - A^2)\phi_k + I\phi_{k+2} = g_{k-1} + g_{k+1} - Ag_k \tag{9}$$

Define $A^{(1)} = 2I - A^2$, $g_k^{(1)} = g_{k-1} + g_{k+1} - Ag_k$.

We have now reduced the solution mesh to only the even-numbered rows (see Figure 3). We could at this point apply the Fourier Transform method to solve for the ϕ_{jk}'s on the even rows, then use (8) to get ϕ_{jk} on the odd rows. This is the procedure usually referred to as "Hockney's Method" [see Hockney, 1965]. However, if m is a power of 2 -- say $m = 2^{\ell+1}$ -- we can perform this reduction process recursively ℓ times until we are left with only one row of unknowns to solve for. Using (9) we can define the recursions

$$A^{(r+1)} = 2I - (A^{(r)})^2, \quad g_k^{(r+1)} = g_{k-2^r}^{(r)} + g_{k+2^r}^{(r)} - A^{(r)}g_k \tag{10}$$

for $k = 2^r(2^r)m - 2^r$, $r = 1(1)\ell$

Each $g_k^{(r+1)}$ may overwrite the previous $g_k^{(r)}$ in computer storage since it will not be needed for the back-substitution phase; thus little additional storage is needed. Each $A^{(r)}$ is a polynomial in A of order 2^r, i.e., $A^{(r)} = P_{2^r}(A)$ and can be easily factored into a sequence of tridiagonal matrices:

$$A^{(r)} = B_1 B_2 B_3 \ldots B_{2^r} \tag{11}$$

Thus all computations involving $A^{(r)}$ can be performed using its factored form, i.e. by solving a sequence of tridiagonal systems, so that all computations involve only sparse matrices.

After ℓ reductions, we have

$$I\phi_0 + A^{(\ell)}\phi_{2^\ell} + I\phi_m = g_{2^\ell}^{(\ell)} \tag{12}$$

but ϕ_o and ϕ_m are known from the boundary conditions, so the $\phi_{j,2^\ell}$'s can be found easily using the factored form of $A^{(\ell)}$. However, it can be shown [see Buzbee, et al., 1970, sec. 10] that the computation of $g^{(r)}$ in (10) is subject to severe round-off error, so this method is not stable for most reasonable values of ℓ. There are various ways of stabilizing this method, generally known as the "Buneman variants on cyclic reduction" or simply "the Buneman algorithms." These algorithms are mathematically identical to the cyclic reduction algorithm just described but are not prone to the round-off error in the computation of the right-hand side of the reduced equations. The right-hand side of (9) may be written as

$$g^{(1)} = A^{(1)}A^{-1} g_k + g_{k-1} + g_{k+1} - 2A^{-1}g_k \tag{13}$$

since $A^{(1)}A^{-1} = 2A^{-1} - A$. Define

$$p^{(1)}_k = A^{-1}g_k \ , \qquad q^{(1)}_k = g_{k-1} + g_{k+1} - 2p^{(1)}_k \tag{14}$$

Then

$$\begin{aligned} g^{(1)}_k &= A^{(1)} p^{(1)}_k + q^{(1)}_k \\ g^{(r)}_k &= A^{(r)} p^{(r)}_k + q^{(r)}_k \end{aligned} \tag{15}$$

From (10), using the identity $(A^{(r)})^2 \equiv 2I - A^{(r+1)}$ we have

$$p^{(r+1)}_k = p^{(r)}_k - (A^{(r)})^{-1} (p^{(r)}_{k-2^r} + p^{(r)}_{k+2^r} - q^{(r)}_k) \tag{16}$$

$$q^{(r+1)}_k = q^{(r)}_{k-2^r} + q^{(r)}_{k+2^r} - 2p^{(r+1)}_k) \tag{17}$$

To compute the last term in (16) we actually solve

$$A^{(r)}(p^{(r)}_k - p^{(r+1)}_k) = p^{(r)}_{k-2^r} + p^{(r)}_{k+2^r} - q^{(r)}_k \tag{18}$$

using the factorized form of A from (11). Thus we never actually compute $(A^{(r)})^{-1}$. After ℓ reductions, we have:

$$A^{(\ell)}\phi_{2^\ell} = A^{(\ell)}p^{(\ell)}_{2^\ell} + q^{(\ell)}_{2^\ell} - (\phi_o + \phi_m)$$

or

$$\phi_{2^\ell} = p^{(\ell)}_{2^\ell} + (A^{(\ell)})^{-1}(q^{(\ell)}_{2^\ell} - \phi_o - \phi_m) \tag{19}$$

To back-solve, we use

$$\phi_{k-2^r} + A^{(r)}\phi_k + \phi_{k+2^r} = A^{(r)}p_k^{(r)} + q_k^{(r)}$$

to get

$$A^{(r)}(\phi_k - p_k^{(r)}) = q_k^{(r)} - (\phi_{k-2^r} + \phi_{k+2^r}) \tag{20}$$

Solving for $(\phi_k - p_k^{(r)})$, again using (11), we then find the ϕ_k's:

$$\phi_k = p^{(r)} + (\phi_k - p_k^{(r)}) \quad \text{for } k = 2^{r-1}(2^r)m-2^{r-1},$$
$$r = \ell(-1)1. \tag{21}$$

This is Buneman's first variant. It consists of:

Step (i): Compute $\{p_k^{(r)}, q_k^{(r)}\}$ using (16)-(18) for $r = 1(1)\ell$.

Step (ii): Back-solve using (20) and (21). This method is stable; however, it requires extra storage for the $p_k^{(r)}$'s. The $q_k^{(r)}$'s require no extra storage, since they can overwrite the g_k's and $q_k^{(r-1)}$'s.

We can improve on this by eliminating $p_k^{(r)}$ between (16) and (17). From (17), we have:

$$p_k^{(r+1)} = 1/2(q_{k-2^r}^{(r)} + q_{k+2^r}^{(r)} - q_k^{(r)}) \tag{22}$$

Substituting (22) into (16) and appropriately rearranging subscripts and superscripts, we get

$$\begin{aligned} q_k^{(r+1)} = {} & q_{k-2^r}^{(r)} - q_{k-2^{r-1}}^{(r-1)} + q_k^{(r)} + q_{k+2^r}^{(r)} \\ & + (A^{(r)})^{-1}[q_{k-3\cdot 2^{r-1}}^{(r-1)} - q_{k-2^r}^{(r)} + q_{k-2^{r-1}}^{(r-1)} - 2q_k^{(r)} \\ & + q_{k+2^{r-1}}^{(r-1)} - q_{k+2^r}^{(r)} + q_{k+3\cdot 2^{r-1}}^{(r-1)}] \end{aligned} \tag{23}$$

for $k = 2^r(2^r)m - 2^r$, with

$$q_k^{(0)} = g_k, \quad q_k^{(1)} = g_{k-1} + g_{k+1} - 2A^{-1}g_k .$$

To back-solve, we use

$$\phi_k = 1/2(q^{(r-1)}_{k-2^{r-1}} + q^{(r-1)}_{k+2^{r-1}} - q^{(r)}_k) - (A^{(r)})^{-1}(\phi_{k-2^r} + \phi_{k+2^r} - q^{(r)}_k) . \tag{24}$$

Of course, we never actually compute $(A^{(r)})^{-1}$, but rearrange (23) and (24) to take advantage of the factorization of $A^{(r)}$ in (11). This is known as "Buneman's second variant."

To summarize,

Step (i): compute $q_k^{(r)}$'s from (23)
Step (ii): Back-solve using (24).

This method requires much less storage than Buneman's first variant requires but takes only a few more floating point operations.

Operation Count: Since $A^{(r)}$ is a polynomial of order 2^r in A, the solution of $q_k^{(r+1)}$ will require the solution of 2^r tridiagonal systems. If A is order n, then $\sim 3n(2^r)$ operations are required for Poisson's equation, since the off-diagonal elements of A are 1. There are $m/2^r-1$ $q_k^{(r)}$'s, so one cycle takes $\sim 3/2$ nm operations, as does one back substitution. For ℓ cycles of reduction and back-substitution, we have a total operation count of $\sim 3mn\ell = 3mn \log_2 m$. (It should be pointed out that m does not have to be a power of 2, but it simplifies the algorithm considerably if it is.)

The operation count for the Fourier Transform method comes primarily from the FFT step. To perform an FFT of length n requires $\sim n \log_2 n$ operations. We must do m of these forward transforms and m back transforms for a total operation count of $\sim 2\, mn \log_2 n$. Another 3mn operations are required for the tridiagonal solution, for a new total of $mn(2\log_2 n + 3)$. So we see that the cyclic reduction and Fourier transform methods are comparable. We can combine the two and get a third method whose operation count is less than either used independently.

Advanced Method 3: Fourier Analysis – Cyclic Reduction (FACR(ℓ)) Algorithm

This method, suggested by Hockney, is often referred to as "Hockney's Method." The FACR(ℓ) algorithm begins with ℓ sweeps of cyclic reduction (where now ℓ is not the same ℓ as used in Method 2, and is yet to be determined), after which we have

$$I\phi_{k-2^\ell} + A^{(\ell)}\phi_k + I\phi_{k+2^\ell} = 1/2\ A(q^{(\ell-1)}_{k-2^{\ell-1}} + q^{(\ell-1)}_{k+2^{\ell-1}} - q^{(\ell)}_k) + q^{(\ell)} \tag{25}$$

But now this can be solved using the Fourier Transform method. Recall that $A^{(\ell)} = P_{2\ell}(A)$ where $P_{2\ell}$ is a polynomial of order 2^ℓ. Now we make the transformation

$$\Phi_k = Q^{-1}\phi_k \;,\quad \overline{q}_k = Q^{-1}q_k \;,\quad k = 2^\ell(2^\ell)m-2^\ell \tag{26}$$

Substituting into (25) and multiplying by Q^{-1} we get

$$\phi_{k-2^\ell} + Q^{-1}A^{(\ell)}Q\phi_k + \phi_{k+2^\ell} = 1/2\; Q^{-1}A^{(\ell)}Q(\overline{q}^{(\ell-1)}_{k-2^{\ell-1}} + \overline{q}^{(\ell-1)}_{k+2^{\ell-1}} - \overline{q}^{(\ell)}_k) + \overline{q}^{(\ell)}_k \tag{27}$$

Which reduces to n independent tridiagonal systems of order $m/2^\ell-1$ since

$$Q^{-1}A^{(\ell)}Q = P_{2\ell}(\Lambda)$$

so that (27) becomes, for the νth component:

$$\Phi_{\nu,k-2^\ell} + P_{2\ell}(\lambda_\nu)\,\Phi_{\nu,k} + \Phi_{\nu,k+2^\ell} = 1/2\; P_{2\ell}(\lambda_\nu)(q^{(\ell-1)}_{k-2^{\ell-1}} + \overline{q}^{(\ell-1)}_{k+2^{\ell-1}} - \overline{q}^{(\ell)}_k) + \overline{q}^{(\ell)}_k \tag{28}$$

for $k = 2^\ell(2^\ell)m-2^\ell$.

These systems are solved using the up-down algorithm as before. The ϕ_k's are found from the inverse transform, for $k = 2^\ell(2^\ell)m - 2^\ell$, and the remaining ϕ_k's are found by back substitution using (24). Hockney, in his original paper, used $\ell = 1$ and later extended the method for arbitrary ℓ. Swarztrauber [1977] has shown that there is an optimum value for ℓ. The cyclic reduction part of this algorithm takes $\sim 3mn\ell$ operations, while the FFT portion takes $\frac{mn}{2^{\ell-1}}\log_2 n$ operations, for a total of $C_\ell = 3mn\ell + 2^{1-\ell}mn\log_2 n$. C_ℓ is minimized at $\ell = \log_2\log_2 n + \log_2(2/3\ \ln 2)$ or $\ell \simeq \log_2\log_2 n - 1$. For this value of ℓ we get a total operation count of $3mn\log_2\log_2 n$.

These three methods are limited to separable elliptic equations. In fact the Fourier Transform method is just a computer implementation of the separation of variables technique. Although we have used Poisson's equation on a rectangle with Dirichlet boundary conditions in y and periodic in x, these methods can be generalized to handle any separable equation, i.e., any equation of the form:

$$a(x)u_{xx} + b(x)u_x + c(x)u + d(y)u_{yy} + e(y)u_y + f(y)u = g(x,y),$$

and to handle Dirichlet, periodic or Neumann boundary conditions in either direction. They are restricted to rectangular regions, or at least to regions that can be mapped onto a rectangle. They have the disadvantage of being rather difficult to program and of requiring a large amount of storage. There are, however, many packaged subroutines in existence to perform these algorithms, so you should never have to program them yourself.

Advanced Method 4: A Block Method

Consider now the most general form of an elliptic equation:

$$a(x,y)\phi_{xx} + b(x,y)\phi_{xy} + c(x,y)\phi_{yy} + d(x,y)\phi_x + e(x,y)\phi_y + f(x,y)\phi = g(x,y) \quad (29)$$

This equation arises in many practical applications in meteorology and oceanography and, in most cases of interest, is not separable. It occurs most often in models seeking normal modes of oscillation in highly variable background density and velocity fields. It cannot be solved using any of the direct methods described thus far. Iterative schemes can handle it, but for most applications, they are too time consuming. We will describe another method, generally called a block method.

In finite difference form, (29) gives rise to the nine-point stencil shown in Figure 4. The matrix form of the finite difference equations is shown schematically in Figure 5. The matrix of the finite difference coefficients, M, can be thought of as a tridiagonal matrix of tridiagonal sub-matrices. We can factor M, just as we did for a general tridiagonal scalar matrix. Let M = LU, where L is lower-triangular and U is upper triangular as in Figure 6, and where the diagonal and sub-diagonal elements of U and L, $\overline{A}_k$ and $\overline{B}_k$, are given by the recursion formula:

$$\overline{A}_{k-1}\,\overline{B}_k = B_k \qquad k = 2(1)m \quad (30)$$

$$\overline{A}_k = A_k - \overline{B}_k\, C_{k-1} \qquad k = 2(1)m \quad (31)$$

and $\overline{A}_1 = A_1$.

On a vector computer (a Class VI machine), these recursions can be performed very efficiently, since A_k, B_k and C_k are tridiagonal. To obtain ϕ_k, we first solve $L\phi = g$ (forward sweep) by

$$\phi_k = g_k - \overline{B}_k\,\phi_{k-1} \qquad k = 2(1)m \quad (32)$$

with $\phi_1 = g_1$.

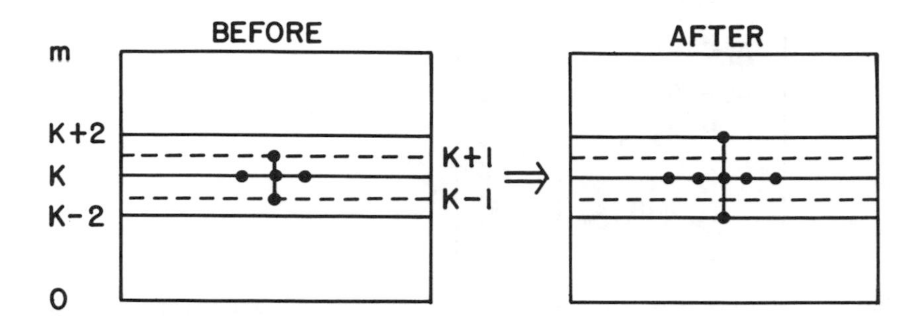

Figure 3. Finite difference stencil before and after one pass of cyclic reduction.

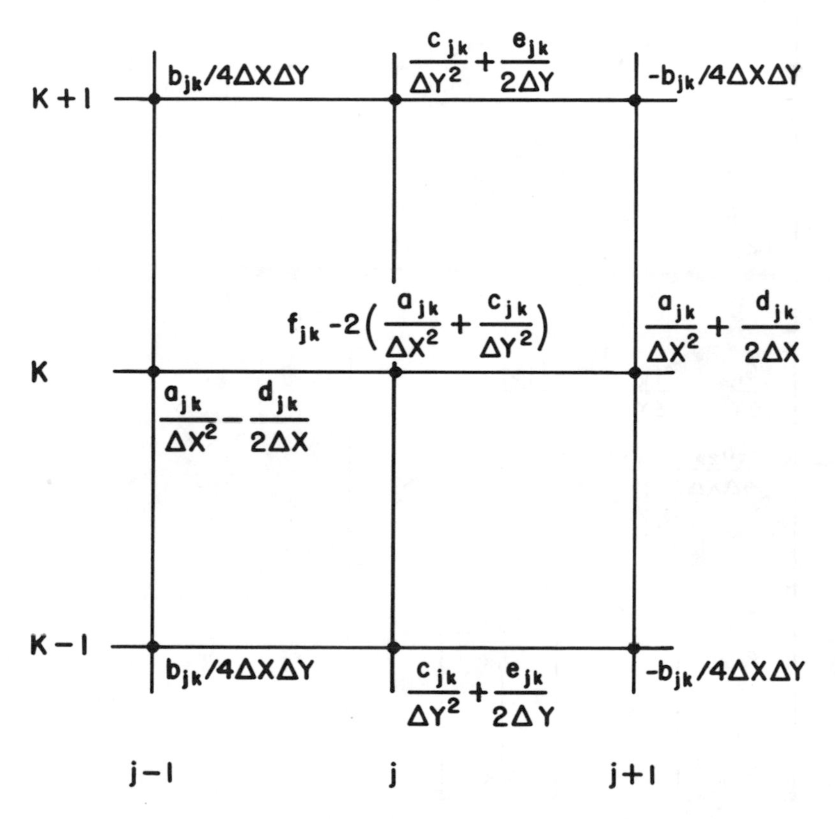

Figure 4. Nine-point finite difference stencil for the general non-separable elliptic equation.

$$M\phi = \begin{bmatrix} A_1 & C_1 & & & \\ B_2 & A_2 & C_2 & & \\ & B_3 & A_3 & C_3 & \\ & & \ddots & \ddots & \\ & & & B_m & A_m \end{bmatrix} \begin{bmatrix} \phi_1 \\ \phi_2 \\ \phi_3 \\ \vdots \\ \phi_m \end{bmatrix} = \begin{bmatrix} g_1 \\ g_2 \\ g_3 \\ \vdots \\ g_m \end{bmatrix}$$

Figure 5a. Matrix form of the finite difference equations for the general elliptic equation

$$A_K = \begin{bmatrix} f_{1,K} - 2\left(\frac{a_{1,K}}{\Delta X^2} + \frac{C_{1,K}}{2\Delta X}\right) & \frac{a_{1,K}}{\Delta X^2} + \frac{d_{1,K}}{2\Delta X} & \\ \frac{a_{2,K}}{\Delta X^2} - \frac{d_{2,K}}{2\Delta X} & f_{2,K} - 2\left(\frac{a_{2,K}}{\Delta X^2} + \frac{C_{2,K}}{\Delta Y^2}\right) & \ddots \\ & \ddots & \ddots \end{bmatrix}_{n \times n}$$

$$B_K = \begin{bmatrix} \frac{C_{1,K}}{\Delta Y^2} - \frac{e_{1,K}}{2\Delta Y} & \frac{-b_{1,K}}{4\Delta X \Delta Y} & \\ \frac{b_{2,K}}{4\Delta X \Delta Y} & \ddots & \ddots \\ & \ddots & \ddots \end{bmatrix}_{n \times n}$$

$$C_K = \begin{bmatrix} \frac{C_{1,K}}{\Delta Y^2} + \frac{e_{1,K}}{2\Delta Y} & \frac{b_{1,K}}{4\Delta X \Delta Y} & \\ \frac{-b_{2,K}}{4\Delta X \Delta Y} & \ddots & \ddots \\ & \ddots & \ddots \end{bmatrix}_{n \times n}$$

$$\phi_K = \begin{bmatrix} \phi_{1,K} \\ \phi_{2,K} \\ \vdots \\ \phi_{n,K} \end{bmatrix} \qquad g_K = \begin{bmatrix} g_{1,K} \\ g_{2,K} \\ \vdots \\ g_{n,K} \end{bmatrix}$$

Figure 5b. Elements of the sub-matrices in Figure 5a

$$L = \begin{bmatrix} I & 0 & \cdots & & \\ \bar{B}_2 & I & 0 & \cdots & \\ & \bar{B}_3 & I & 0 & \cdots \\ & & \ddots & \ddots & \\ & & & \bar{B}_m & I \end{bmatrix}$$

$$U = \begin{bmatrix} \bar{A}_1 & C_1 & 0 & \cdots & \\ 0 & \bar{A}_2 & C_2 & 0 & \cdots \\ & 0 & \bar{A}_3 & C_3 & \cdots \\ & & \ddots & \ddots & \ddots \\ & & & & \bar{A}_m \end{bmatrix}$$

Figure 6. Lower-upper (L-U) decomposition of matrix M

Then $U\phi = \phi$ is solved (backward sweep) by

$$A_m \phi_m = \phi_m$$

$$A_k \phi_k = \phi_k - C_k \phi_{k+1} , \quad k = m - 1(-1)1 \tag{33}$$

where ϕ_k on the right hand side of (33) comes from (32). Again, (32) and (33) can be performed very efficiently on a vector computer such as a CRAY or a CDC Cyber 205, whereas on a scalar or sequential computer they would be very time consuming. This method suffers, alas, from the same difficulty as the other direct methods; it is restricted to regions that map nicely onto a rectangle. It can handle very general boundary conditions, from simple Dirichlet to fully mixed derivative (or oblique derivative) boundary conditions. It is very fast and accurate on a vector machine, but has the additional disadvantage of being very machine-specific, because it uses what are called "hardware vector instructions" in the matrix recursion formulas to take advantage of the great speed of the vector processors. As before, there are canned subroutines available from NCAR (The National Center for Atmospheric Research) that will perform the computations for you, so you never have to program the method yourself. This method is general in that it will solve <u>any</u> elliptic equation, either separable or nonseparable.

In summary, we have seen four of the most efficient and most commonly used direct elliptic solvers. They are all much faster than iterative schemes (usually an order of magnitude faster), except for perhaps multi-grid methods. Their primary disadvantages are their large storage requirements and their restriction to very simple geometries.

APPLICATION OF MULTI-LEVEL TECHNIQUES TO THE STOMMEL PROBLEM WITH IRREGULAR BOUNDARIES

T. G. Jensen
Geophysical Institute
Department of Physical Oceanography
University of Copenhagen

ABSTRACT. Multigrid solution techniques for elliptic PDE's are presented and applied to a simple Stommel-type ocean model. Two general multigrid strategies are outlined: the Correction Scheme (CS), and the Full Approximation Scheme (FAS), and the choices of smoothers, injections and interpolations between different grids are discussed. The problem of handling irregular boundaries is discussed in some detail. A section outlines a method to include islands in ocean models, when a stream function is used to describe the flow. Finally, test results comparing the efficiency of SOR computations and multigrid methods are shown using a time-dependent Stommel model on a spherical grid. Solutions with rectangular and irregular coastlines are presented. The multigrid method is faster in the initial phases of a spin up process for a rectangular ocean up to a factor of 5 and for irregular coastlines up to a factor of 3.

1. INTRODUCTION

In early atmospheric large-scale circulation models, the non-divergent barotropic vorticity equation was used for prediction. Also, the first simple wind-driven models of ocean circulation were based on conservation of vorticity. In more recent studies, e.g. Bryan [1969], the same equations play a central role as a prediction equation for the barotropic mode of the velocity field. The equation is solved by successive relaxations, a method which is very time consuming.

The objective of the present study is to show that the relatively new multigrid technique can be used as an alternative and to test whether correct results can be obtained when irregular coastlines are introduced. Secondly, the efficiency compared to SOR is investigated.

2. THE MULTIGRID METHOD

A traditional solution method of elliptic partial differential

J. J. O'Brien (ed.), Advanced Physical Oceanographic Numerical Modelling, 87–110.

equations is by means of iterations. Successive relaxations such as Gauss-Seidel or Successive Overrelaxation (SOR) are very commonly used to solve elliptic equations in irregularly shaped domains where direct solvers such as Fourier Transform methods are not applicable.

The multigrid method as a fast solver was developed in the 1970s. It is especially applicable to discrete elliptic boundary value problems. However, a large number of problems involving the solution of PDEs or systems of PDEs may benefit from the multigrid technique. For a comprehensive review of multigrid methods in general, reference is made to Hackbusch and Trottenberg [1982].

The idea of using multigrids in an adaptive algorithm was originated by Brandt [1972]. Here the Multi-Level Adaptive Technique (MLAT) was introduced. This allows the grid choice to be determined by the computer program, depending on the convergence rates or the spatial variability of the forcing. In the following, a short presentation of the multigrid method will be given.

An iterative process using relaxation is effective in smoothing out errors which have scales of just a few grid points, but it is ineffective in smoothing out errors on larger scales. This has been demonstrated by Fourier analysis of the errors

$$V^k = U^k - u^k$$

where U^k is an approximate solution to the problem in some stage of the iteration and U^k is the exact solution.

This means that when the convergence rate is slow, it is due to large-scale errors. The idea of the multigrid approach is to use a coarser grid to speed up the smoothing of the errors which have large scales on the fine grids. The method will be outlined below.

Let a discrete elliptic boundary value problem in a two-dimensional domain with boundary conditions be given, where

$$L^k U^k = F^k \qquad (1)$$

is its discretization on a uniform grid k with grid distance h. L^k is a linear difference operator, U^k the solution, and F^k the forcing function. The assumption of linearity of L^k is not essential to the multigrid method. Also the dimensions of the domain may be larger than two and a non-uniform grid can be applied.

As mentioned previously, the error at any stage in an iteration process can be written

$$V^k = U^k - u^k \qquad (2)$$

which satisfies the residual equation

$$L^k V^k = r^k \qquad (3)$$

where the residual r^k is

$$r^k = F^k - L^k u^k. \qquad (4)$$

The differential operator for the residual equation is only the same as in the original equation in the case where L^k is linear.

Assume that an approximate solution or initial guess has been obtained on a grid k. Denote the next coarser grid by k-1, having a grid distance larger than h. Normally, the so-called "standard coarsening" is used where the grid distance on k-1 is 2 h. The error at grid k obeys the residual equation above, so this equation can be solved faster on grid k-1.

Let I_k^{k-1} be an interpolation operation from k to k-1. One example is simple injection.

The residual equation on grid k-1 yields now

$$L^{k-1}\, V^{k-1} = I_k^{k-1}\, r^k \tag{5}$$

where the L^{k-1} is a discretization of the differential operator on grid k-1, and the right-hand side is calculated from the approximate solution.

After the last equation is solved on the coarser grid, the solution V^{k-1} or an approximate solution v^{k-1} can be interpolated back to grid k,

$$v^k = I_{k-1}^k\, v^{k-1} \tag{6}$$

where I_{k-1}^k is an interpolation operator from grid k-1 to grid k. A new approximation to the solution is then

$$\tilde{u}^k = u^k + v^k. \tag{7}$$

Since the residual equation on grid k-1 can be solved by iteration, a still coarser grid k-2 can be introduced, and in this manner the process is done recursively using coarser and coarser grids until a solution can quickly be found on a very coarse grid.

Thus the multigrid method involves smoothing by relaxation, fine-to-coarse grid transfers of residuals, smoothing of errors, and finally, error correction of fine grid solutions.

The process described above is called the Correction Scheme (CS) by Brandt [1977]. A similar process is the Full Approximation Scheme (FAS), suggested by Brandt. Using this scheme, nonlinear differential operators can be used without difficulty.

Brandt introduces the function

$$U^{k-1} = I_k^{k-1} + V^{k-1}$$

where U^{k-1} is the solution to a differential equation,

$$L^{k-1}\, U^{k-1} = F_*^{k-1} \tag{8}$$

where F_*^{k-1} is a modified forcing.
An approximate residual equation is

$$L^{k-1}(I_k^{k-1}u^k + V^{k-1}) - L^{k-1}(I_K^{k-1}u^k) = I_k^{k-1}r^k. \quad (9)$$

In the linear case this reduces to (5). The problem is then to solve for U^{k-1} the equation

$$L^{k-1}U^{k-1} = L^{k-1}(I_k^{k-1}u^k) + I_k^{k-1}r^k,$$

or

$$L^{k-1}U^{k-1} = F_*^{k-1}$$

where the modified forcing is

$$F_*^{k-1} = L^{k-1}(I_k^{k-1}u^k) + I_k^{k-1}(F^k - L^k u^k). \quad (10)$$

As seen, in this case an equation for the full solution u^{k-1}, instead of equation for the residual, is solved on the coarse grid. However, the forcing has been corrected by the truncation error on grid k relative to grid
k-1.

The correction of the fine grid solution u^k can be found by calculating the error on the coarse grid,

$$v^{k-1} = u^{k-1} - I_k^{k-1}u^k \quad (11)$$

then interpolating v^{k-1} to grid k by

$$v^k = I_{k-1}^k v^{k-1}, \quad (12)$$

and calculating a new fine grid solution

$$\tilde{u}^k = u^k + v^k. \quad (13)$$

As with the CS scheme, the large-scale errors are smoothed by this process while small-scale errors eventually introduced by the interpolation from coarse to fine grid are easily smoothed on the fine grid.

The following notation for the different operations is used:

O(n_2) - smoothing by relaxation; n_2 iterations;
\ - fine-to-coarse grid transfer of residuals (CS) or solution (FAS);
/ - coarse-to-fine grid interpolation and error correction;
□ - exact solution calculated.

In a case with 3 grids, the combinations in Figure 1 are termed V-cycles and W-cycles for obvious reasons.

Figure 1.

For the V-cycle this indicates that n_1 iteration steps are made by relaxation; then, n_2 relaxation steps are made on a coarser grid before solving exactly on the coarsest grid. After this, returns are made as shown. A cycle is one step in a multigrid iteration, and the cycles may be repeated successively until a satisfactory solution is obtained.

However, a more efficient calculation can be obtained if the process is initated using coarse grid approximations as a first guess to the solution to the finer grids. Combined with V- or W-cycles, this technique is called the Full Multigrid Method.

An example is a calculation path as shown in Figure 2

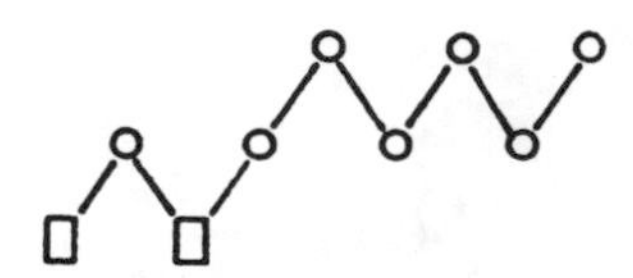

Figure 2. Full Multigrid Method.

where only a few relaxation steps are taken on the finest grids.

3. AN EXAMPLE OF AN OCEAN MODEL

To test the multigrid method on an ocean model with irregular boundaries, a simple Stommel-type model has been chosen. The physics are the same as in Stommel's model, except that this model is time-dependent, and spherical coordinates are used. The advantage of the time-dependent model is that inclusion of nonlinear terms and higher order friction is relatively simple.

Assuming the rigid lid approximation, the vertically integrated equations of motion for an incompressible fluid of constant density ρ_o may be written in vector form as

$$\frac{\partial\vec{v}}{\partial t} + 2\vec{\Omega} \times \vec{v} + (\frac{\vec{v}}{H} \cdot \vec{\nabla})\vec{v} = - \frac{H}{\rho_o} \vec{\nabla}p - gH \frac{\vec{r}}{r} + \frac{\vec{\tau}}{\rho_o} - R\vec{v} \quad (14)$$

$$\vec{\nabla} \cdot \vec{v} = 0 \quad (15)$$

where $\vec{v}$ is the volume transport per unit length. The centrifugal force has been included in $\vec{g}$. The depth is $H(\lambda,\phi)$. $\vec{\Omega}$ is the angular rotation vector of the earth, p the pressure, g the acceleration of gravity, and $\vec{r}$ a radial vector. A frictional term is proportional by the constant R to the volume transport.

The boundary condition applied at the sea surface is

$$\frac{\partial \vec{v}}{\partial z} = \vec{\tau} \tag{16}$$

where $\vec{\tau}$ is the wind stress vector.

For a slow viscous ocean the nonlinear term may be neglected. Assuming this and taking the curl of (14) gives for the radial component

$$\frac{1}{H}\frac{\partial}{\partial t}(\vec{\nabla} \times \vec{v})_r - \frac{1}{H^2}\vec{\nabla}H \times \frac{\partial \vec{v}}{\partial t} + \frac{2}{H}((\vec{v}\cdot\vec{\nabla})\vec{\Omega})_r -$$

$$- \frac{2}{H^2}(\vec{\nabla}H \times ((\vec{v}\cdot\vec{\nabla})\vec{\Omega}))_r = (\vec{\nabla} \times \vec{\tau}/\rho_o) - R(\vec{\nabla} \times \vec{v})_r. \tag{17}$$

A stream function ψ is introduced, so

$$\vec{v} = \vec{k} \times \vec{\nabla}\psi \tag{18}$$

where $\vec{k}$ is a radial unit vector.
In spherical coordinates this is

$$v = \frac{1}{a\cos\phi}\frac{\partial\psi}{\partial\lambda} \qquad u = -\frac{1}{a}\frac{\partial\psi}{\partial\phi} \tag{19}$$

where a is the radius of the earth and ϕ and λ the latitude and the longitude, respectively. v and u are the meridional and zonal transport components since the transport field is assumed non-divergent. For an ocean of constant depth, equation (17) can be written in spherical coordinates as

$$(\frac{\partial}{\partial t} + R)\nabla^2\psi + \frac{2\Omega}{a^2}\frac{\partial\psi}{\partial\lambda} = \vec{\nabla} \times \vec{\tau}' \tag{20}$$

where $\tau' = \tau/\rho_o$.

The equation is solved in a closed ocean. A necessary condition is that there is no flow normal to the boundary. This is accomplished by using

$$\psi = \text{constant at the boundaries.}$$

This assumption allows the tangential velocity component to be very large in the vicinity of the boundary. This is often referred to as the free-slip condition.

4. METHOD OF SOLUTION

Equation (20) is a time-dependent non-homogeneous elliptic PDE of the second order with variable coefficients. It will be solved using a finite difference representation in time and space variables. The time integration is made using the leap frog scheme with a Euler forward scheme used initially and occassionally to filter out computational modes.

The difference scheme becomes

$$\frac{1}{\Delta t^*} L\,(\psi^{n+1} - \psi^{m}) + RL\psi^{m} + \frac{2\Omega}{a^2}\frac{1}{2h}(\psi^{n}_{j+1,k} - \psi^{n}_{j-1,k})$$

$$= F(j,k,n) \qquad (21)$$

where L is a finite difference operator for ∇^2 in spherical coordinates and F(j,k,n) is the forcing, i.e. the curl of the wind stress. The grid spacing in radians is h.

For the leap frog scheme,

$$\Delta t^* = 2\Delta t \quad \text{and} \quad m = n - 1,$$

and for the Euler scheme

$$\Delta t^* = \Delta t \quad \text{and} \quad m = n.$$

The equation can be rewritten

$$L\psi^{n+1} = \Delta t^* F(j,k,n) + (1 - R\Delta t^*)\, L\psi^{m}$$

$$- \frac{\Omega}{a^2}\frac{\Delta t^*}{h}(\psi^{n}_{j+1,k} - \psi^{n}_{j-1,k}) \equiv f_n(j,k). \qquad (22)$$

The right-hand side depends on previous timesteps, so ψ^{n+1} is determined by solving a Poissons equation.

This structure can be used for any explicit model where the stream function has to be calculated. The physics will in most cases be included in the forcing f_n (j,k).

However, when bottom topography is included, the left-hand side will not be as simple, but instead, as seen from (17), will be of the more general form

$$\nabla^2\psi^{n+1} + a(\lambda,\phi)\frac{\partial\psi^{n+1}}{\partial\lambda} + b(\lambda,\phi)\frac{\partial\psi^{n+1}}{\partial\phi} = f_n(\lambda,\phi). \qquad (23)$$

Similar equations arise if some terms in the equation are handled implicitly.

In each timestep (22) has to be solved.

The differential operator L is in spherical coordinates

$$L\psi = \frac{1}{a^2}\left[\frac{1}{\cos^2\phi}\cdot\frac{1}{h^2}(\psi_{j+1,k} + \psi_{j-1,k} - 2\psi_{j,k})\right. \tag{24}$$

$$+ \frac{1}{h^2}(\psi_{j,k+1} + \psi_{j,k-1} - 2\psi_{j,k})]$$

$$\left. -\tan\phi\,\frac{1}{2h}(\psi_{j,k+1} - \psi_{j,k-1})\right]$$

This gives an iteration scheme that will be convergent if

$$\gamma = \frac{1}{2(1 + 1/\cos^2\phi)}\left[\frac{2}{\cos^2\phi} + \left|1 - \frac{h\tan\phi}{2}\right|\right.$$

$$\left. + \left|1 + \frac{h\tan\phi}{2}\right|\ \right] \leqq 1$$

If $\left|\frac{h\tan\phi}{2}\right| \leqq 1$, we have $\lambda = 1$ otherwise

$$\lambda = \frac{\frac{h\tan\phi}{2} + \frac{1}{\cos^2\phi}}{(1 + \frac{1}{\cos^2\phi})} > 1 \tag{25}$$

Care should be taken when a multigrid method is used, since the coarsest grids can be divergent for iteration processes, especially for high latitudes.

The iteration was done by using SOR and different multigrid cycles.

For the SOR iteration an acceleration parameter α determined by

$$\alpha = \frac{2}{1 + \sqrt{1 - r^2}}, \tag{26}$$

where $r = 1/2\,(\cos\frac{\pi}{N} + \cos\frac{\pi}{M})$ and N,M is the number of gridpoints in each coordinate. The expression for α holds only for large N,M.

5. MULTIGRID ITERATIONS

Three different multigrid strategies have been used: W-cycles, a full multigrid method, and an adaptive technique (MLAT). The FAS scheme was used in all cases, as was standard coarsening, so that all points in a coarse grid are also in all finer grids. When W-cycles were used, a fixed number of relaxations were done on each grid before shifts to coarser or finer grids. The MLAT structure is seen in Figure 3. This structure worked very fast for regular boundaries. However, when irregular coastlines were introduced, the best criterion for changing grids seemed to be time-dependent.

Instead of a fixed combination of V- and W-cycles, (Full

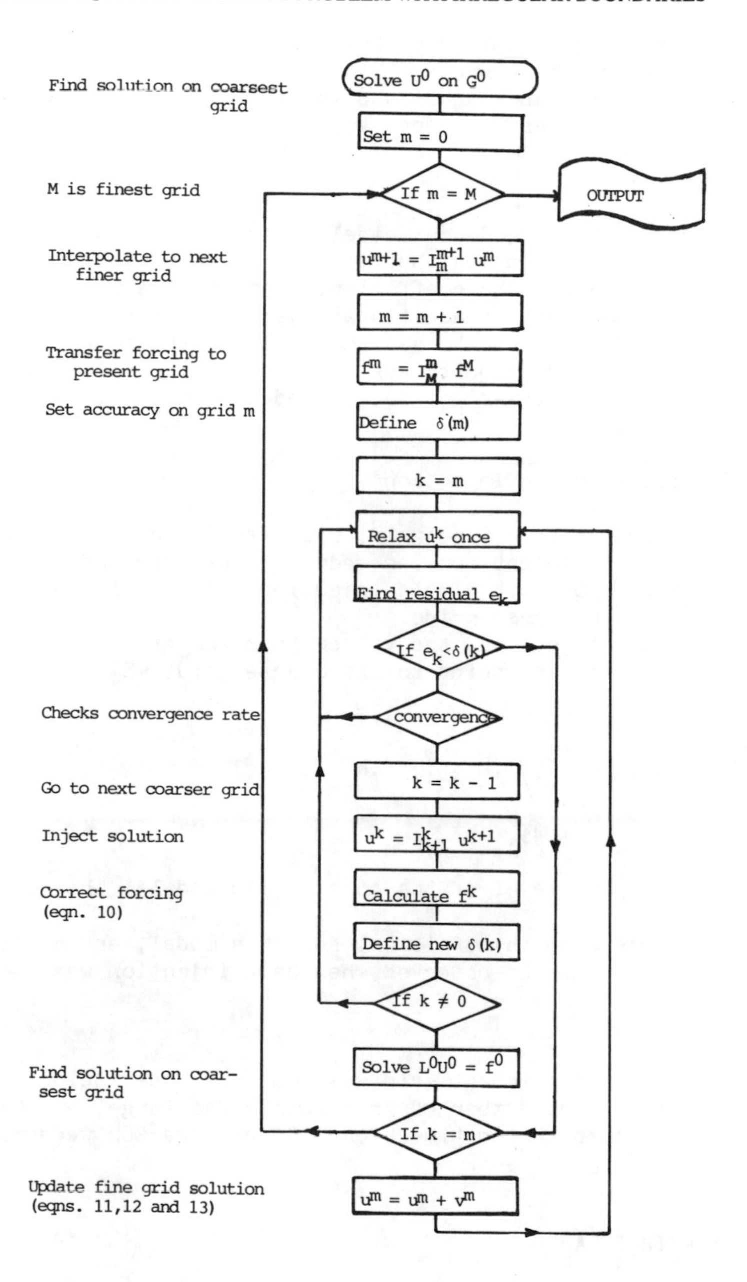

Figure 3. Program structure for MLAT (after Brandt, 1977).

multigrid method) was applied to this case. The structure was found using MLAT for some time steps. However, 2-W-cycles were the most efficient solution. When an island was introduced, a combination of several V- and W-cycles was necessary for an accurate solution.

5.1. Smoothing

As a smoothing operator, Gauss-Seidel relaxation is generally the best for a Poissons equation, but in this case, because of the spherical coordinates, the coefficients vary with ϕ . Point Gauss-Seidel can be used, and if additional iterations are done in the region where the coefficients vary most, i.e., the northern part of the basin, the method is quite effective. However, line Guass-Seidel relaxation in the east-west direction made a more efficient and general program.

5.2. Interpolation and Injection

Going from a coarse to a fine grid, bilinear interpolation was used. This is sufficient for 2nd order equation [Brandt 1977]. Tests were made with both simple injection and half-injection when going from fine to coarse grids.

With simple injection, the values from the fine grid are transferred to the same point in the coarse grid, while for the half injection

$$\psi^{l}_{j,k} = \frac{1}{8} (\psi_{j+1,k} + \psi_{j-1,k} + \psi_{j,k+1} + \psi_{j,k-1} + 4\psi_{j,k})^{l+1} \tag{27}$$

where l denotes the level of the coarse grid and l+1 the next fine grid.

Simple injection is sufficient for this model, and no difference in the solution could be observed when half injection was used.

5.3. Solvers

The solution on the coarsest grid was found by Gauss-Seidel relaxation or, if the number of grid points was large, by SOR or line-SOR. The computation times for SOR and line-SOR were roughly the same.

6. OCEAN WITH ISLANDS

The solution of the vorticity equation with islands may be solved by applying a method used by Kamenkovitch [1962]. This method was also used by Bryan [1969] in his formulation of a model for the world ocean.

The boundary condition, i.e., that there is no flow

perpendicular to a solid wall, implies

$$\psi = \text{constant.}$$

In the case where there are no islands, the constant can be chosen to be zero. When islands are present, the constant will generally be different for each island, and is an unknown constant.

So the boundary condition is

$$\psi = 0 \text{ (side walls)}, \ \psi = \mu_r, \ r = 1,2, \ldots Q$$

when there are Q islands in the ocean.

For the time-dependent problem, μ_r is a function of time.

The unknown constants μ_r can be found by taking the circulation along a closed contour around each island r.

In the present case the equation of motion can be written

$$\frac{\partial u}{\partial t} - fv = -\frac{1}{\rho_o}\frac{\partial p}{\partial x} + \tau^{\lambda} - Ru$$

$$\frac{\partial v}{\partial t} + fu = -\frac{1}{\rho_o}\frac{\partial p}{\partial y} + \tau^{\phi} - Rv \tag{28}$$

where $f = 2\Omega\sin\phi$, $dx = a\cos\phi d\lambda$, and $dy = ad\phi$ has been introduced. Taking the line integral counter-clockwise along a closed contour gives

$$\frac{\partial}{\partial t}\oint udx + vdy = \oint F^{(x)}\, dx + F^{(y)}\, dy,$$

where $F^{(x)} = fv + \tau^{\lambda} - Ru$ and $F^{(y)} = -fu + \tau^{\phi} - Rv$. (29)

Introducing the stream function, defined previously, gives

$$\frac{\partial}{\partial t}\oint \frac{1}{a}\left(-\frac{\partial\psi}{\partial\phi}dx + \frac{1}{\cos\phi}\frac{\partial\psi}{\partial\lambda}dy\right) = \oint F^{(x)}\, dx + F^{(y)}\, dy \tag{30}$$

and

$$F^{(x)} = f\frac{1}{a\cos\phi}\frac{\partial\psi}{\partial\lambda} + \tau^{\lambda} + R\frac{1}{a}\frac{\partial\psi}{\partial\phi}$$

$$F^{(y)} = f\frac{1}{a}\frac{\partial\psi}{\partial\phi} + \tau^{\phi} - R\frac{1}{a\cos\phi}\frac{\partial\psi}{\partial\lambda}.$$

According to Bryan [1969], the solution of the stream function can be written

$$\psi(t) = \psi_o(t) + \sum_{r=1}^{Q}\mu_r(t)\,\psi_r \tag{31}$$

where ψ_r is the solution to the Laplace equation

$$\nabla^2 \psi_r = 0$$

with the boundary conditions $\psi_r = 1$ along the perimeter of island r and $\psi_r = 0$ along all other boundaries. The function ψ_r is independent of time.

In Bryan [1969] ψ_o is the solution to the vorticity equation with the boundary condition $\psi_o = 0$ at all boundaries.

For an efficient computation, especially when SOR relaxation is used, a good initial guess for the solution of the next timestep must be made. Since ψ and ψ_o generally are quite different, either two separate fields for the stream function should be stored, or recalculation of ψ_o^n as an initial guess for calculation ψ_o^{n+1} is necessary.

This is unnecessary if ψ_o is solved according to the boundary condition

$$\psi_o = 0 \text{ at the side walls, and}$$

$$\psi_o = \psi^n \text{ at all island points.}$$

In this way, μ_r is not the value of ψ along the coasts of the islands, but just the correction from the previous timestep.

Introducing finite differences in time and for step n+1

$$\psi^{n+1} = \psi_o^{n+1} + \sum_{r=1}^{Q} \mu_r^{n+1} \psi_r$$

gives us Q equations of Q variables μ_r to be determined. Here is one equation written for r=1:

$$\frac{1}{\Delta t^*} \oint_1 \sum_{r=1}^{Q} \frac{\mu_r}{a} \left(- \frac{\partial \psi_r}{\partial \phi} dx + \frac{1}{\cos\phi} \frac{\partial \psi_r}{\partial \lambda} dy\right) = I_1(t), \qquad (32)$$

where

$$I_1(t) = \oint_1 \frac{1}{a} \left(\frac{\partial \psi_o}{\partial \phi} - \frac{\partial \psi^m}{\partial \phi}\right) dx + \frac{1}{a \cos\phi} \left(\frac{\partial \psi^m}{\partial \lambda} - \frac{\partial \psi_o}{\partial \lambda}\right) dy$$

$$+ \Delta t^* \oint_1 F^{(x)} dx + F^{(y)} dy.$$

For the Euler forward scheme $\Delta t^* = \Delta t$ and $m = n$ and the Coriolis and friction term is evaluated at timestep n in the F terms above. In the case of leap frog, $\Delta t^* = 2\Delta t$, $m = n-1$ and the Coriolis and frictional terms are taken to time steps n and n-1, respectively.

Note that the coefficients to μ_r on the left-hand side are independent of time. In the case of just one island, it is especially simple since

$$\mu_r(t) = \frac{I_1(t)}{I_r} \tag{33}$$

where I_r is the integral constant on the left-hand side.

The inclusion of an island is not more difficult when multigrid techniques are used. The ψ_r fields can be evaluated by multigrid or SOR. Since ψ_r only has to be calculated once, computation time is not essential.

The ψ_o field is in every time step determined by a multigrid technique.

7. IRREGULAR BOUNDARIES ON A MULTIGRID

The ocean is divided into domains so that if two points at the same latitude are ocean points, all points between are also ocean points. Figure 4 shows such a division into domains.

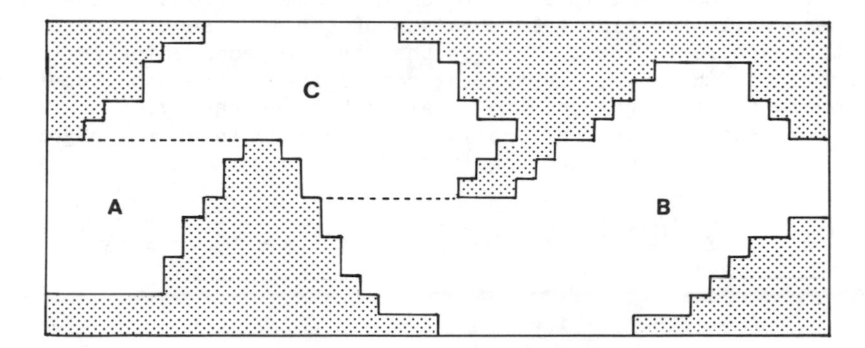

Figure 4. Division of ocean basin into three domains.

The division can be done in several ways, and it is a way of administering the variable boundaries. The actual division chosen is not critical to the method. The domains are chosen according to the east-west direction since the storage allocation of a two-dimensional matrix defined in a FORTRAN program is arranged in a similar way.

For each domain, two vector variables are stored, one containing the start address, i.e., the first ocean point near the western vector variable boundary and one vector variable, where the addresses of the last ocean point are stored.

Each domain is scanned from north to south or vice versa depending on the start and end addresses of the domain. For some ocean basins, it might be preferable to choose the divisions according to a similar criterion applied in the north-south direction which might reduce the number of domains. In that case, the north-south variable should be represented by the first index in the matrix of the stream function to obtain efficient computation speed.

The stream function in this model is stored in a two-dimensional array. In very complex-shaped basins, a lot of unused storage might be allocated in the computer, and another approach may be preferable.

The use of multiple grids is most convenient using a rectangular matrix where the number of grid intervals is a power of 2.

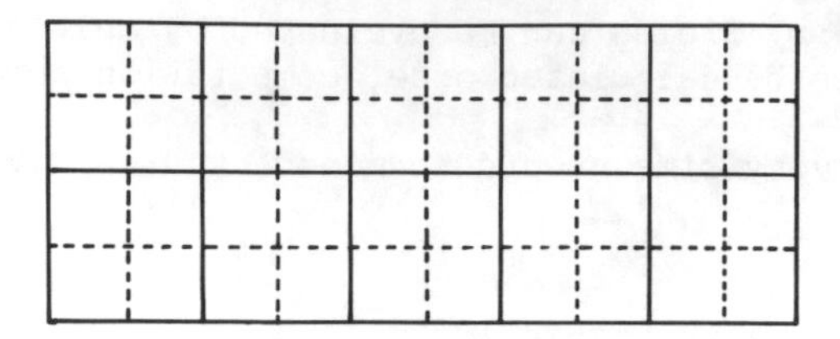

Figure 5. Two-grid standard coarsening.

Figure 5 displays how a fine grid is placed relative to the next coarser grid. The grids have every second point in common in every line. The coarse grid consists of the intersections of the full lines, while the fine grid consists of the intersection of both the dotted and full lines. This is the standard coarsening previously mentioned. However, even if the points are in the same location, storage for each grid has to be allocated separately.

To store a field of one variable defined in an m x n grid with M coarser grids, the space required is

$$\sum_{k=1}^{M} \left(\frac{m-1}{2^k} + 1\right)\left(\frac{n-1}{2^k} + 1\right) \leq \frac{1}{3}(m-1)(n-1) + (m-1) + (n-1) + 1. \qquad (34)$$

Using the last expression for dimensioning a one-dimensional array containing all the coaser grids, the space requirements are fulfilled, no matter how many coarse grids are actually introduced.

If M = 5 and m x n = 65 x 33, the extra space for coarse grid variables is 770, while the formula allocates 779 floating point variables.

When irregular coasts are introduced, the distance to the coast may vary in each direction, as seen from Figure 6.

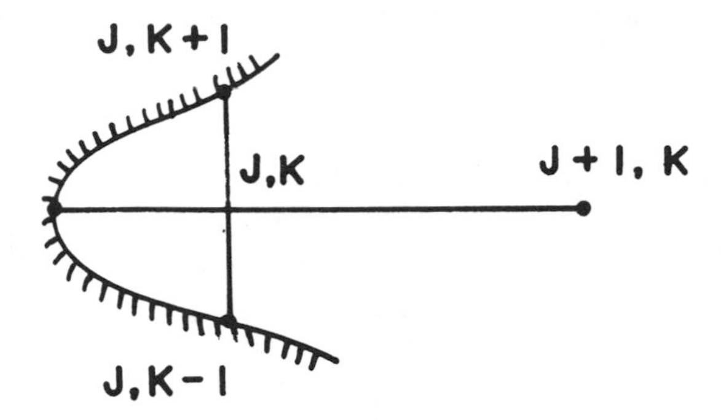

Figure 6. Variable grid distance along coast.

For a central difference approxmation to $\frac{\partial\psi}{\partial x}$, the formula to be used for the point j,k is

$$\frac{\psi_{j+1,k} - \psi_{j,k}}{h_E} \cdot \frac{h_W}{h_W+h_E} + \frac{\psi_{j,k} - \psi_{j-1,k}}{h_W} \frac{h_E}{h_W+h_E} \tag{35}$$

and for $\frac{\partial\psi}{\partial y}$

$$\frac{\psi_{j,k+1} - \psi_{j,k}}{h_N} \frac{h_S}{h_N+h_S} + \frac{\psi_{j,k} - \psi_{j,k-1}}{h_S} \frac{h_N}{h_N+h_S}. \tag{36}$$

As seen, for $h_N = h_S$ and $h_E = h_W$ the usual central difference expression is obtained.

For an operator of the form

$$a_1(x,y) \frac{\partial^2\psi}{\partial x^2} + a_2(x,y) \frac{\partial^2\psi}{\partial y^2} ,$$

the 5-point Shortley-Weller approximation should be used:

$$2\left[a_1(x,y) \left(\frac{\psi_{j-1,k}}{h_W(h_W+h_E)} + \frac{\psi_{j+1,k}}{h_E(h_W+h_E)}\right) + \right.$$
$$+ a_2(x,y)\left(\frac{\psi_{j,k-1}}{h_S(h_N+h_S)} + \frac{\psi_{j,k+1}}{h_N(h_N+h_S)}\right) - \tag{37}$$
$$\left. - \left(\frac{a_1(x,y)}{h_W h_E} + \frac{a_2(x,y)}{h_N h_S}\right)\psi_{j,k}\right].$$

Applying this to the Poissons equation in spherical coordinates yields

$$\left[\frac{1}{h_W h_E \cos^2\phi} + \frac{1}{h_N h_S} + \tan\phi \left(\frac{h_N}{h_S(h_N+h_S)} - \frac{h_S}{h_N(h_N+h_S)} \right) \right] \psi_{j,k} =$$

$$\left(\frac{2}{h_W(h_W+h_E)\cos^2\phi} \psi_{j-1,k} + \frac{2}{h_E(h_W+h_E)\cos^2\phi} \psi_{j+1,k} + \right.$$

$$+ \left(\frac{1}{h_S(h_N+h_S)} + \tan\phi \frac{h_N}{h_S(h_N+h_S)} \right) \psi_{j,k-1} + \qquad (38)$$

$$+ \left(\frac{1}{h_N(h_N+h_S)} - \tan\phi \frac{h_S}{h_N(h_N+h_S)} \right) \psi_{j,k+1} - a^2 F(\lambda,\phi).$$

Since h_N, h_S, h_W, and h_E are different for every point along the coast, much additional programming and computational efforts have to be made with this formulation. However, minor errors made in coarse grids tend to be corrected while going to finer grids, so in this study another simpler approach has been attempted. The errors may, however, slow down the computation speed significantly.

When the irregular coastlines are resolved on an equidistant fine grid and coarse grid, the coastlines on the two grids will not be the same everywhere.

Three approaches have been tested:

I. The simplest is to make the ocean wider for the coarse grid whenever necessary. As seen in Figure 7, the coarse grid ocean has a shore line indicated by the crossed area, while the finer grid has the whole shaded area; i.e., the solution on the coarse grid will be the same whether the shaded area is land or ocean.

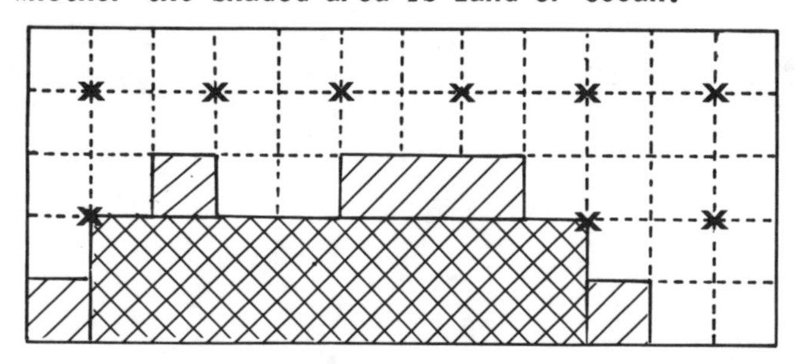

Figure 7. Different coastlines for two grids.

The disadvantage of using this method is that the coarse grid ocean is always wider than the fine grid ocean, depending on the grid arrangement, which in a spin-up case tends to give larger numerical values for the stream function and thus less accuracy. Since the error made is of the same scale as the forcing, it may require many iterations of a finer grid to smooth it out, and no gain in computation speed will be obtained.

II. A "best fit" of the coarse grid to a non-resolvable coastline is generally better. A weighted average coastline is used for the coarse grid, so narrow peninsulas or fjords are ignored. Here the disadvantage is the more complex structure of an automatic calculation routine of the coarse grid boundaries.

III. Finally, a coarse grid containing only interior points, that is, points where all surrounding points are also ocean points, may be chosen. The injection from fine grids should, however, include all ocean points on the coarse grid, so the boundary condition of this "interior ocean" is non-zero. However, errors of large-scale are still introduced in some cases.

When the strategy of handling the boundaries is as outlined above, it is necessary to have different start and end addresses for the iteration process for each grid. The coarse boundaries defined in II and III both worked well, while the simple approach of I was too crude in complex shaped basins.

8. TEST RESULTS

The ability of the multigrid approach to produce correct results was tested using the linear ocean model when a constant wind stress was applied to an initially motionless sea. The wind stress is similar to that used in the classical study by Stommel [1948] with zero wind stress curl at the northern and southern boundaries and one maximum in the interior, and independent of the longitude, i.e.,

$$\tau_x = \tau_o \cos\left(\frac{\pi}{M} \cdot \phi\right), \quad o \leqq \phi \leqq M \tag{39}$$

$$\tau_y = 0$$

where M is the scale of the ocean basin in the north-south direction in degrees and ϕ the latitude.

Tests were made with an ocean rectangular in degrees, an ocean with an irregular coastline, and, finally, an ocean with irregular coastlines and one island. The coastlines chosen correspond crudely to the border of the continental shelf of the North Atlantic between 48^oN and 80^oN. The parameters used in the model are found in

TABLE I

Model Parameters	
Radius of the Earth:	6,371.305 km
Earth rotation:	$7.29 \cdot 10^{-5}$ s^{-1}
Amplitude of wind stress:	0.2 N/m^2
Linear friction coefficient:	$1.0 \cdot 10^{-6}$ s^{-1}
Resolution:	$1^o \times 1^o$
Number of gridpoints:	65 x 33
Timestep:	1.5 - 6 hours

The computation is stable with a larger timestep, but then phase errors become significant.

In the case of a rectangular sea, a very efficient calculation could be obtained with a self-adaptive algorithm. A typical computational route calculated by the program for a four grid method is shown in figure 8 below.

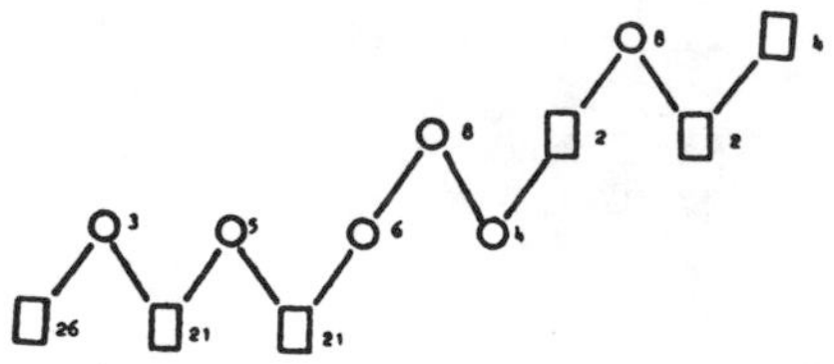

Figure 8. MLAT-calculation path.

Here the number of iterations on each level is shown. Convergence is obtained whenever a square symbol is used. As the diagram shows, four grids were used.

The criterion for changing grids and for convergence is outlined here. An iteration is convergent when

$$r_{max} < \delta(m) \cdot |\bar{\psi}| \ ,$$

where r_{max} is the maximal residual, $\delta(m)$ is a prescribed parameter which can depend on the level m, and $|\bar{\psi}|$ is the mean of the absolute value of the stream function.

In the present case, a value of $\delta(m) = 0.0001 \cdot 2^{(M-m)}$ where M is the total number of grids was used.

During the process of solving for grid m, a solution for (m-1) with the accuracy $\delta(m-1)$ has already been found. When returning to the coarse (m-1)-grid later, a convergence criterion of $\delta'(m-1) = 0.2\ \delta(m)$ is used for the coarse grid. The change to a coarser grid is made when the convergence speed is too slow. If $\bar{e}_k$ is the maximal residual of the previous iteration, a shift is made when

$$e_k > \eta \bar{e}_k \ ,$$

where e_k is the maximal residual of the last iteration and η a parameter. A value of η in the range of 0.7 - 0.9 worked well. However, during time itegration the optimal value can change.

The efficiency of the MLAT to line-SOR obtained at a Cyber 760 is shown in Table II.

TABLE II

Timestep	SOR iterations	CPU-ratio (MLAT/SOR)
1	97	5.6
10	53	2.8
20	40	2.2
30	31	1.7

The timestep in the calculation was 1.5 hours.

As the ocean is spun up, the solution from one timestep to the next is nearly identical. The solution for timestep n-1 was used as an initial guess for timestep n. This is essential to ensure efficient computation with SOR, while the MLAT, starting with the coarsest grid, does not benefit much from this approach. This indicates that a more efficient calculation may be obtained if the coarser grids are skipped after some time.

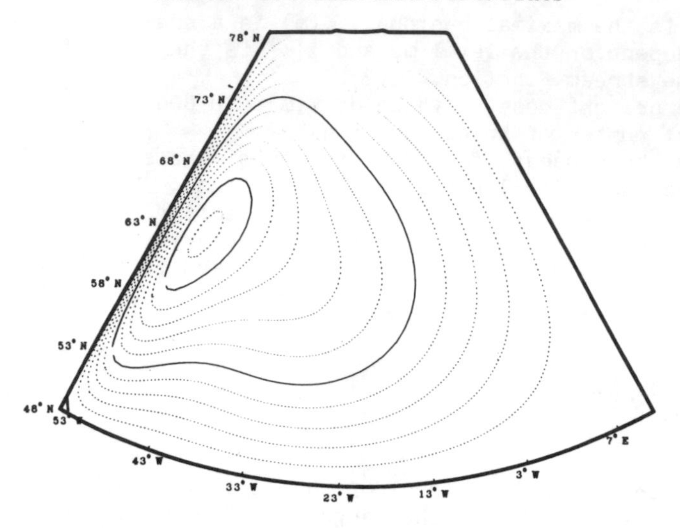

Figure 9. Ocean with regular boundaries.

A result of a 20 day integration using a timestep of 6 hours is displayed in Figure 9. The maximum of the stream function is initially found in the middle of the ocean and moves westward to form a boundary current. The maximum transport during the spin up estimated from the Sverdrup relation and using the parameters as presented in Table I, is about 50 Sv for an ocean covering 64° in longitude. As seen from Figure 12, the numerical calculation is in agreement with this result.

When irregular coastlines were introduced, problems with the self-adaptive algorithm arose. Sometimes the iterations were inefficient and the corrections of the residuals on the coarse grid insufficient; so the iteration process was trapped between two grid levels. This could be due to the method of handling the irregular boundaries on coarse grids.

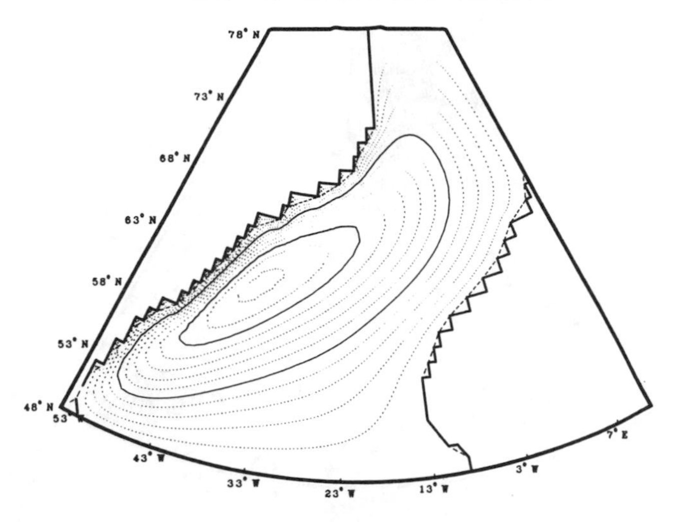

Figure 10. Ocean with irregular boundaries.

Another simpler approach was used. Two W-cycles with line-relaxation using 3 grids worked well. The efficiency compared to line-SOR is, however, somewhat lower than for the rectangular sea. Figure 13 shows the CPU-time used for SOR for $\delta(M)$, being 0.1%, 0.5%, and 1% compared with the time used for a multigrid calculation. The multigrid solution was closest to the 0.1% SOR solution. However, the maximum value of ψ was about 1% too large after 80 timesteps.

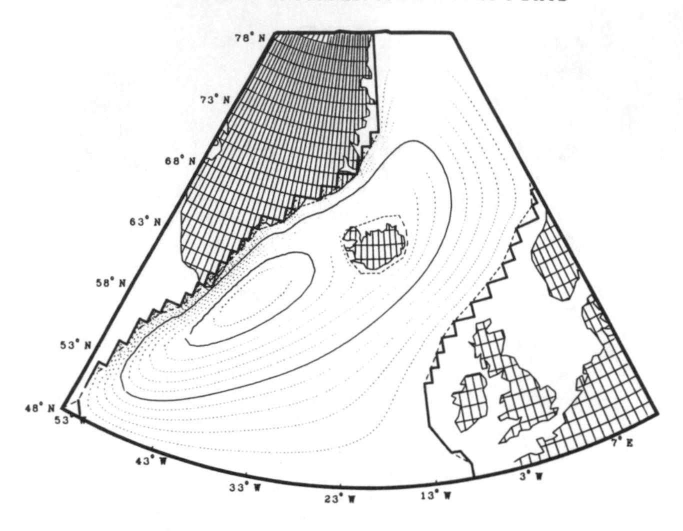

Figure 11. Ocean with one island.

When an island was introduced, the computation times of the SOR and 2-W cycles were the same. Other multigrid schemes with combined W and V cycles were tested, but no faster computation has yet been found.

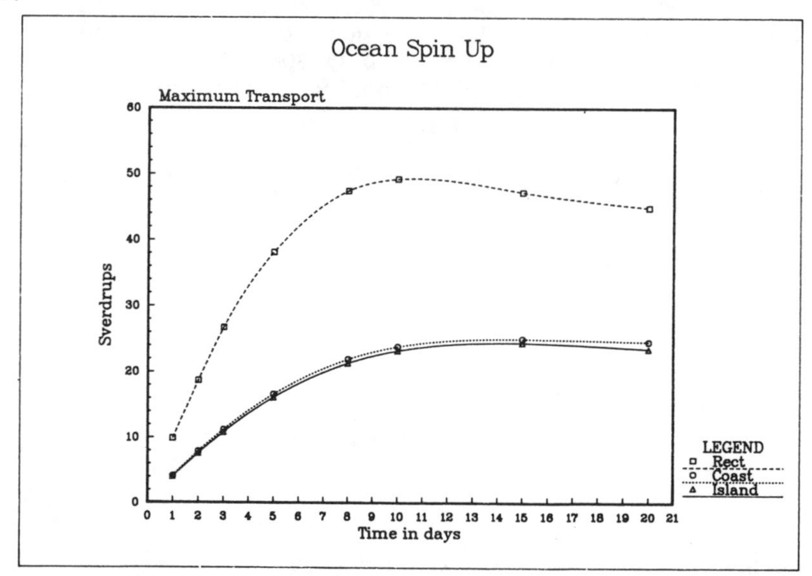

Figure 12. Maximum transport as function of time. Broken line: rectangular ocean. Dotted line: irregular coastline. Full line: with island.

Physically, the solution with the island included is not very different from the calculation without the island (Figure 10 and Figure 11). Only in the vicinity of the island is the flow modified. Also, the maximum transport is only slightly decreased, as seen from Figure 12. This is expected as the condition of free-slip boundaries was imposed.

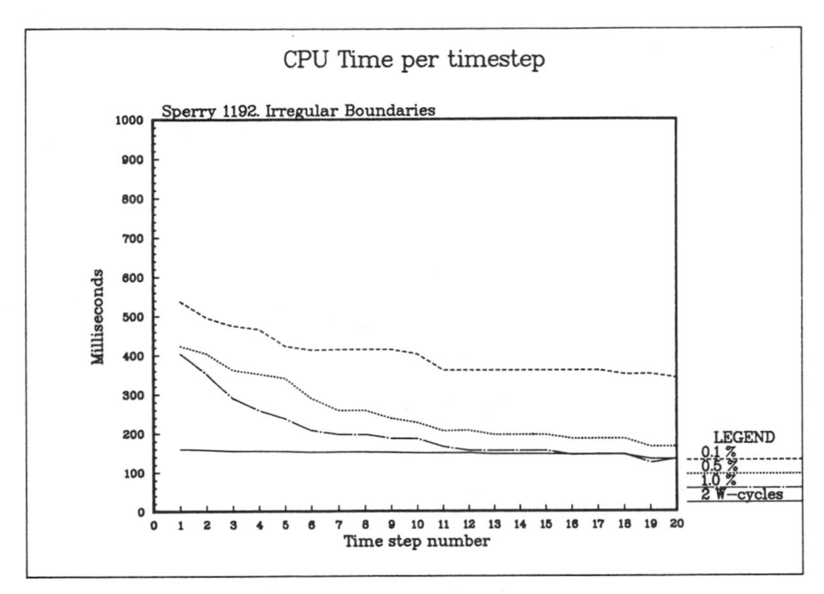

Figure 13. CP-time in milliseconds used per time step. Broken lines: SOR with different accuracies, 0.1%, 0.5% and 1%. Full line: 2-W cycle multigrid.

9. DISCUSSION AND CONCLUSIONS

The multi-grid technique has been shown to work when the boundaries are resolved by the coarsest grid used, as demonstrated by the example with a rectangular ocean. When the coefficients in the spherical Poissons equation vary, line relaxation should be used as a smoother.

The attempt of the present investigation, i.e., to avoid a strict computation by using different boundaries for different grids, has been shown to work, though the efficiency of the multigrid method has to be improved. However, the efficiency compared to SOR will be better when a finer grid is introduced. The present model had only about 1000 ocean points left when the irregular coastline was introduced.

The poor performance after the island was implemented may be caused by the fact that the maximum of the stream function is near the island boundary which is not modelled correctly on the coarsest grid.

A NUMERICAL APPROACH TO EQUATORIAL OCEANIC WAVE-MEAN FLOW INTERACTIONS

J. A. Proehl, M. J. McPhaden and L. M. Rothstein
School of Oceanography, WB-10
University of Washington
Seattle, Washington 98195

ABSTRACT. An efficient numerical method is employed to investigate the interaction of equatorial trapped waves with vertically and meridionally sheared zonal jets. The model is formulated with the Equatorial Undercurrent in mind but has general applicability. The governing differential equation is derived and expanded using central differences into finite difference form. The linear system of equations leads to a coefficient matrix of block-tridiagonal form which is efficiently solved using a direct method. The direct method involves an optimized Gauss elimination and backsubstitution. For long Kelvin waves in the absence of mean flow, the analytic solution is obtained numerically with a high degree of accuracy. A diagnostic test is derived for use in cases where the analytic solution is unknown. The model is then applied to situations in which waves are superimposed on a background flow containing a critical surface where the flow speed matches the phase speed of the wave. For these cases, the waves are damped and absorbed in a frictional layer surrounding the critical surface. For speeds characteristic of the Undercurrent, all but the first few baroclinic Kelvin modes encounter critical surfaces which may prevent the establishment of higher baroclinic modes in the equatorial ocean.

1. INTRODUCTION

In recent years much insight has been gained into the dynamics of the equatorial ocean. Past modelling efforts have been concerned with either the pure linear wave theory in the absence of flow or the steady state dynamics. The details of many of the models, including reduced gravity and continuously stratified models, have been recently discussed in an excellent review article by McCreary [1985]. One aspect of equatorial dynamics which has yet to be fully clarified is the effect of the intense, narrow equatorial current system on the linear equatorial wave modes. The Equatorial Undercurrent, being the most intense, is of particular interest. This eastward jet residing in the pycnocline, reaches speeds in excess of 100cm/s. It is geostrophically balanced in the cross stream direction and possesses a vertical scale of the order of

J. J. O'Brien (ed.), Advanced Physical Oceanographic Numerical Modelling, 111–126.

100m and a meridional scale of the order of 150km. In most studies the equatorial current field is either ignored or assumed passive as far as the wave response is concerned. This assumption simplifies the dynamics to where analytical treatments are possible. In the case of the Undercurrent this is likely to be an extremely weak assumption. The presence of critical surfaces in the Undercurrent will strongly affect the wave propagation. A critical surface or level is the surface upon which the Doppler shifted phase velocity, c-U, vanishes. The critical layer, associated with the critical surface, is a frictional layer in which, due to the smallness of c-U, damping is greatly enhanced.

The interaction of equatorial waves with oceanic mean flows has received some attention. Complementary studies by McPhaden and Knox [1979] and Philander [1979] investigated the effects of meridionally sheared jets on zonal wave propagation in two layered systems. The two layered approach prohibits the inclusion of realistic vertical shears in the mean flow. Both studies evaded the possibility of critical layers and instabilities by restricting attention to eastward phase speeds everywhere greater than the background flow. The study of McPhaden and Knox [1979] limited attention to the Kelvin and inertia-gravity wave range of dispersion space. They reported significant Doppler shifting of both wave types. The case of the Kelvin wave showed only slight distortion of the latitudinal structure, whereas the distortion of the zonal velocity of the inertia-gravity waves was substantial. Distortion of the pressure field of the inertia-gravity wave was less significant. Philander [1979] included consideration of the Rossby waves. For background flows lacking meridional shear, short Rossby waves which normally propagate westward were swept downstream,resulting in eastward phase propagation. For flows possessing meridional shear, this eastward propagation was eliminated though significant Doppler shifting remained. In both of these studies, the governing equations lead to an ordinary differential equation in y which is sufficiently complex to require a numerical treatment.

Analogous numerical problems were first investigated in the atmosphere over a decade ago [Holton, 1970; Lindzen, 1970]. The study of Lindzen [1970] included continuous stratification, but only vertical shear was considered. Holton [1970], however, included both meridional and vertical shears in his geostrophically-balanced zonal-wind field. In his study an analytic form was chosen for the wind profile, and damping was in the form of Rayleigh friction and Newtonian cooling. The primary results of that work were: 1) waves approaching a critical surface undergo a contraction of both meridional and vertical scales; 2) Kelvin wave energy is absorbed at the height of the critical surface at the equator even though the critical surface may vary with latitude; and 3) momentum exchange between the waves and mean flow is due to distortion of the wave field by the mean shear. Subsequent to these numerical studies, it was realized that the zonal wind field was sufficiently "slowly varying" [Lindzen, 1971; 1972] so that the equations admit solutions based on WKBJ theory. Comparison with the WKBJ solutions validated the results obtained in the earlier finite difference studies and therefore demonstrated that critical layer dynamics can be treated in this way. In recent years, advances have been made which depend upon the background fields possessing sufficiently slowly varying vertical structure. For this case, flow fields possessing arbitrary meridional structure can be treated using analytical procedures, ordinary perturbation methods, or Hermite spectral methods depending upon the type of flow field [Boyd, 1979]. For more recent reviews of the pertinent literature see Holton [1975; 1979] or Dunkerton [1980].

The present study assumes continuous stratification and allows for both vertical and meridional shear in the zonal background flow. Due to the presence of the intense, narrow, highly sheared equatorial jet and the non-separability of the resulting elliptic partial differential equation, a numerical treatment is dictated. In recent years, much progress has been made in the direct numerical solution of linear systems arising from elliptic partial differential equations. This paper is intended to set up the proposed problem and to describe the solution procedure. A test of accuracy, performed on a case where the analytic solution is known, is used to calibrate a diagnostic test for cases when the analytic solution is unknown. Only preliminary results will be offered supporting the strength of the numerical model employed.

2. MODEL FORMULATION

We begin by specifying an inviscid zonal jet which is in geostrophic balance with the continously stratified background density field. The dynamics for the perturbations are assumed hydrostatic, incompressible, viscid, and Boussinesq. For a background state temporally and zonally invariant, we seek solutions of the form $e^{i(kx-\sigma t)}$. The governing equations for the perturbations, posed on an infinite equatorial β-plane and linearized about the background flow are

$$-i\sigma u + ikUu + v\frac{\partial U}{\partial y} + w\frac{\partial U}{\partial z} - \beta yv = -ikp - \kappa u, \qquad (1a)$$

$$-i\sigma v + ikUv + \beta yu = -\frac{\partial p}{\partial y} - \kappa v, \qquad (1b)$$

$$iku + \frac{\partial v}{\partial y} + \frac{\partial w}{\partial z} = 0, \qquad (1c)$$

$$\rho_0\frac{\partial p}{\partial z} = -g\rho, \qquad (1d)$$

$$-i\sigma\rho + ikU\rho + v\frac{\partial \rho_b}{\partial y} + w\frac{\partial \rho_b}{\partial z} = -\kappa\rho. \qquad (1e)$$

The total density is given by $\rho_T(x,y,z,t) = \rho_0 + \rho_b(y,z) + \rho(x,y,z,t)$ and ρ_0 has been absorbed into the definition of pressure. The background state satisfies

$$\beta yU = -\frac{\partial P}{\partial y} \quad \text{and} \quad \rho_0\frac{\partial P}{\partial z} = -g\rho_b. \qquad (2)$$

For the special case U=0, the equations are separable and their solutions, in the inviscid limit, are the well-known free equatorial waves [Moore and Philander, 1977]. In an unbounded ocean, these waves possess Hermite function dependence with latitude and vertical normal mode dependence with depth. They also have the properties that they form complete orthogonal sets in both y and z respectively. In addition, the odd numbered meridional modes possess odd symmetry in v and even symmetry in u, w, p, and ρ around the equator. The converse holds true for even numbered modes. This will be important in the specification of boundary conditions.

In general we will be interested in the interaction of all types of equatorially trapped waves with background flow fields including Kelvin, mixed Rossby-gravity, inertia-gravity, and Rossby waves. But for the present study we focus on the gravest meridional mode, the Kelvin wave. We also choose to restrict attention to long Kelvin waves at the annual period. Therefore, we make the longwave approximation reducing (1b) to the geostrophic balance

$$\beta y u = -\frac{\partial p}{\partial y}. \tag{1b'}$$

This eliminates from consideration short zonal wavelength Rossby waves, the mixed Rossby-gravity wave and high frequency inertia-gravity waves. This is not a requirement of the numerical model but a simplifying assumption.

The Kelvin wave solution, in the absence of background flows, is given by

$$v \equiv 0 \quad ; \quad u = \frac{1}{c}\, p = \Pi^{\frac{1}{4}} e^{-\{\beta y^2/2c\}} F(z) \tag{3}$$

where F(z) is the solution to the vertical Stürm-Liouville problem

$$\frac{d^2F}{dz^2} + \left[\frac{N(z)}{c}\right]^2 F = 0$$
$$\text{subject to } \frac{dF}{dz} = 0 \text{ at } z=0,-H \tag{4}$$

with eigenvalue 1/c and N, the frequency of buoyancy oscillations for the background state, defined through

$$N^2 \equiv \frac{-g}{\rho_0}\frac{d\rho_b}{dz}. \tag{5}$$

Cross differentiating (2) to eliminate pressure, differentiating with respect to z and integrating with respect to y and using (5) gives

$$N^2(y,z) = N^2(\infty,z) + \int_y^\infty \left[\beta y \frac{\partial^2 U}{\partial z^2}\right] dy. \tag{6}$$

At the northern boundary, we must specify N^2 as function of depth. For the present problem we chose a constant N^2 value of $10^{-4}s^{-2}$ which corresponds to a vertical temperature gradient of approximately 1°C per 100m. For constant N, F(z) has the form of a pure sinusoid with wavenumber c/N. The Kelvin wave has a natural meridional scale, $\alpha=(c/\beta)^{\frac{1}{2}}$, as is evident from (3).

Nondimensionalizing and eliminating in favor of p from (1) leads to the second order partial differential equation

$$A\frac{\partial^2 p}{\partial y^2} + B\frac{\partial^2 p}{\partial y \partial z} + C\frac{\partial^2 p}{\partial z^2} + D\frac{\partial p}{\partial y} + E\frac{\partial p}{\partial z} + Fp = 0, \qquad (7)$$

where

$$A = 1,$$

$$B = \frac{2\beta y}{N^2}U_z,$$

$$C = \frac{\beta y}{N^2}[\beta y - U_y],$$

$$D = -\frac{2}{y} - \frac{1}{S}S_y - \frac{\beta y}{SN^2}U_z S_z + \frac{\beta y}{N^2}U_{zz} - \beta y N^{-4} N_z^2 U_z,$$

$$E = 2\left(\frac{\beta y}{N^2}\right)^2 U_z U_{zz} - \frac{1}{S}\left(\frac{\beta y}{N^2}\right)^2 U_z S_z + \beta^2 y^2 \left(\frac{1}{N^2}\right)_z [S + \frac{2}{N^2}U_z^2] + $$

$$+ \beta y\left(\frac{1}{N^2}U_z\right)_z - \frac{\beta}{N^2}U_z\left(\frac{y}{S}S_y + 1\right),$$

$$F = \frac{\beta^2 y^2 S}{(c-U+ir)}\left(\frac{\beta y}{SN^2}U_z S_z - \frac{\beta y}{N^2}U_{zz} - \beta y U_z\left(\frac{1}{N^2}\right)_z + \left(\frac{1}{S}S_y - \frac{1}{y}\right)\right).$$

Subscripts refer to differentiation, and S and r are defined by

$$S \equiv 1 - N^{-2}U_z^2 - [\beta y]^{-1}U_y \qquad ; \qquad r = \frac{\kappa}{k} \ .$$

The form of the boundary conditions and method of solution depend upon the sign of the discriminant, Γ, defined by

$$\Gamma \equiv B^2 - 4AC = -[\beta y N^{-1}]^2 S. \qquad (8)$$

Evaluating the discriminant for velocity and density data from the Hawaii-Tahiti shuttle experiment (Figure 1a,b) we find that the discriminant (Figure 1c) is everywhere negative with the exception of the equator (y=0) and a region near the surface south of the equator in the South Equatorial Current. Assuming this is typical of the Equatorial Undercurrent system, the governing equation is essentially elliptic and we must specify p, its normal derivative or a linear combination of both on the bounding contour.

From a solution standpoint, the primary differences in the various equatorial wave types are their symmetry properties and equatorial trapping scales. The full 2-D problem is posed on an infinite strip in y centered on the equator with the condition that all dependent variables decay to zero for large $|y|$. However, for symmetric background configurations of the type we will consider, the symmetry properties of equatorial waves allow the problem to instead be posed on a semi-infinite strip. The equatorial trapping condition at minus infinity is then replaced by a symmetry condition at the equator which depends upon the meridional mode in question. For the Kelvin wave these conditions are

$$p = 0 \qquad \text{as } y \to \infty, \qquad (9a)$$

$$\frac{\partial p}{\partial y} = 0 \qquad \text{at } y=0. \qquad (9b)$$

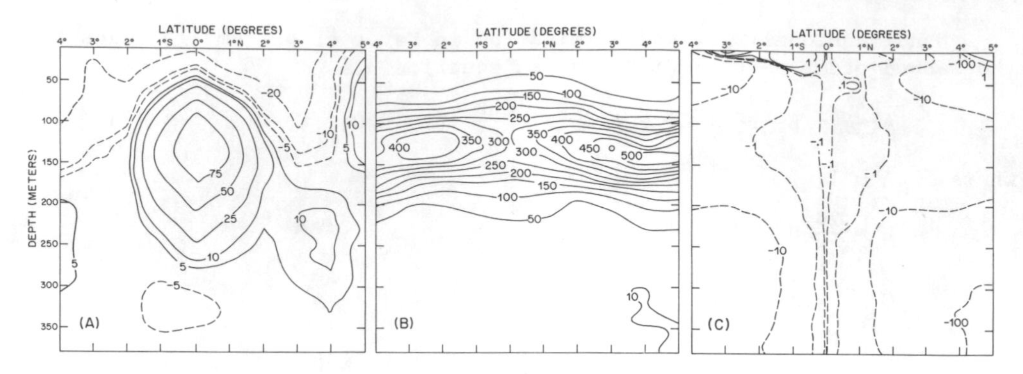

Figure 1. Background fields from the Hawaii-Tahiti Shuttle experiment, (a) Zonal velocity (cm/s) (b) Squared buoyancy frequency (N^2) ($\times 10^6 s^{-2}$) (c) Discriminant

In the physical problem, the ocean is primarily forced through a wind stress acting upon its upper surface. However, for the present we are interested only in the interaction of waves with mean flows, so the forcing mechanism is not of crucial importance. The interesting question of the wind forced response of the ocean in the presence of background flows will be considered later. Therefore, instead of specifying a stress at the surface, we specify a vertical velocity which is proportional to p_z. This may be thought of as an Ekman pumping velocity at the base of the mixed layer (analogous to a plunger in the case of a laboratory experiment). To investigate a particular wave type, we specify that the vertical velocity have meridional, zonal, and temporal dependence which matches the structure and dispersion characteristics of the wave. In the Kelvin wave case, the vertical velocity has the Gaussian form given in (3). Since we have chosen to fix the frequency at one cycle per year, varying the zonal phase speed of the wave is equivalent to varying the zonal wavenumber. The bottom boundary condition is one of no mass flux through the lower boundary. Thus the surface and bottom boundary conditions are

$$\frac{\partial p}{\partial z} = e^{-\{\beta y^2/2c\}} \qquad \text{at } z=0, \tag{9c}$$

$$\frac{\partial p}{\partial z} = 0 \qquad \text{at } z=-H. \tag{9d}$$

3. NUMERICAL CONSIDERATIONS

3.1. Methods

A large variety of methods have been developed for the numerical solution of elliptic partial differential equations [Birkhoff, 1972]. Finite difference, finite element, and spectral methods have all been employed with success. The three approaches differ in their basic philosophy. Finite difference methods replace the continuous differential equation with a discrete difference equation. Finite element methods reformulate the differential equation as a variational principle. The spectral methods expand the unknown solution in terms of a predetermined basis set and employ orthogonality relations. All three approaches result in a linear algebraic system of equations. For certain choices of basis set (spectral) and minimizing scheme (finite element) the three approaches are equivalent at discrete grid points. For rectangular domains, in which we are working, the finite difference methods are the simplest to program and yet yield sufficiently accurate results. In addition, in the presence of strong shear, the mode coupling and deformation of the wave field that occurs make spectral methods less useful in that a large number of terms must be retained in the series. Therefore, we chose to employ a finite difference approach.

The finite difference methods employed on elliptic systems are generally one of two types, iterative or direct. The iterative methods attempt to find solutions by successive correction of an initial "guess" until a sufficiently accurate solution is obtained. This is done by solving the finite difference equation for the dependent variable at a point in terms of its surrounding values. Then at each iteration a new value for the solution at this point is found from the old values at the surrounding ones. The iterative methods are simpler to program and in many problems can be faster and more accurate than direct methods. However, this usually can not be shown *a priori*. For coefficient matrices which are diagonally dominant, usually the case for elliptic problems, iterative methods are guaranteed to converge to the correct solution. In contrast, the direct methods attempt to solve the linear system directly. They usually take advantage of the form and sparseness of the coefficient matrix to reduce it to upper triangular form, whereupon the solution is obtained through backsubstitution. For a direct method to be useful, full advantage must be taken of the sparseness of the coefficient matrix. For example, classic Gaussian elimination cannot be used in problems of reasonable size due to the large storage necessary for coefficients and the accumulation of significant roundoff errors resulting from the large number of operations. For a more detailed discussion of both iterative and direct methods see Forsythe and Wasow [1960], Varga [1962], Issacson and Keller [1966] or Dorr [1970]. For application to atmospheric problems see Haltiner and Williams [1980].

For the present problem, the coefficient matrix is not diagonally dominant. Therefore convergence of the iterative methods, though possible, is not assured. Solutions were sought using both the Gauss-Seidel method and successive overrelaxation (SOR) with little success. More sophisticated iterative techniques were not considered. As a result of these failures, we concentrated on using direct methods to solve the difference equation. In order to understand the procedure it is best to consider the form of the coefficient matrix. Therefore we begin by discretizing the domain in y and z such that

$$y_i = \delta y(i-\tfrac{1}{2}) \quad \text{and} \quad z_j = \delta z(j-\tfrac{1}{2})$$

and expand (7) using centered differences into finite difference form.

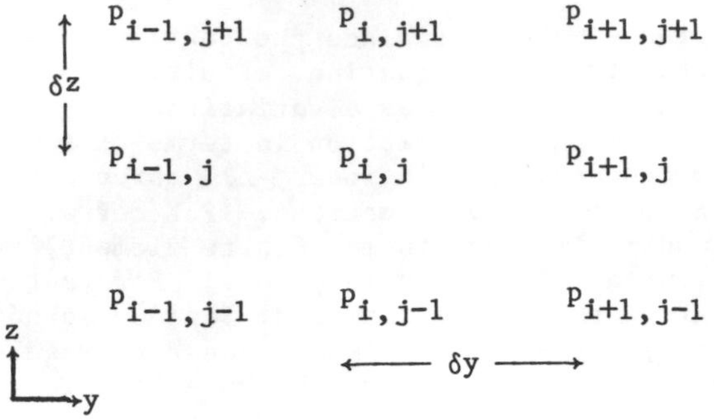

$p_{i-1,j+1}$ $p_{i,j+1}$ $p_{i+1,j+1}$

δz

$p_{i-1,j}$ $p_{i,j}$ $p_{i+1,j}$

$p_{i-1,j-1}$ $p_{i,j-1}$ $p_{i+1,j-1}$

z ← δy →

y

Figure 2. Grid arrangement for the 9-point stencil

The centered difference expansion employs "nearest neighbor" points and neglects terms which are proportional to the y or z stepsize squared and thus are second order accurate. The above grid is used for convenience in expressing derivative boundary conditions. On this grid all boundaries, except the northern where pressure is specified, lie between grid points. Therefore, the derivative conditions at the boundary can easily be expressed in central difference form using grid values ½ step inside and outside the boundary. Due to the presence of the cross derivative term, the centered differences require a nine point stencil (Figure 2) to evaluate. Evaluating for every grid point leads to NY*NZ equations for the pressure at each of the NY*NZ grid points. To construct the coefficient matrix in the most compact form which conserves symmetry, a linear mapping is performed with a new index, L, defined by

$$L = NY(j-1) + i \qquad i=1,2...NY \;\; \& \;\; j=1,2...NZ.$$

This mapping was chosen since, in general, NY is much less than NZ and therefore the bandwidth of the coefficient matrix is minimized. Then for grid point i,j (= L), the finite difference equation can be written

$$\begin{aligned} &Q_L p_{L-NY-1} + R_L p_{L-NY} - Q_L p_{L-NY+1} + S_L p_{L-1} + T_L p_L + U_L p_{L+1} - \\ &- Q_L p_{L+NY-1} + V_L p_{L+NY} + Q_L p_{L+NY+1} = [\text{Forcing} + \text{BC}]_L . \end{aligned} \tag{10}$$

From this it can be seen that the equation is the normal linear matrix equation, $AX=B$ where A is the coefficient matrix, X the solution vector, and B is the vector containing forcing and/or boundary conditions. Furthermore, the coefficient matrix is of block (or banded)-tridiagonal form with bandwidth 2(NY+1). The bandwidth is defined as the maximum number of elements between the first and the last nonzero elements in any row of the coefficient matrix. Perhaps the simplest way of visualizing the coefficient matrix is as a square tridiagonal matrix of order NZ whose elements are square tridiagonal matrices of order NY. The matrix in symbolic form is

$$A = \begin{bmatrix} C_1 & D_1 & \cdot & \cdot & & 0 \\ B_2 & C_2 & D_2 & \cdot & & \\ \cdot & B_3 & C_3 & D_3 & \cdot & \\ & \cdot & \cdot & \cdot & \cdot & \cdot \\ & & \cdot & B_{NZ-1} & C_{NZ-1} & D_{NZ-1} \\ 0 & & & \cdot & B_{NZ} & C_{NZ} \end{bmatrix} = [B_j, C_j, D_j]_{NZxNZ}$$

Employing the compact notation on the extreme right hand side, the tri-diagonal matrices B_j, C_j and D_j are then

$$B_j = [Q_i, R_i, -Q_i]_{NYxNY}|_j,$$

$$C_j = [S_i, T_i, U_i]_{NYxNY}|_j,$$

$$D_j = [-Q_i, V_i, Q_i]_{NYxNY}|_j$$

with the Q, R, S, T, U, and V_i's being the coefficients in (10) and i and j refer to the y and z dependent grid positions, respectively.

It is just this form which lends itself to efficient solution by direct methods. The approach that we have chosen was used successfully by McCreary and Chao [1985] in the study of trapped waves on the continental shelf, and they have supplied us with their code for the solver. The procedure is essentially a Gaussian elimination with operations only performed for non-zero elements. This reduces the operations count from $NZ^3NY^3/3$ of the classical elimination to $NZ*NY^3$ and the memory requirements from NZ^2NY^2 to $2NZ*NY^2$. Operationally, passes are made through the matrix eliminating elements below the main diagonal. This can be done efficiently by using the knowledge of the form of the coefficient matrix. Since this is a banded matrix, elimination can only generate non-zero elements between the two original bands allowing the savings in memory. Another direct method which has been employed on this type of problem is the block reduction method [Isaacson and Keller, 1966; Lindzen and Kuo, 1969]. This method uses a generalization of the tridiagonal solver (or two pass method) on the matrix form of the coefficient matrix. However, one drawback to this method is that it requires NZ inversions of matrices of order NY for a variable coefficient problem.

3.2. Accuracy

By subtracting the finite difference equation (FDE) from the partial differential equation (PDE) and taking the limit of small stepsizes, one can easily show that the truncation error tends to zero. Therefore, one can conclude that the FDE is consistent with the PDE in question. Consistency assures that as the stepsizes approach zero we still approxi-

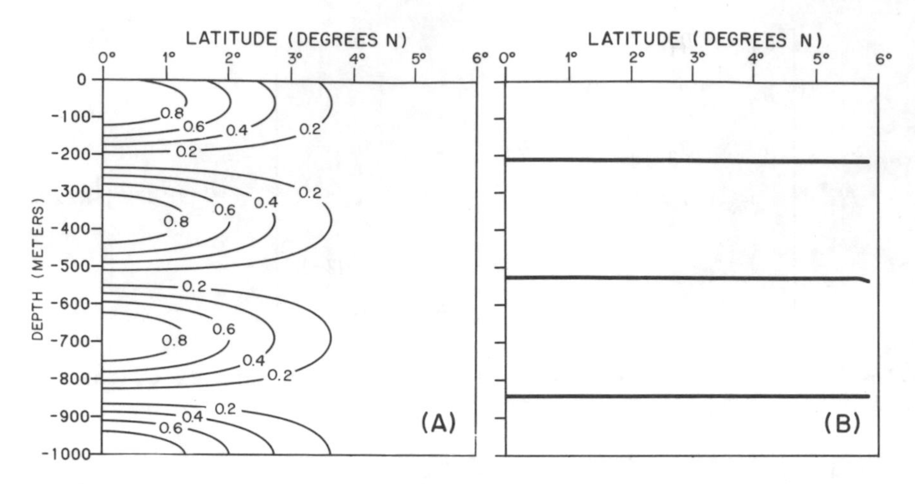

Figure 3. Pressure field solution for U=0cm/s, Inviscid (NY=40, NZ=131) (a) Amplitude (b) Phase.

mate the required PDE. Consistency, along with stability, implies convergence of the numerical solution to the true one as the stepsizes decrease. The truncation error of the FDE is also a convienient estimate of the expected error of the finite difference (FD) approximation. The truncation error, τ, is given by

$$\tau \equiv \frac{\delta y^2}{6}\left[\frac{A}{2}\frac{\partial}{\partial y} + B\frac{\partial}{\partial z} + D\right]\frac{\partial^3 p}{\partial y^3} + \frac{\delta z^2}{6}\left[\frac{C}{2}\frac{\partial}{\partial z} + B\frac{\partial}{\partial y} + E\right]\frac{\partial^3 p}{\partial z^3} + \text{H.O.T.}$$

In order to address the actual accuracy of the numerical method, we first solve (7) for an inviscid, no mean flow case with c specified as 100cm/s. For the results presented here, the domain chosen extends from the equator to 6°N and has a depth of 1000m. This domain, although considerably shallower than the real ocean, is sufficiently deep to address the questions which will be of interest later. In addition, the northern boundary is, necessarily, placed at a finite value of y, where p is set to zero. However, at the northern boundary, the amplitude of the forcing at the surface has decreased to less than 2% of its equatorial value. Therefore, in placing the northern boundary at 6°N we have introduced roughly 2% error in that boundary condition. For other cases, when we vary the phase speed, we allow the domain to expand or contract in the latitudinal direction in direct proportion to the equatorial decay scale, α.

For testing the accuracy of the numerical scheme, various choices of grid resolution were used. To test the sensitivity to meridional resolution, cases were run with NZ=131 for NY=30,40, and 50 points. Likewise, to test sensitivity to the vertical resolution, cases were run with NY=40 for NZ=61,101,131, and 161. Figure 2 shows the amplitude and phase of the numerical solution in the inviscid, no flow case for NY=40 and NZ=131. The phase speed of the forcing is specified as 100cm/s, slightly away from the resonant value of 106cm/s for the third baroclinic mode. The meridional structure is Gaussian as in (3) and a standing

TABLE I: Error dependence on grid resolution as a percentage of the wave amplitude. (No background flow)

Vertical Resolution (NY=40)	Maximum	RMS
Number of points in z		
61	8.5	2.2
101	6.9	1.9
131	6.6	1.8
161	6.4	1.7
Latitudinal Resolution (NZ=131)		
Number of points in y		
30	11.5	3.0
40	6.6	1.8
50	4.2	1.2

wave is set up in the vertical with phase jumps of 180° across nodal lines. In Table I, we have computed the maximum and the root mean square difference between the analytic and numerical solution over the entire domain. The result is quoted as a percent of the amplitude of the analytic solution.

There is little change in the accuracy of the solution for the differing vertical resolutions. The no mean flow case is nearly as well resolved using 61 grid points as 161 in this direction. Indeed, for the three highest resolutions there is essentially no change in the solution. This is due to the fact that the mode in Figure 2 has a vertical wavelength of $2\pi c/N$ or roughly 628 meters. For the coarsest vertical resolution (NZ=61) there are approximately 38 grid points per wavelength. However, in the presence of background flow the vertical wavelength is instead proportional to c-U, to leading order, and the resolution requirements clearly increase. Therefore, further consideration of vertical resolution will be postponed until we consider a mean flow.

While it is apparent that all resolutions in z for the no mean flow case are more than sufficient, this is not true for y. From Table I we see that changing the number of grid points in the y direction from 30 to 50 causes a decrease by a factor of 3 in the error. The improvement seen is more drastic from 30 points to 40 than from 40 to 50. It happens that these errors are decreasing roughly as the y stepsize squared in agreement with the estimate of the truncation error. Since a decrease is not seen for the vertical direction, clearly the first term in the truncation error is dominant under these conditions. Unfortunately, the practical constraints on the resolution in y are more stringent than in z since the scheme that we are employing requires $NZ*NY^3$ operations. Doubling the resolution in y, for example, requires four times the computation of a doubling in z. However, to investigate a critical layer case, we expect that the z resolution will be more crucial. Hence, we chose to work with NY=40 rather than NY=50 since it was felt that the increase in computer time associated with increasing resolution to 50 points outweighed the few percent increase in accuracy. The spacing for NY=40 corresponds to a dimensional stepsize in the meridional direction of roughly 16km for the mode in Figure 2.

A direct test for accuracy in the presence of mean flows is impossible because we do not have, in general, analytical solutions available to us. One must therefore rely on the consistency and stability of the numerical scheme to be assured of a convergent solution. However, as an additional means of addressing accuracy, especially for mean flow cases, a second diagnostic test was performed. From (1) and from the numerical pressure field we calculate the remaining dynamic variables (velocity and density), using only four of the five remaining governing equations. The remaining equation is satisfied if and only if the numerical pressure field is a solution to the PDE. It is simplest to use the continuity equation for the remaining equation. It should be noted that error is incurred in calculating the dynamic variables since differentiation is involved.

The test uses a pointwise form of the continuity equation (1c) to compare in a rms sense the residual of the finite difference form of (1c) to the individual terms therein. In the first two columns of Table II we present the results of this test for the inviscid, no mean flow case. These results are quoted as the percent of each term of the continuity equation that the residual represents. These results are in agreement with the accuracy test of Table I. In particular, the same meridional dependence of the error is evident, although the proportionality to the square of the stepsize is lost. Additionally, the insensitivity to the vertical resolution remains. The v field we obtain is an order of magnitude smaller than the error in the other two velocity

TABLE II: Continuity equation residual as a percentage of the individual terms.

	U=0cm/s		U=150cm/s	
Vertical Resolution (NY=40)	$\lvert ku \rvert$	$\lvert w_z \rvert$	$\lvert ku \rvert$	$\lvert w_z \rvert$
Number of points in z				
61	1.26	1.27	8.88	9.48
101	0.89	0.89	5.78	6.01
131	0.86	0.87	4.85	5.03
161	0.87	0.87	4.51	4.66
Latitudinal Resolution (NZ=131)				
Number of points in y				
30	1.12	1.12	5.23	5.35
40	0.86	0.87	4.85	5.03
50	0.72	0.73	4.72	4.93

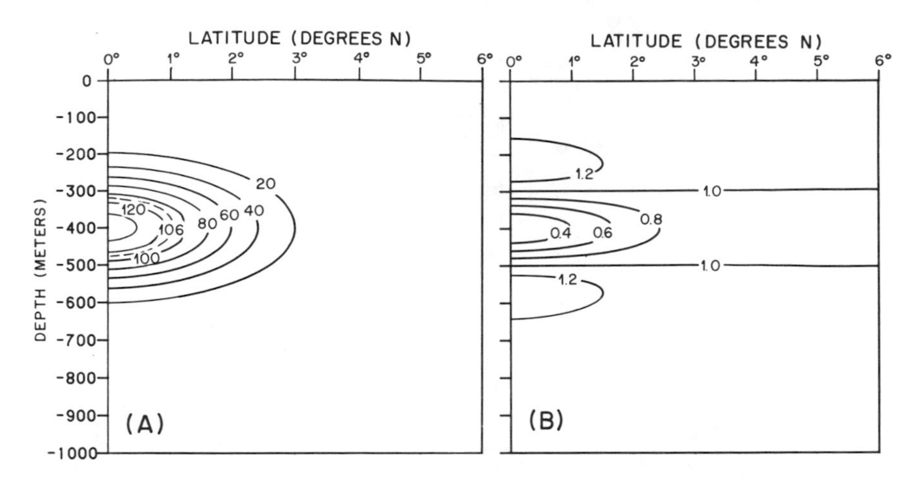

Figure 4: Imposed background fields (a) Zonal velocity (cm/s) (b) N^2 ($\times 10^4 s^{-2}$)

components. However, for this case the v component of the flow should be identically zero.

When a realistically sheared zonal background flow is present, we in general cannot solve (7) analytically. Thus, to determine the accuracy of the present method, we must resort to the traditional technique of increasing resolution until numerical convergence is attained. The flow considered is an eastward jet of amplitude 150cm/s in which a critical surface exists for the third baroclinic mode. The flow is assumed to have Gaussian meridional structure, centered at 400m depth with decay scales of 150km and 100m with latitude and depth, respectively. The background fields (U and N^2) are shown in Figure 4 with the 106cm/s contour, corresponding to the third baroclinic mode phase speed, dashed. Due to the presence of the critical surface, friction is necessarily included.

We can consider a simple scale analysis to obtain lower limits on the vertical resolution necessary to resolve the critical layer (since an analytic form is chosen for the Undercurrent, there is no constraint imposed upon the grid in order that the background shear flow be accurately represented). For Newtonian cooling and Rayleigh friction, the region over which damping is important is that for which the damping, r, is comparable to the Doppler shifted phase speed, c-U. A Taylor series expansion of c-U around the critical surface leads to

$$\Delta z \sim \left(\frac{2\kappa U}{\sigma U_z}\right), \qquad \Delta y \sim \left(\frac{2\kappa U}{\sigma U_y}\right).$$

For damping used herein, corresponding to an e-folding time of 5 years, Δz is O(10m) thick and Δy is O(15km) wide. For a basin depth of 1000m and width of 600km, this dictates that at least 100 points be used in z and 40 points be used in y in order to assure that grid points fall

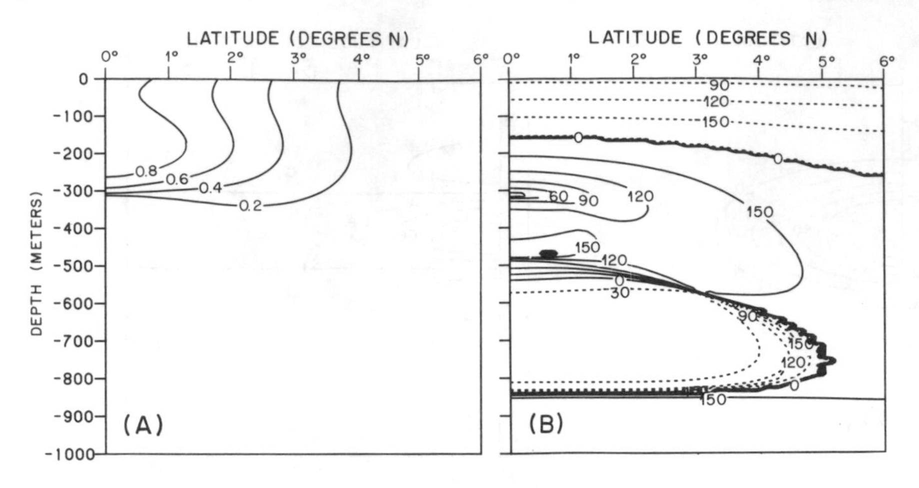

Figure 5: Pressure solution for U=150cm/s, $\kappa=(5 \text{ years})^{-1}$ (a) Amplitude (b) Phase

within the entire extent of the critical region. The requirement on the y resolution occurs at the core of the Undercurrent where the critical surface is oriented vertically and tangential to the direction of wave propagation. Therefore, this is likely an overly strict requirement on the y resolution. These spacings do not resolve the fine details of the critical layer but merely assures that its effects will be felt strongly by the waves. The test of accuracy was performed for the same choices of resolution as in the no flow case and the results are presented in the last two columns of Table II.

For runs made with critical layers, numerical convergence was attained by a resolution of NZ=131 and NY=40 points with dimensional stepsizes of approximately 8m and 16km, respectively. This convergence is also apparent in the continuity balance, in that little change is seen in increasing resolution from 131 to 161 in z or from 40 to 50 in y. The solution for NY=40 and NZ=131 is shown in Figure 5. Significant amplitude is seen to be confined to the region above the critical layer at the equator similar to the findings of Holton [1970]. The phase is continuous and increasing upward, indicating downward energy propagation into the critical layer where it is dissipated and absorbed in the background flow. Clearly, the presence of an intense background flow will have a strong dissipative impact on vertically standing Kelvin modes. For realistic amplitudes of the Undercurrent, all but the lowest of these modes encounter a region of critical flow over at least a part of their meridional extent. This is especially true when one considers that the equatorial decay scale (being proportional to c) decreases with increasing vertical mode number. This causes the critical region to extend over a greater portion of the wave region, further enhancing the absorption.

4. CONCLUSIONS

We have applied a direct finite difference solution technique to the problem of oceanic equatorial waves in the presence of realistically sheared background flows. In the long wave limit, the governing second order partial differential equation is of elliptic type and is efficiently solved using finite differences. The resulting finite difference equation leads to a coefficient matrix which is of block-tridiagonal form. The numerical algorithm takes full advantage of the sparseness of the coefficient matrix in reducing it to upper diagonal form. The method is relatively fast and accurate for strong flow cases requiring high vertical and meridional resolution. This is crucial for consideration of the large variety of problems of interest for which this technique may be applicable, not only in equatorial oceanography but in more general physical oceanography as well.

REFERENCES

Birkhoff, G., The Numerical Solution of Elliptic Equations, Regional Conference Series in Applied Math., 11, SIAM Publications, Philadelphia, 1972.

Boyd, J.P., The effects of latitudinal shear on equatorial waves: Parts I and II, J Atmos Sci., 35, 2236-2267, 1979.

Dorr, F.W., The direct solution of the discrete Poisson equation on a rectangle, SIAM Review, 12(2), 248-263, 1970.

Dunkerton, T., A Lagrangian mean theory of wave, mean-flow interaction with applications to nonacceleration and its breakdown, Rev of Geophys and Space Phys., 18(2), 387-400, 1980.

Forsythe G.E. and W.R. Wasow, Finite-Difference Methods for Partial Differential Equations, John Wiley, New York, 1960.

Haltiner, G.J. and R.K. Williams, Numerical Prediction and Dynamic Meteorology, John Wiley and Sons, New York, 1980.

Holton, J.R., The influence of mean wind shear on the propagation of Kelvin waves, Tellus, 22, 186-193, 1970.

Holton, J.R., The Dynamic Meteorology of the Stratosphere and Mesosphere, Meteor Monogr., No. 37, Amer Meteor Soc., 1975.

Holton, J.R., Equatorial wave-mean flow interaction: a numerical study on the role of latitudinal shear, J Atmos Sci., 36(6), 1030-1040, 1979.

Issacson, E. and H.B. Keller, Analysis of Numerical Methods, John Wiley and Sons, New York, 1966.

Lindzen, R.S., Internal equatorial planetary-scale waves in shear flow, J Atmos Sci., 27, 394-407, 1970.

Lindzen, R.S., Equatorial planetary waves in shear: Part I, J Atmos Sci., 28, 609-622, 1971.

Lindzen, R.S., Equatorial planetary waves in shear: Part II, J Atmos Sci., 29, 1452-1463, 1972.

Lindzen, R.S. and H-L. Kuo, A reliable method for the numerical integration of a large class of ordinary and partial differential equations, Mon Wea Rev., 97(10), 732-734, 1969.

McCreary, J.P. and S-Y Chao, Three-dimensional shelf circulation along

an eastern ocean boundary, J Mar Res., 43, 13-36, 1985.
McCreary, J.P., Modeling equatorial ocean circulation, Ann Rev Fluid Mech., 17, 359-409, 1985.
McPhaden, M.J. and R.A. Knox, Equatorial Kelvin and inertio-gravity waves in zonal shear flow, J Phys Ocean., 9, 263-277, 1979.
Moore, D.W. and S.G.H. Philander, Modeling of the tropical ocean circulation, In The Sea, 6, edited by E.D. Goldberg, Wiley-Interscience, New York, 319-361, 1977.
Philander, S.G.H., Equatorial waves in the presence of the Equatorial Undercurrent, J Phys Ocean., 9, 254-262, 1979.
Varga, R.S., Matrix Iterative Analysis, Prentice-Hall, Englewood Cliffs, New Jersey, 1962.

THE DIFFUSIVE PROBLEM

J. J. O'Brien
Mesoscale Air-Sea Interaction Group
The Florida State University
435 Oceanography/Statistics Building
Tallahassee, FL 32306-3041

1. INTRODUCTION

The ocean contains motions on many scales. In all numerical models it is necessary to parameterize the effect of unresolved scales of motion by some smoothing, usually in the form of Laplacian friction. This leads to equations which are both hyperbolic and parabolic. In this chapter we study simple diffusion equations in order to learn how to formulate the diffusive part of the problem and how to investigate the linear numerical stability of finite difference schemes. All methods are not stable.

2. THE SIMPLE ONE-DIMENSIONAL DIFFUSIVE EQUATION

$$\frac{\partial u}{\partial t} = K \frac{\partial^2 u}{\partial x^2}, \qquad K > 0 \text{ and constant.} \tag{1}$$

The simplest finite difference scheme is

$$u_j^{n+1} = u_j^n + \frac{\Delta t K}{(\Delta x)^2} [u_{j+1}^n + u_{j-1}^n - 2u_j^n] \tag{2}$$

This has a truncation error $O(\Delta t) + O(\Delta x^2)$.

Exercise:

If $K = \Delta x^2/6\Delta t$, then the scheme is $O(\Delta t^2) + O(\Delta x^4)$. Prove it.

Equation (1) states physically that u should decrease smoothly in time. If we multiply (1) by u and use the chain rule on the right side, we obtain

$$\frac{1}{2}\frac{\partial u^2}{\partial t} = uK\frac{\partial^2 u}{\partial x^2} = K\frac{\partial}{\partial x}\left(u\frac{\partial u}{\partial x}\right) - K\left(\frac{\partial u}{\partial x}\right)^2$$

Integrate over all x, say $x \in [0,1]$

J. J. O'Brien (ed.), Advanced Physical Oceanographic Numerical Modelling, 127–144.

$$\frac{\partial}{\partial t} \int_0^1 u^2/2 \, dx = K \, u\frac{\partial u}{\partial x} \Bigg|_{x=0}^{x=1} - K \int_0^1 \left(\frac{\partial u}{\partial x}\right)^2 dx$$

The first term on the right hand side is called the "dispersion" or "diffusion" term; the second is the "dissipation".

If we have cyclic boundaries, u or $\frac{\partial u}{\partial x}$ vanishes at boundaries

$$\frac{\partial}{\partial t} E(t) = -K \int_0^1 \left(\frac{\partial u}{\partial x}\right)^2 dx < 0$$

Let us call E(t) energy or variance of u. The relationship says that E(t) must decrease monotonically with time. Therefore any finite difference solution of (1) must exhibit this same property. It also demonstrates that K must be positive for Laplacian friction.

Another way to understand the physics of this simple problem is to assume that there is a known solution

$$u(x,t) = U(t) \, e^{i\ell x} \tag{3}$$

Substitution into (1) proves immediately that (3) is at least a local solution of (1) without regard to boundary conditions. If ℓ is arbitrary, then U(t) satisfies

$$\frac{\partial U}{\partial t} + K\ell^2 U = 0$$

which has a solution

$$U = c \exp\left[- K\ell^2 t\right]$$

This explicitly states that the time-dependent solution of (1) must decay in time for every wavelike solution! The solution decays exponentially in time and the e-folding scale is a quadratic function of wave number, ℓ.

Since we expect any physical solution to remain bounded in time, we must require any finite difference solution to remain bounded in time.

Therefore, for any finite-difference approximation to be called numerically stable, the solution should consist of Fourier components whose amplitudes do not grow unbounded in time. We shall call a scheme stable if all Fourier components remain bounded with time. It is unstable if even one component (or mode) is not bounded.

3. JOHN VON NEUMANN NECESSARY CONDITION FOR STABILITY

Consider the simple diffusion equation in the finite difference form

$$u_j^{n+1} = u_j^n + \hat{K}[u_{j+1}^n + u_{j-1}^n - 2u_j^n] \tag{4}$$

where $\hat{K} = K\Delta t/\Delta x^2$. Assume

$$u_j^n = U_n \exp [ikj\Delta x]$$

and substitute in (1). Use of Euler's formula yields

$$U_{n+1} = (1 + 2\hat{K}(\cos k\Delta x - 1))U_n$$

Define an amplification factor, G, such that

$$U_{n+1} = GU_n \quad \text{where } G = 1 + 2\hat{K}(\cos k\Delta x - 1)$$

In this somewhat trivial case G is a scalar. However, in later, more general examples G will be a matrix. The simple difference equation has as a solution

$$U_n = G^n U_0$$

Since the initial condition U_0 is arbitrary, U_n is bounded for large n if $|G| \leq 1$. If G is a matrix, then we require the spectral radius of G (say $\sigma(G)$) to be less than or equal to 1.

$\sigma(G) \leq 1$ is the Von Neumann necessary condition for *linear* computational stability.

For physical problems with growing solutions, we need to allow for growth. Therefore it is permissible for $|G| = 1 + O(\Delta t)$ but not faster. Indeed, let

$$U_{n+1} \equiv \lambda\, U_n$$

$|\lambda|$ is the amplification factor

$$|U_{n+1}| = |\lambda|\,|U_n|$$

For the method to be stable, we require U_n to be bounded after n time steps

$$|U_n| = |\lambda|^n |U_0| < B$$

Therefore

$$n \ln|\lambda| < \ln(B/|U_0|) \equiv B'$$

Since $n = t/\Delta t$, the necessary condition for stability

$$\ln|\lambda| < B'\Delta t/t$$

If we require boundedness for a finite time, then

$$\ln|\lambda| \leq O(\Delta t)$$

Let $$|\lambda| \equiv 1 + \delta$$

$$\ln(1+\delta) \simeq \delta + \text{higher order terms}$$

Therefore,

$$\delta < O(\Delta t)$$

or

$$|\lambda| \leq 1 + O(\Delta t)$$

This is the Von Neumann necessary condition for linear computational stability. In cases where no physical growth is anticipated, the practical stability criterion to use is

$$|\lambda| \leq 1$$

For our special case for the diffusion equation, $|G| \leq 1$ if $\hat{K} < 1/2$. Therefore we say that the scheme is conditionally stable if

$$\frac{K\Delta t}{\Delta x^2} \leq \frac{1}{2} \tag{5}$$

It is important to note that $k\Delta x = \pi$ is the most unstable wave; i.e., if we violate (5), the waves of length $2\Delta x$ will grow fastest. (See Figures 1a and 1b)

4. CENTERED TIME DIFFERENCE FOR DIFFUSION EQUATION

Suppose we use the leap-frog finite difference scheme

$$u_j^{n+1} = u_j^{n-1} + 2\hat{K}\,[u_{j+1}^n + u_{j-1}^n - 2u_j^n]$$

The reader can show that the truncation error is now $O(\Delta t^2 + \Delta x^2)$ and is, therefore, more accurate than the forward time difference. However, we can use the stability test to show that the scheme is unconditionally unstable!

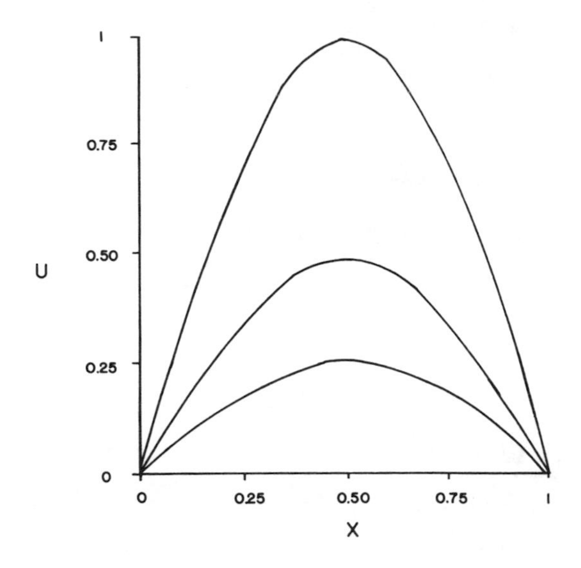

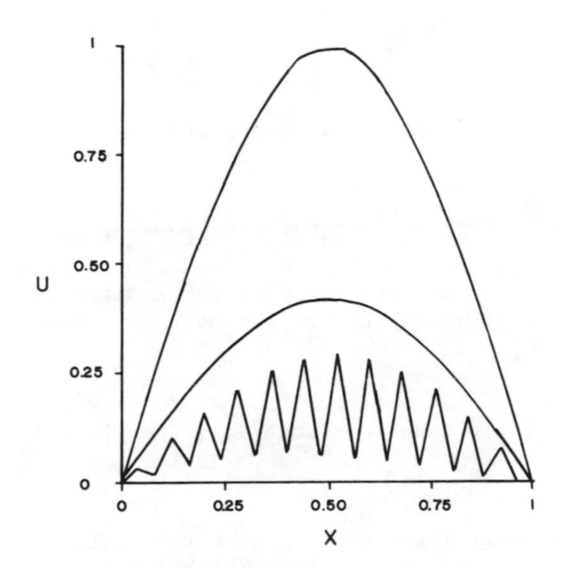

Figures 1a and 1b. The solution of the diffusion equation (1) with $u(x,0) = \sin(\pi x)$, $u(0,t) = 0$. The curves are at $n = 0, 50, 100$ for $\hat{K} = 0.45$ and 0.55 which exceeds the value 0.5. Note that the $2\Delta x$ wave is the most unstable wave.

Let

$$u_j^n = U_n \exp [ikj\Delta x]$$

to obtain

$$U_{n+1} = U_{n-1} + 4\hat{K} [\cos k\Delta x - 1] U_n \tag{6}$$

This is a three-term difference equation which can be solved by introducing the trivial equation $U_n = U_n$. We then have a set of two equations

$$\begin{bmatrix} U_{n+1} \\ U_n \end{bmatrix} = \begin{bmatrix} 4\hat{K}(\cos k\Delta x - 1) & 1 \\ 1 & 0 \end{bmatrix} \begin{bmatrix} U_n \\ U_{n-1} \end{bmatrix}$$

Define the amplification matrix

$$G = \begin{bmatrix} 4K(\cos k\Delta x - 1) & 1 \\ 1 & 0 \end{bmatrix} \tag{7}$$

The scheme is stable if $\sigma(G) \leqslant 1$.

Exercise:

Show that this is not true. Hence, this centered-difference scheme is always unstable!

There is a nice trick for handling the analysis of three time level schemes such as (6). Observe that we want to find the amplification factor, G, which satisfies $U_{n+1} = GU_n$. Since $G = 0$ is not useful, G^{-1} must exist and thus $U_{n-1} = G^{-1}U_n$. Substitution of these 2 equations into (6) immediately yields the desired equation for G. This trick will be used frequently in the next few chapters.

Exercise:

Show that the characteristic equation for (7) is identical to (8).

$$G = G^{-1} + 4\hat{K}(\cos k\Delta x - 1) \tag{8}$$

We shall use this general approach to test the stability of many finite difference schemes even for problems which are not parabolic. However, we can only find G for linear problems. For non-linear problems, it is necessary to linearize and apply the Von Neumann technique locally. Experience indicates that if a non-linear problem is linearly unstable, it is unstable.

There is some important vocabulary related to linear computational stability.

Definition: If $\sigma(G) = 1$, we say the scheme is <u>neutral</u>
If $\sigma(G) < 1$, we say the scheme is <u>stable</u>. The stability may be conditional or unconditional.
If $\sigma(G) > 1$, the scheme is <u>unstable</u>.
$\sigma(G)$ is the spectral radius of a matrix G or the modulus of a scalar G. These conditions must be true for any wavenumber k, if the scheme is to be called stable.

5. FINITE DIFFERENCE SCHEMES FOR PARABOLIC PDE'S

Consider the simplest diffusion equation

$$\frac{\partial u}{\partial t} = K \frac{\partial^2 u}{\partial x^2} \qquad K = \text{const} > 0$$

$$u = u(x,t)$$

$$u(x,0) = u_0(x) \qquad \text{and is prescribed.}$$

Define the operator δ_2 such that

$$\delta_2 u_j = u_{j+1} + u_{j-1} - 2u_j$$

A general four or six point difference equation is

$$u_j^{n+1} - u_j^n = \frac{\Delta t K}{(\Delta x)^2} [\theta \delta_2 u_j^{n+1} + (1 - \theta) \delta_2 u_j^n]$$

where $\theta \geqslant 0$. $\theta = 0$ gives the usual four point formula (2)

$\theta = 1/2$ gives the trapezoidal rule (centered time difference called the Crank-Nicholson Method).

The truncation error is $O(\Delta t^2) + O(\Delta x^2)$ when $\theta = 1/2$.

If $\theta = 0$, we have an <u>explicit</u> system to solve; meaning we can solve for u^{n+1} without any terms on the right hand side depending on u_j^{n+1} for any j.

If $\theta \neq 0$, we have an <u>implicit</u> system to solve, and it may require a matrix technique like the tridiagonal algorithm.

In the following, let $\hat{K} = K\Delta t/\Delta x^2$ and e be the truncation error. The stick diagrams in the following lists are called the 'finite difference stencil' (adapted from Richtmyer & Morton [1967]).

5.1. Brief Outline of Methods for Diffusion Equation

5.1.1. Method 1. ($\theta = 0$)

$$u_j^{n+1} = u_j^n + \hat{K}\delta_2 u_j^n , \qquad e = O(\Delta t) + O(\Delta x^2)$$

This scheme is explicit and stable if $\hat{K} \leqslant 1/2$ as Δt, $\Delta x \to 0$.

5.1.2. Method 2. ($\theta = 1/2$)

$$u_j^{n+1} = u_j^n + \frac{\hat{K}}{2} [\delta_2 u_j^n + \delta_2 u_j^{n+1}], \quad e = O(\Delta t^2) + O(\Delta x^2)$$

This is the Crank-Nicholson scheme. It is implicit and always stable.

5.1.3. Method 3. ($\theta = 1$)

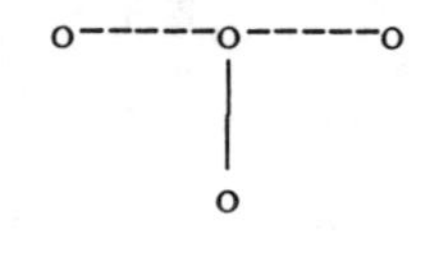

$$u_j^{n+1} = u^n + \hat{K}\ \delta_2 u_j^{n+1} \qquad e = O(\Delta t) + O(\Delta x^2)$$

This is implicit and always stable.

5.1.4. Method 4.

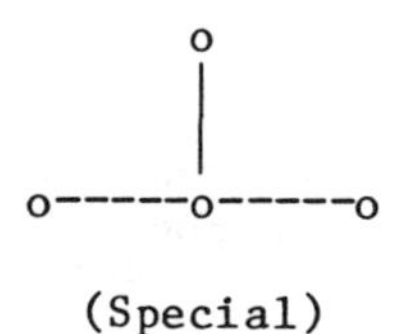

(Special)

Same as Method 1 except $\hat{K} = 1/6$, $e = O(\Delta t^2) + O(\Delta x^4)$

This scheme is explicit and stable.

5.1.5. Method 5.

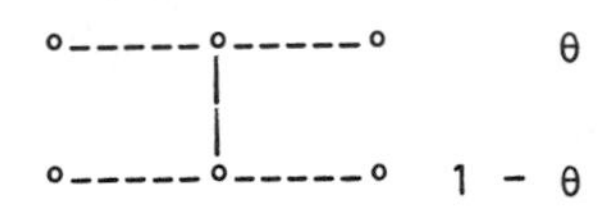

$$u_j^{n+1} = u_j^n + \hat{K}\,[\theta\,\delta_2\,u_j^{n+1} + (1-\theta)\,\delta_2\,u_j^n]$$

If $0 < \theta < 1/2$, the scheme is stable if $\hat{K} < (2-4\theta)^{-1}$. If $1/2 \leq \theta \leq 1$, the scheme is always stable. Methods 1 through 4 are special cases of 5.

Exercise:

Verify that $e = O(\Delta t^2 + \Delta x^2)$ if $\theta = 1/2$ and if $\theta = 1/2(1-1/6\hat{K})$.

5.1.6. Method 6.

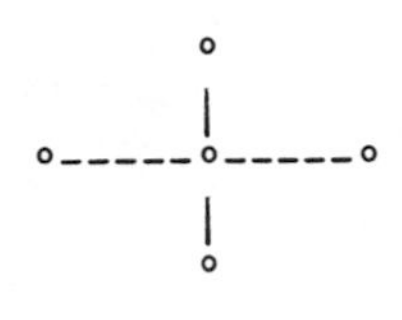

$$u_j^{n+1} = u_j^{n-1} + 2\hat{K}\,\delta_2\,u_j^n$$

This scheme is *always* unstable. This is the centered scheme considered in 4.4.

5.1.7. Method 7.

DuFort-Frankel [1953]

$$u_j^{n+1} = u_j^{n-1} + 2\hat{K}\,[u_{j+1}^n + u_{j-1}^n - u_j^{n-1} - u_j^{n+1}]$$

$$e = O(\Delta t^2) + O(\Delta x^2) + O\left((\Delta t/\Delta x)^2\right)$$

This scheme is implicit and always stable. The u_j^{n+1} term on right may be brought to the left yielding an explicit formula. The DuFort-Frankel scheme is widely used in oceanography. However it has an inherent problem. This scheme is not consistent unless $(\Delta t/\Delta x)$ goes to zero at the same rate that Δt, Δx 0. In cases with time steps large compared to $\hat{K} = 1/2$, the solution behaves like a hyperbolic problem with the governing equation

$$\frac{\partial u}{\partial t} = - \left(\frac{\Delta t}{\Delta x} \right)^2 K \frac{\partial^2 u}{\partial t^2} + K \frac{\partial^2 u}{\partial x^2}$$

However, in many problems $K(\Delta t)^2 \ll \Delta x^2$, and the scheme behaves properly. It is convenient to use the Dufort-Frankel method in many oceanography problems that are two and three dimensional.

There are numerous other schemes involving three time levels. They are too rarely used to reproduce here.

6. TWO-DIMENSIONAL DIFFUSIVE PROBLEMS

Let us consider the linear diffusive equation

$$\frac{\partial u}{\partial t} = K \left(\frac{\partial^2 u}{\partial x^2} + \frac{\partial^2 u}{\partial y^2}\right) \tag{9}$$

If $u(x,y,t) = u(\ell\Delta x, m\Delta y, n\Delta t) = u^n_{\ell,m}$, the usual second order finite difference approximation is

$$u^{n+1}_{\ell,m} = u^n_{\ell,m} + \frac{K\Delta t}{\Delta x^2}\left(u^n_{\ell+1,m} + u^n_{\ell-1,m} - 2\,u^n_{\ell,n}\right) + \frac{K\Delta t}{\Delta y^2}\left(u^n_{\ell,m+1} + u^n_{\ell,m-1} - 2u^n_{\ell,n}\right) \tag{10}$$

Let us investigate the linear computational stability of this equation. Let

$$u^n_{\ell,m} = U_n\, e^{i\ell j\Delta x}\, e^{imk\Delta y} \tag{11}$$

Substitution of (11) in (10) yields

$$U_{n+1} = U_n + [2K\Delta t/\Delta x^2(\cos j\Delta x-1) + 2K\Delta t/\Delta y^2\,(\cos k\Delta y-1)]U_n$$

For $j\Delta x = k\Delta y = \pi$, the amplification factor is

$$G = 1 - 4K\Delta t/\Delta x^2 - 4K\Delta t/\Delta y^2$$

or

$$G = 1 - 4K\Delta t\left(\frac{1}{\Delta x^2} + \frac{1}{\Delta y^2}\right)$$

This scheme is stable if

$$K\Delta t\left(\frac{1}{\Delta x^2} + \frac{1}{\Delta y^2}\right) < \frac{1}{2}$$

In the usual physical problem $\Delta x = \Delta y$, and thus

$$K\Delta t/\Delta x^2 < \frac{1}{4}$$

Thus the time step is more restrictive for a two-dimensional problem than for a one-dimensional problem. The reader must also realize that this is a practical restriction on the time step. We expect that K is chosen based on the physics of the problem; Δx should be chosen on the basis of the scales of variability that one is modelling. Practically, Δx is frequently chosen on the basis of storage availability or economics of computing. Hence, for most conditionally stable finite difference schemes, the stability criterion is a constraint on the time step!

7. ADVANCED SECTION: TURBULENCE PARAMETERIZATIONS IN OCEANOGRAPHY AND METEOROLOGY

On the length scales where rotation is important, we know relatively little about turbulence in the ocean or atmosphere. Numerous studies in air-sea interaction and boundary layer meteorology have given us insight into the vertical distribution of eddy diffusivities but little understanding of parameterizing horizontal turbulence. Various simple and complex approaches have been taken. Many atmospheric modelers use numerical damping or occasional spatial filtering to remove noise from numerical solutions. In this section we will review some of the explicit damping mechanisms used in large scale models. Some friction is always needed in numerical models to damp unwanted noise which arises from numerical errors or non-linear interactions.

7.1. Bottom Friction

The simplest form of friction is linear bottom friction or Rayleigh friction.

$$\frac{\partial u}{\partial t} = -\alpha u + \text{other physics} \; ; \; \alpha > 0 \qquad (12)$$

The inverse of α is an e-folding scale for damping. It damps all wavenumbers at the same rate.

Exercise:

Show that forward time differencing is stable if $\alpha\Delta t \leqslant 2$ and that leap frog is unconditionally unstable.

In many ocean problems, quadratic bottom friction is used. Numerous flow studies in streams, rivers and the laboratory show that the bottom friction is proportional to the square of velocity. Usually the equation takes the form

$$\frac{\partial u}{\partial t} = -cu^2/H + \text{other physics} \quad , \quad c > 0 \tag{13}$$

The drag coefficient is non-dimensional; H is a characteristic depth - sometimes the depth of the fluid and sometimes the depth at which u is measured. We cannot check stability of this equation because it is non-linear. However, if we employ a "local" approximation in time and space, (2) reduces to (1) for linear stability analysis.

7.2. Laplacian Friction

The most common approximation for turbulence is Laplacian friction. Most researchers believe that eddy dissipation can be modelled like Fickian diffusion with a constant viscosity, such as:

$$\frac{\partial u}{\partial t} = K \frac{\partial^2 u}{\partial x^2} + \text{other physics} \tag{14}$$

This is the problem we have investigated in detail in this chapter. It is important to recall that each wavenumber, k, is damped according to

$$u(x,t) \approx U(t,k) = U_o e^{-k^2 Kt}$$

The smallest waves, $L_{min} = \frac{2\pi}{2\Delta x}$, are damped fastest.

7.3. Biharmonic Friction

Many oceanographers have found that they cannot afford to use very small grid distances (high resolution) and have invented new ways to damp the smallest scales in numerical solutions. The simplest is bi-harmonic friction in the form

$$\frac{\partial u}{\partial t} = -A\nabla^4 u + \text{other physics} \quad , \quad A > 0$$

The necessity for the negative sign is easily seen if we substitute a single harmonic,

$$u(x,t) = U(t,k,\ell)e^{ikx} e^{i\ell y}$$

which has the "local" solution

$$U(t,k,\ell) = U_o e^{-A(k^2+\ell^2)t}$$

In order for biharmonic friction to be dissipative, the negative sign is imperative. This parameterization of friction is called "scale selective" because it damps the waves according to k^4 instead of k^2. Consequently the rate of damping decreases faster for longer waves for biharmonic friction than Laplacian friction. It still is an open question whether it is a good approximation for oceanic turbulence.

If we consider the equation

$$\frac{\partial u}{\partial t} = -A\nabla^4 u$$

and use second-order finite differences with $\Delta x = \Delta y$, we obtain the weights shown in Figure 2.

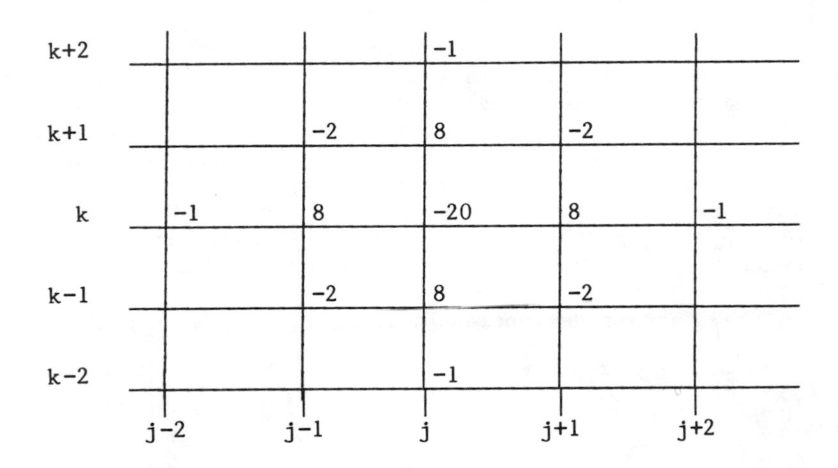

Figure 2. The stencil is very wide. If a boundary is at $(j-1)\Delta x$, or, in other terms, if the nearest interior point is $j\Delta x$, then we need to specify u_{j-1}, $(u_x)_{j-1}$ and $\nabla^2 u_{j-1}$ as appropriate boundary conditions.

Exercise:

Use the von Neumann stability analysis with a forward time difference and $\Delta x = \Delta y$ to show that

$$\frac{A\Delta t}{\Delta x} \leq \frac{1}{128}$$

for stability.

7.4. Non-Linear Diffusivity Based on Turbulence Theory

The inclusion of eddy diffusivities is based on a parameterization of the turbulence at the smallest scales in a particular oceanic model. If the scales are small enough so that the flow is not significantly dominated by rotation, we may propose that the turbulence is three dimensional; if rotation is important, we might propose that geostrophic turbulence [Charney, 1976] is important. In each case there is a formulation which has found success in various models. As a preliminary *caveat*, we state that these nonlinear formulations are expensive to include in a model. If we want to investigate the linear physics of a particular situation, we should not include the following parameterizations. If we are trying to predict a real ocean flow, they are suggested as useful parameterizations.

7.5. Three-Dimensional Turbulence

We envision a three-dimensional turbulence problem in which the largest scales are forced; the intermediate scales are an inertial subrange and the smallest scales (smaller than $2\Delta x$) are viscously dominated. The Kolmogoroff hypothesis states that in the inertial subrange, the kinetic energy spectrum, $E(k)$ depends only on wavenumber, k, and the rate of energy dissipation, ε.

The concept of using this hypothesis for eddy diffusivities was developed by Smagorinsky [1963] and Lilly [1967]. It has been used with success by Smagorinsky [1963], Deardorff [1970, 1971] and O'Brien [1971].

In the standard derivation of the -5/3 law, we hypothesize

$$E(k) = Ck^{\alpha}\varepsilon^{\beta}$$

where

$$\varepsilon = \nu \, D_{ij} \, D_{ij}$$

The molecular viscosity is ν and D_{ij} is the deformation tensor

$$D_{ij} = 1/2\left(\frac{\partial u_i}{\partial x_j} + \frac{\partial u_j}{\partial x_i}\right)$$

where the velocity, $\underline{v}$ has components u_i; C is the Kolmogoroff constant. Dimensional analysis yields

$$E(k) = Ck^{-5/3} \, \varepsilon^{2/3}$$

The hypothesis for eddy diffusivity, K, is that K depends only on k and ε

$$K = ck^{\alpha}\varepsilon^{\beta}$$

Dimensional analysis yields

$$K = ck^{-4/3}\,\varepsilon^{1/3}$$

We only want to use K for horizontal diffusivity where

$$\varepsilon = KD^2$$

and

$$D^2 = (D_T^2 + D_S^2)$$

where

$$D_T = \frac{\partial u}{\partial x} - \frac{\partial v}{\partial y} \quad \text{stretching deformation}$$

$$D_S = \frac{\partial v}{\partial x} + \frac{\partial u}{\partial y} \quad \text{shearing deformation}$$

The solution for K is

$$K = c^{3/2}\,k^{-2}D$$

The viscosity depends on the square of the wavelength of the flow. To implement the approximation, we assign the smallest scale to wavenumber

$$k = 2\pi/2\Delta x$$

which yields

$$K = c'(\Delta x)^2 D$$

where c' is an empirical constant. If T is the quantity being diffused, K is used in the form

$$\frac{\partial T}{\partial t} = \nabla\cdot(K\nabla T) + \text{other physics}$$

Second-order derivatives are recommended, and D is calculated from the flow field at each time step from previous values.

7.6. Two-Dimensional Turbulence

We hypothesize that the turbulence at scales near $2\Delta x$ is dominated by geostrophic turbulence or 2-dimensional turbulence theory. Leith [1968], Crowley [1968] and Haney [1975] have used this idea successfully in various ocean circulation simulations. In this 2-D case, the kinetic energy spectrum, E(k), depends on wavenumber, k, and the rate of enstrophy dissipation, η. Enstrophy is mean square vorticity and η is

$$\eta = \nu\nabla\zeta \cdot \nabla\zeta$$

where the vertical component of vorticity, ζ, is

$$\zeta = \underline{k} \cdot \nabla\times\underline{V} = \frac{\partial v}{\partial x} - \frac{\partial u}{\partial y}$$

The dimensional analysis

$$E(k) = Ck^{\alpha}\dot{\eta}^{\beta}$$

yields the −3 law in wavenumber

$$E(k) = Ck^{-3}\eta^{2/3}$$

For a turbulent eddy viscosity formulation, we propose that K_2 depends only on wavenumber, k, and rate of enstrophy dissipation, η.

$$K_2 = ck^{\alpha}\eta^{\beta}$$

Dimensional analysis yields

$$K_2 = ck^{-2}\eta^{1/3}$$

As in the previous case, we solve for K_2, letting $k = 2\pi/2\Delta x$, and obtain

$$K_2 = c'(\Delta x)^3|\nabla\zeta|$$

The "constant," c', must be determined by numerical experiment. The magnitude of the gradient of vorticity is expensive to calculate. The coefficient is used in the equation for a variable, T, as

$$\frac{\partial T}{\partial t} = \frac{\partial}{\partial x}(K_2 \frac{\partial T}{\partial x}) + \frac{\partial}{\partial y}(K_2 \frac{\partial T}{\partial y}) + \text{other physics}$$

The eight nearest neighbors of a grid point are involved in determining K_2. This formulation for eddy diffusivity has the advantage that we induce substantial dissipation in those regions where the gradient of vorticity is large (strong currents) and no dissipation in regions where geostrophy dominates. We obtain dissipation where it is essential but not where it is not part of the essential physics.

7.6.1. An anisotropic nonlinear eddy viscosity. In many rotating stratified simulations of atmospheric and oceanic convection, symmetry in one horizontal direction is presumed. Computer efficiency and availability also dictate that the horizontal mesh spacing Δx exceed the vertical spacing Δz. We expect, *a priori*, that the eddy viscosities appropriate to these scales will not be isotropic. To prevent nonlinear computational instability, we anticipate that $K_H > K_v$ where these are the horizontal and vertical eddy viscosities, respectively.

If $Q = \zeta^2/2$ is the enstrophy, then the local enstrophy dissipation rate η is

$$\eta = K_H\left(\frac{\partial\zeta}{\partial x}\right)^2 + K_v\left(\frac{\partial\zeta}{\partial z}\right)^2 .$$

Following Leith [1968], if we assert that K_H and K_v depend only on wavenumber and η, the dimensionally consistent forms for K_H and K_v are

$$K_H = ck_x^{-2}\,\eta^{1/3} ,$$

$$K_v = ck_z^{-2}\,\eta^{1/3} ,$$

where some generality has been sacrificed by allowing c to appear in both equations. The wavenumbers are chosen to be the largest resolved by the grid, $k_x = 2\pi/2\Delta x$, $k_z = 2\pi/2\Delta z$. Eliminating η in the above, we obtain

$$K_H = \gamma(\Delta x)^3\left[\left(\frac{\partial\zeta}{\partial x}\right)^2 + \frac{(\Delta z)^2}{(\Delta x)^2}\left(\frac{\partial\zeta}{\partial z}\right)^2\right]^{1/2} , \qquad (15)$$

$$K_v = \frac{(\Delta z)^2}{(\Delta x)^2}K_H$$

where γ contains all the constants. It is possible that γ may eventually be estimated from turbulence theory. If $\Delta z = \Delta x$, (2) reduces to Leith's [1968] formulation. Kasahara (personal communication) has used c = 0.5 successfully in the horizontal eddy viscosity prescription for the NCAR general circulation model. Crowley [1968] reports a value of c = 3.7, but there is apparently some uncertainty about his definition of c. The reader must see numerical experimentation to determine an appropriate value of c. However, whereas most fluid simulation codes will integrate satisfactorily with a wide range of constant eddy viscosities, this author's experience [O'Brien, 1971] is that non-linear viscosity constants may only be varied by a factor of two for acceptable results.

In many simulations of mesoscale convection, $\Delta x \gg \Delta z$, the formula reduces to the simple equations

$$K_H = \gamma(\Delta x)^3\left|\frac{\partial\zeta}{\partial x}\right| ,$$

$$K_v = \gamma(\Delta z)^2\Delta x\left|\frac{\partial\zeta}{\partial x}\right| .$$

These are easily calculated on a grid.

These suggestions are no panacea for parameterizing the turbulent flow unresolved in two-dimensional simulations of rotating convective elements. We expect qualitatively better integrations than with constant eddy viscosities, but more sophisticated prescriptions based on physical understanding of microscale turbulence are required.

8. REFERENCES

Charney, J., Geostrophic Turbulence, J. Atmos. Sci., 28, 1087-1095, 1977.

Crowley, W. P., A global numerical model: Part 1, J. Comp. Phys., 3, 111-147, 1968.

Deardorff, J. W., A three-dimensional numerical investigation of the idealized planetary boundary layer, Geophys. Fluid Dynamics, 1, 377-410, 1970.

Deardorff, J. W., On the magnitude of the subgrid scale eddy coefficient, J. Comp. Phys., 7, 120-133, 1971.

DuFort, E. C., and Frankel, S. P., Stability conditions in the numerical treatment of parabolic differential equations, Math Tables and Other Aids to Computations, 7, 135-152, 1953.

Haney, R., The relationship between the grid size and the coefficient of nonlinear lateral eddy viscosity in numerical ocean circulation models, J. Comp. Phys., 19, 257-266, 1975.

Leith, C. E., Two dimensional eddy viscosity coefficients, Proc. WMO/IUGG Symposium on Numerical Weather Prediction, Tokyo, Japan, Nov. 26-Dec. 4, 1968, I-41-I-44, 1968.

Lilly, D. K., The representation of small-scale turbulence in numerical simulation experiments, Proc. IBM Scientific Computing Symposium on Environmental Sciences, 195-210, 1967.

O'Brien, J. J., A two-dimensional model of the wind-driven North Pacific, Inves. Pesq., 35(1), 331-349, 1971.

Ogura, Y., Convection of isolated masses of a buoyant fluid, J. Atmos. Sci., 19, 492-502, 1962.

Phillips, N. A., An example of nonlinear computational instability, The Atmosphere and The Sea in Motion: Rossby Memorial Volume, Rockefeller Institute Press, New York, 501-504, 1959.

Richtmyer, R. D., and Morton, K. W., Difference Methods for Initial Value Problems, 406 pp., Interscience, New York, 1967.

Smagorinsky, J., General circulation experiments with the primitive equations, I. The basic experiment, Mon. Wea. Rev., 91, 99-164, 1963.

SMALL-SCALE PARAMETERIZATION IN LARGE-SCALE OCEAN MODELS

A. E. Gargett
Institute of Ocean Sciences
P.O. Box 6000
9860 West Saanich Road
Sidney, B.C.
Canada V8L 4B2

ABSTRACT. The parameterization of sub-grid scale processes, a necessity in present numerical ocean models, is generally accomplished in terms of eddy diffusion coefficients, usually taken as constants. This note examines two non-constant eddy diffusivities which have recently been suggested for processes associated with the diapycnal transport of mass: internal wave induced diffusion considered by Gargett and Holloway [1984] and boundary mixing considered by Armi [1979]. Two points are made: first that both parameterizations are very model-dependent, requiring careful attention to assumptions incorporated in each; second that both processes, as parameterized, lead to diapyncal mass diffusivities which have inverse power-law dependences upon N, $K_d \propto N^{-p}$, $p > 0$. It is noted that spatial gradients of diffusion coefficients act like velocities in advection-diffusion balances of conservative scalars, and hence may strongly affect ocean property distributions. In addition, vertical gradients of K_d enter the geostrophic vorticity equation, and hence may have dynamical effects. The actual importance of spatially variable diffusivities to numerical model results should be explored. For a proper assessment, diffusivities should be incorporated as acting along and across isopycnals, instead of locally horizontally and vertically.

1. INTRODUCTION

Despite advances in computing power, it is now, and in the foreseeable future still will be, necessary to parameterize sub-grid-scale "mixing" processes. The present generation of ocean models uses horizontal and vertical eddy diffusivities, all generally assumed constant in both space and time, to describe the effects on momentum and mass fields of complex processes which are not explicitly resolved by the models. Depending upon the model scale, it may be necessary to parameterize effects of eddies, nonlinear Rossby waves, unstable convection, double-diffusive convection and "ordinary" three-dimensional turbulent mixing. The subject of appropriate parameterizations is correspondingly vast, and nothing like a complete survey will be attempted in this note.

J. J. O'Brien (ed.), Advanced Physical Oceanographic Numerical Modelling, 145–154.

While restricting discussion to diapycnal (across-isopycnal) mixing of mass, an area with which I am most familiar, I will try to make two major points.

The first point is that suggested parameterizations are presently very model-dependent, depending not only upon the particular mixing mechanism being considered, but also upon specific assumptions about that process. Until more observational evidence accumulates, it is necessary to be aware of the level of assumptions built into a parameterization before incorporating it into a model. This point will be illustrated with two suggestions for parameterization of K_d, the diapycnal mass diffusivity: the model of Gargett and Holloway [1984] for diapycnal diffusivity associated with internal wave breaking and that of Armi [1979] for the diapycnal diffusivity due to boundary mixing.

The second point is that apparently insignificant changes in parameterization of K_d may have significant dynamical effects, a point illustrated with a simple vertical advective-diffusive model of the ocean interior. Even if these dynamical effects prove secondary (relative to isopycnal (along-isopycnal) processes), the distributions of passive scalar properties may be dominated by pseudo-velocities associated with spatial gradients of both diapycnal and isopycnal diffusivities.

Finally, I will urge modellers to incorporate non-constant diapycnal diffusivities into basin-scale models, in an attempt to discover whether this effect really can produce significant changes in model predictions. For a proper assessment of the importance of diapycnal processes, however, it is essential that the models incorporate diapycnal and isopycnal diffusivities via a local density-coordinate rotation of the diffusion tensor [Solomon, 1971; Redi, 1982], or else formulate the model in isopycnal coordinates [Bleck and Boudra, 1981].

2. TWO PARAMETERIZATIONS FOR K_d, THE DIAPYCNAL DIFFUSIVITY OF DENSITY

In the following brief discussions I hope not to promote particular parameterizations, but to illustrate the dependence of present parameterizations upon the specific mechanisms of mixing envisaged and upon further assumptions made about characteristics of a particular mechanism. The goals are to increase awareness that parameterizations of sub-grid processes are process-related, hence must be carefully selected with particular applications in mind and to increase the degree of attention given to the assumptions, frequently numerous, which underlie any parameterization.

2.1. K_d Resulting From Internal Wave Breaking

Gargett and Holloway [1984] considered a stably-stratified (but double-diffusively stable) system in which diapycnal mixing is brought about by instabilities associated with an internal wave field. Having thus defined the type of mixing which is being parameterized, they made the following assumptions:

(i) the kinetic energy of the system resides dominantly in internal wave motions;

(ii) the wave kinetic energy is in a statistically steady state (hence an appropriate averaging time for wave quantities must include many generation events for the wave field);

(iii) both the divergence $\partial\overline{(pu_i)}/\partial x_i$ of the pressure (p) - velocity (u_i) correlation and horizontal shear-stress terms, such as $\overline{(uv\partial u/\partial y)}$ where u, v are the horizontal velocity components and x, y are horizontal coordinates, make negligible contributions to the kinetic energy balance;

(iv) the wave field exists near a critical condition for stability of a stratified shearing flow: in particular,

$$Ri_w \quad \frac{N^2}{\overline{\left(\frac{\partial u}{\partial z}\right)^2_w}} \sim O(1)$$

where Ri_w is a "wave Richardson number" formed with the mean Väisälä frequency $N \equiv (g\rho_o^{-1}\bar{\rho}_z)^{\frac{1}{2}}$ and the "wave" shear $(\partial u/\partial z)^2_w$ (discussion of how this should be defined is found in Gargett and Holloway [1984]);

(v) the dissipative processes are characterized by a small flux Richardson number

$$R_f \equiv \frac{-g\rho_o^{-1}\overline{(\tilde{\rho}w)}}{\overline{\left(uw\,\frac{\partial u}{\partial z}\right)}} << 1 \quad ,$$

the ratio of kinetic energy change resulting from transport by the vertical velocity w of density fluctuations $\tilde{\rho}$ to that resulting from vertical stress-shear terms.

Of the above set of assumptions, all but (iii) have at least some observational support from oceanic and/or laboratory measurements (for details, see Gargett and Holloway [1984]). From the kinetic energy equation under these assumptions, they derive

$$\varepsilon = T_o\left[\overline{u^2}\;\overline{w^2}\overline{\left(\frac{\partial u}{\partial z}\right)^2_w}\right] \tag{1}$$

where T_o is a non-dimensional triple correlation coefficient and ε is the rate of decay of kinetic energy due to molecular viscosity. From assumption (v) above, $\overline{(\partial u/\partial z)^2_w} \sim N^2$, so the functional form for ε, hence K_d, depends upon the N-dependence of the internal wave velocity variances $\overline{u^2}$ and $\overline{w^2}$ in (1). By the definition $K_d \equiv -\overline{\tilde{\rho}w}/(\partial\bar{\rho}/\partial z)$ for K_d and that given above for R_f,

$$K_d = \frac{R_f}{1-R_f}\,\varepsilon N^{-2} \quad ,$$

hence the same conclusion holds for K_d. Gargett and Holloway [1984] identified two extreme cases.

Case (1): Internal waves of a single frequency have $\overline{u^2} \sim N$ and $\overline{w^2} \sim N^{-1}$, hence $\varepsilon \sim N^{+1}$ and $K_d \sim N^{-1}$.

Case (2): The broad-band oceanic internal wave model spectrum GM79 [Munk, 1981], in which bandwidth as well as amplitude varies with N, has $\overline{u^2} \sim N$ and $\overline{w^2} \sim$ constant, hence $\varepsilon \sim N^{-3/2}$ and $K \sim N^{-1/2}$.

A collection of oceanic measurements of ε, shown as a function of N in Figure 1, and the observed narrow-band nature of vertical velocity frequency spectrum [Pinkel, 1981] suggest that the first case may best describe the oceanic internal wave field. If so,

$$K_d \simeq a_o N^{-1} \tag{2}$$

where Gargett [1984] has estimated an oceanic value for the "constant" $a_o \simeq 1 \times 10^{-3}\ cm^2s^{-2}$, based on the slope of the line relating ε and N in Figure 1 and a value of $R_f \simeq 0.2$. Oceanic diapycnal diffusivities predicted by (2) thus range from a value of order $0.2\ cm^2s^{-1}$ in the main thermocline where $N \sim 6 \times 10^{-3}\ s^{-1}$, to a value of order $2\ cm^2s^{-1}$ in the deep ocean where $N \sim 0.5 \times 10^{-3}\ s^{-1}$. The latter value compares well with values of the order of $2\text{-}3\ cm^2s^{-1}$ derived by Hogg et al. [1982] and Saunders [1984] from temperature budget studies of semi-enclosed deep ocean basins.

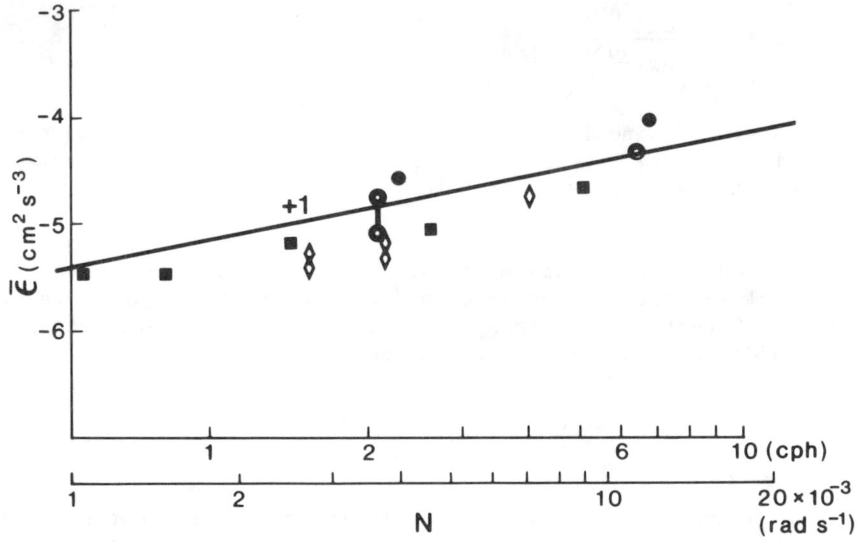

Figure 1. Values of $\bar{\varepsilon}$, averaged kinetic energy dissipation rate, as a function of Väisälä frequency N: most points are averaged over 2-3 km of data. Measurements were taken in the North Atlantic subtropical gyre (◎ Gargett and Osborn [1981]), the eastern North Pacific off Vancouver Island (△ Lueck, Crawford and Osborn [1983]), and the western North Pacific near the Kuroshio (■ Moum [1984]).

2.2. K_d Resulting From Boundary Mixing

Starting with the idea that most diapycnal mixing takes place when fluid encounters the solid boundaries of the ocean, Armi [1979] expresses the average buoyancy flux across depth z as the sum of interior and boundary contributions:

$$A(z)\ \overline{K_d(z)\frac{\partial\rho(z)}{\partial z}} = A_i(z)\ \overline{K_i(z)\left(\frac{\partial\rho(z)}{\partial z}\right)_i} + A_b(z)\ \overline{K_b(z)\left(\frac{\partial\rho(z)}{\partial z}\right)_b}$$

where A(z) is total area, $A_i(z)$ interior area, and $A_b(z)$ boundary area at depth z, and the overbar denotes averaging at depth z. The interior contribution is found to be negligible relative to the boundary contribution, hence

$$\overline{K_d(z)} \simeq \frac{A_b(z)}{A(z)}\ \overline{K_b(z)} \equiv A_r(z)\ \overline{K_b(z)}\ , \tag{3}$$

provided the following assumptions hold:

(i) vertical property gradients at the boundary are similar to those in the interior, i.e. $\partial\rho(z)/\partial z \sim (\partial\rho(z)/\partial z)_i \sim (\partial\rho(z)/\partial z)_b$

(ii) the ratio of interior to boundary areas $A_i(z)/A_b(z)$ is of 0(100) and is z-independent

(iii) the interior diffusivity $\overline{K_i(z)}$ is much smaller than the boundary diffusivity $K_b(z)$

$$\overline{K_i(z)} \sim 0(10^{-2}\ cm^2 s^{-1}) << 0(10^2\ cm^2\ s^{-1}) \sim \overline{K_b(z)}\ .$$

With additional assumptions that:

(iv) the ocean density field is steady-state,

(v) and the averaged vertical velocity w(z) at depth z is negligible,

Armi writes the equation for the mean density field as

$$\frac{\partial}{\partial z}\left(\overline{K_d(z)}\ \frac{\partial\rho}{\partial z}\right) = 0\quad .$$

Using expression (3) for $\overline{K_d(z)}$, and a final assumption that

(vi) the boundary diffusivity $\overline{K_b}$ is independent of depth, this reduces to the relation

$$A_r(z)N^2(z) \sim C\quad , \tag{4}$$

where C is a constant, independent of depth.
From (3),

$$\overline{K_d(z)} \simeq A_r(z)\ \overline{K_b} \simeq C\overline{K_b}N\quad , \tag{5}$$

where a value of 0.04 [(c.p.h.)^2] for the constant C was determined by a comparison of observed North Atlantic N (z) and area ratio $A_r(z)$ (since he expressed $A_r(z)$ as a percent (%), the constant given by Armi is 4[%(c.p.h.)^2], incorrectly quoted as 4[%(c.p.h.)]). With this value, and the order of magnitude estimate of $K_b \sim 10^2\ cm^2s^{-1}$ taken above in (iii), expression (5) predicts that K_d increases from a value of order 10 cm^2s^{-1} at 2 km, to one of order 60 cm^2s^{-1} at 5 km, the depth range over which Armi expects this model to hold. While such magnitudes are much larger than present upper ocean estimates based upon a variety of techniques [Garrett, 1979; Gargett, 1984], there are few reliable estimates from this depth range.

3. MODEL EFFECTS OF VARIABLE DIFFUSIVITIES

As noted by Armi [1979], ocean property distributions may be sensitive to the presence of spatial gradients in the diffusivities which characterize mixing processes. This may be illustrated by

$$\left(u - \frac{dk}{dx}\right) \frac{dS}{dx} - k\frac{d^2S}{dx^2} = 0 \quad ,$$

the one-dimensional steady advective-diffusive equation for a conservative property S. Here the (negative) gradient of the diffusivity k appears as a diffusive pseudo-velocity $u_d \equiv -dk/dx$ which acts, in addition to the true fluid velocity u, to "advect" the mean scalar gradient dS/dx. Using a point source of salt at the eastern boundary of a numerical ocean basin and spatially inhomogeneous and/or anisotropic horizontal eddy diffusivites, Armi and Haidvogel [1982] produced tongue-like distributions, similar to observations of the Mediterranean salt tongue in the North Atlantic, with zero flow field. Although we presently lack a quantitative parameterization for horizontal eddy diffusivities based upon detailed knowledge of oceanic eddy kinetic energy distributions and/or directional anisotropy associated with Rossby wave propagation, it seems clear from these exploratory simulations that horizontal pseudo-velocities may play a significant role in determining basin-scale distributions of scalar properties in the ocean. Thus it is important to continue efforts to develop an appropriate description of horizontal scalar transport by eddies and nonlinear Rossby waves for incorporation into global ocean models.

Another intriguing possibility is that the diffusive pseudo-velocity associated with vertical variation of the diapycnal mass diffusivity may be comparable in magnitude to actual vertical velocities, and hence may affect dynamics of models as well as distributions of scalars. This was pointed out by Gargett [1984] in the context of a very simple model for the mass balance of the ocean interior, the vertical advective-diffusive balance given by

$$w\,\frac{\partial \rho}{\partial z} - \frac{\partial}{\partial z}\left(K_d\frac{\partial \rho}{\partial z}\right) = 0 \quad .$$

This may be rewritten as

$$\frac{w + w_d}{K_d} = \frac{\dfrac{\partial^2 \rho}{\partial z^2}}{\dfrac{\partial \rho}{\partial z}} \quad (6)$$

where $w_d \equiv -\partial K_d/\partial z$ is the vertical pseudo-velocity associated with vertical variation of K_d. If w and K_d are assumed constant (hence $w_d \equiv 0$), (6) is the abyssal recipe of Munk [1966]. However if K_d is a function of z, then in general (6) implies that the true fluid vertical velocity w is also a function of z, a fact which may be of considerable dynamical significance. By the geostrophic vorticity equation

$$v = \frac{f}{\beta}\frac{\partial w}{\partial z} \quad ,$$

the direction of the meridional geostrophic velocity v depends upon the sign of $\partial w/\partial z$: if $\partial w/\partial z$ is positive (negative), v is poleward (equatorward). If the mean density profile of the deep ocean is nearly exponential in character, use of any diffusivity of the form $K_d = a_p N^{-p}$ where a_p is independent of z and $p > 0$, in (6) yields a vertical velocity profile characterized by $\partial w/\partial z < 0$, associated with equatorward meridional flows. With the particular expression for K given by (2), the resulting profile of w(z) was derived by Gargett [1984] who found that the magnitude calculated for w in the deep ocean ($z \sim 4$ km) is close to the canonical value of 1×10^{-5} cm s^{-1} calculated by Munk [1966].

The simple balance given by (6) may not actually determine the field of vertical velocity in the ocean of course, since advection parallel to sloping isopycnals also contributes to the total vertical velocity at any point. Nevertheless it is instructive to discover that the addition of stratification to the constant-density ocean of Stommel and Arons [1960] can, via the simplest model of the density field balance and a density-dependent diffusivity K_d, reverse the implied direction of meridional flow in the interior.

4. APPLICATION OF VARIABLE DIFFUSIVITIES IN BASIN-SCALE MODELS

Before microscale researchers spend a great deal more time and effort attempting to further refine parameterizations of the diapycnal diffusivities associated with various processes, it would be useful to discover what effects, if any, are observed when variable diffusivities of the kind suggested in section 2 are introduced into basin-scale models. Are there observable dynamical effects? Where are effects on property distributions most noticeable? Meaningful answers to such questions require that existing models incorporate the general belief that physical processes producing fluxes which are parameterized by eddy diffusivities act along and across isopycnals, rather than in directions of local horizontal and vertical coordinate axes [McDougall and Church, 1985].

At present, basin-scale models assume a diffusivity tensor which is assumed to have the form, in geographic coordinates,

$$K_{ij} = K_H \begin{bmatrix} 1 & 0 & 0 \\ 0 & 1 & 0 \\ 0 & 0 & \text{"}\gamma\text{"} \end{bmatrix} \tag{7}$$

where "γ" $\equiv K_v/K_H \sim 0(10^{-7})$ is a small parameter describing the ratio of vertical (K_v) to horizontal (K_H) diffusivities. As noted for the two-dimensional case by Solomon [1971] and the three-dimensional case by Redi [1982], a physically more appropriate form is, in isopycnal coordinates,

$$\overset{I}{K}_{ij} = K_\rho \begin{bmatrix} 1 & 0 & 0 \\ 0 & 1 & 0 \\ 0 & 0 & \gamma \end{bmatrix} \tag{8}$$

where $\gamma \equiv K_d/K_\rho \sim 0(10^{-7})$ is the small ratio of diapycnal (K_d) to isopycnal (K_ρ) diffusivities. Transforming (8) to local geographic coordinates, using the local slope of density surfaces, gives

$$\overset{G}{K}_{ij} = \frac{K_\rho}{(1+\delta^2)} \begin{bmatrix} 1 + \frac{(\rho_y^{\,2}+\gamma\rho_x^{\,2})}{\rho_z^{\,2}} & (\gamma - 1)\frac{\rho_x\rho_y}{\rho_z^{\,2}} & (\gamma - 1)\frac{\rho_x}{\rho_z} \\ (\gamma - 1)\frac{\rho_x\rho_y}{\rho_z^{\,2}} & 1 + \frac{(\rho_x^{\,2}+\gamma\rho_y^{\,2})}{\rho_z^{\,2}} & (\gamma - 1)\frac{\rho_y}{\rho_z} \\ (\gamma - 1)\frac{\rho_x}{\rho_z} & (\gamma - 1)\frac{\rho_y}{\rho_z} & \gamma + \frac{(\rho_x^{\,2}+\rho_y^{\,2})}{\rho_z^{\,2}} \end{bmatrix} \tag{9}$$

where $\delta^2 = (\rho_x^{\,2} + \rho_y^{\,2})/\rho_z^{\,2}$. The diffusivity tensor is no longer diagonal, and the full diffusive flux divergence of scalar S appears as

$$\sum_{i=1}^{3} \frac{\partial}{\partial x_i}\left(\sum_{j=1} \overset{G}{K}_{ij} \frac{\partial S}{\partial x_j}\right) \text{ rather than } \sum_{i=1}^{3} \frac{\partial}{\partial x_i}\left(K_{ii} \frac{\partial S}{\partial x_i}\right) \text{ which}$$

results if (7) is used.

As an example of a significant difference between the formulations (7) and (9), consider the x-component of the horizontal diffusive flux of density ($S = \rho$). Using (9), this flux is given by

$$[F_\rho]_x = K^G_{11}\rho_x + K^G_{12}\rho_y + K^G_{13}\rho_z = \gamma K_\rho \rho_x \quad .$$

Since γ is small, this expression is much smaller than the x-component flux $K_H\rho_x \simeq K_\rho\rho_x$ which is found when (7) is used. Thus if rotation of the diffusion tensor is not employed, models framed in a geographic coordinate system will contain inappropriately large horizontal diffusive fluxes of mass in regions where isopycnals are sloped, i.e. ρ_x and/or ρ_y

are non-zero. In addition, it can be shown that a large horizontal diffusivity acting across sloping isopycnals produces false diapycnal fluxes of tracers (although not of mass) which can mask the effects, hence the importance of truly diapycnal processes on large-scale tracer distributions.

5. SUMMARY

My aims in this short note have been extremely limited. I wanted to make modellers aware of the tentative nature of suggested parameterizations of the diapycnal diffusivity for mass, to point out that significant effects are associated with eddy diffusivities which have spatial gradients, and to suggest that a reasonable next step would be to explore model sensitivity to simple spatially-variable diffusivities, using appropriate methods to incorporate the isopycnal/diapycnal nature of mixing processes in the sea.

6. ACKNOWLEDGEMENTS

I would like to thank Trevor McDougall for forcing me to pay attention to the serious model repercussions of not using local coordinate rotation of the diffusivity tensor, Greg Holloway for suggestions and comments on the first draft of this paper, and Jim O'Brien for inviting an outsider to a very informative and enjoyable study institute.

REFERENCES

Armi, L., Effects of variations in eddy diffusivity on property distributions in the oceans, J. Mar. Res., 37(3), 515-530, 1979.

Armi, L., and D.B. Haidvogel, Effects of variable and anisotropic diffusivities in a steady-state diffusion model, J. Phys. Oceanogr., 12, 785-794, 1982 .

Bleck, R. and D.B. Boudra, Initial testing of a numerical ocean circulation model using a hybrid (quasi-isopycnic) vertical coordinate, J. Phys. Oceanogr., 11, 755-770, 1981.

Gargett, A.E., Vertical eddy diffusivity in the ocean interior, J. Mar. Res., 42, 359-393, 1984.

Gargett, A.E., and G. Holloway, Dissipation and diffusion by internal wave, breaking, J. Mar. Res., 42, 15-27, 1984.

Gargett, A.E., and T.R. Osborn, Small-scale shear measurements during the Fine and Microstructure Experiment (FAME), J. Geophys. Res., 86, 1929-1944, 1981.

Garrett, C., Mixing in the ocean interior, Dyn. Atmosph. Oceans, 3, 239-265, 1979.

Hogg, N., P. Biscaye, W. Gardner, and W.J. Schmitz, Jr., On the transport and modification of Antarctic Bottom Water in the Vema Channel, J. Mar. Res., 40 (Suppl.), 231-263, 1982.

Lueck, R.G., W.R. Crawford, and T.R. Osborn, Turbulent dissipation over the continental slope off Vancouver Island, J. Phys. Oceanogr., 13, 1809-1818, 1983.

McDougall, T., and J.A. Church, Pitfalls with the numerical representation of isopycnal and diapycnal mixing, J. Phys. Oceanogr., in press.

Moum, J.N., Measurements of velocity microstructure in the central equatorial and western Pacific Ocean. PhD Thesis, Dept. of Oceanography, Univ. of British Columbia, Vancouver, B.C., Canada. 1984.

Munk, W.H., Abyssal recipes, Deep-Sea Res., 13, 707-730, 1966.

Munk, W., Internal Waves and Small-Scale Processes. in The Evolution of Phys. Oceanography: Scientific Papers in Honour of Henry Stommel, edited by B.A. Warren and C. Wunsch, MIT Press, 264-291, 1981.

Pinkel, R., Observations of the near-surface internal wave field, J. Phys. Oceanogr., 11, 1248-1257, 1981.

Redi, M.H., Oceanic isopycnal mixing by coordinate rotation, J. Phys. Oceanogr., 12, 1154-58, 1982.

Saunders, P.M., Benthic boundary layer IOS observational programme; Discovery Gap measurements, March 1984. I.O.S. Rep No. 180, Wormley, U.K., 1984.

Solomon, H., On the representation of isentropic mixing in ocean circulation models, J. Phys. Oceanogr., 1, 233-234, 1971.

Stommel, H., and A.B. Arons, On the abyssal circulation of the world ocean - II An idealized model of the circulation pattern and amplitude in oceanic basins, Deep-Sea Res., 6, 217-233, 1960.

TIME INTEGRATION SCHEMES

J. J. O'Brien
Mesoscale Air-Sea Interaction Group
The Florida State University
Tallahassee, Florida 32306-3041

1. INTRODUCTION

It is basically incorrect to consider only the time-dependent character of a partial differential equation, since the space differencing and time differencing are closely related for stability. However, when low-order spectral studies are integrated, the space dependence of the dependent variables has been explicitly chosen *a priori*, and only a set of coupled nonlinear ordinary differential equations need to be solved. There is a great body of literature on the numerical solution of ordinary differential equations. Here we will consider some of the simplest schemes.

As a simple oceanographic example, we consider the model of Pollard and Millard [1970]. They chose a simple upper ocean to compare observed variations in horizontal velocity. The basic hypothesis was that wind stress variations were responsible for the local variations in velocity. A simple linear bottom drag is introduced to allow momentum to leak to the deeper ocean. Define $w = u + iv$ be the horizontal upper ocean velocity. The model equation is

$$\frac{dw}{dt} = -ifw - Kw + \tau/\rho H \tag{1}$$

where f is the local Coriolis parameter, $K(s^{-1})$ is the drag coefficient; $\tau = \tau^x + i\tau^y$ is the observed estimate of wind stress and H is the depth of the mixed layer. This equation contains the resonant inertial frequency, $2\pi/f$. It is physically constructive to derive the energy equation for this model. The kinetic energy in the upper ocean is $E = \rho H(u^2+v^2)/2$; its equation is

$$\frac{dE}{dt} = -2KE + (u\tau^x + v\tau^y)$$

The term, 2KE, represents the rate of loss of kinetic energy to the lower ocean. The last term represents the input *or* loss of kinetic

J. J. O'Brien (ed.), Advanced Physical Oceanographic Numerical Modelling, 155–163.

energy to the atmosphere. The sign depends on the direction of the stress vector with respect to the ocean velocity vector. We shall return to numerical approximations of this model shortly.

2. SOME SIMPLE TIME DIFFERENCING SCHEMES

Let u be a vector governed by the equation

$$\frac{\partial u}{\partial t} = G(u,x,y) \tag{2}$$

where any spatial derivatives in x or y have been replaced by finite differences or removed by appropriate wavenumber expansions. Therefore x and y are only parameters.

Integrate (1) over $t \in [(n-p)\Delta t,(n+1)\Delta t]$

$$u_k^{n+1} = u_k^{n-p} + \int_{t=(n-p)\Delta t}^{t=(n+1)\Delta t} G\, dt$$

The integral can be approximated

$$u_k^{n+1} = u_k^{n-p} + \Delta t \sum_{j=0}^{q+1} a_j G_k^{n-j+1}$$

where $q \geq p$.

The weights a_j may be chosen in a variety of ways. If $q = p$ we have an ordinary Newton-Cotes integration rule. If $a_0 \neq 0$, the scheme is *implicit*. (Why?) If $a_0 = 0$, the scheme is *explicit*.

Several common schemes follow.

2.1. Forward difference (Euler)

$$u_k^{n+1} = u_k^n + \Delta t\, G_k^n$$

2.2. Backward difference

$$u_k^{n+1} = u_k^n + \Delta t\, G_k^{n+1}$$

2.3. Trapezoidal Rule (Crank-Nicholson Scheme for Diffusion Equation)

$$u_k^{n+1} = u_k^n + \frac{\Delta t}{2} \left(G_k^{n+1} + G_k^n\right)$$

2.4. Leapfrog or Midpoint Rule

$$u_k^{n+1} = u_k^{n-1} + 2\Delta t\ G_k^n$$

2.5. Simpson's Rule (or Milne Corrector)

$$u_k^{n+1} = u_k^{n-1} + \frac{2\Delta t}{6}\left(G_k^{n-1} + 4G_k^n + G_k^{n+1}\right)$$

2.6. Adam-Bashforth

$$u_k^{n+1} = u_k^n + \Delta t\ \left(\frac{3}{2}\ G_k^n - \frac{1}{2}\ G_k^{n-1}\right)$$

2.7. Milne Predictor

$$u_k^{n+1} = u_k^{n-3} + \frac{4\Delta t}{3}\left[G_k^n - G_k^{n-1} + G_k^{n-2}\right]$$

2.8. Predictor-Corrector

If we use an explicit scheme to achieve a guess for u_k^{n+1} and an implicit scheme for the final result, where the explicit guess is used on the right hand side to make the second step explicit the two step process is called a Predictor-Corrector scheme.

2.8.1. General Predictor-Corrector: The *predictor step* is

$$\text{Let } u_k^* = u_k^{n-p} + \Delta t \sum_{j=1}^{q+1} a_j\ G_k^{n+1-j}$$

The second, *corrector step* is

$$u_k^{n+1} = u_k^{n-p} + \Delta t \sum_{j=1}^{q+1} b_j\ G_k^{n+1-j} + \Delta t\ b_0 G_k^{n+1}(u_k^*)$$

where the coefficients, $a_j \neq b_j$, in general. Note there is no a_0 in the predictor step and G_k^{n+1} is evaluated using the guess vector u_k^*, not the answer. Of course, the corrector step may be repeated several times.

2.8.2. Simplest Predictor-Corrector:

1st step (Predictor)

$$u_k^* = u_k^n + \Delta t\ G_k^n \qquad \text{(forward difference)}$$

or

$$u_k^* = u_k^{n-1} + 2\Delta t \, G_k^n \quad \text{(leap frog)}$$

2nd step (Corrector).

$$u_k^{n+1} = u_k^n + \frac{\Delta t}{2}[G_k^n + G(u_k^*)^{n+1}]$$

This is sometimes called the Heun scheme in meteorology. If the 2nd step is a backward step, it is the more common Matsuno scheme.

Exercise:

Prove that if $G(u,x,y)$ is linear in u, then the predictor step is not necessary, since the corrector step becomes explicit instead of implicit. Excellent articles on time differencing schemes are Lilly [1965], Kurihara [1965], Young [1968] and Baer and Simons [1970]. In realistic numerical problems in oceanography and meteorology we cannot afford to save the solution at many old (or earlier) time steps since we do not have the computer storage space. In addition the space truncation errors are many times greater than the time differencing errors. Therefore we generally select simple time differencing schemes.

3. THE OSCILLATION EQUATION

The simple oscillation equation is a very useful model for understanding time differencing schemes. This equation is

$$\frac{\partial u}{\partial t} = i\omega u \quad ; \quad \omega \text{ is real} \tag{3}$$

In the Pollard and Millard model we retrieve this equation if the damping $(K = 0)$ and the forcing $(\tau = 0)$ are zero. The simple advective flow equation

$$\frac{\partial v}{\partial t} + c\frac{\partial v}{\partial x} = 0$$

can be reduced to (3) if we consider a single wavenumber, k. If $v(x,t) = u(t)e^{ikx}$, the equation for u is

$$\frac{\partial u}{\partial t} = -ikcu \qquad \text{where} \quad -kc = \omega$$

If $|\omega\Delta t| < 1$, $|G| = 1$; hence the scheme is neutral. However this scheme has two roots: the root for which G goes to 1 (+ sign) as $\Delta t \to 0$ is called the *physical mode;* the other root (-sign) is called the *computational mode*. When using leapfrog, it is very important to surpress the computational mode, which can lead to unwanted time splitting or $2\Delta t$ oscillations.

3.5. Heun Scheme

This is the first of two predictor corrector schemes we consider for the oscillation equation.

$$u^* = u^n + i\omega\Delta t\ u^n \quad \text{(forward step)}$$

$$u^{n+1} = u^n + \frac{i\omega\Delta t}{2}(u^n + u^*) \quad \text{(modified trapezoidal)}$$

We can find G by eliminating u*

$$u^{n+1} = u^n + \frac{i\omega\Delta t}{2}[2u^n + i\omega\Delta t\ u^n]$$

or

$$G = 1 + i\omega\Delta t - \frac{(\omega\Delta t)^2}{2}$$

$$|G|^2 = 1 + \frac{(\omega\Delta t)^4}{4}$$

Again, the result is stable by the von Neuman criterion but physically unstable, since the oscillation equation does not allow growth. The growth is small, $O(\Delta t^2)$.

3.6. Matsuno Scheme

$$u^* = u^n + i\omega\Delta t\ u^n \quad \text{(forward scheme)}$$

$$u^{n+1} = u^n + i\omega\Delta t\ u^*$$

or

$$G = 1 + i\omega\Delta t - (\omega\Delta t)^2$$

$$|G|^2 = 1 - (\omega\Delta t)^2 + (\omega\Delta t)^4$$

If $|\omega\Delta t| < 1$, $|G| < 1$; $|\omega\Delta t| < 1$ is a realistic condition since the frequency, ω, is the only physics in our simple problem. The Matsuno scheme is widely used in meteorology since it selectively damps higher frequencies. This is seen in Figure 5.1.

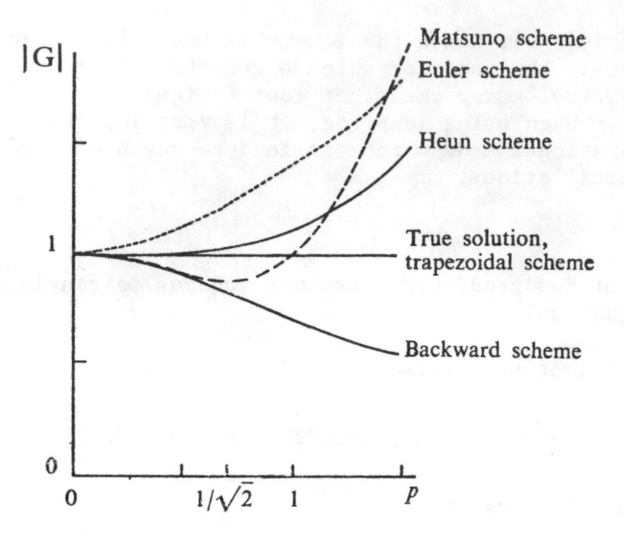

Figure 5.1 The amplification factor as a function of $\rho \equiv \omega\Delta t$ for five two level schemes and for the true solution.

3.7. Lax-Wendroff Scheme

This scheme is an important idea which is used in fluid dynamics. It is rarely illustrated for a single frequency problem. We will return to it later in more complicated problems. The Taylor series expansion in time for u(t) is

$$u^{n+1} = u^n + \Delta t \frac{\partial u}{\partial t} + \frac{(\Delta t)^2}{2} \frac{\partial^2 u}{\partial t^2} + \text{other terms}$$

We substitute for both u_t and u_{tt} from the differential equation, and obtain the Lax-Wendroff scheme

$$u^{n+1} = u^n + i\omega\Delta t u^n - \frac{(\omega\Delta t)^2}{2} u^n \tag{4}$$

The amplification factor is

$$G = 1 + i\omega\Delta t - (\omega\Delta t)^2/2$$

$$|G|^2 = 1 + (\omega\Delta t)^4$$

If $|\omega\Delta t|<1$, then $|G|\leq 1 + O(\Delta t^2)$. The scheme is stable in the von Neuman sense. We can interpret the result physically by recognizing that (4) may be considered an approximation of

$$\frac{du}{dt} = i\omega u - Ku$$

where $K = \omega^2\Delta t/2$. We call K an *artificial* or *computational viscosity*.

3.8. Semi-Implicit Method

When there are two or more frequencies involved in a problem, we might consider different methods for each frequency. Kwizak and Robert [1971] considered the oscillation problem when

$$\omega = \omega_s + \omega_f$$

where $\omega_s<\omega_f$; the "slow" wave has frequency, ω_s; the fast wave has frequency, ω_f. We treat the low frequency explicitly and the high frequency implicitly; the approximation for (1) is

$$u^{n+1} = u^{n-1} + i\Delta t[2\omega_s u^n + \omega_f (u^{n+1} + u^{n-1})]$$

If $\omega_s = 0$, the scheme is completely *implicit*, if $\omega_f = 0$, the scheme is *explicit*. The amplification factor is

$$G = G^{-1} + i2\omega_s\Delta t + i\omega_f\Delta t(G + G^{-1})$$

or

$$(1 - i\omega_f\Delta t)G^2 - i2\omega_s\Delta tG - (1 + i\omega_f\Delta t) = 0$$

A simple analysis shows that if

$(\omega_s\Delta t)^2 \leq 1 + (\omega_f\Delta t)^2$, then $|G| = 1$, and the scheme is stable.

To use this scheme to practical advantage in geophysical problems, it is necessary to partition any fluid flow problem into two parts. All terms in the model equations which govern the propagation of the fastest moving waves are treated implicitly and the remainder explicitly. As an example, consider the shallow water wave equations in one dimension for a rotating fluid

$$\frac{\partial u}{\partial t} + u\frac{\partial u}{\partial x} + g\frac{\partial h}{\partial x} - fv = 0$$

$$\frac{\partial h}{\partial t} + u\frac{\partial h}{\partial x} + h\frac{\partial u}{\partial x} = 0$$

$$\frac{\partial v}{\partial t} + u\frac{\partial u}{\partial x} + fu = 0$$

The fastest moving waves are external gravity waves which propagate at speed $(gh)^{1/2}$.

Let $h(x,t) = H + h'(x,t)$ where H is the mean depth of the fluid. Then the terms $g\frac{\partial h'}{\partial x}$ and $H\frac{\partial u}{\partial x}$ are treated implicitly and the remainder explicitly. The details are left to the reader except for the comment that the semi-implicit scheme yields complicated equations, but the payoff comes from large time steps. This idea has been used by O'Brien and Hurlburt [1972] for a coastal upwelling model.

4. THE MIXED OSCILLATION-DAMPING EQUATION

The Pollard and Millard model contains an oscillation and a damping term. The best way to handle the combination of effects is to use a mixture of schemes. Let us consider leapfrog for the Coriolis term and forward for the bottom drag

$$w^{n+1} = w^{n-1} - if(2\Delta t)w^n - 2\Delta t K w^{n-1} + \tau/\rho H$$

For stability, we ignore the forcing term and find

$$G = G^{-1} - i2f\Delta t - 2\Delta t K G^{-1}$$

If $(K + |f|)\,\Delta t \leqslant 1$ then $|G| \leqslant 1$. In essence we are recommending leapfrog for the oscillation or wave part of the problem and a forward time step for bottom friction or diffusion terms. We can, of course, consider *implicit* methods for either or both types of physics. These always lead to more complicated sets of finite difference equations. If the terms made implicit are linear, we have the rich methods of Chapter 3 to help us. If the implicit parts are nonlinear, we have a serious mathematical problem.

Three time level methods such as leapfrog are conditionally stable and neutral for oscillation problems but we have to store two complete time levels. Explicit forward difference methods have to be dissipative if they are stable but only require storage of one complete time level.

5. REFERENCES

Baer, F., and T. J. Simmons, Computational stability and time truncation of coupled nonlinear equations with exact solutions, Mon. Wea. Rev., 98, 665-679, 1970.

Kurihara, Y., On the use of implicit and iterative methods for the time integration of the wave equation, Mon. Wea. Rev., 93, 33-46, 1965.

Kwizak, M., and A. J. Robert, A semi-implicit scheme for a grid-point atmospheric model of the primitive equations, Mon. Wea. Rev., 99, 32-36, 1971.

Lilly, D. L., On the computational stability of numerical solutions of the time-dependent non-linear geophysical fluid dynamics problems, Mon. Wea. Rev., 93, 11-26, 1965.

O'Brien, J. J., and H. E. Hurlburt, A numerical model of coastal upwelling, J. Phys. Oceanogr., 2, 14-26, 1972.

Pollard, R. T., and R. C. Millard, Jr., Comparison between observed and simulated wind-generted inertial oscillations, Deep-Sea Research, 17, 813-821, 1970.

Young, J. A., Comparative properties of some time differencing schemes for linear and nonlinear oscillations, Mon. Wea. Rev., 96, 357-364, 1968.

THE HYPERBOLIC PROBLEM

J. J. O'Brien
Mesoscale Air-Sea Interaction Group
The Florida State University
Tallahassee, FL 32306-3041

1. INTRODUCTION

Before we proceed to realistic ocean modelling problems, it is useful to have a knowledge of finite difference schemes for simple hyperbolic partial differential equations (PDE).

2. HYPERBOLIC EQUATIONS

The general linear second-order PDE is

$$au_{xx} + 2bu_{xy} + cu_{yy} + 2du_{x} + 2eu_{y} + fu = h(x,y) \quad (1)$$

If $b^2 - ac > 0$, then (1) is hyperbolic. We can eliminate the u_{xy} term by the transformation

$$y - \lambda_1 x = \xi + \eta$$

$$y - \lambda_2 x = \xi - \eta$$

where λ_1, λ_2 are the roots of $a\lambda^2 - 2b\lambda + c = 0$. The transformed equation is

$$u_{\xi\xi} - u_{\eta\eta} + 2Du_{\xi} + 2Eu_{\eta} + Fu = H(\xi,\eta) \quad (2)$$

This is the canonical form of a second order hyperbolic equation. A commonly encountered hyperbolic equation is the wave equation

$$u_{tt} - c^2 u_{xx} = g(x)$$

which can be written as the system of first order equations (if c is a constant).

J. J. O'Brien (ed.), Advanced Physical Oceanographic Numerical Modelling, 165–186.

$$u_t = cv_x + tg$$

$$v_t = cu_x$$

In fluid dynamics we frequently encounter systems of hyperbolic equations. Suppose we have a system of equations for the vector

$$u(x,t) = [u_1, u_2, \ldots, u_n]^T$$

such as

$$\frac{\partial u_1}{\partial t} = a_{11}\frac{\partial u_1}{\partial x} + a_{12}\frac{\partial u_2}{\partial x} + \ldots + a_{1n}\frac{\partial u_n}{\partial x}$$

$$\frac{\partial u_2}{\partial t} = a_{21}\frac{\partial u_1}{\partial x} + \ldots$$

$$\vdots$$

$$\frac{\partial u_n}{\partial t} = a_{n1}\frac{\partial u_1}{\partial x} + \ldots + a_{nn}\frac{\partial u_n}{\partial x}$$

which can be written in matrix form

$$\frac{\partial u}{\partial t} = A\frac{\partial u}{\partial x}$$

Suppose that A has n distinct *real* eigenvalues $\{\lambda_i\}$; then the system is hyperbolic and there exists a nonsingular matrix, P, of eigenvectors such that

$$D = PAP^{-1}$$

is a diagonal matrix whose diagonal elements are $\{\lambda_i\}$. Let $w = Pu$, then

$$w_t = (PAP^{-1})\ Pu_x = Dw_x$$

or

$$\frac{\partial w_1}{\partial t} = \lambda_1\frac{\partial w_1}{\partial x}$$

$$\vdots$$

$$\frac{\partial w_n}{\partial t} = \lambda_n\frac{\partial w_n}{\partial x}$$

where each equation is a simple hyperbolic equation; thus the simplest model equation for linear hyperbolic problems is the one-dimensional advection equation

$$u_t + Au_x = 0 \tag{3}$$

with initial condition $u(x,0) = f(x)$ for $-\infty < x < \infty$. The solution is well-known; i.e., $u(x,t) = f(x - At)$.

3. FINITE DIFFERENCE SCHEMES FOR ONE-DIMENSIONAL HYPERBOLIC EQUATIONS

There are numerous methods for handling hyperbolic problems. Several frequently encountered schemes follow.

The model problem is

$$\frac{\partial u}{\partial t} = -A \frac{\partial u}{\partial x}$$

where $u(x,t)$ is real, A is a real constant; define $a = A\Delta t/\Delta x$, $\alpha = k\Delta x$, where Δt is the time step and Δx is the grid spacing and k is a wavenumber. The dimensionless number, a, is called the Courant number.

Method 1

Unstable (forward time step and centered in space)

```
    0
    |
0--0--0
```

$$u_j^{n+1} = u_j^n - \frac{a}{2}(u_{j+1}^n - u_{j-1}^n)$$

$$G = 1 - i\, a \sin \alpha$$

This scheme is always unstable because $|G| \geq 1$. This is an important result. For a diffusive problem, forward in time and centered in space is conditionally stable, but for the hyperbolic prblem it is unconditionally unstable.

Method 2: Upstream Differencing

$$u_j^{n+1} = u_j^n - a \left\{ \begin{array}{ll} u_{j+1}^n - u_j^n, & \text{if } a < 0 \\ u_j^n - u_{j-1}^n, & \text{if } a > 0 \end{array} \right\} \tag{4}$$

$$G = 1 - a \left\{ \begin{array}{ll} e^{i\alpha} - 1, & a < 0 \\ 1 - e^{-i\alpha}, & a > 0 \end{array} \right\}$$

Stencil

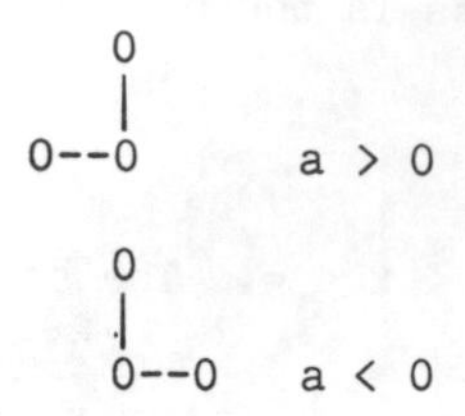

This scheme is stable if $|a| \leq 1$. It is very popular in meteorology and oceanography because of the ease of implementation. However, this author strongly opposes the use of upstream differencing because it is a very dissapative finite difference scheme. This is demonstrated by considering the case, $a > 0$ and adding a particular zero to (4)

$$u_j^{n+1} = u_j^n - a(u_j^n - u_{j-1}^n) + \frac{a}{2}(u_{j+1}^n - u_{j+1}^n)$$

The last term is obviously zero. By rearranging, we can obtain

$$u_j^{n+1} = u_j^n - \frac{a}{2}(u_{j+1}^n - u_{j-1}^n) + \frac{a}{2}(u_{j+1}^n + u_{j-1}^n - 2u_j^n)$$

The first two terms on right are the *unstable* scheme; the last term is a second-order approximation to a diffusion equation! If we use *upstream* differencing, we are actually solving the differential equation

$$\frac{\partial u}{\partial t} + A\frac{\partial u}{\partial x} = K\frac{\partial^2 u}{\partial x^2}$$

with the computational viscosity, $A\Delta x/2$, $A > 0$, and an unconditionally unstable approximation for the advection term. We have added enough artificial friction to damp the unstable part of the problem. In practice, $A\Delta x/2$ represents a value for eddy viscosity that no one would attribute physically to a large scale flow problem in meteorology or oceanography. In fact, things are humorous. There are several published papers with real diffusive terms for which the authors "learned" that the value of eddy viscosity did not effect their solution, but they had used upstream differencing for which $A\Delta x$ was at least an order of magnitude larger than their prescribed eddy viscosity! Naturally, we should not give the references here.

Exercise: Derive the computational viscosity for the case $a < 0$.

Method 3: Diffusion Scheme (Friedrich's scheme)

$$u_j^{n+1} = \frac{1}{2}(u_{j+1}^n + u_{j-1}^n) - \frac{a}{2}(u_{j+1}^n - u_{j-1}^n) \tag{5}$$

$$G = \cos\alpha - i\, a \sin\alpha$$

Stable if $|a| < 1$.

For the time derivative, if we use $(u_j^{n+1} - u_j^n)/\Delta t$ the scheme will be unstable. To avoid this problem, we replace u_j^n by the average between u_{j+1}^n and u_{j-1}^n so that the stencil becomes:

```
    0
    |
   / \
  0   0
```

$$|G|^2 = \cos^2\alpha + a^2\sin^2\alpha \; : \text{ if } |a| \leq 1 \text{ then } |G| \leq 1, \text{ i.e. } \frac{|A|\Delta t}{\Delta x} < 1$$

This scheme is the first one we have seen which is staggered in time-space. As for the upstream scheme, we can consider the discretization to be from

$$\frac{\partial u}{\partial t} = -A\frac{\partial u}{\partial x} + K\frac{\partial^2 u}{\partial x^2} \tag{6}$$

which we can rewrite in the discrete form:

$$u_j^{n+1} = u_j^n - \frac{a}{2}(u_{j+1}^n - u_{j-1}^n) + \frac{K\Delta t}{2(\Delta x)^2}(u_{j+1}^n + u_{j-1}^n \; 2u_j^n)$$

The advection part is the unstable scheme and the diffusive part is the conditionally stable forward difference. In this diffusive scheme, the artificial viscosity is $K = (\Delta x)^2/\Delta t$.

Method 4: Leapfrog

This is one of the most popular schemes in oceanography and meteorology. We choose to use a centered-in-time, centered-in-space scheme.

$$u_j^{n+1} = u_j^{n-1} - a(u_{j+1}^n - u_{j-1}^n)$$

The stencil is

```
        0
        |
  0----)--0
        |
        0
```

The amplification factor is

$$G = G^{-1} - a\,2i \sin \alpha$$

or

$$G = i\,a \sin \alpha \pm (-a^2 \sin^2 \alpha + 1)^{1/2}$$

If $a^2 \sin^2 \alpha \leq 1$, the radical is real and $|G| = 1$. There are several points to notice. If $|a| > 1$, the most unstable wave is $\alpha = \pi/2$ or $k = 2\pi/4\Delta x$. If we violate the stability condition, the four Δx wave will grow fastest. Recall in the diffusive problem that the two Δx wave was the most unstable.

Definition: The ratio, $A\Delta t/\Delta x$ is called the Courant Number. The condition $A\Delta t/\Delta x \leq 1$ for stability is called the CFL Condition after Courant, Friedrich and Lewy. Physically the CFL condition states that useful information must propagate less than one Δx in time Δt.

Method 5: Lax-Wendroff Scheme for the advection equation

Consider

$$\frac{\partial u}{\partial t} = -A \frac{\partial u}{\partial x}$$

If A is constant, the second derivative in time is

$$u_{tt} = -Au_{xt} = A^2 u_{xx}$$

A *Taylor Series* expansion in time yields

$$u_j^{n+1} = u_j^n + \Delta t \frac{\partial u}{\partial t} + \frac{\Delta t^2}{2} + \frac{\partial^2 u}{\partial t^2} \frac{\Delta t^3}{6} + \frac{\partial^2 u}{\partial t^3} + \ldots$$

or

$$u_j^{n+1} = u_j^n - A\Delta t\, u_x + \frac{\Delta t^2 A^2}{2} u_{xx} + 0\,(\Delta t^3)$$

Let $a = \frac{A\Delta t}{\Delta x}$, then the finite difference scheme is

$$u_j^{n+1} = u_j^n - \frac{a}{2}[u_{j+1}^n - u_{j-1}^n] + \frac{a^2}{2}[u_{j+1}^n - 2u_j^n + u_{j-1}^n] + O(\Delta t^3)$$

The linear stability analysis for this scheme is instructive; let

$$u_j^n = U_n\ e^{ikj\Delta x}$$

$$G = 1 - i\ a \sin k\Delta x + a^2[\cos k\Delta x - 1]$$

$$|G|^2 = 1 + 2a^2(\cos\theta - 1) + a^4(\cos\theta - 1)^2 + a^2 \sin^2\theta$$

$$= 1 - a^2(1 - a^2)(1 - \cos\theta)^2$$

If $|a| < 1$, then $a^2(1 - a^2) \leq 1/4$

and $(1 - \cos\theta)^2 \leq 4$

therefore $|G|^2 \leq 1$ if $|a| < 1$.

Suppose we were solving the following equation:

$$\frac{\partial u}{\partial t} = -A\frac{\partial u}{\partial x} + K\frac{\partial^2 u}{\partial x^2}$$

if

$$\frac{a^2}{2} = \frac{K\Delta t}{(\Delta x)^2}$$

then

$$\frac{a^2\Delta t^2}{2(\Delta x)^2} = \frac{K\Delta t}{(\Delta x)^2}$$

Thus $K = A^2\Delta t/2$ is the Artificial Viscosity induced by the Lax-Wendroff scheme.

Method 6: The Two-Step Lax-Wendroff Scheme

For primitive equation models, it is awkward to use the Lax-Wendroff scheme because we have to differentiate the equations in time. With non-linear problems, this leads to complicated spatial derivatives. There is a two step version of this idea which is particularly useful for nonlinear problems.

For step one, use the diffusing time step (Method 2); for step two, use leapfrog (Method 4). It is important to realize that we do *not* consider the solution at odd time steps a solution of our equation.

Step 1: Diffusing Time Step

$$u_j^{n+1} = \frac{1}{2}(u_{j+1}^n + u_{j-1}^n) - \frac{A\Delta t}{2\Delta x}(u_{j+1}^n - u_{j-1}^n)$$

Step 2: Leap-Frog

$$u_j^{n+2} = u_j^n - \frac{A\Delta t}{\Delta x}(u_{j+1}^{n+1} - u_{j-1}^{n+1})$$

The two-step method is best suited for 2-D problems and non-linear problems.
The stencil is

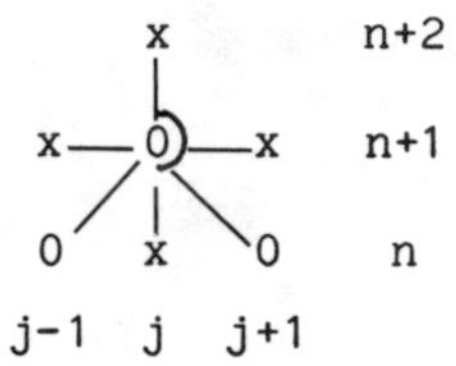

```
        x        n+2
        |
   x————0————x   n+1
       /|\
      / | \
     0  x  0     n

   j-1  j  j+1
```

The circles are the first step; the crosses are the second step. Note we have to solve for *all* points at time level, n+1, before finding solution at (n+2)Δt.

We can show that the two-step method is equivalent to the one-step method. Let us eliminate the dependence on (n+1)

$$u_j^{n+2} = u_j^n - a[\frac{1}{2}u_{j+2}^n + \frac{1}{2}u_j^n - \frac{a}{2}(u_{j+2}^n - u_j^n)$$

$$- \frac{1}{2}u_j^n - \frac{1}{2}u_{j-2}^n + \frac{a}{2}(u_j^n - u_{j-2}^n)]$$

$$u_j^{n+2} = u_j^n - \frac{a}{2}(u_{j+2}^n - u_{j-2}^n) + \frac{a^2}{2}(u_{j+2}^n - 2u_j^n + u_{j-2}^n)$$

Therefore this is the same as the one-step method if Δx is (2Δx) and Δt is (2Δt).

Space-time staggered version

There is an interesting variant of the two-step method called the space-time staggered version. Examination of the stencil for the two-step method suggests that we might calculate the first step for odd j at time, n+1, and at even j for time levels, n, n+2. The new stencil is

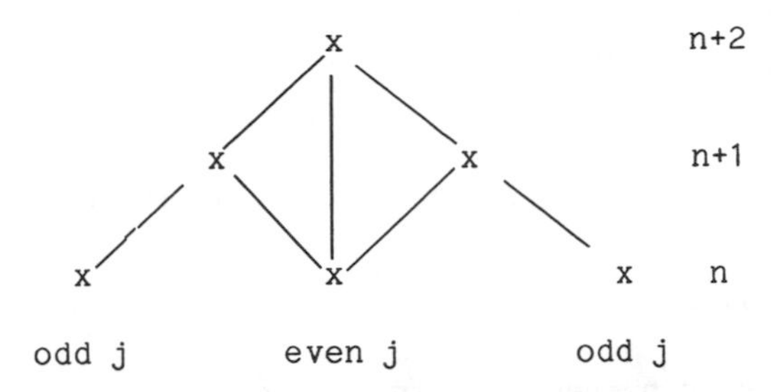

The equations are

$$u_{j+1}^{n+1} = \frac{1}{2}(u_j^n + u_{j+2}^n) - \frac{a}{2}(u_{j+2}^n - u_j^n) \quad ; \quad j(\text{odd})$$

$$u_j^{n+2} = u_j^n - a(u_{j+1}^{n+1} - u_{j-1}^{n+1}) \quad ; \quad j(\text{even})$$

Observe that we have thrown away half the space points for each time level. We note that the scheme is said to be staggered in time and space.

Method 7: Implicit Scheme

We write:

$$u_j^{n+1} = u_j^n - \frac{a}{2}\left\{\frac{u_{j+1}^n + u_{j+1}^{n+1}}{2} - \frac{u_{j-1}^n + u_{j-1}^{n+1}}{2}\right\}$$

For stability, let $u_j^n = U_n e^{ikj\Delta x}$,

$$U_{n+1} = U_n - \frac{a}{2}\left[\frac{2i\sin\alpha}{2} U_{n+1} + \frac{2i\sin\alpha}{2} U_n\right]$$

Then $|G|^2 = (1 + \frac{a^2\sin^2\alpha}{4})/(1 + \frac{a^2\sin^2\alpha}{4}) = 1$

This scheme is neutral and stable. In practice we never use implicit schemes for advection problems in oceanography because the equations are nonlinear.

Method 8: Quasi-Lagrangian

This idea is an important concept which is not used as frequently as one would anticipate. For the simple linear advection equation, it is easy to understand. The distance, $A\Delta t$ is the spatial distance that a parcel will travel in time, Δt, to reach a point, $j\Delta x$. In finite difference form we might approximate our model equation with

$$\frac{u_j^{n+1} - u_j^n}{\Delta t} = -A\left(u_j^n - u_{(j-a)}^n\right)/A\Delta t$$

where $u_{(j-a)}^n = u(n\Delta t, (j-A\Delta t)\Delta x)$. The above equation reduces to

$$u_j^{n+1} = u^n(j-a)$$

Unless $A\Delta t$ is an integer number of grid distances, Δx, we do *not* know the right-hand side. However, we can use an interpolating polynomial to determine it.

In the two-dimensional problem,

$$u_t + Au_x + Bu_y = 0 \ ,$$

we have two distances, $A\Delta t$, and $B\Delta t$ which define a position at some distance from $(j\Delta x, k\Delta y)$. However, the quasi-Lagrangian scheme still reduces to

$$u_{j,k}^{n+1} = u(n\Delta t, j\Delta x - A\Delta t, k\Delta y - B\Delta t)$$

In simple terms, the one term on the right-hand side is the approximation to the entire advection term. Of course, we have to perform a two-dimensional interpolation of $u_{j,k}^n$ to obtain the right-hand side. The scheme is stable if we use the nearest neighbors for the interpolation. The proof is left as an exercise for the reader.

4. STABILITY FOR THE GRAVITY WAVE EQUATIONS

When waves exist in the solution of hyperbolic partial differential equations, the linear stability of the usual second-order finite difference schemes will be controlled by the phase speed of the fastest moving wave. If C is the phase speed, we will find that

$$\frac{C\Delta t}{\Delta x} < O(1)$$

If advection is included, then the Doppler-shifted speed,

$|U| + C$ governs the stability, i.e.,

$$(|U| + C)\frac{\Delta t}{\Delta x} < O(1)$$

In this section we investigate the linear stability of a sequence of problems leading to the shallow water equations. Let us consider simple one-dimensional flow for gravity waves. The differential equations are

$$\frac{\partial u}{\partial t} = -g\frac{\partial h}{\partial x}$$

$$\frac{\partial h}{\partial t} = -H\frac{\partial u}{\partial x}$$

Using simple leapfrog and centered differences, we can write

$$u_j^{n+1} = u_j^{n-1} - \frac{\Delta t}{\Delta x}\, g(h_{j+1}^n - h_{j-1}^n)$$

$$h_j^{n+1} = h_j^{n-1} - \frac{\Delta t}{\Delta x}\, H(u_{j+1}^n - u_{j-1}^n)$$

Define $\lambda = \Delta t/\Delta x$ and $C^2 = gH$. Let

$$u_j^n = U_n\ e^{ikj\Delta x}$$

$$h_j^n = h_n\ e^{ikj\Delta x}$$

Define $\alpha = k\Delta x$, and after substituting, we obtain

$$U_{n+1} = U_{n-1} - g\lambda 2i\sin\alpha\ h_n$$

$$h_{n+1} = h_{n-1} - H\lambda 2i\sin\alpha\ U_n$$

If we rewrite the U_{n+1} equation for U_{n+1} for U_{n+2} and U_n and subtract, we can substitute the equation for $h_{n+1} - h_n$. The result is

$$U_{n+2} - 2U_n + U_{n+2} = -2i\lambda C\sin\alpha(-2i\lambda C\sin\alpha U_n) \qquad (9)$$

If an amplification factor, G, exists such that

$$U_{n+2} = GU_n$$

Note only every second time step is involved in (9). The equation for G is

$$G^2 - (2 - 4\lambda^2C^2\sin^2 \alpha)G + 1 = 0$$

or

$$G = 1 - 2\lambda^2\sin^2 \alpha \pm i[1 - (1-2\lambda^2C^2\sin^2\alpha)^2]^{1/2}$$

If the radical is real, then $|G| = 1$. This condition will be realized if $C^2\lambda^2\sin^2\alpha \leq 1$, or $C^2\lambda^2 \leq 1$, which implies that

$$\frac{C\Delta t}{\Delta t} \leq 1$$

It should be noted that if $C\lambda$ exceeds one, then the wavenumber scales near $\alpha = \pi/2$ will be the most unstable. These are the smallest resolved scales,

$$k = 2\pi/4\Delta x$$

The instability will manifest itself in a numerical solution by exponential growth of $4\Delta x$ waves.

4.1. CFL for inertial gravity waves

If we consider the one-dimensional problem

$$\frac{\partial u}{\partial t} = fv - g\frac{\partial h}{\partial x}$$

$$\frac{\partial v}{\partial t} = -fu$$

$$\frac{\partial h}{\partial t} = -H\frac{\partial u}{\partial x}$$

and use leapfrog and centered differences, we obtain after defining

$$u_j^n = U_n\ e^{ikj\Delta x},\ v_j^n = V_n\ e^{ikj\Delta x},\ h_j^n = h_n\ e^{ikj\Delta x}$$

$$\alpha = k\Delta x$$

$$U_{n+1} - U_{n-1} = 2\Delta tf\ V_n - 2i\lambda g\sin\alpha\ h_n$$

$$V_{n+1} - V_{n-1} = -2\Delta tf\ U_n$$

$$h_{n+1} - h_{n-1} = -2iH\sin\alpha\, U_n$$

Elimination of h_n and V_n yields

$$U_{n+2} - 2U_n + U_{n-2} = -4f^2\,\Delta t^2\, U_n - 4gH\lambda^2\,\sin^2\alpha\, U_n.$$

Following our standard procedure, we define

$$U_{n+2} = GU_n$$

which leads to

$$G^2 - [2-(2f\Delta t)^2 - 4C^2\lambda^2\sin^2\alpha]G + 1 = 0$$

or

$$G = [1 - 2(f\Delta t)^2 - 2C^2\lambda^2\sin^2\alpha]$$
$$\pm\, i\left[1 - [1 - 2(f\Delta t)^2 - 2C^2\lambda^2\sin^2\alpha]^2\right]^{1/2}$$

Again, if the radical is real, then $|G| = 1$ for all α. This is true if

$$(f\Delta t)^2 + C^2\lambda^2\sin^2\alpha \leqq 1$$

or

$$|f\Delta t| < 1 \text{ and } \frac{C\Delta t}{\Delta t} \leqq [1 - (f\Delta t)^2]^{1/2}$$

We observe that the Coriolis term, $f\Delta t$, reduces the CFL condition below unity. In practice, the time step is a small fraction of the inertial period $2\pi/f$. and this is not a serious problem.

4.2. C.F.L. for Two-Dimensional Flow

The linearized shallow water equations on an f-plane are

$$\frac{\partial u}{\partial t} = fv - g\frac{\partial h}{\partial x}$$

$$\frac{\partial v}{\partial t} = -fv - g\frac{\partial h}{\partial y}$$

$$\frac{\partial h}{\partial t} = -H\left(\frac{\partial u}{\partial x} + \frac{\partial v}{\partial y}\right)$$

If we use leapfrog and centered spatial finite differences and we define

$$u^n_{j,k} = U_n e^{i\ell j\Delta x} e^{imk\Delta y}, \quad v^n_{j,k} = V_n e^{i\ell j\Delta x} e^{imk\Delta y},$$

$$h^n_{j,k} = h_n e^{i\ell j\Delta x} e^{imk\Delta y}$$

the system reduces to

$$U_{n+1} - U_{n-1} = 2f\Delta t\, V_n - 2ig\lambda_x \sin\alpha\, h_n$$

$$V_{n+1} - V_{n-1} = -2f\Delta t\, U_n - 2ig\lambda_y \sin\beta\, h_n$$

$$h_{n+1} - h_{n-1} = -2i\lambda_x H \sin\alpha\, U_n - 2i\,\lambda_y H \sin\beta\, V_n$$

where $\lambda_x = \Delta t/\Delta x$, $\lambda_y = \Delta t/\Delta y$, $\alpha = \ell\Delta x$, and $\beta = m\Delta y$.
In order to eliminate V and h in favour of U, we write the first two equations for U_{n+1}, U_n and V_{n+2}, V_n to eliminate h_n. These are

$$U_{n+2} - 2U_n + U_{n-2} = 2f\Delta t(V_{n+1} - V_{n-1})$$

$$- 2i\lambda_x\, g\sin\alpha(-2i\lambda_x H\sin\alpha\, U_n - 2i\lambda_y H\sin\beta\, V_n)$$

$$V_{n+2} - 2V_n + V_{n-2} = -2f\Delta t(U_{n+1} - U_{n-1})$$

$$- 2i\lambda_y\, g\sin\beta(-2i\lambda_x H\sin\alpha\, U_n - 2i\lambda_y H\sin\beta\, V_n)$$

Using straightforward linear algebra, we can now eliminate V_n from these two equations. If we define an amplification factor

$$U_{n+2} = GU_n$$

we can derive a fourth-order polynomial in G, which fortunately has two roots of magnitude unity. The reduced polynomial is

$$G^2 + [-2 + 4(f\Delta t)^2 + 4C^2\lambda_x^2\sin^2\alpha + 4C^2\lambda_y^2\sin^2\beta]G + 1 = 0$$

or

$$G^2 - 2AG + 1 = 0$$

If $(1-A^2)\geq 0$, then $|G| = 1$ for all α and β; in other words, the solution is stable for all wavenubers ℓ and m. Here

$$A = 1 - 2(f\Delta t)^2 - 2\lambda_x^2C^2\sin^2\alpha - 2\lambda_y^2C^2\sin^2\beta$$

The stability condition is satisfied if

$$(f\Delta t)^2 + C^2\lambda_x^2\sin^2\alpha + C^2\lambda_y^2\sin^2\beta \leq 1$$

for the shortest waves, $\alpha = \beta = \pi/4$,

$$(f\Delta t)^2 + C^2\lambda_x^2 + C^2\lambda_y^2 \leq 1.$$

For the longest waves, $\alpha = \beta = 0$, the $|f\Delta t| \leq 1$.
If $\Delta x = \Delta y$

$$\frac{C^2\Delta t^2}{\Delta x^2} \leq \frac{1 - (f\Delta t)^2}{2}$$

Again we note that the Coriolis term, $f\Delta t$, reduces the CFL condition below unity. It is more important to observe that if $f = 0$,

$$\frac{C\Delta t}{\Delta x} \leq \frac{\sqrt{2}}{2}$$

In a two-dimensional system a wave can have a velocity of C in both spatial directions and thus a vector speed of $\sqrt{2}C$. The linear stability must account for this. As in the one-dimensional case, we observe that if CFL is violated, we expect the $4\Delta x$ and $4\Delta y$ waves to grow fastest.

5. CFL FOR TWO-DIMENSIONAL GRAVITY WAVES WITH ADVECTION

The linear primitive equations without rotation are

$$u_t + Au_x + Bu_y = -gh_x$$

$$v_t + Av_x + Bu_y = -gh_y$$

$$h_t + Ah_x + Bh_y = -D(u_x + v_y)$$

Let us use second order approximations in time and space and surpress, j, k, n, unless it is indexed

Define: $\lambda_x = \Delta t/\Delta x$, $\lambda_y = \Delta t/\Delta x$, $\alpha = \ell\Delta x$, $\beta = m\Delta y$

$$u^{n+1} = u^{n-1} - A\lambda_x(u_{j+1} - u_{j-1}) - \beta\lambda_y(u_{k+1} - u_{k-1})$$

$$- g\lambda_x(h_{j+1} - h_{j-1})$$

$$v^{n+1} = v^{n-1} - A\lambda_x(v_{j+1} - v_{j-1}) - B\lambda_y(u_{k+1} - u_{k-1})$$

$$-g\lambda_x(h_{k+1} - h_{k-1})$$

$$h^{n+1} = h^{n-1} - A\lambda_x(h_{j+1} - h_{j-1}) - B\lambda_y(h_{k+1} - h_{k-1})$$

$$- D\lambda_x(u_{j+1} - u_{j-1}) - D\lambda_y(v_{k+1} - v_{k-1})$$

Use our standard linear stability technique and eliminate the spatial dependence

$$(u,v,h) = (U,V,H)e^{imj\Delta x}\, e^{i\ell k\Delta y}$$

The new equations are

$$U_{n+1} = U_{n-1} - A\lambda_x 2i\sin\alpha U - B\lambda_y 2i\sin\beta U - g\lambda_x 2i\sin\alpha H$$

$$V_{n+1} = V_{n-1} - A\lambda_x 2i\sin\alpha V - B\lambda_y 2i\sin\beta V - g\lambda_y 2i\sin\beta\alpha H$$

$$H_{n+1} = H_{n-1} - A\lambda_x 2i\sin\alpha H - B\lambda_y 2i\sin\beta H - D\lambda_x 2i\sin\alpha U$$

$$- D\lambda_y 2i \sin \beta V$$

This looks like a difficult algebraic problem. However, the solution is straightforward. Proceed as follows; eliminate H from the first and third equations <u>and</u> the second and third equations; then eliminate either U or V from the two equations; after using

$$U_{n+1} = GU_n ,$$

we obtain a second-order equation, $G^2 + 2iRG - 1 = 0$, whose analysis is left to the reader and

$$G^4 + 4iRG^3 - 2SG^2 - 4iRG + 1 = 0$$

where

$$R = A\lambda_x\sin\alpha + B\lambda_y\sin\beta$$

$$S = 1 + 2(A\lambda_x\sin\alpha + B\lambda_y\sin\beta)^2 - 2gD\lambda_x^2 \sin^2\alpha - 2gD\lambda_y^2 \sin^2\beta$$

Let $iX = G$ and obtain

$$X^4 + 4RX^3 + 2SX^2 + 4RX + 1 = 0$$

The product of the roots must be unity, and if one, $|X|<1$ is less than unity, another must be greater in absolute value.

For stability all roots must have magnitude unity. The reader can complete the details. The general condition for stability becomes

$$|A|\lambda_x + |B|\lambda_y + [gD(\lambda_x^2 + \lambda_y^2)]^{1/2} \leq 1$$

We can reduce this to the famous CFL condition by letting $\Delta x = \Delta y$, $\alpha = \beta = \pi/4$, $A = B$ and $U = (A^2 + B^2)^{1/2}$, to obtain

$$[U + (gD)^{1/2}]\ \Delta t/\Delta x \leq \frac{1}{\sqrt{2}}$$

In physical terms, the Doppler-shifted gravity wave may not propagate a greater distance then $\sqrt{2}\ \Delta x$ in time Δt.

Exercise:

Add the Coriolis terms to the primitive equations with advection and determine the stability condition.

6. COMBINED ADVECTIVE-DIFFUSIVE PROBLEMS

In Chapter 4 we demonstrated that the diffusion equation is unstable if we use a leapfrog scheme. In this chapter, we learned that a forward time step is unstable for the advection equation. In most oceanographical problems we must include some dffusion in the problem; hence it is necessary to combine several time-differencing schemes. The simplest is leapfrog for advection and forward for diffusion. Suppose we have the equation

$$\frac{\partial q}{\partial t} + A\frac{\partial q}{\partial x} = K\frac{\partial^2 q}{\partial x^2} \tag{10}$$

We might use

$$q_j^{n+1} = q_j^{n-1} - \frac{A\Delta t}{\Delta x}\left[q_{j+1}^n - q_{j-1}^n\right]$$
$$+ \frac{2K\Delta t}{(\Delta x)^2}\left[q_{j+1}^{n-1} + q_{j-1}^{n-1} - 2q_j^{n-1}\right]$$

Note that the diffusion term is evaluated at $(n-1)\Delta t$. Each term is stable by itself; if $K = 0$, then the scheme is stable if $|A|\Delta t/\Delta x \leq 1$; if $A = 0$, then the scheme is stable if $K\Delta t/\Delta x^2 < 1/4$. The factor, 1/4 ,arises because of the $2\Delta t$ time step. In the general case, we get a modified condition. The amplification factor is

$$G = G^{-1} - 2i(A\Delta t/\Delta x)\sin\alpha + 4(K\Delta t/\Delta x^2)(\cos\alpha - 1)G^{-1}$$

or

$$G^2 + 2i(A\Delta t/\Delta x)\sin\alpha G - (1 + 4(K\Delta t/\Delta x^2)(\cos\alpha - 1) = 0$$

The resulting necessary and sufficient condition of numerical stability is

$$\frac{A^2\Delta t^2 + 4K\Delta t}{\Delta x^2} \leq 1 \ .$$

One notes that for either A=0 or K=0, the stability conditions of the individual advective and diffusive schemes are recovered, but also that imposing each condition separately is insufficient.

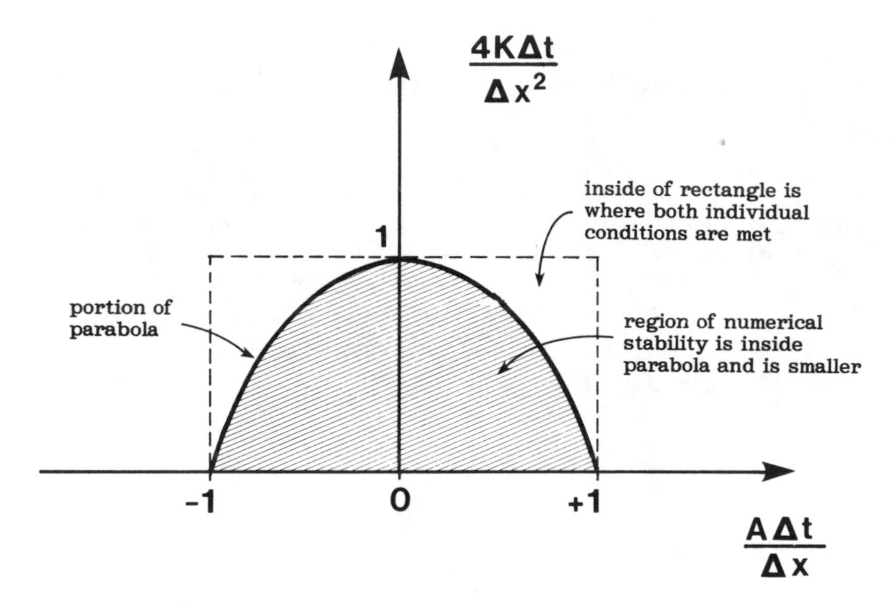

Figure 1. The diagram shows the stability region for the advection-diffusive problem.

We obtain the surprising conclusion that adding explicit diffusion actually reduces the maximum time step allowed for advection. This is not a serious problem in oceanography, since we normally have

$$\frac{K\Delta t}{(\Delta x)^2} \ll \left(\frac{C\Delta t}{\Delta x}\right) \ , \ C > 0$$

for most practical problems.

Exercise:

Derive the linear stability condition for 2-D advection and diffusion using forward-in-time for diffusion and centered-in-time for advection.

Many ocean modellers use Dufort-Frankel for the diffusion term; (10) would be approximated by

$$q_j^{n+1} = q_j^{n-1} - \frac{A\Delta t}{\Delta x}\left[q_{j+1}^n - q_{j-1}^n\right]$$

$$+ \frac{2K\Delta t}{(\Delta x)^2}\left[q_{j+1}^n + q_{j-1}^n - q_j^{n+1} - q_j^{n-1}\right]$$

We have combined a conditionally stable advection scheme $\left(|A|\frac{\Delta t}{\Delta x} < 1\right)$ with an unconditionally stable scheme for diffusion. In this one dimensional case, it can be shown that only the CFL condition need be satisfied. However, in the two-dimensional case, there is a more stringent restriction (Cushman-Roisin, 1984).

Several authors suggest the use of the unstable forward-in-time, centered-in-space (FTCS) advection scheme when diffusion is present. Clancy [1981] derived the necessary and sufficient conditions for stability and suggested the use of this scheme in ocean modelling; (10) is written

$$q_j^{n+1} = q_j^n - A\Delta t/(2\Delta x)(q_{j+1}^n - q_{j-1}^n)$$

$$+ K\Delta t/\Delta x^2(q_{j+1}^n + q_{j-1}^n - 2q_j^n)$$

Define: $\hat{k} = K\Delta t/\Delta x^2$; $a = |A|\Delta t/\Delta x$. The amplification factor is

$$G = 1 - ia\sin\alpha \; 2\hat{k}(1 - \cos\alpha)$$

Clancy derives the two conditions for stability. $|G| \leq 1$ iff

$$K\Delta t/\Delta x \leq 1/2$$

and

$$\frac{|A|\Delta x}{2K} \leq 1$$

The FTCS scheme is not recommended for ocean models, in spite of the enthusiasm of several authors.

7. NONLINEAR STABILITY

A large body of empirical evidence exists which demonstrates that inviscid nonlinear hyperbolic equations will become unstable after many time steps even when the linear stability is not violated. If one attempts an integration of a hyperbolic model with initial conditions which are well resolved at low wavenumber, one finds that after some time, variance will appear on small scales ($4\Delta x$ - $2\Delta x$). The amplitude of the "noise" will grow slowly at first but eventually, at an unpredictable time, it will grow exponentially. This renders the solution useless. Phillips [1959] is credited with the first analytical example which explains this phenomenon. Richtmyer [1963] provided another example which we will reproduce here. Robert et al., [1970] generalized the previous examples.

We approach the example with these preconceptions. In a linear problem, no Fourier mode can interact with any other mode. However when the equations are nonlinear (or have nonconstant coefficients) we expect the modes to interact and create variance in scales which have no variance initially.

Second, we recognize that we have a bandlimited wavenumber space. A uniform grid in Cartesian coordinates can only have wavelengths, $k \in [0, 2\pi/2\Delta x]$. If any nonlinear interaction should produce variance in scales, $k > 2\pi/2\Delta x$, the grid cannot resolve this energy, and it will be folded into some low wavenumber. Let us arbitarily call $k < 2\pi/4\Delta x$ low wavenumber and $2\pi/4\Delta x < k < 2\pi/2\Delta x$ high wavenumber. We expect *apriori* that even though all initial energy is low wavenumber, nonlinear interactions will eventually provide variance (or energy) at high wavenumbers.

The example of Richtmyer [1963] inspects the stability of a nonlinear problem which has variance at wavelengths, 0, $4\Delta x$ and $2\Delta x$. The model is

$$\frac{\partial u}{\partial t} = -u\frac{\partial u}{\partial x} \tag{11}$$

where, as in the von Neumann linear analysis, we will not be concerned with any boundary condition, but only perform a local analysis. Let us approximate (1) with a leapfrog scheme

$$u_j^{n+1} = u_j^{n-1} - \frac{\lambda}{2}[(u_{j+1}^n)^2 - (u_{j-1}^n)^2] \tag{12}$$

where $\lambda = \Delta t/\Delta x$. It will be useful to write (12) as

$$u_j^{n+1} = u_j^{n-1} - \frac{\lambda}{2}[(u_{j+1}^n + u_{j-1}^n)(u_{j+1}^n - u_{j-1}^n)] \tag{13}$$

We can show that an exact solution for (13) is

$$U_j^n = C_n \cos\pi j/2 + S_n \sin\pi j/2 + U_n \cos\pi j + V \tag{14}$$

We can identify (C_n, S_n) as the amplitudes of a wave with length, $4\Delta x$, and U_n as the amplitude of a wave with length, $2\Delta x$ and V as a low wavenumber component with zero wavenumber. When we substitute (14) into (13) we obtain relationships among the amplitudes

$$C_{n+1} - C_{n-1} = +\lambda S_n(U_n - V)$$

$$S_{n+1} - S_{n-1} = \lambda C_n(U_n + V)$$

$$U_{n+1} = U_{n-1}$$

The last equation says U_n may take on different initial values -- say A and B -- for the odd and even time steps. If we eliminate S_n from the first equation, we obtain

$$C_{n+2} - 2C_n + C_{n-2} = +4\lambda^2(A + V)(B - V)C_n$$

Is this equation stable in the von Neumann sense? Define $C_{n+2} = GC_n$ and derive

$$G^2 - 2\big(1 + 2\lambda^2(A + V)(B - V)\big)G + 1 = 0$$

the roots are

$$G = 1 + 2\lambda^2(A + V)(B - V) \pm$$

$$\pm \left[1 + \big(1 + 2\lambda^2(A + V)(B - V)(b - V)\big)^2\right]^{1/2}$$

If the radical is imaginary, then $|G| = 1$; this requires the coefficient

$$-1 \leq \lambda^2(A + V)(B - V) \leq 0 \tag{15}$$

Obviously this is only possible if $|A| < V$ and $|B| < V$. This is violated when the amplitude of the $2\Delta x$ wave is large. In this case the $4\Delta x$ wave will grow exponentially and the scheme is unstable.

In numerical models of the ocean or atmosphere, it is important to damp out the smallest space scales in order to control nonlinear instability. This may be done with an explicit eddy viscosity (Chapter 4) or with a dissipative finite difference method.

8. REFERENCES

Clancy, R. M., A note on the finite differencing of the advection-diffusion equation, Mon. Wea. Rev., 109, 1807-1809, 1981.

Cushman-Roisin, B., Analytical, linear stability criteria for the leap-frog, Dufort-Frankel method, J. Comp. Phys., 53, 227-239, 1984.

Phillips, N. A., An example of nonlinear computational instability, in The Atmosphere and the Sea in Motion, edited by Bert Bolin, pp. 501-504, Rockerfeller Institute Press, New York, 1959.

Richtmyer, R. D., A survey of differene methods for non-steady fluid dynamics. NCAR Technical Note 63-2, 25 pp., 1963.

Robert, A. J., and J. P. Gerrity, Jr., On partial difference equations in mathematical physics, Mon. Wea. Rev., 98, 1-6, 1970.

FINITE-DIFFERENCE FORMULATION OF A WORLD OCEAN MODEL

A. J. Semtner, Jr.
National Center for Atmospheric Research
Boulder, Colorado 80307-3000
USA

ABSTRACT. The finite-difference formulation of a multi-level primitive equation ocean model is described. The formulation is suitable for irregular geometry such as that of the world ocean. The second-order, energetically consistent techniques of Bryan [1969] and Takano [1974] are followed.

1. INTRODUCTION

This paper describes the finite-difference formulation of a multi-level primitive equation ocean model. The historical development of such models and the rationale for many of the numerical choices of this model are described in another paper within this volume [Semtner, 1986]. The model is formulated in spherical coordinates for oceans of arbitrary coastline, bottom topography, and connectedness. For the most part, the second-order, energetically consistent space differencing scheme of Bryan [1969] is used. With regard to the vertically averaged flow, the methods of Takano [1974] for a semi-implicit treatment of the beta term and for hole relaxation of islands are generalized to the case of variable bottom topography. Also discussed are the options for a five-point or nine-point relaxation algorithm for obtaining the vertically averaged flow, and for semi-implicit treatment of the pressure term for the baroclinic mode. The model was programmed in 1974 to be highly efficient on scientific computers which achieve a substantial increase in computing speed through vector processing. Most of the material presented here is from unpublished documentation of that model code [Semtner, 1974]. The model is still in widespread use, and it has recently been updated for Cyber-type machines by Cox [1984]. Although many additional physical and numerical improvements have been developed for specific applications (e.g., treatments of open boundaries, the North Pole, and isopycnal mixing as well as the direct solution of the vertically averaged flow in regular domains), these will not be discussed in this paper.

J. J. O'Brien (ed.), Advanced Physical Oceanographic Numerical Modelling, 187–202.

2. THE BASIC EQUATIONS IN DIFFERENTIAL FORM

2.1. Governing equations

We take the earth to be a sphere of radius a, rotating with angular speed Ω. A spherical coordinate system is used, with λ, ϕ, and z representing longitude, latitude, and height. The ocean is contained between the surface $z = 0$ and the bottom $z = -H(\lambda, \phi)$. Seven variables specify the physical condition of the ocean: three velocity components (u, v, and w), pressure p, density ρ, potential temperature T, and salinity S. We assume that the ocean is incompressible and behaves as a horizontally isotropic fluid with constant eddy viscosity coefficients μ and A_M in the vertical and horizontal directions respectively. We also assume that sub-grid scale transfers of heat and salt are similarly describable using constant eddy diffusivity coefficients κ and A_H. Several standard approximations are used to simplify the governing equations: the thin-shell approximation, on account of the shallowness of the ocean relative to the earth's radius; the hydrostatic assumption, because of the scale of the motions being considered; the Boussinesq approximation, because of the relatively small variations in density; and the neglect of terms involving w in the horizontal momentum equations, on the basis of scale analysis. With respect to these assumptions and approximations, the governing equations are as follows:

$$\frac{\partial u}{\partial t} + Lu - \frac{uv\tan\phi}{a} - fv = -\frac{1}{\rho_0 a\cos\phi}\frac{\partial p}{\partial\lambda} + \mu\frac{\partial^2 u}{\partial z^2} + A_m\left\{\nabla^2 u + \frac{(1-\tan^2\phi)u}{a^2} - \frac{2\sin\phi}{a^2\cos^2\phi}\frac{\partial v}{\partial\lambda}\right\} \tag{1}$$

$$\frac{\partial v}{\partial t} + Lv + \frac{u^2\tan\phi}{a} + fu = -\frac{1}{\rho_0 a}\frac{\partial p}{\partial\phi} + \mu\frac{\partial^2 v}{\partial z^2} + A_m\left\{\nabla^2 v + \frac{(1-\tan^2\phi)v}{a^2} + \frac{2\sin\phi}{a^2\cos^2\phi}\frac{\partial u}{\partial\lambda}\right\} \tag{2}$$

$$\frac{\partial p}{\partial z} = -\rho g \tag{3}$$

$$\frac{1}{a\cos\phi}\frac{\partial u}{\partial\lambda} + \frac{1}{a\cos\phi}\frac{\partial}{\partial\phi}(v\cos\phi) + \frac{\partial w}{\partial z} = 0 \tag{4}$$

$$\frac{\partial T}{\partial t} + LT = \kappa\frac{\partial^2 T}{\partial z^2} + A_H\nabla^2 T \tag{5}$$

$$\frac{\partial S}{\partial t} + LS = \kappa\frac{\partial^2 S}{\partial z^2} + A_H\nabla^2 S \tag{6}$$

$$\rho = \rho(T, S, p)\,. \tag{7}$$

In the above, we have the advection operator,

$$L(\alpha) = \frac{1}{a\cos\phi}\frac{\partial}{\partial\lambda}(u\alpha) + \frac{1}{a\cos\phi}\frac{\partial}{\partial\phi}(v\alpha\cos\phi) + \frac{\partial}{\partial z}(w\alpha) \tag{8}$$

the horizontal Laplacian operator,

$$\nabla^2\alpha = \frac{1}{a^2\cos^2\phi}\frac{\partial^2\alpha}{\partial\lambda^2} + \frac{1}{a^2\cos\phi}\frac{\partial}{\partial\phi}\left(\frac{\partial\alpha}{\partial\phi}\cos\phi\right) \tag{9}$$

and the Coriolis parameter,

$$f = 2\Omega\sin\phi\,. \tag{10}$$

Equation (7) represents the nonlinear equation of state. Accurate polynomial approximations to equation (7) which are suitable for numerical modelling have been developed by Bryan and Cox [1972] and by Friedrich and Levitus [1972].

A large-scale hydrostatic model, such as the present one, cannot handle convection explicitly. A simple convective adjustment is therefore applied to remedy any situation of static instability. The mechanics of this adjustment are described in the section on finite differencing.

2.2. Boundary Conditions

The configuration of the ocean is defined by specifying the depth field $H(\lambda,\phi)$. Northern and southern boundaries are assumed to be closed. East and west boundaries may be either closed or cyclically continuous. Otherwise, arbitrary specifications of the coastline, bottom topography, and connectedness are allowed.

At the ocean bottom, fluxes of momentum, heat, and salt are taken to be zero:

$$\left.\begin{aligned} \rho_0\mu\frac{\partial}{\partial z}(u,v) &= 0 \\ \rho_0\kappa\frac{\partial}{\partial z}(T,S) &= 0 \end{aligned}\right\} \quad \text{at} \quad z = -H(\lambda,\phi)\ . \tag{11}$$

(Alternatively, the momentum flux could be specified by a quadratic drag law.) At the ocean surface, wind stress and surface fluxes of heat and salt are prescribed from observations or from an atmospheric model:

$$\left.\begin{aligned} \rho_0\mu\frac{\partial}{\partial z}(u,v) &= (\tau^\lambda,\tau^\phi) \\ \rho_0\kappa\frac{\partial}{\partial z}(T,S) &= (F^T,F^S) \end{aligned}\right\} \quad \text{at} \quad z = 0. \tag{12}$$

(Precipitation minus evaporation in the real ocean is handled in terms of the equivalent salt flux for the model ocean by neglecting the small amount of water mass transfer at the surface.)

At lateral walls, a no-slip condition $(u, v) = 0$ imposed, and no flux of heat or salt is allowed. For the computational purpose of allowing a longer model timestep, the surface of the ocean is assumed to be a horizontal rigid lid, so that

$$w = 0 \quad \text{at} \quad z = 0\,. \tag{13}$$

This assumption removes high frequency external gravity waves from the system without seriously affecting the lower frequency motions which contain most of the energy. At the ocean bottom, the flow is required to parallel the slope, i.e.,

$$w = -\frac{u}{a\cos\phi}\frac{\partial H}{\partial \lambda} - \frac{v}{a}\frac{\partial H}{\partial \phi} \quad \text{at} \quad z = -H(\lambda, \phi)\,. \tag{14}$$

2.3. Prognostic Equations

Although the rigid lid assumption (13) allows a more efficient calculation, it complicates the solution of the time-dependent equations. We note from the hydrostatic relation (3) that pressure at any depth z is composed of a pressure p_s at the rigid lid and a hydrostatic part, i.e.,

$$p(z) = p_s + \int_z^0 g\rho\, dz'\,.$$

No simple prediction equation for the pressure p_s at $z = 0$ is available, but p_s can be eliminated from the equations by differentiation. It is removed from equations (1) and (2) by application of $\frac{\partial}{\partial z}$ and substitution of the hydrostatic relation to obtain

$$\frac{\partial}{\partial t}\left(\frac{\partial u}{\partial z}\right) - f\frac{\partial v}{\partial z} = \frac{g}{\rho_0 a\cos\phi}\frac{\partial \rho}{\partial \lambda} + \frac{\partial}{\partial z}(G^\lambda) \tag{15}$$

$$\frac{\partial}{\partial t}\left(\frac{\partial v}{\partial z}\right) + f\frac{\partial u}{\partial z} = \frac{g}{\rho_0 a}\frac{\partial \rho}{\partial \phi} + \frac{\partial}{\partial z}(G^\phi)\,. \tag{16}$$

Here G^λ and G^ϕ represent the collected contributions of the nonlinear and viscous terms in equations (1) and (2). The above equations allow prediction of the vertical shear of velocity, so that a further specification of some vertically independent velocity components is needed. The vertically *averaged* velocity components $(\overline{u}, \overline{v})$ can be written in terms of a volume transport streamfunction ψ for the vertically *integrated* flow:

$$\overline{u} = \frac{1}{H}\int_{-H}^0 u\, dz = -\frac{1}{Ha}\frac{\partial \psi}{\partial \phi} \tag{17}$$

$$\overline{v} = \frac{1}{H}\int_{-H}^0 v\, dz = \frac{1}{Ha\cos\phi}\frac{\partial \psi}{\partial \lambda}\,. \tag{18}$$

The existence of ψ is guaranteed by the nondivergent nature of the vertically integrated flow, i.e.,

$$\frac{1}{a\cos\phi}\frac{\partial}{\partial\lambda}\left(\int_{-H}^{0} u\,dz\right)+\frac{1}{a\cos\phi}\frac{\partial}{\partial\phi}\left(\cos\phi\int_{-H}^{0} v\,dz\right)=0$$

This relation can be verified by integrating the continuity equation (4) and applying conditions (13) and (14).

A prediction equation for ψ which does not include the unknown surface pressure p_s is obtained by forming the vertical averages of equations (1) and (2) and taking the curl of these. In terms of ψ and p_s, the vertically averaged equations are as follows:

$$-\frac{1}{Ha}\frac{\partial}{\partial\phi}\left(\frac{\partial\psi}{\partial t}\right)-\frac{f}{Ha\cos\phi}\frac{\partial\psi}{\partial\lambda}=-\frac{1}{a\rho_0\cos\phi}\frac{\partial p_s}{\partial\lambda}-\frac{g}{Ha\rho_0\cos\phi}\int_{-H}^{0}\int_{z}^{0}\frac{\partial\rho}{\partial\lambda}dz'\,dz+\frac{1}{H}\int_{-H}^{0}G^{\lambda}\,dz \tag{19}$$

$$\frac{1}{Ha\cos\phi}\frac{\partial}{\partial\lambda}\left(\frac{\partial\psi}{\partial t}\right)-\frac{f}{Ha}\frac{\partial\psi}{\partial\phi}=-\frac{1}{a\rho_0}\frac{\partial p_s}{\partial\phi}-\frac{g}{Ha\rho_0}\int_{-H}^{0}\int_{z}^{0}\frac{\partial\rho}{\partial\phi}dz'\,dz+\frac{1}{H}\int_{-H}^{0}G^{\phi}\,dz\,. \tag{20}$$

Applying the $curl_z$ operator, defined by

$$curl_z(q_1,q_2)=\frac{1}{a\cos\phi}\left[\frac{\partial q_2}{\partial\lambda}-\frac{\partial}{\partial\phi}(q_1\cos\phi)\right]$$

and simplifying by eliminating a factor of $1/(a^2\cos\phi)$, we get

$$\begin{aligned}\left[\frac{\partial}{\partial\lambda}\left(\frac{1}{H\cos\phi}\frac{\partial^2\psi}{\partial\lambda\,\partial t}\right)+\frac{\partial}{\partial\phi}\left(\frac{\cos\phi}{H}\frac{\partial^2\psi}{\partial\phi\,\partial t}\right)\right]-\left[\frac{\partial}{\partial\lambda}\left(\frac{f}{H}\frac{\partial\psi}{\partial\phi}\right)-\frac{\partial}{\partial\phi}\left(\frac{f}{H}\frac{\partial\psi}{\partial\lambda}\right)\right]\\ =-\left[\frac{\partial}{\partial\lambda}\left(\frac{g}{\rho_0 H}\int_{-H}^{0}\int_{z}^{0}\frac{\partial\rho}{\partial\phi}dz'\,dz\right)\right.\\ \left.-\frac{\partial}{\partial\phi}\left(\frac{g}{\rho_0 H}\int_{-H}^{0}\int_{z}^{0}\frac{\partial\rho}{\partial\lambda}dz'\,dz\right)\right]\\ +\left[\frac{\partial}{\partial\lambda}\left(\frac{a}{H}\int_{-H}^{0}G^{\phi}\,dz\right)-\frac{\partial}{\partial\phi}\left(\frac{a\cos\phi}{H}\int_{-H}^{0}G^{\lambda}\,dz\right)\right]\,.\end{aligned} \tag{21}$$

Equation (21) is a prediction equation for ψ, which requires inversion of a second-order differential operator to obtain the function. It is necessary to specify boundary conditions for this inversion. In general, the domain of ψ will be a multiply connected region whose boundary consists of a primary continent and several islands. (For example, in the case of the world ocean, the primary continent might be chosen to consist of Eurasia and the Americas, whereas the "islands" might be Antarctica, Australia, and New Zealand.) The value of ψ must be spatially constant along each individual coastline, in order to have vanishing normal velocity. On the chosen continent, ψ is held constant in time as well; but on the islands, ψ varies in response to the changing circulation. In order to predict the time change of ψ on the islands, we use the method of Takano [1974], generalized to the case of variable bottom topography. We require that the surface pressure p_s be a single-valued function, in the sense that a line integral of ∇p_s around the coastline of each island should vanish. By integrating equations (19) and (20) around each island and applying this condition, the following equation is obtained to predict the change of ψ on the island:

$$\oint \left(-\frac{\cos\phi}{H} \frac{\partial^2 \psi}{\partial \phi\, \partial t}\, d\lambda + \frac{1}{H\cos\phi} \frac{\partial^2 \psi}{\partial \lambda\, \partial t}\, d\phi \right)$$

$$= -\oint \frac{g}{\rho_0 H} \left[\left(\int_{-H}^{0} \int_{z}^{0} \frac{\partial \rho}{\partial \lambda}\, dz'\, dz \right) d\lambda + \left(\int_{-H}^{0} \int_{z}^{0} \frac{\partial \rho}{\partial \phi}\, dz'\, dz \right) d\phi \right] \tag{22}$$

$$+ \oint \left[\left(\frac{a\cos\phi}{H} \int_{-H}^{0} G^{\lambda}\, dz \right) d\lambda + \left(\frac{a}{H} \int_{-H}^{0} G^{\phi}\, dz \right) d\phi \right] .$$

In the above, the fact that ψ is spatially constant along the coastline eliminates any contribution from the Coriolis terms.

It should be noted that the pressure p_s at all points in the domain can be obtained using line integration of equations (19) and (20) from a fixed point and then normalizing to make the area-averaged pressure equal to zero. Thus, the surface pressure is a diagnostic quantity which can be recovered for purposes of analysis.

3. FINITE-DIFFERENCE EQUATIONS

The ocean to be studied is approximated by a collection of boxes, each having horizontal dimensions corresponding to increments $\Delta\lambda$ and $\Delta\phi$ in longitude and latitude. Boxes of varying thickness are stacked downward from the surface until the bottom is reached. The kth box down has thickness Δk. Points at which various quantities are computed are distributed in a staggered fashion through the lattice of boxes. Temperature and salinity points reside at the centers of the boxes, and horizontal vector velocity points reside at the midpoints of the vertical edges. The coastline is defined as the boundary between boxes having zero levels and those having nonzero levels; u and v are identically zero on this boundary and similarly along the bathymetric contours defined by the vertical grid levels. The vertical velocity points used for the calculation of T and S are at the centers of the horizontal faces, while those used for the calculation of u and v are at the endpoints of the vertical edges. The streamfunction ψ is defined at horizontal points overlying the T and S points.

Given the values of a variable g at adjacent points with longitude $\lambda - \frac{\Delta\lambda}{2}$ and $\lambda + \frac{\Delta\lambda}{2}$, we define difference and average values at the midpoint between them as follows:

$$\delta_\lambda g = \frac{g\left(\lambda + \frac{\Delta\lambda}{2}\right) - g\left(\lambda - \frac{\Delta\lambda}{2}\right)}{\Delta\lambda} \tag{23}$$

$$\overline{g}^\lambda = \frac{g\left(\lambda + \frac{\Delta\lambda}{2}\right) + g\left(\lambda - \frac{\Delta\lambda}{2}\right)}{2} . \tag{24}$$

Similar operations are defined with respect to the coordinates ϕ, z, and t. In the vertical differencing of quantities at box centers, a separation $\Delta' k = (\Delta k + \Delta(k-1))/2$ is used. In time differencing an interval $2\Delta t$ is used, i.e.,

$$\delta_t g = \frac{g(t + \Delta t) - g(t - \Delta t)}{2\Delta t} .$$

The finite differencing scheme of Bryan [1969] is used because it is both simple and energetically consistent. No fictitious sources of energy arise in the nonlinear exchanges of kinetic energy or in the conversion of potential energy to kinetic energy. Other quantities such as mass, heat, and salt are also conserved, apart from specified fluxes at the boundaries. For completeness, the proof of these properties will be included in Section 4. The basic means by which consistency is achieved is the use of averaging operators whenever values of a variable are needed at points other than the regular grid points.

The differencing for T, S, and w at T, S points is presented first. The advection operator L of Eq. (8) has as its discrete counterpart the operator L^*, such that

$$L^*(\alpha) = \frac{1}{a\cos\phi}\delta_\lambda\left(\overline{u}^\phi\overline{\alpha}^\lambda\right) + \frac{1}{a\cos\phi}\delta_\phi\left(\overline{v}^\lambda\overline{\alpha}^\phi\cos\phi\right) + \delta_z(w\overline{\alpha}^z) . \tag{25}$$

The continuity equation can be written as

$$L^*(1) = 0 . \tag{26}$$

The vertical velocity is computed using this relation plus the condition that $w = 0$ at $z = 0$. The heat equation is written as

$$\delta_t T = -L^*T + \kappa\delta_z(\delta_z T) + A_H\left[\frac{1}{a^2\cos^2\phi}\delta_\lambda(\delta_\lambda T) + \frac{1}{a^2\cos\phi}\delta_\phi(\delta_\phi T\cos\phi)\right] \tag{27}$$

taking into account appropriate boundary conditions, and the salt equation is similar. In the above, the advective terms are computed for $T(t)$, whereas the diffusive terms are computed for $T(t - \Delta t)$. Thus the model has a leapfrog time step for advective processes and a forward time step for diffusive processes.

We now discuss the differencing scheme to obtain the velocity. We first remark that the ocean depth for a vertical column of u, v points is taken as the minimum of the four surrounding depths for columns of T, S points (measured by summing the (Δk)s in each column). The depth between two columns of u, v points is defined as the maximum of the two depths at T, S columns on either side. We also note that u or v at any point may be decomposed into an internal mode plus an external (vertically averaged) mode. Thus

$$(u,v) = (u',v') + (\overline{u},\overline{v}) \tag{28}$$

where

$$\sum_{k=1}^{kz}(u',v')\,\Delta k = (0,0) \tag{29}$$

for a column with a total of kz levels, and where

$$(\overline{u},\overline{v}) = \left(-\frac{1}{Ha}\delta_\phi(\overline{\psi}^\lambda) \quad , \quad \frac{1}{Ha\cos\phi}\delta_\lambda(\overline{\psi}^\phi)\right) . \tag{30}$$

To aid in constructing an advection operator L^{**} at u, v points, we define two auxiliary velocities

$$\begin{aligned} U &= \overline{u'}^\lambda - \frac{1}{Ha}\delta_\phi\,\psi \\ V &= \overline{v'}^\phi + \frac{1}{Ha\cos\phi}\delta_\lambda\,\psi . \end{aligned} \tag{31}$$

In terms of these, we let

$$L^{**}(\alpha) = \frac{1}{a\cos\phi}\delta_\lambda(U\overline{\alpha}^\lambda) + \frac{1}{a\cos\phi}\delta_\phi(V\overline{\alpha}^\phi\cos\phi) + \delta_z(w\overline{\alpha}^z) . \tag{32}$$

The continuity equation is

$$L^{**}(1) = 0 \tag{33}$$

which allows the determination of w in u, v columns after setting $w = 0$ at $z = 0$. It can be shown that

$$w \cong -\frac{u}{a\cos\phi}\,\delta_\lambda H - \frac{v}{a}\,\delta_\phi H \qquad \text{at} \qquad z = -H . \tag{34}$$

We now must construct prediction equations for u' and v'. One way of doing this is to render equations (15) and (16) into finite difference form so that $\delta_z u$ and $\delta_z v$ are predicted. Then u' at level m could be calculated by the relation

$$u'_m = -\sum_{k=1}^{m}(\delta_z u\,\Delta' k) + \frac{1}{H}\sum_{k_2=1}^{kz}\left[\sum_{k_1=1}^{k_2}(\delta_z u\,\Delta' k_1)\right]\Delta k_2 \tag{35}$$

with a similar relation for v'. A simpler method for predicting u' and v' is to make finite-difference versions of equations (1) and (2) rather than of equations (15) and

(16). By simply ignoring the unknown contribution made by the surface pressure p_s, equations for preliminary velocities u^* and v^* at time $t + \Delta t$ are obtained. These preliminary velocities are in error by amounts that are independent of depth, so that no final error occurs in calculating u' by the rule

$$u'_k = u^*_k - \frac{1}{H} \sum_{k=1}^{kz} u^*_k \, \Delta k \, . \tag{36}$$

Before the prediction equations for u^* and v^* are exhibited, some remarks should be made about the semi-implicit treatment of two terms. The Coriolis term in equation (1) is written as $-f[v^{\ell-1} + \alpha(v^* - v^{\ell-1})]$, where superscripts in ℓ indicate the time level of the variable and α lies in the interval (0,1). By a choice of $\alpha \geq 1/2$, a longer time step is allowable because of the filtering of inertial waves and barotropic Rossby waves. The hydrostatic component of pressure p_h is also treated semi-implicitly so that internal waves may also be filtered to some extent. Thus p_h at level m is given by

$$p_h = g \sum_{k=1}^{m} \overline{[\rho^{\ell-1} + \gamma(\rho^{\ell+1} - \rho^{\ell-1})]}^{z} \, \Delta' k \tag{37}$$

where the $\overline{(\cdot)}^{z}$ operator is employed for values of $k \geq 2$. The value of $\rho^{\ell+1}$ is obtainable from $T^{\ell+1}$ and $S^{\ell+1}$, which are predicted using u^{ℓ} and v^{ℓ} in the advective terms.

To obtain prediction equations for the preliminary velocities u^* and v^* at time level $\ell + 1$, we render equations (1) and (2) into finite-difference form and put all terms not involving $(u^* - u^{\ell-1})$ or $(v^* - v^{\ell-1})$ on the right sides of the equations. The resulting equations are:

$$\begin{aligned} \frac{u^* - u^{\ell-1}}{2\Delta t} - f\alpha(v^* - v^{\ell-1}) &= F^{\lambda} \\ \frac{v^* - v^{\ell-1}}{2\Delta t} + f\alpha(u^* - u^{\ell-1}) &= F^{\phi} \end{aligned} \tag{38}$$

where

$$\begin{aligned} F^{\lambda} = &- L^{**}(u^{\ell}) + \frac{\tan\phi \, u^{\ell} v^{\ell}}{a} + f v^{\ell-1} - \frac{1}{\rho_0 a \cos\phi} \delta_{\lambda}(\overline{p_h}^{\phi}) \\ &+ \mu \delta_z(\delta_z u) + A_M \left\{ \frac{1}{a^2 \cos^2\phi} \delta_{\lambda}(\delta_{\lambda} u) + \frac{1}{a^2 \cos\phi} \delta_{\phi}(\delta_{\phi} u \, \cos\phi) \right. \\ &\left. + \frac{1 - \tan^2\phi}{a^2} u - \frac{2 \sin\phi}{a^2 \cos^2\phi} \delta_{\lambda} v \right\} \end{aligned} \tag{39}$$

and

$$F^{\phi} = -L^{**}(v^{\ell}) - \frac{\tan\phi\, u^{\ell}u^{\ell}}{a} - fu^{\ell-1} - \frac{1}{\rho_0 a}\delta_{\phi}(\overline{p}_h^{\lambda})$$
$$+ \mu\delta_z(\delta_z v) + A_M\left\{\frac{1}{a^2\cos^2\phi}\delta_{\lambda}(\delta_{\lambda}v) + \frac{1}{a^2\cos\phi}\delta_{\phi}(\delta_{\phi}v\cos\phi)\right.$$
$$\left. + \frac{1-\tan^2\phi}{a^2}v + \frac{2\sin\phi}{a^2\cos^2\phi}\delta_{\lambda}u\right\}. \quad (40)$$

From the above, it follows that

$$(u^*, v^*) = \left(u^{\ell-1} + 2\Delta t\frac{F^{\lambda} + (2\Delta t\,\alpha f)F^{\phi}}{1+(2\Delta t\,\alpha f)^2},\ v^{\ell-1} + 2\Delta t\frac{F^{\phi} - (2\Delta t\,\alpha f)F^{\lambda}}{1+(2\Delta t\,\alpha f)^2}\right). \quad (41)$$

The internal mode velocities u' and v' are then obtained by equation (36).

To form a finite-difference equation analogous to equation (21) for the streamfunction, we first write down the difference equations for (1) and (2):

$$\frac{u^{\ell+1} - u^{\ell-1}}{2\Delta t} - f\alpha(v^{\ell+1} - v^{\ell-1}) = \frac{-1}{\rho_0 a\cos\phi}\delta_{\lambda}(\overline{p}_s^{\phi}) + F^{\lambda} \quad (42)$$

$$\frac{v^{\ell+1} - v^{\ell-1}}{2\Delta t} + f\alpha(u^{\ell+1} - u^{\ell-1}) = \frac{-1}{\rho_0 a}\delta_{\phi}(\overline{p}_s^{\lambda}) + F^{\phi}. \quad (43)$$

The terms F^{λ} and F^{ϕ} are as defined in equations (39) and (40). We then apply a sequence of finite-difference operations in exact analogy to the derivation of equation (21). Specifically, the following operators are used in turn:

(i) vertical averaging by

$$\frac{1}{H}\sum_{k=1}^{kz}(\cdot)\,\Delta k;$$

(ii) expression of averaged velocities in terms of ψ using equation (30);

(iii) the finite-difference curl, defined by

$$curl_z(q_1, q_2) = \frac{1}{a\cos\phi}\left[\delta_{\lambda}(\overline{q_2}^{\phi}) - \delta_{\phi}(\cos\phi\,\overline{q_1}^{\lambda})\right] \quad (44)$$

(iv) multiplication by $2\Delta t$ and setting $\psi^{\ell+1} - \psi^{\ell-1} = D$.

The resulting equation is

$$\frac{1}{a\cos\phi}\left\{\delta_\lambda\overline{\left[\frac{1}{Ha\cos\phi}(\delta_\lambda\overline{D}^\phi)\right]}^\phi+\delta_\phi\overline{\left[\frac{\cos\phi}{Ha}(\delta_\phi\overline{D}^\lambda)\right]}^\lambda\right\}$$

$$-\frac{2\Delta t\,\alpha}{a\cos\phi}\left\{\delta_\lambda\overline{\left[\frac{f}{Ha}(\delta_\phi\overline{D}^\lambda)\right]}^\phi-\delta_\phi\overline{\left[\frac{f}{Ha}(\delta_\lambda\overline{D}^\phi)\right]}^\lambda\right\} \tag{45}$$

$$=\frac{2\Delta t}{a\cos\phi}\left\{\delta_\lambda\overline{\left[\frac{1}{H}\sum_{k=1}^{kz}F^\phi\,\Delta k\right]}^\phi-\delta_\phi\overline{\left[\frac{1}{H}\sum_{k=1}^{kz}F^\lambda\,\Delta k\right]}^\lambda\right\}.$$

The first expression in the curly brackets above involves values of D at nine points. It is a generalization to the topographic case of the nine-point Laplacian developed by Haney [1971]. It differs from the five-point Laplacian used by Bryan [1969], which is essentially the following:

$$\frac{1}{a\cos\phi}\left\{\delta_\lambda\left[\frac{1}{\overline{H}^\phi a\cos\phi}\delta_\lambda D\right]+\delta_\phi\left[\frac{\cos\phi}{\overline{H}^\lambda a}\delta_\phi D\right]\right\}. \tag{46}$$

The nine-point formulation may be preferable to the five-point one because of the consistency with which all terms in equation (45) are obtained by operations on the basic equations (42) and (43). But in certain cases, the nine-point scheme may be more susceptible to computational noise [Takano, personal communication]. One may choose which of these expressions to use in a given problem.

The second expression in curly brackets in equation (45) is a generalized beta term. Although it appears to involve nine values of D, it actually involves only four, namely those defined immediately to the north, south, east, and west of a central point.

Regardless of which Laplacian operator is used, equation (45) specifies an algebraic relation at each gridpoint, involving a value D_0 defined at that point and n values $D_1,\ldots,D_n$ at surrounding points. The equation can be abbreviated in the form

$$\sum_{i=1}^{n}c_i(D_i-D_0)=E. \tag{47}$$

We can solve the system of such equations at all grid points by the method of successive over-relaxation. Given values $D_0, D_1,\ldots,D_n$, a new central value is estimated as

$$D_0^{new}=D_0+\frac{\delta}{(\sum_{i=1}^{n}c_i)}\left[\sum_{i=1}^{n}c_i(D_i-D_0)-E\right]. \tag{48}$$

In the above, δ is an over-relaxation parameter which speeds convergence. The over-relaxation technique is equivalent to solving a time-dependent heat equation by forward time stepping, and limitations on the size of δ are related to stability considerations involving the coefficients c_i. For the five-point Laplacian operator, the largest allowable value of δ depends on whether immediate updating within the array D is carried out ($\delta < 2$ if so, $\delta \leq 1$ if not). In the nine-point case, acceptable values of δ have been found empirically to lie below 1/2. In obtaining a solution of acceptable accuracy using over-relaxation, care must be taken in choosing a first guess and in carrying out sufficient iterations [Killworth and Smith, 1984].

It remains to discuss the computation of the streamfunction on islands. Rather than construct a finite difference version of equation (22) directly, we use an indirect approach which is based on a finite-difference form of Stokes' theorem. This theorem applies to any area A covered by a collection of grid boxes and having a perimeter P of box edges. If arbitrary values of two fields q_1 and q_2 are defined at the corners of the grid boxes, the following can be shown to hold

$$\begin{aligned}\sum_A\sum \frac{1}{a\cos\phi}[\delta_\lambda(\overline{q_2}^{\phi}) - \delta_\phi(\cos\phi\,\overline{q_1}^{\lambda})]a^2\cos\phi\,\Delta\lambda\,\Delta\phi \\ = \sum_P(\overline{q_1}^{\lambda}\,a\cos\phi\,\Delta\lambda + \overline{q_2}^{\phi}\,a\Delta\phi)\,.\end{aligned} \tag{49}$$

To compute the value of the island streamfunction, a line integral of the vertically averaged momentum equations is required. The curl of those equations is already available in equation (45). By virtue of the Stokes theorem above, we can equivalently take the area sum of equation (45). (We can arbitrarily set the values of the vertically averaged velocity and the vertically averaged forcing to be zero at the interior corners of boxes; then the area sum will pick up nonzero contributions only from boxes on the margin of the area.) The resulting area sum gives an algebraic relation between the value of D for an island and all the values of D immediately surrounding the island. This relation can be solved simultaneously with equation (45) at each point in the ocean. The successive over-relaxation procedure (48) is essentially the same as before, except that n is now the number of points immediately adjacent to an island.

A remark should be made about the method of convective adjustment. Since virtually nothing is known about oceanic convection, a simple ad hoc method can be used. At each time step, each layer of water is given the opportunity to mix with water immediately above and/or below it. Mixing of two layers is carried out only when the upper layer is found upon adiabatic displacement to the lower layer to have a greater density than the lower layer.

Another remark concerns the time differencing. Leapfrog differencing in time tends to cause "splitting" of results at alternate time steps, which must be remedied in some way. One can use the method of Bryan [1969], whereby a forward time step of length Δt is taken periodically to eliminate one branch of the solutions. However, an Euler-backward time step is preferable, since less high-frequency noise is excited than in the forward case.

A final remark about time stepping is in order. A model can be programmed to allow timesteps of different lengths on the baroclinic, barotropic, and density equations. This artificial device has been found useful in speeding convergence to a final state in steadily forced, viscous oceans. Computer time is reduced by taking a longer time step for the slowly adjusting fields of temperature and salinity. A discussion of this method and some justification for its use in certain nonsteady problems is given in Killworth et al. [1984].

4. ENERGETIC CONSISTENCY

The differencing scheme of Bryan [1969] is designed to conserve global integrals of a number of linear and quadratic quantities, along the general lines outlined by Arakawa [1966]. In particular, volume integrals of the following quantities are preserved by the spatial differencing of the advective processes: mass, heat, salt, variance of temperature, variance of salinity, and total energy. It should be noted that the momentum advection operator L^{**} does not preserve enstrophy (mean square vorticity). The conservation of this quantity may be important for high resolution ocean studies, and it can be included following the approach of Arakawa [1972].

Conservation of mass follows from the finite difference version of the continuity equation, which is used to obtain values of w from u and v. We recapitulate an argument of Bryan [1969] to show that heat, salt, and the variances of temperature and salinity are conserved. Let q be either T or S, and have values $q_1, q_2, \dots q_N$ at centers of grid boxes with volumes $\alpha_1, \dots, \alpha_N$ respectively. We wish to show that the finite difference analogue of the advection equation,

$$\frac{\partial q}{\partial t} + \nabla \cdot (\vec{v} q) = 0 \tag{50}$$

is such that the quantities

$$\sum_{n=1}^{N} \alpha_n q_n \qquad \text{and} \qquad \sum_{n=1}^{N} \alpha_n q_n^2$$

are preserved in time. The spatially finite differenced form of the above equation in the model can be written as

$$\alpha_n \frac{dq_n}{dt} + \sum_{i=1}^{6} A_n^i V_n^i \left(\frac{q_n + q_n^i}{2} \right) = 0 \tag{51}$$

where A_n^i is the area of the ith face of the nth grid box, V_n^i is a normal velocity defined at the center of that face, and q_n^i is the value at the center of the grid box adjacent to that face. The time change of the volume integral of q is

$$\frac{d}{dt} \sum_{n=1}^{N} \alpha_n q_n = \sum_{n=1}^{N} \alpha_n \frac{dq_n}{dt} = - \sum_{n=1}^{N} \sum_{i=1}^{6} A_n^i V_n^i \left(\frac{q_n + q_n^i}{2} \right) . \tag{52}$$

We note that contributions associated with a common face between two boxes will cancel out because of the essential "symmetry" of $\frac{q_n+q_n^i}{2}$ and the essential "antisymmetry" of V_n^i. Also, at lateral boundaries the normal velocity is zero. Consequently,

$$\frac{d}{dt}\sum_{n=1}^{N}\alpha_n q_n = 0\,. \tag{53}$$

The time change of the volume integral of q^2 is

$$\begin{aligned}\frac{d}{dt}\sum_{n=1}^{N}\alpha_n q_n^2 &= 2\sum_{n=1}^{N}\alpha_n q_n \frac{dq_n}{dt}\\ &= -\sum_{n=1}^{N} q_n^2\left(\sum_{i=1}^{6} A_n^i V_n^i\right) - \sum_{n=1}^{N}\sum_{i=1}^{6} A_n^i V_n^i q_n q_n^i\,.\end{aligned} \tag{54}$$

The first term vanishes by continuity of mass in each box, and the second term vanishes by the "symmetry" of $q_n q_n^i$ and "antisymmetry" of V_n^i, with $V_n^i = 0$ on boundaries.

It should be noted that equation (52) will have nonzero V_n^i at lateral boundaries of the lattice of momentum gridpoints when applied to $q = u$ and $q = v$. Thus momentum is not conserved. Experience has shown that few problems arise from this in highly geostrophic systems.

It remains to show that total energy is conserved in the model. Equation (52) can be applied to $q = u$ and $q = v$ successively to obtain total interior conservation of $u^2 + v^2$ by advection. (Note that at lateral boundaries a no-slip condition $q_n^i = 0$ is needed, since the corresponding V_n^i may be nonzero.) Also the metric and Coriolis terms cannot affect kinetic energy (when they are treated explicitly). Total energy will be conserved (apart from frictional effects) if any change in the total kinetic energy through work by the horizontal pressure gradient forces is accompanied by an equal and opposite change in the total potential energy. The volume integral of the rate of work by the meridional component of the pressure gradient force is given by

$$\begin{aligned}&\sum_{ijk} v_{ij}\frac{1}{a}\delta_\phi \overline{p}^\lambda a^2 \cos\phi_j\,\Delta\lambda\,\Delta\phi\,\Delta k\\ &= \frac{a\Delta\lambda}{2}\sum_{ijk} v_{ij}\cos\phi_j(p_{i+1j+1} + p_{ij+1} - p_{i+ij} - p_{ij})\,\Delta k\,.\end{aligned}$$

We can use the fact that $v = 0$ at the margins of the grid to rewrite the above as

$$-\frac{a\Delta\lambda}{2}\sum_{ijk} p_{ij}[(v_{ij} + v_{i-1j})\cos\phi_j - (v_{ij-1} + v_{i-1j-1})\cos\phi_{j-1}]\,\Delta k\,.$$

This is simply

$$-\sum_{ijk} p_{ij} \frac{1}{a \cos \phi_j} \delta_\phi (\cos \phi \, \overline{v}^\lambda) a^2 \cos \phi_j \, \Delta\lambda \, \Delta\phi \, \Delta k \, .$$

A similar relation can be derived for the zonal component of the pressure gradient force. By combining the two relations and using the finite difference form of the continuity equation, one obtains that the total rate of work by the horizontal pressure gradient forces is

$$\sum_{ijk} p_{ijk} \frac{w_{ijk} - w_{ijk+1}}{\Delta k} a^2 \cos \phi_j \, \Delta\lambda \, \Delta\phi \, \Delta k \, .$$

We now note that

$$\begin{aligned} &\sum_{k=1}^{kz} p_k \left(\frac{w_k - w_{k+1}}{\Delta k} \right) \Delta k \\ &= \sum_{k=1}^{kz} p_k w_k - \sum_{k=2}^{kz+1} p_{k-1} w_k \\ &= -\sum_{k=2}^{kz} w_k \left(\frac{p_{k-1} - p_k}{\Delta' k} \right) \Delta' k \end{aligned}$$

where the last equality follows from the fact that $w_1 = w_{kz+1} = 0$ at T, S points. Thus the rate of work by the horizontal pressure gradient forces is

$$-\sum_{ijk} w_{ijk} \frac{p_{ijk-1} - p_{ijk}}{\Delta' k} a^2 \cos \phi_j \, \Delta\lambda \, \Delta\phi \, \Delta' k$$

but this is simply the rate of change of potential energy due to buoyancy forces.

We note that no fictitious energy is generated in nonlinear exchanges between the internal and external modes of motion. The finite difference equations (42) and (43) are energy conserving as far as advection is concerned. These equations are satisfied by $(u, v) = (\overline{u} + u', \overline{v} + v')$. Any change of energy of the internal mode due to nonlinear exchanges must therefore be compensated by a similar but opposite change in external mode energy. Thus the model formulation is fully energy conserving apart from prescribed physical dissipation and small time-truncation error.

Acknowledgments. The author wishes to thank Ms. Eileen Boettner for assistance in preparing the manuscript. The National Center for Atmospheric Research is sponsored by the National Science Foundation.

REFERENCES

Arakawa, A., 1966: Computational design for long-term numerical integration of the equations of fluid motion. *J. Comput. Phys.*, **1**, 119–143.

Arakawa, A., 1972: Design of the UCLA general circulation model. *Numerical Simulation of Weather and Climate*, Tech. Rept. No. 7, Department of Meteorology, University of California, Los Angeles, 116 pp.

Bryan, K., 1969: A numerical method for the study of the circulation of the world ocean. *J. Comput. Phys.*, **4**, 347–376.

Bryan, K., and M.D. Cox, 1972: An approximate equation of state for numerical models of ocean circulation. *J. Phys. Oceanogr.*, **2**, 510–514.

Cox, M.D., 1984: A primitive equation three-dimensional model of the ocean. *GFDL Ocean Group Tech. Rept. No. 1*, GFDL/NOAA, Princeton University, Princeton, 250 pp.

Friedrich, H., and S. Levitus, 1972: An approximation to the equation of state for sea water, suitable numerical ocean models. *J. Phys. Oceanogr.*, **2**, 514–517.

Haney, R.L., 1971: "A numerical study of the large-scale response of an ocean circulation to surface heat and momentum flux." UCLA Ph.D. Thesis, 191 pp.

Killworth, P.D., J.M. Smith, and A.E. Gill, 1984: Speeding up ocean circulation models. *Ocean Modelling*, No. 56, 1–4.

Killworth, P.D., and J.M. Smith, 1984: Gradual instability of relaxation extrapolation schemes. *Dyn. Atmos. Oceans*, **8**, 185–213.

Semtner, A.J., 1974: An oceanic general circulation model with bottom topography. *Numerical Simulation of Weather and Climate*, Tech. Rept. No. 9, Department of Meteorology, University of California, Los Angeles, 99 pp.

Semtner, A.J., 1986: History and methodology of modelling the circulation of the world ocean. *Proceedings of the NATO Advanced Study Institute on Advanced Physical Oceanographic Numerical Modelling*, D. Reidel Publishing Co., Dordrecht.

Takano, K., 1974: A general circulation model for the world ocean. *Numerical Simulation of Weather and Climate*, Tech. Rept. No. 8, Department of Meteorology, University of California, Los Angeles, 47 pp.

QUASIGEOSTROPHIC MODELLING OF EDDY-RESOLVED OCEAN CIRCULATION

William R. Holland
National Center for Atmospheric Research
Boulder Colorado, 80307-3000
USA

ABSTRACT. A discussion is given of the development and application of quasi-geostrophic models of ocean circulation. Several applications are shown, including simple basin models, North Atlantic models, and several limited area models that illustrate the flexibility and power of QG eddy-resolving models for understanding ocean dynamics.

1. Introduction

Numerical modelling of large scale ocean circulation, including mesoscale eddies as an integral aspect of the oceanic general circulation, involves many compromises. These compromises include aspects of both the physical and numerical choices needed to actually carry out a three-dimensional, time-dependent calculation of the evolving, often turbulent, fluid motion.

One choice in the possible array of "physics" is the quasigeostrophic (henceforth QG) model. As suggested by the emphasis in the above paragraph on the word compromise (probably the key word in reference to oceanic numerical modelling!), there are both advantages and disadvantages to the use of such models and indeed all models. While QG models make basic assumptions that limit the applicability of the model to certain regions or features in the ocean (assumptions such as small Rossby number, small topographic relief, linearization around a constant static stability), these models also have a numerical efficiency and a physical simplicity that allow very large, high resolution numerical experiments to be run. Moreover, the relatively simple physics (compared to primitive equation models) allows an in-depth understanding of the eddy-mean flow interactions that are fundamental to understanding the general circulation of the ocean. In the end, it will be necessary to make use of a wide variety of models with different types of physics and numerics, resolution and forcing, from the most simple to "complete" models of the World Ocean, in order to unravel the host of complex questions facing the physical oceanographer.

J. J. O'Brien (ed.), Advanced Physical Oceanographic Numerical Modelling, 203–231.

2. A Historical (But Limited and Personal) Perspective

In the early 1970's, I set out to do some three-dimensional tracer problems, using the primitive equation model developed at GFDL. Even at that time water mass property distributions, including passive tracers, were the topic of the day, since it was well understood that such studies were needed to explain deep water formation and thermocline penetration problems (the WOCE problem). A 19-layer, three degree horizontal resolution model was used [Holland, 1971] to look at "Ocean Tracer Distributions: Part I, Preliminary Numerical Experiments." I have yet to do Part II ! The horizontal grid size was just too big to enable us to choose subgrid scale diffusivities for heat, salt and momentum (as well as tracers) that were reasonable. The oceans were diffusively dominated and not very realistic. Even a crude comparison between the model results and ocean property distributions showed the inadequacy of the model resolution. This was basically due to the lack at that time of computing power needed to reduce the grid size so as to adequately model the three-dimensional physics of even a single ocean basin. Even these simple calculations took hundreds of computer hours on the "supercomputer" of the day, a UNIVAC 1108 (approximately 100 times slower than the present CRAY-1).

It became apparent in the early 70's that (i) eddies were important in the ocean (the catch word is "ubiquitous") and (ii) numerical models with the kinds of resolution I thought I would need for tracer studies would become turbulent. Thus the idea arose that we should no longer try to parameterize (perhaps badly) the effects of eddies but let them take on the their rightful role of fluxing momentum, heat and salt, as well as passive tracers, by making explicit diffusivities quite small in numerical calculations. This would require very fine horizontal resolutions to resolve the mesoscale eddy fields that would spontaneously occur due to instability processes. If we could take this to the limit (and our models had the right physics in other respects), then the general circulation of the ocean would become a statistical concept. Not only would $\overline{u}$, $\overline{T}$, $\overline{S}$ be needed to give a complete picture of the ocean, but also $\overline{u'T'}$, $\overline{u'v'}$, eddy kinetic energy, and so forth, that is the relevant eddy statistics, would be needed.

The obvious importance of mesoscale transience led in the early 70's to the development of the Mid-Ocean Dynamics Experiment and later in the decade to the POLYMODE observational program. As part of these programs, designed by oceanographers to understand the role of eddies in the oceanic general circulation, models of eddy resolved ocean circulation were developed. The first study of this kind, by Liang B. Lin and myself, was carried out in 1973-1974 making use of a small basin, primitive equation model developed for the purpose [Holland and Lin, 1975a,b]. This model was a two-layer, adiabatic, rigid lid model driven by wind forcing alone. It was the closest model we could think of to provide a baroclinic extension of the work done by Veronis [1966a,b] for a barotropic ocean.

Veronis' calculations were steady while the Holland-Lin model was strongly transient. Mesoscale eddies, due to baroclinic instability, grew to finite amplitude and fundamentally altered the mean flow in the baroclinic case. In calculations with multiple gyres, the free eastward jet also became unstable, and barotropic instability processes also played a role. Figure 1 shows one such calculation. Eddies are a vigorous part of the final statistical equilibrium, even though the wind forcing is steady. In any case, these numerical experiments were still quite costly in computer resources due to the presence of internal gravity waves and we had to use small basins. Typically, with a 20 km grid necessary to resolve the eddies, a 2000 km square basin was the best we could do. This led us in the next few years to develop models (QG) that filtered out the gravity waves, thus allowing a speed-up of a factor of ten in doing such fine resolution calculations.

3. Quasigeostrophic Model Equations

The so-called primitive equations have been presented in great detail elsewhere in this volume (see Semtner's Chapter). Here we shall briefly discuss the assumptions involved in deriving the quasigeostrophic equations that are an approximation to the complete set. Since the subject has a long history, going back to Charney's [1949] pioneering work and to Phillips' [1951] construction of an atmospheric numerical model, we shall only sketch the broad outlines here. The reader is referred to the excellent discussion in Pedlosky's [1979] book and to the references therein for a complete history of the subject.

Following Pedlosky [1979], we shall make use of the shallow water equations to illustrate the process of derivation of the QG analogue of those equations. The shallow water equations are

$$\begin{aligned} u_t + uu_x + vu_y - fv &= -g\eta_x \\ v_t + uv_x + vv_y + fu &= -g\eta_y \\ h_t + (uh)_x + (vh)_y &= 0 \end{aligned} \tag{1}$$

where the Coriolis parameter f = constant, η is the surface height perturbation, and h is the net thickness of the fluid layer. If D is the constant mean thickness of the fluid layer and h_B is the variable bottom topography perturbation, then

$$h = D + \eta - h_B. \tag{2}$$

We focus our attention on time scales which are long compared to f^{-1}, and choose scales that characterize the magnitudes of the variables:

$$\begin{aligned} x, y &= L(x', y') \\ t &= Tt' \\ u, v &= U(u', v') \\ \eta &= N_o\eta'. \end{aligned} \tag{3}$$

Plugging these into (1) and (2), and making the assumption that

$$\epsilon = \frac{U}{fL} << 1$$

we find that the nonlinear acceleration terms in (1) are small compared to the Coriolis terms, and that

$$N_o = \frac{fUL}{g} = \epsilon \frac{f^2 L^2}{g}$$

Also

$$h = D\left[1 + \epsilon \frac{L^2}{R^2}\eta' - \frac{h_B}{D}\right] \quad (4)$$

where $R = (gD)^{1/2}/f$, the Rossby radius of deformation. If $L/R \sim 1$ or less, the departure of the layer thickness from its value in the absence of motion is $0(\epsilon)$. Then the equations (1)-(2) can be written

$$\begin{aligned} &\epsilon u_t + \epsilon(uu_x + vu_y) - v = -\eta_x \\ &\epsilon v_t + \epsilon(uv_x + vv_y) + u = -\eta_y \\ &\epsilon F\eta_t + \epsilon F(u\eta_x + v\eta_y) - u(\tfrac{h_B}{D})_x - v(\tfrac{h_B}{D})_y + (1 + \epsilon F\eta - \tfrac{h_B}{D})(u_x + v_y) = 0. \end{aligned} \quad (5)$$

Here $F = (f^2L^2/gD) = (L/R)^2$ is assumed to be order one. Since $\epsilon << 1$, we expand in an asymptotic series $u = u_o + \epsilon u_1 + ...$, and balance the various powers of ϵ to give

$$v_o = \eta_{ox}, \quad u_o = -\eta_{oy}, \quad u_{ox} + u_{oy} = 0, \quad (6)$$

i.e., the lowest order fields are geostrophic and nondivergent. Now, if we assume $h_B/D \sim \epsilon$ (that is $h_B/D = \epsilon\eta_B$ where $\eta_B \sim 0(1)$), then the order ϵ equations are

$$\begin{aligned} &u_{ot} + u_o u_{ox} - v_o u_{oy} - v_1 = -\eta_{1x} \\ &v_{ot} + u_o v_{ox} + v_o v_{oy} + u_1 = -\eta_{1y} \\ &F\{\eta_{ot} + u_o\eta_{ox} + v_o\eta_{oy}\} - u_o\eta_{Bx} - v_o\eta_{By} + (u_{1x} + v_{1y}) = 0. \end{aligned} \quad (7)$$

The order zero fields are solved for by forming the vorticity equation

$$\frac{d\zeta_o}{dt} = \frac{\partial \zeta_o}{\partial t} + u_o\zeta_{ox} + v_o\zeta_{oy} = -(u_{1x} + v_{1y})$$

where

$$\zeta_o = v_{ox} - u_{oy} = \nabla^2 \eta_o.$$

Then

$$\frac{d}{dt}(\nabla^2 \eta_o - F\eta_o + \eta_B) = 0. \tag{8}$$

This is a statement that potential vorticity is conserved.

Note that producing a valid quasigeostrophic approximation has required several assumptions. These are

$\epsilon << 1$: small Rossby number

$\frac{h_B}{D} << 1$: slope of bottom cannot be large

$\frac{\eta}{D} << 1$: slope of perturbed density surfaces must be small

(this is especially important for 3 – D cases to be discussed below).

The disadvantages of the QG model then are: we may have gotten rid of "physics" we would like to keep (thermally forced cases not easily treated, gravity wave response is lost, equatorial problems cannot be done since $f \rightarrow 0$); isopycnal surfaces cannot be allowed to substantially deviate from level surfaces, i.e., surfacing is not allowed; and the Rossby number must be small. The advantages of the QG model are that the computations are quite efficient (because there are no gravity waves), the dynamics are relatively simple (important for understanding), and adiabatic problems (no heat diffusion) can be treated.

These results for the shallow water equations can be extended to the full three-dimensional primitive equations in spherical geometry [Pedlosky, 1979, pp. 340-345]. Then the statement of the conservation of potential vorticity becomes

$$\frac{D}{Dt}\left[\nabla^2\psi + \frac{\partial}{\partial z}\left(\frac{1}{S}\frac{\partial\psi}{\partial z}\right) + \beta y\right] = W \tag{9}$$

where ψ is three dimensional quasigeostrophic streamfunction, W represents the nonconservative effects of forcing and friction, $S = S(z)$ is a static stability parameter determined by the mean background stratification, and β is the local rate of change of the Coriolis parameter f. This equation forms the basis for the QG model developed by Holland [1978] to study eddy-resolved ocean circulation problems of the kind begun earlier with the Holland-Lin primitive equation model.

The quasigeostrophic model formulation with N arbitrary layers is a straightforward extension of the two-layer case described by Holland [1978]. Here we shall present the form of the equations in which the vertical discretization has already been done. The horizontal discretization and the form of the finite difference equations will not be discussed here.

The governing equations are the vorticity and interface height perturbation equations and the thermal wind relation:

$$\frac{\partial}{\partial t} \nabla^2 \psi_k = J(f + \nabla^2 \psi_k, \psi_k) + \frac{f_o}{H_k}(w_{k-1/2} - w_{k+1/2}) + F_k + T_k \quad : k = 1 \text{ to } N$$

$$\frac{\partial}{\partial t} h_{k+1/2} = J(h_{k+1/2}, \psi_{k+1/2}) + w_{k+1/2} \quad : k = 1 \; to \; N-1 \tag{10}$$

$$h_{k+1/2} = \frac{f_o}{g'_{k+1/2}}(\psi_{k+1} - \psi_k).$$

Whole number subscripts (k) denote the vertical layers (k increasing downward) in which the quasigeostrophic streamfunction is defined (nominally at the center of each of the layers) while fractional subscripts (k+1/2) denote the interfaces between layers where vertical velocity and interface height perturbations are defined (Figure 2). The variables are the quasigeostrophic streamfunction (ψ_k) with horizontal velocity components ($u = -\psi_y, v = \psi_x$), the interface height perturbation ($h_{k+1/2}$), positive upward, and the vertical velocity ($w_{k+1/2}$), also positive upward. The horizontal coordinates are x (eastward) and y (northward), the Coriolis parameter is $f = f_o + \beta y$, and the mean layer thicknesses are H_k. The values of f_o and β are chosen to represent typical midlatitude gyre values. The basic background vertical stratification is written in terms of the reduced gravity $g' = g\Delta\rho_{k+1/2}/\rho_o$, where $\Delta\rho_{k+1/2}$ is the (positive) density difference between layers k+1 and k. Frictional effects, written symbolically in Equation (1) as F_k, have been parameterized in different ways in various calculations—as lateral friction of the Laplacian or biharmonic kind [Holland, 1978], in which $F_k = A_2 \nabla^4 \psi_k$ or $F_k = -A_4 \nabla^6 \psi_k$ respectively. In addition, in many experiments, F_k includes a bottom friction, $-\varepsilon \nabla^2 \psi_N$, when $k = N$ (the bottom layer). Note that the effect of the wind forcing T_1, equal to curlτ/H_1, produces an Ekman pumping stretching tendency in the upper layer that is equivalent to a body force acting on the upper layer. The T_k for $k > 1$ are zero. Also note that a gentle bottom slope can be consistently included in the QG framework so that the influence of variable depth can be examined, at least within the limitations of quasigeostrophy. This is taken care of by the bottom boundary conditions on w, i.e., $w_{N+1/2} = J(\psi_N, H_B)$, where $H_B(x, y)$ is the variable bottom topography (positive upward). At the sea surface, $w_{1/2} = 0$. The advective velocities at the interfaces, needed in Equation (10), are calculated from a weighted average of the velocities in the layers, i.e., $\psi_{k+1/2} = \alpha_{k+1/2}\psi_{k+1} + (1 - \alpha_{k+1/2})\psi_k$, where $\alpha_{k+1/2} = H_k/(H_{k+1})$.

These equations can be written in potential vorticity form

$$\frac{DQ_k}{Dt} = T_k + F_k \tag{11}$$

where

$$Q_k = \nabla^2 \psi_k + f + \frac{f_o}{H_k}(h_{k+1/2} - h_{k-1/2}).$$

For consistency in this equation, $h_{1/2} = 0$ and $h_{N+1/2} = H_B$, the variable bottom topography. Potential vorticity is conserved except for wind forcing and frictional effects.

Note that, although not included here, thermal forcing can be easily added (within the restrictions of QG physics) to the above model. Also note that auxilliary momentum conditions are needed in multiply-connected geometry [McWilliams, 1977]. The model numerics used in the following examples make use of finite differences in space and time, include conservation properties for energy, potential vorticity, and potential enstrophy (Q^2), and allow a timestep of several hours (a factor of 10 greater than that used by Holland and Lin).

4. Applications #1 - Simple Basins

A large number of studies using QG models have now been carried out in simple rectangular basins with steady wind forcing. These have provided valuable insight into a number of key problems in ocean circulation theory. These include (a) the role of eddies in the oceanic general circulation [Holland, 1978]; b) the nature of the instability processes going on in and near ocean jets [Haidvogel and Holland, 1978; Holland and Haidvogel, 1980]; (c) a comparison of QG and primitive equation model results [Semtner and Holland, 1978]; (d) the nature of potential vorticity balances and the implications for gyre scale equilibration [Harrison and Holland, 1981; Rhines and Holland, 1979; Holland and Rhines, 1980; Holland et al., 1984]; (e) the nature of Gulf Stream penetration into the ocean interior [Holland and Schmitz, 1985]; and (f) thermally forced QG models [Rhines et al., 1985]. In addition, a number of more general discussions have been published [Holland et al., 1983; Holland, 1985] and a Technical Note describing the detailed numerics of the QG model is presently being written by J.C. Chow and W.R. Holland.

Given this large quantity of published literature, here we will present only a few examples of recent calculations. The reader is referred to the above literature for a more complete perspective on QG ocean modelling. In addition there are numerous other works that either make use of or discuss results from QG ocean models.

Let us examine some recent results from an eight-layer model calculation in a simple rectangular domain. The model ocean is taken as 3600 km east-west, 2800 km north-south, and 5000 meters deep. The horizontal resolution is 20 km

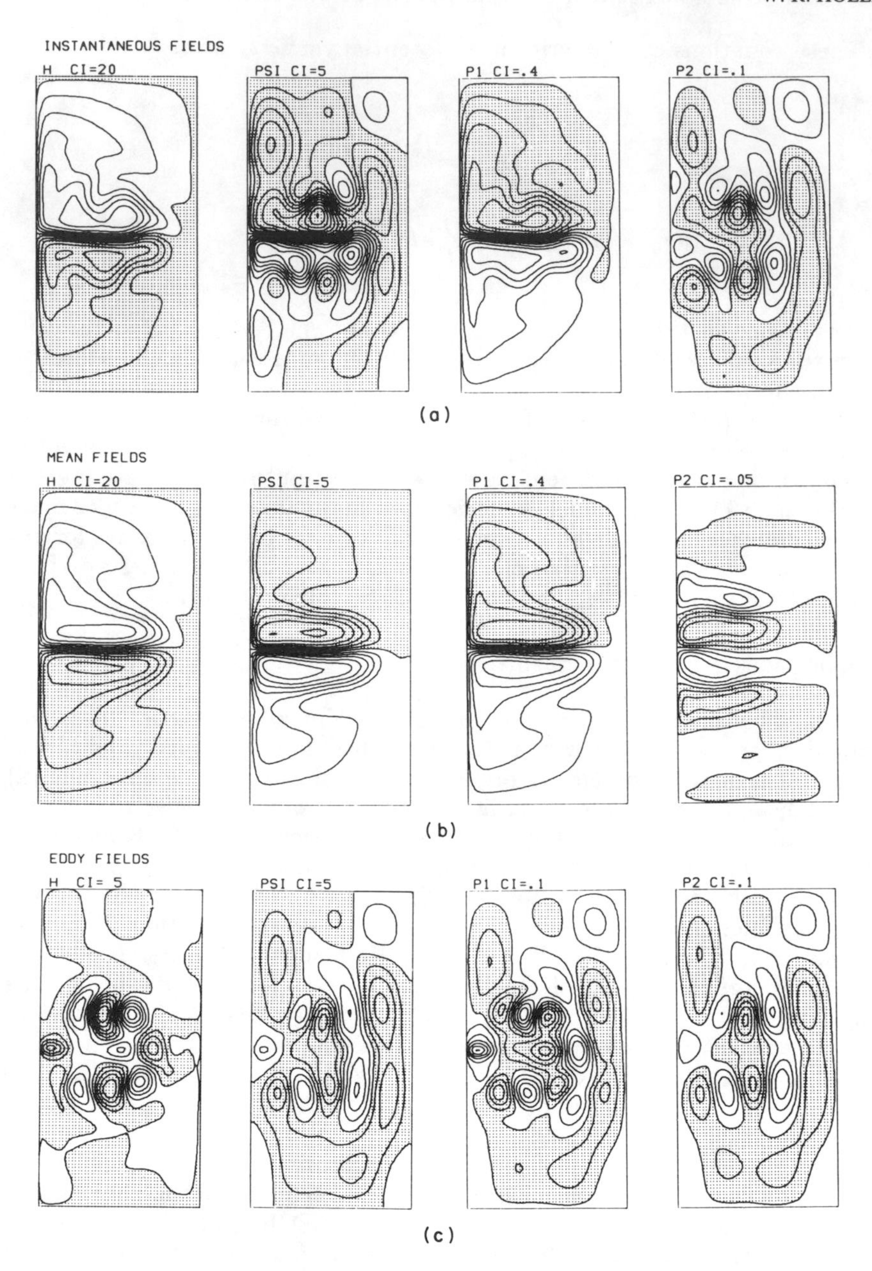

Figure 1. The patterns of motion in a double gyre, primitive equation ocean model [Holland and Lin, 1975]: the interface heights, streamfunctions, layer one pressures, and layer two pressures for (a) instantaneous, (b) mean, and (c) eddy fields.

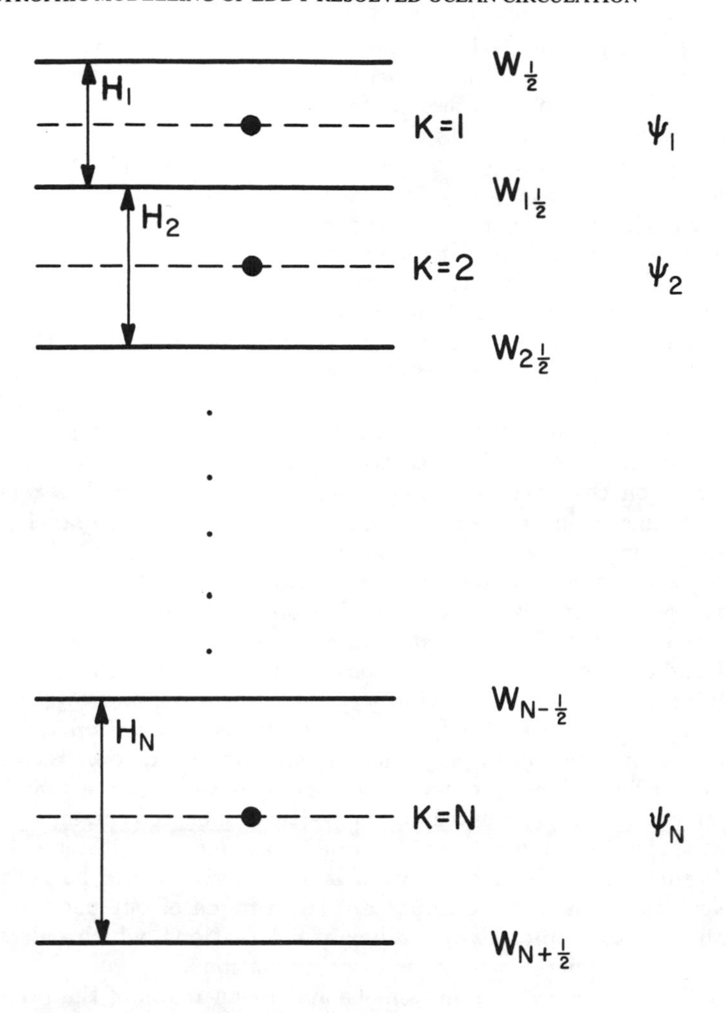

Figure 2. Schematic diagram of the vertical structure of a multilayer QG model.

and the vertical depth range is divided into eight layers of varying thickness, changing from 300 m near the surface to 1300 m at depth. The vertical stratification is exponential, with parameters in approximate agreement with stratification in the North Atlantic subtropical gyre.

Figure 3 shows the QG streamfunctions at three levels (of the eight) at two different instants in time (a = day 1860, b = day 1920). The levels are at 150 m (ψ_1), 850 m (ψ_3), and 1750 m (ψ_5) respectively. Note the strong meandering processes that lead to Gulf Stream Ring production. At day 1860 a warm core ring is just breaking off from the Gulf Stream. Sixty days later, that same ring has moved westward several hundred kilometers. Note also the nature of the gyre structure apparent in these plots. At the surface (top panels), a strongly transient Sverdrup gyre is apparent. Instabilities in the Gulf Stream and also on the flanks of the gyres give rise to an important eddy field throughout the basin. At mid-depths (middle panel), the basin scale of flow has been replaced by a tight recirculation in the vicinity of the eastward jet. Eddies are most energetic near the stream and on the north and south flanks of the gyre with a relative minimum in eddy energy in between. At greater depths (bottom panels), only the eddy signal is apparent; the mean flow is very nearly obscured by the eddy field.

Figure 4 shows the time averaged streamfunctions for these same fields, and Figure 5 shows the time averaged perturbation height fields (the equivalent of temperature) at depths of 300 m, 1050 m, and 2000 m. The vertical structure of these wind and eddy driven gyres is apparent as shown by the streamfunction in the subtropical gyre. The broad wind gyre at the surface narrows rapidly with depth, such that the southern edge of the westward circulation is further and further north with depth until only the very narrow, barotropic recirculation is left. The height field gives a somewhat different picture. (Note that the height field is proportional to ψ_z, related to the vertical <u>shear</u> of currents.) The gyre structure as shown by the height field is much broader at depth, suggesting that while the barotropic mode of circulation is quite narrow, the baroclinic modes are not nearly so. This may be important since much of our perspective about gyre structure is based upon dynamic height calculations, which reflect only the baroclinic modes associated with density perturbations.

Finally, Figure 6 shows instantaneous and mean maps of the potential vorticity Q at a middle depth in the ocean (850 m, level 3). As discussed earlier (and see Holland et al. [1984]; Rhines and Young [1982]) this quantity is quasi-conservative, especially at mid-depth where very weak laterial friction is the only non-conservative term in the potential vorticity equation. Vigorous eddy mixing (as is apparent in Figure 6a) leads to a homogenization of potential vorticity over vast regions of the ocean gyres, a process that is reflected in many regions of the ocean [Holland et al., 1984].

In addition to the above theoretical studies, comparisons of model results with observations have played an important part in model refinement and in identifying important issues regarding the physics of the Gulf Stream System [Schmitz and Holland, 1982; Schmitz et al., 1982; Holland, 1985]. This work is currently being extended with models of much higher vertical resolution than heretofore to examine the vertical structure of mean and eddy fields in the Gulf Stream and Kuroshio. As illustrations of this kind of comparison, Figures 7 through 11 show several observational/model intercomparisons that are currently being examined. Figures 7 and 8 show meridional sections of mean zonal flow and eddy kinetic energy in the eight layer numerical experiment shown above; Figures 9 and 10 show similar sections [Richardson 1983, 1985] of mean zonal flow and eddy kinetic energy based upon a whole host of observations along 55W crossing the Gulf Stream. The correspondence is by no means exact but it is clear that vertical/horizontal structure in the numerical experiment has many features in common with the data in both mean and eddy quantities, including approximately correct ratios of surface to deep mean and eddy currents in the intense flow, and similar meridional structure in terms of eastward and westward (recirculating) mean flows. Figure 11 shows an additional comparison between the data at 55W (in the Gulf Stream) and a similar point in the intense eastward flow of another eight layer model calculation (with somewhat different parameters than the previous one). In this case the vertical structure of eddy kinetic energy is found to be remarkably similar between observations and model, suggesting strongly that we are on the right track regarding the eddy/mean flow interactions that give rise to the intense eddy field in the western North Atlantic.

5. Applications #2 - North Atlantic Basin

To date, most numerical studies of eddy resolved ocean circulation have been highly idealized with respect to the geometry of the ocean basin in question. There are many good reasons for this. For one thing, simpler situations are a vital and necessary part of understanding the more complex ones. When confronting questions about detailed comparisons with observations, however, one must ask whether or where dynamically similar regions of an idealized basin can be found in the western North Atlantic. Or, turning the question around, where in an idealized basin should one seek to compare observations along 55W (or any other place)? Thus, as models more faithfully reproduce observed features, we are driven toward more faithful inclusion in our models of basin shape, bottom topography, and boundary conditions (e.g., wind stress, buoyancy flux, inflow and outflow across the boundaries of local domains, etc.). Such models, with various physical and geometrical factors successively put in or taken out, allow us to ascertain which features are key to understanding the dynamics and which are not.

In the past several years, oceanographers have begun to develop models of the North Atlantic (and other basins) that have somewhat "realistic" geometry. In the model #1 results shown below [Holland, 1983], the horizontal resolution is 1/4° of latitude and longitude, the depth is constant, and the eastern boundary is simplified to a straight north-south coastline. Figure 12 shows the mean $\overline{\psi_i}$ for a three-layer model driven by the mean annual Bunker wind stress. Figure 13 shows the patterns of eddy kinetic energy; and Figure 14 shows the time averaged potential vorticity in the middle layer. Finally, Figure 15 shows a time sequence of instantaneous upper layer streamfunction to show something of the time dependence.

These experiments have a realistic flavor to them. Gulf Stream meandering produces warm and cold core rings; mesoscale eddy energy has about the right amplitude and structure; the mean flows have about the right strength. Careful and thorough comparisons with observations are still underway but we are on the verge of having a true general circulation model of the North Atlantic basin in which we can test our ideas about realistic oceanic flows.

The experiments above have a major discrepancy associated with them; the regions far from the Gulf Stream and the southern flank of the gyre, particularly the eastern half of the basin, have too little eddy energy. The strong energy regions to the west do not seem to effectively radiate eastward to account for eastern basin energy levels. This suggests that transient forcing, not present in the above calculations, may be necessary.

Figures 16 and 17 show calculations in a new North Atlantic model (#2 with eastern boundary geometry and transient wind forcing now included). Figure 16 shows the time average layer streamfunction when only mean wind forcing is included. In Figure 17 the model is forced only by the transient wind component; the mean forcing is omitted (temporarily) to clarify the nature of the transient response itself. Hellerman's monthly mean winds (annual average removed) are used for this purpose.

The results in Figure 17 show a strong baroclinic response at the annual period (the forcing has one- to twelve-month periods). First baroclinic mode Rossby waves with an east-west wave length of about 500 km and a much longer north-south wave length, propagate westward from the eastern boundary. The main thermocline moves up and down (with nearly annual frequency) by about 5 to 10 meters and associated "eddy" currents are a few cm/sec in amplitude. Note that the shape of the eastern boundary of the basin is evident in the westward propagating wave field almost all the way to the western boundary! While these eddy signals are considerably less strong than those near the Gulf Stream (when mean wind forcing is also included) they do constitute the major signal in the eastern regions.

At this point considerable effort is needed to understand these various competing energy sources as well as to examine more complex models with actual bottom topography and even more realistic wind forcing. We are, however, well on our way toward simulations of the geographical structure of variability in mid-latitude gyres and can hope that the models can then serve adequately as a testing groud for dynamical rationalization of the large-scale general circulation of the ocean.

6. Applications #3 - Limited Area Models

Even though basin scale studies can be made with horizontal resolution as fine as 1/4 degree in latitude and longitude, for some purposes even higher resolution

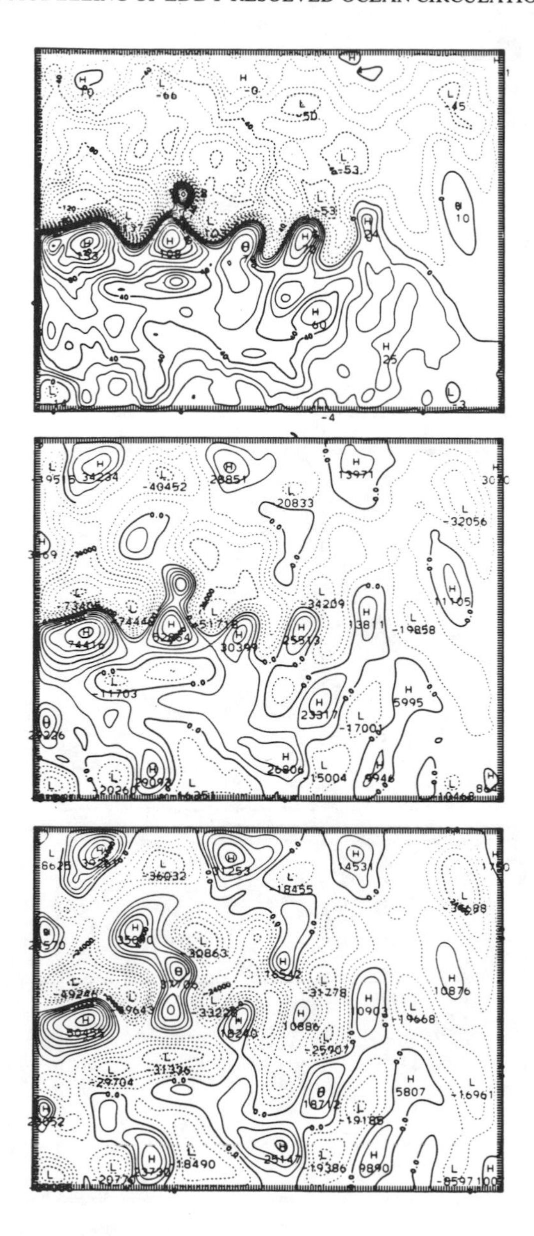

Figure 3a. The QG streamfunctions at three levels (150 m, 850 m, 1750 m) in an eight layer model: day 1860. Note the strong meandering and the breakoff of a warm core ring.

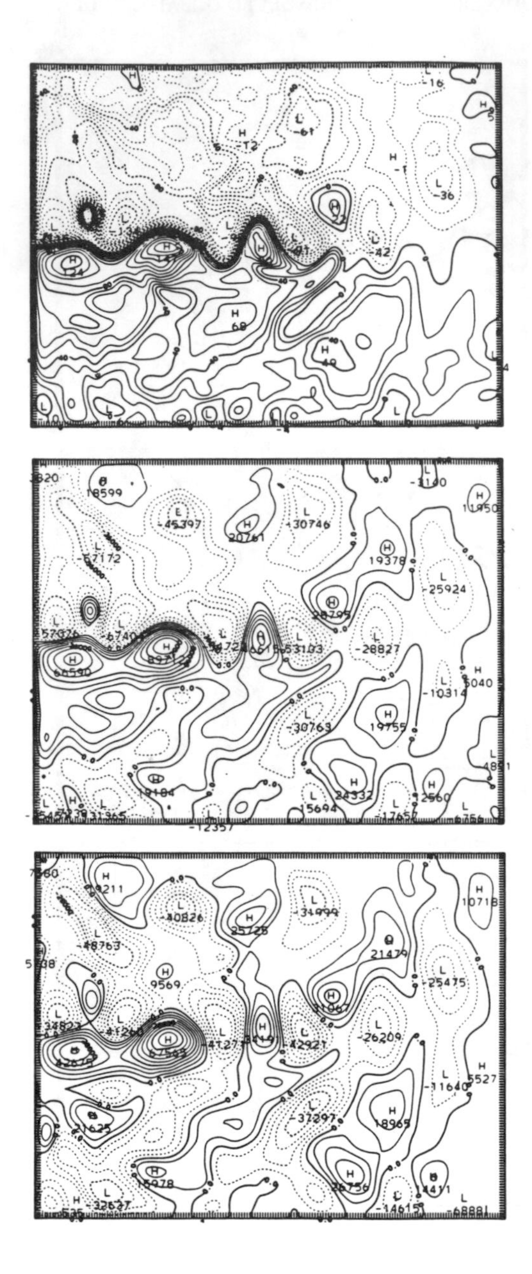

Figure 3b. The QG streamfunctions at three levels (150 m, 850 m, 1750 m) in an eight layer model: day 1920. The ring has propagated westward several hundred kilometers in this 60 day period.

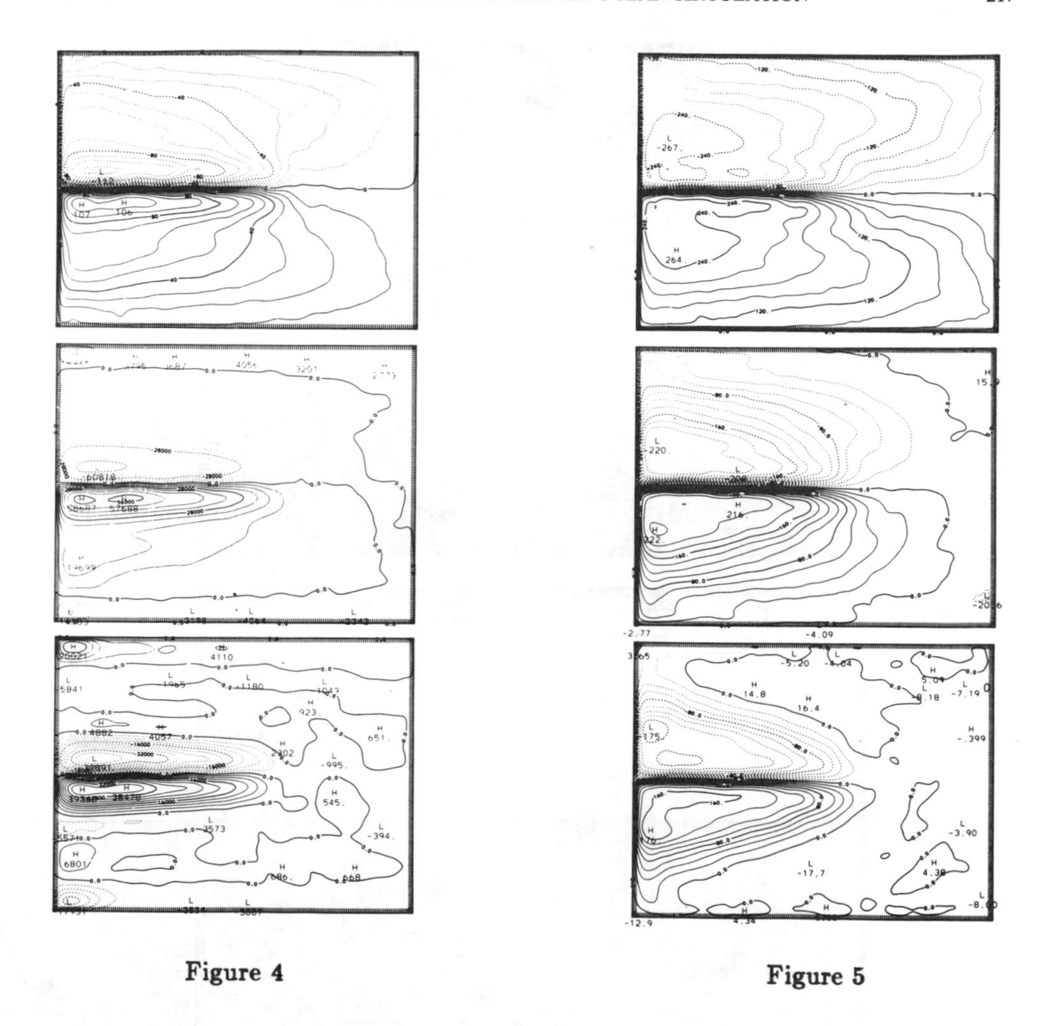

Figure 4. The time averaged streamfunctions at the same levels as in Figure 3. Note the Sverdrup gyre at the surface, the two scales of recirculation at mid-depths, and the tight recirculation alone in the deep ocean.

Figure 5. The time averaged height perturbation fields at levels 300 m, 1050 m, and 2000 m. These are the interfaces at the bottom of the layers in Figure 3. The baroclinic flow is much broader than the barotropic recirculation even in the deep ocean.

Figure 6. Maps of instantaneous (top) and mean (bottom) potential vorticity Q at 850 m. Potential vorticity is rapidly mixed from the fringes of the homogeneous region as well as being advected in by the western boundary current.

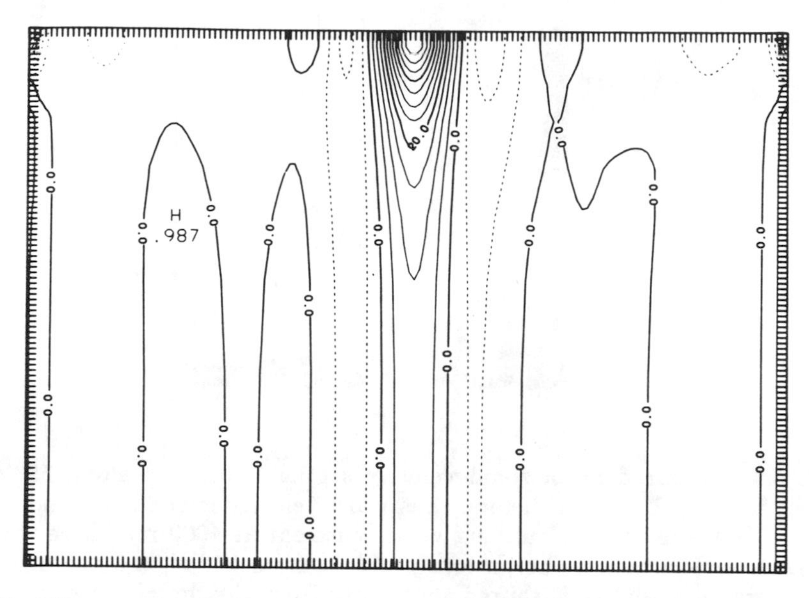

Figure 7. The mean zonal velocity in a north-south section at mid-longitude in the basin for the experiment shown in Figure 3. The contour interval is $5cm\ s^{-1}$, showing a surface jet maximum of $55cm\ s^{-1}$, a deep eastward flow of about $8cm\ s^{-1}$, and deep westward recirculating flows of $6cm\ s^{-1}$.

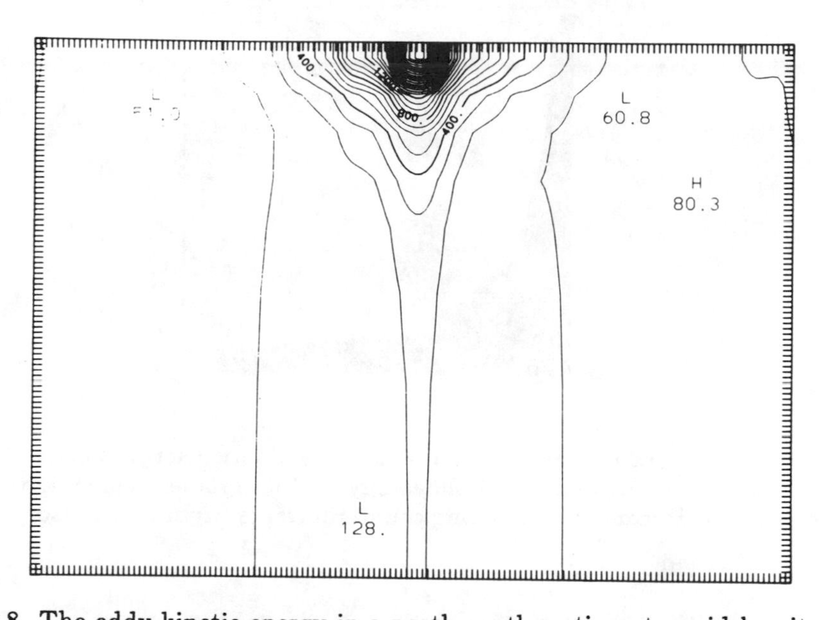

Figure 8. The eddy kinetic energy in a north-south section at a mid-longitude in the basin for the experiment shown in Figure 3. The entire depth range—0 to 5000 m—is shown. The contour interval is $100cm^2s^{-2}$, the surface maximum is about $2900cm^2s^{-2}$, and the bottom maximum is about $160cm^2s^{-2}$.

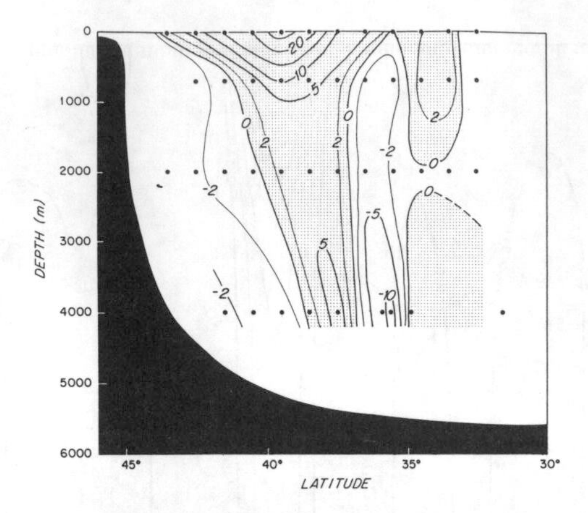

Figure 9. Contoured mean zonal velocity section ($cm\ s^{-1}$) along 55°W from drifters, floats and current meters. Eastward velocity is shaded. Dots indicate centers of boxes used in calculating velocity except at 4000 m, where they show current meter locations. The bottom profile is from 55°W; the average bottom profile between 50-60°W is shifted southward from this by about one degree in latitude (from Richardson [1985]).

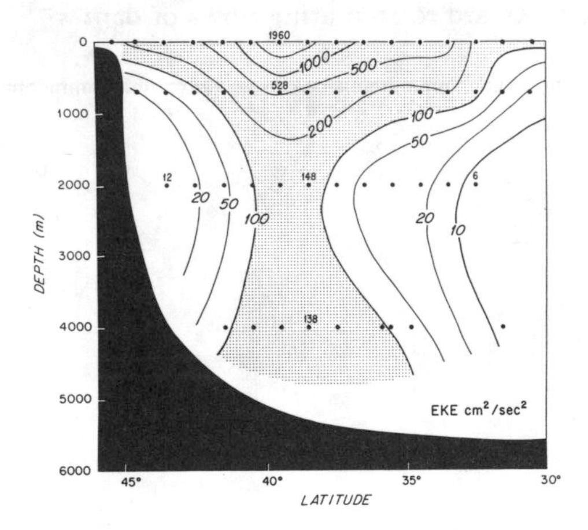

Figure 10. Contoured section along 55°W of eddy kinetic energy (per unit mass). Units are $cm^2 s^{-2}$. High eddy kinetic energy and its gradient coincide with the deep mean Gulf Stream and bounding countercurrents (from Richardson [1983]).

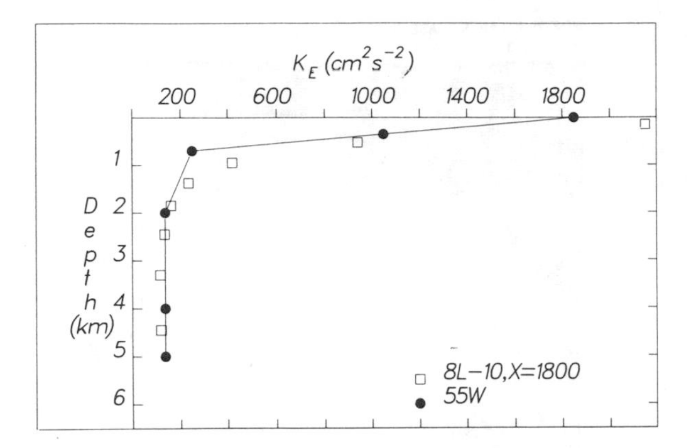

Figure 11. The vertical distribution of eddy kinetic energy. The observations (solid circles) are taken at 55°W at the latitude of the mean Gulf Stream. The model results (open squares) are from an eight layer numerical experiment in the intense eastward jet 1800 kilometers from the western boundary.

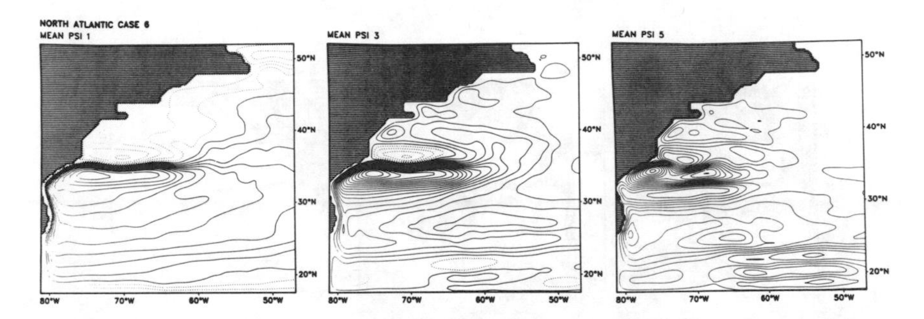

Figure 12. Mean quasigeostrophic streamfunctions at the three levels in the North Atlantic Basin Model #1: 150 m, 650 m, and 3000 m. Only the western half of the domain is shown.

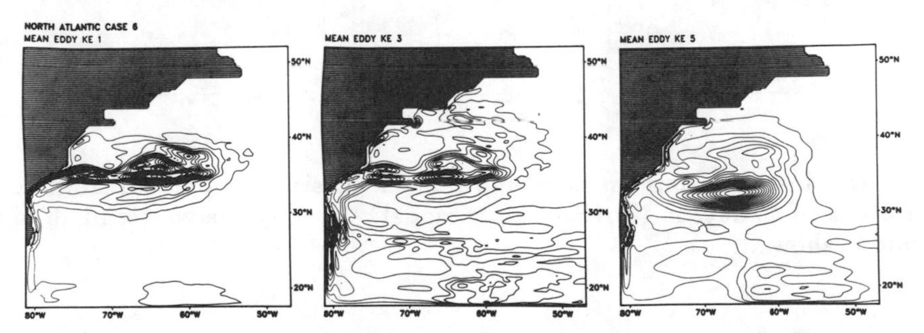

Figure 13. Mean eddy kinetic energy at three levels in the North Atlantic model #1: 150 m, 650 m, and 3000 m. Only the western half of the domain is shown.

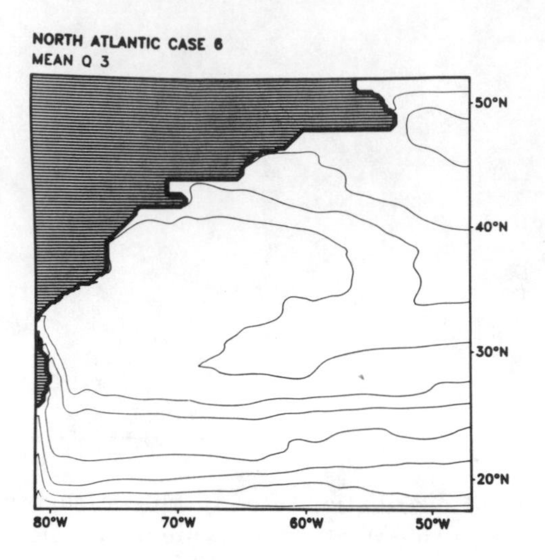

Figure 14. Mean potential vorticity in the middle layer (at 650 m) of the North Atlantic model #1. Only the western half of the domain is shown.

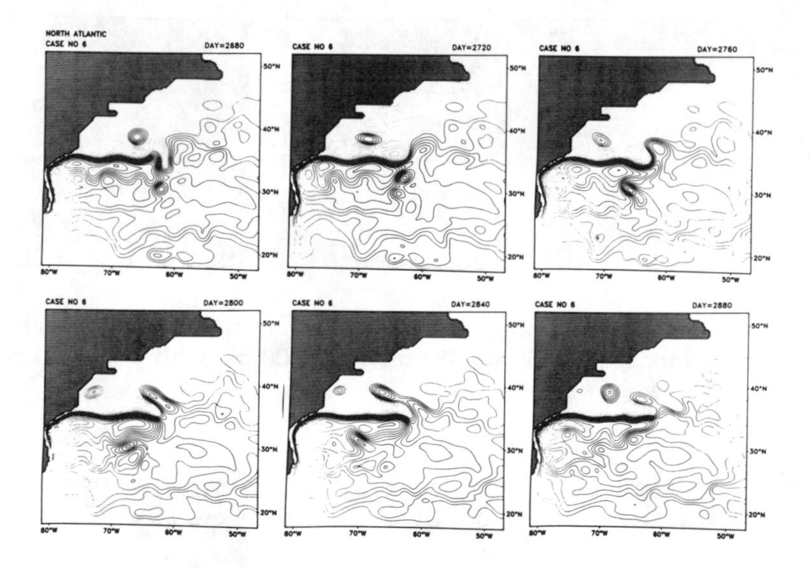

Figure 15. Instantaneous maps of upper layer quasigeostrophic streamfunction at intervals of 40 days in North Atlantic model #1. Only the western half of the domain is shown.

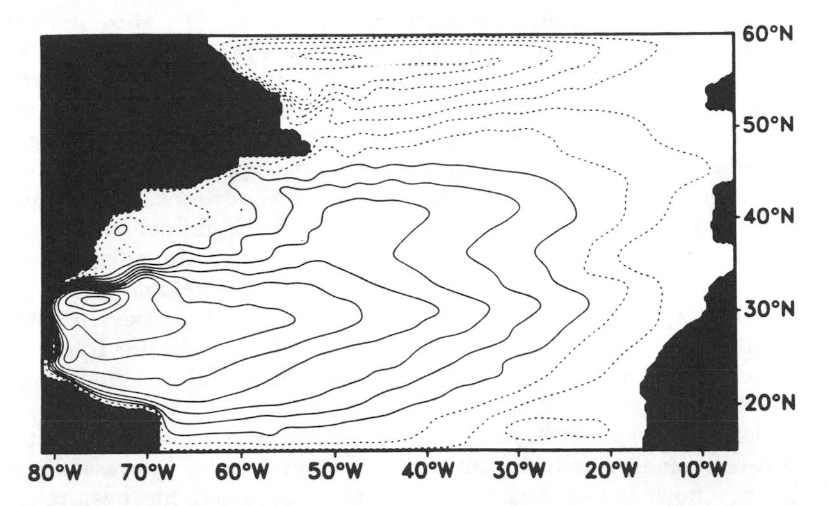

Figure 16. The time averaged, upper layer streamfunction from three-layer, quasi-geostrophic model #2 of the North Atlantic basin. The model ocean is forced with the mean Hellerman wind stress, the horizontal resolution is 1/4°, and the instantaneous model ocean (not shown) is strongly transient with Gulf Stream meandering, ring formation, and mesoscale eddies present.

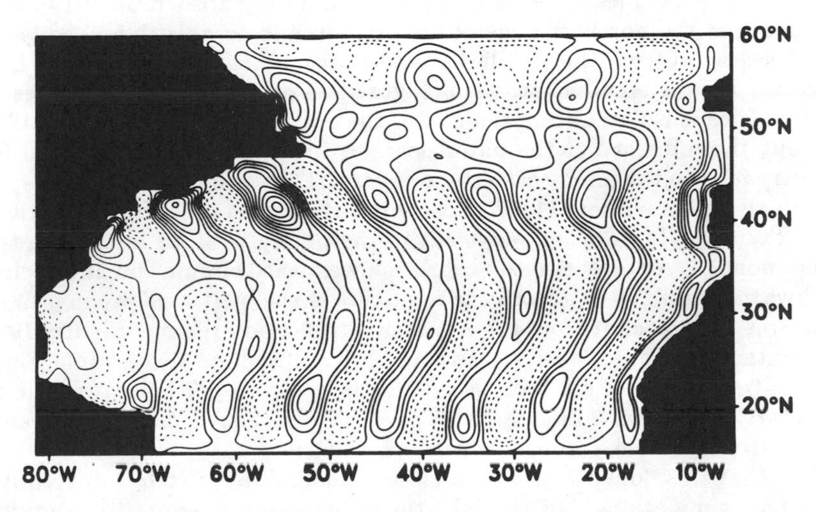

Figure 17. An instantaneous picture of the perturbation of the main thermocline depth for North Atlantic ocean basin model #2 driven by Hellerman's seasonal winds only (no mean forcing). Annual period, first baroclinic mode Rossby waves make up the primary response.

will be needed. Unfortunately, given present computers, the above models cannot effectively be used with much higher resolution—the computational expense is just too high. Therefore, a new generation of eddy-resolving limited area models (ELAMs) that can successfully handle open boundaries is being developed; these will allow us to telescope in on local regions of a larger domain. Here I will show some examples from my own work of such an approach.

Regional ocean models have "open" boundary conditions. These conditions are required both to allow a "parameterization" of the influence of the rest of the ocean on the region of interest and to provide a technique for allowing the radiation or movement of features generated within the region out of it. This subject has a long history but remains a very active area of research with many unsolved problems (see the chapter in this volume by Lars Petter Röed).

For some problems, closed boundaries might be useful to simulate a region that is in reality open. Then, part of the model domain is dedicated to special forcing regions and "sponge layers" with enhanced friction to recirculate water and absorb eddy energy radiated into it from the region of interest. Figures 18 and 19-20 show flows in two models where such an approach has been taken. The first is a barotropic model of the formation region of the Gulf Stream, i.e., the Caribbean/Gulf of Mexico complex, and the second is a baroclinic model of the region south of South Africa where the Agulhas Current actively spins off intense ring-like features that drift westward into the South Atlantic. In Figure 18, the eastern half of the domain is not realistic; it includes a simple wind-curl like forcing designed to provide a known (westward) inflow to the Caribbean basin. This region also includes enhanced friction to smoothly absorb the very nonlinear Gulf Stream at the northern edge of the domain and to absorb any eddy energy radiated eastward from the Antilles Island region when the flow passes between the islands. For the Agulhas problem (Figures 19-20), the region of interest south of South Africa is isolated from the rest of the Southern Ocean (by closed boundaries), but the influence of the outer ocean is incorporated by a special forcing region very near to the phony eastern boundary. The "wind curl" there is specially chosen to give a known zonal flow whose interaction with the African continent we want to examine. In addition, adjacent to the phony western boundary, there is another sponge layer intended to absorb the westward circulation and recirculate it southward, as well as to absorb the remnants of the westward propagating ring-like features that reach it from the Agulhas retroflection region. Figure 19 shows the instantaneous upper layer velocity and streamfunction fields while Figure 20 shows the trajectories of six "simulated" floats in the upper layer over the course of five years. Note that some particles pass into the South Atlantic Ocean while others "retroflect" back into the South Indian Ocean.

Note that this "open region in a closed domain" approach is not unlike laboratory flow simulations. In the lab, the experimentor wants to examine flow locally in his apparatus; he has a "test section" in which he wants to examine the fluid behavior. Upstream and downstream he might require pumps and baffles to produce the kind of local flow environment he wants to study. These numerical

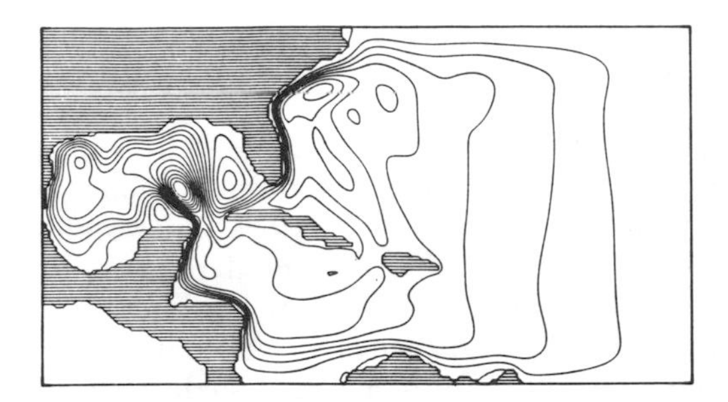

Figure 18. An instantaneous picture of the barotropic streamfunction in a limited area model of the southwestern North Atlantic including the Caribbean and Gulf of Mexico portions of the Gulf Stream system. Model resolution is 1/4° and the ocean is forced in the eastern part of the domain by a uniform wind stress curl. Flows across the northern and eastern boundaries could be specified in this model but have not been in this calculation. Loop Current meandering and ring separation occur even in this simple barotropic case.

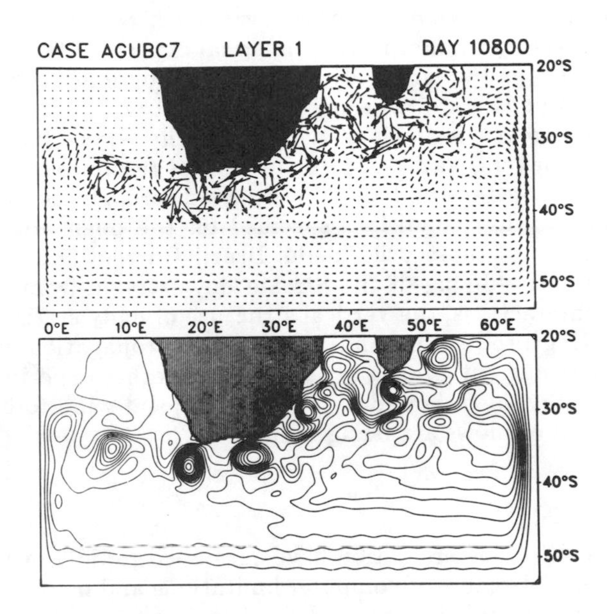

Figure 19. A baroclinic (three layer) model of the Agulhas Retroflection region with variable bottom topography. A closed domain is used to simulate the flow in a region that is actually open. Near the eastern boundary is an artificial "forcing" to establish a zonal flow of known horizontal and vertical structure while near the western boundary, a frictional sponge layer is included to recirculate mass and absorb Agulhas Rings that drift westward. Upper panel: velocity vectors. Lower panel: upper layer streamfunction.

models have similar requirements, often needing experimentation, judicious tinkering, and compromise to achieve the flows wanted in the "test section" of the model.

For other problems, open boundary conditions can be used. Radiation conditions, that is sophisticated extrapolation techniques, coupled perhaps with some choice of sponge layers, can in some cases work quite well. The best choice of such techniques depends upon the nature of the dynamics, whether dominated by advection, by wave-like "radiation", or by very nonlinear vortex-like propagation across the boundary. At the present time, this is more an art than a science but experience should lead to successful handing of such open boundary problems in the next few years.

Figures 21 and 22 show two examples with open boundaries. In the first, a barotropic model of the confluence region for the Brazil and Malvinas/Falklands Currents in the Southwest Atlantic Ocean, the east and west boundaries have a specified distribution of the streamfunction ψ. This fixes the inflow through the Drake Passage as well as the Brazil Current from the north. The outflow region is also specified at the eastern boundary. The vorticity is specified at inflow points and a simple extrapolation scheme, the gradient of vorticity set equal to zero, is used at outflow points. Note that these inflow/outflow conditions (amplitude and locations) could be specified from observations if they were available or, as done here, can be varied in alternative numerical experiments to study how the internal dynamics depend upon the details of these boundary conditions.

The final example (Figure 22) shows some recent results from a limited area model of the Gulf Stream, from Cape Hatteras to the Grank Banks. The model is baroclinic (three layers) and has $1/6°$ horizontal resolution. In the example shown the depth is constant and there is no wind forcing; the flow is "driven" by the inflow and outflow conditions specified on the open boundary. Vorticity conditions similar to those used in the Brazil Current simulation are used. Note the vigorous meandering of the Stream downstream from Hatteras. The eddy/mean flow interactions, the vertical structure of eddy statistics, the nature of Gulf Stream Ring production, and role of Ring propagation and interaction with the Stream are currently being investigated. In other experiments, the roles of bottom topography, open boundary conditions, local wind forcing, and more complex inflows and outflows are being examined.

7. Conclusions

Models of large scale ocean circulation have developed rapidly over the course of the past two decades. Because of computer limitations and because it is probably the best way to understand the ocean, it has been necessary to develop a whole hierarchy of models, from the most simple analytic ones to very complex, quite complete models of the World Ocean. Most importantly, this array of models has begun to unravel the quite complex physical processes at work in the ocean.

One of the models that has proven quite useful in the last decade is based upon quasigeostrophic dynamics. This has been true because these models are

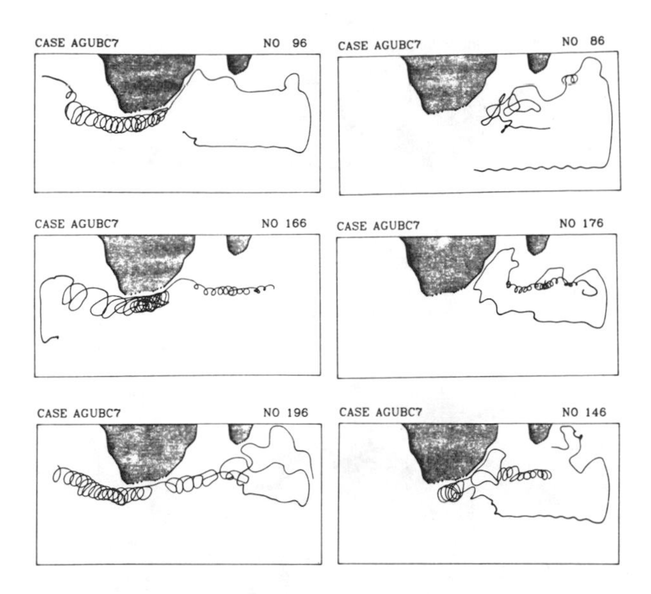

Figure 20. Simulated trajectories for six water parcels in the upper layer of the Agulhas model calculation. The "floats" are started at different locations, and a five year history of their tracks is shown. Note that some (floats 86, 146, and 176) return to the Indian Ocean while others (floats 96, 166, and 196) pass into the South Atlantic.

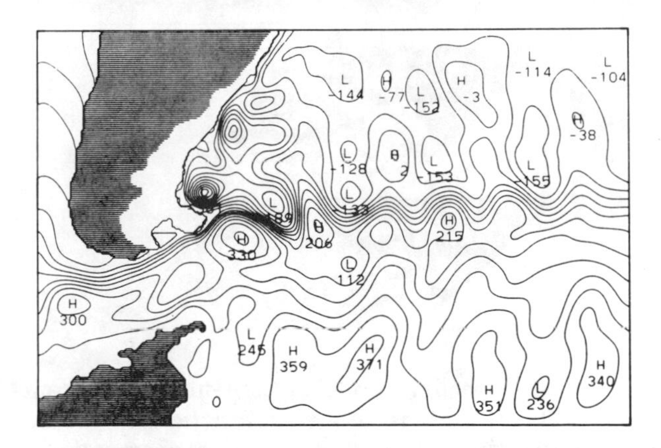

Figure 21. A barotropic Limited Area Model of the southwestern sector of the South Atlantic Ocean showing the confluence region of the Brazil and Malvinas/Falklands Currents. The domain is open with simple boundary conditions on the open boundary chosen: the streamfunction is specified everywhere, and the vorticity is given on the inflow and extrapolated from the interior at the outflow points.

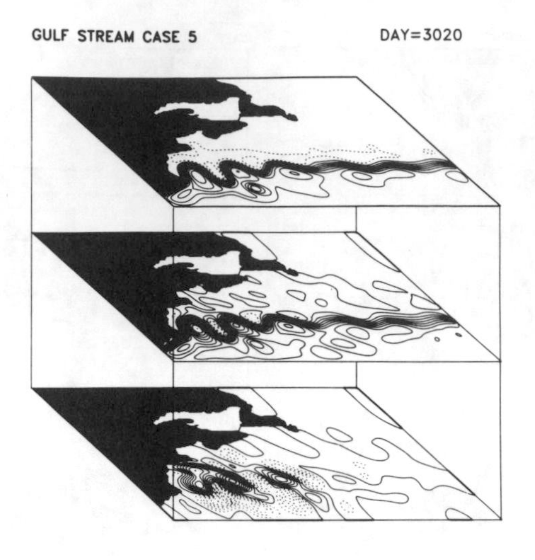

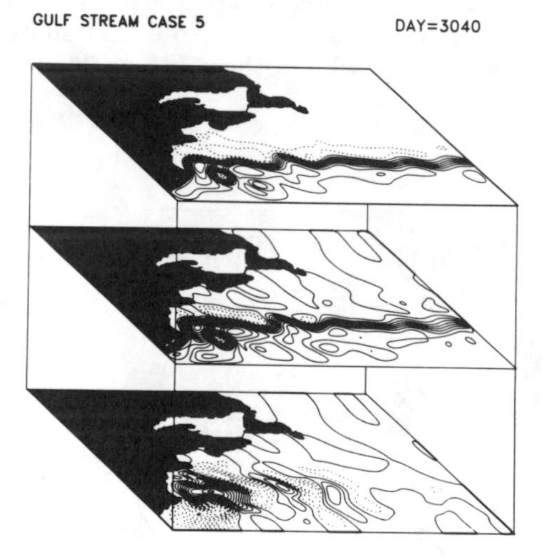

Figure 22. Results from a baroclinic model of the Gulf Stream from Cape Hatteras to the Grand Banks. The instantaneous streamfunctions at the three levels in the model are shown for two times 20 days apart in a perspective view. Note the strong meandering and the Cold Core Ring south of the Stream. The model has open boundaries on its south and east boundaries. No wind forcing is present and the depth is constant in this calculation.

computationally efficient and because even though dynamically rather simple, they seem to capture the essential dynamical processes responsible for the internal instabilities that give rise to vigorous mesoscale transience in mid-latitude oceans.

This class of models is now being used to attack problems for local regions of the ocean with very high horizontal and vertical resolutions, resolutions needed to adequately capture the relevant dynamical processes occurring in the turbulent ocean. The "representation" of the influence of the rest of the ocean upon these local regions through appropriate boundary conditions becomes a major new problem in ocean circulation modelling, and much more work is needed both to learn how to drive the local region with "observed" conditions and to properly let waves and eddies generated inside the domain pass out of it in a realistic fashion.

Finally I would like to stress that the nature of the art/science of ocean modeling is a process of judicious choice of numerics and physics coupled with careful, calculated compromises forced upon us by computer power and ocean complexity. It is already clear that the ocean modelling community stands on the threshold of being able to faithfully reproduce real ocean flows—for the first time. With the arrival of new, more powerful computers and a new generation of well trained oceanographers of the modeling persuasion, we should see this culminate in an ability to sort out the complex of physical processes actually occurring in the oceans.

8. Acknowledgements. The National Center for Atmospheric Research is sponsored by the National Science Foundation.

REFERENCES

Charney, J.G., On a physical basis for numerical prediction of large-scale motions in the atmosphere, J. Meteor., 6, 371-385, 1949.

Haidvogel, D. B., and W.R. Holland, The stability of ocean currents in eddy-resolving general circulation models, J. Phys. Oceanogr., 8, 393-413, 1978.

Harrison, D. E. and W.R. Holland, Regional eddy vorticity transport and the equilibrium vorticity budgets of a numerical model ocean circulation, J. Phys. Oceanogr., 11, 190-208, 1981.

Holland, W.R., Ocean tracer distributions. I. A preliminary numerical experiment, Tellus, 23, 371-392, 1971.

Holland, W.R., The role of mesoscale eddies in the general circulation of the ocean: Numerical experiments using a wind-driven quasigeostrophic model, J. Phys. Oceanogr., 8, 363-392, 1978.

Holland, W.R., Simulation of midlatitude variability, in The Role of Eddies in the General Ocean Circulation, Proceedings Hawaiian Winter Workshop, University of Hawaii, January 5-7, 1983.

Holland, W.R., Simulation of mesoscale ocean variability in midlatitude gyres, in Atmospheric and Oceanic Modeling–Volume 28A of Advances in Geophysics, Academic Press, Orlando, 1985.

Holland, W.R., and L.B. Lin, On the origin of mesoscale eddies and their contribution to the general circulation of the ocean. I. A preliminary numerical experiment, J. Phys. Oceanogr., 5, 642-657, 1975a.

Holland, W.R., and D.B. Haidvogel, A parameter study of the mixed instability of idealized ocean currents, Dyn. Atmos. & Oceans, 4, 185-215, 1980.

Holland, W.R., and L.B. Lin, On the origin of mesoscale eddies and their contribution to the general circulation of the ocean. II. A parameter study, J. Phys. Oceanogr., 5, 658-669, 1975b.

Holland, W.R. and P.B. Rhines, An example of eddy induced ocean circulation, J. Phys. Oceanogr., 10, 1010-1031, 1980.

Holland, W.R., and W.J. Schmitz, Jr., On the zonal penetration scale of model midlatitude jets, J. Phys. Oceanogr., 15, 1859-1875, 1985.

Holland, W.R., T. Keffer, and P.B. Rhines, Dynamics of the oceanic general circulation: the potential vorticity field, Nature, 308, 698-705, 1984.

McWilliams, J.C., A note on a consistent quasigeostrophic model in a multiply-connected domain, Dyn. Atmos. Oceans, 1, 427-441, 1977.

Pedlosky, J., Geophysical Fluid Dynamics, 624 pp., Springer-Verlag, New York, 1979.

Phillips, N.A., A simple three-dimensional model for the study of large-scale extratropical flow patterns, J. Meteor., 8, 381-394, 1951.

Rhines, P.B., and W.R. Holland, A theoretical discussion of eddy-driven mean flows, Dyn. Atmos. & Oceans, 3, 289-325, 1979.

Rhines, P.B. and W.R. Young, Homogenization of potential vorticity in planetary gyres, J. Fluid Mech., 122, 347-367, 1982.

Rhines, P.B., W.R. Holland, and J.C. Chow, Experiments with buoyancy-driven ocean circulation, NCAR Tech Note, in press, 1985.

Richardson, P.L., A vertical section of eddy kinetic energy through the Gulf Stream System, J. Geophys. Res., 88, 2705-2709, 1983.

Richardson, P.L., Average velocity and transport of the Gulf Stream near 55W, J. Mar. Res., 43, 83-111, 1985.

Schmitz, W.J., Jr., and W.R. Holland, A preliminary comparison of selected numerical eddy-resolving general circulation experiments with observations, J. Mar. Res., 40, 75-117, 1982.

Schmitz, W.J., Jr., P.P. Niiler, R.L. Bernstein, and W.R. Holland, Recent long-term moored instrument observations in the Western North Pacific, J. of Geophys. Res., 87, 9425-9440, 1982.

Semtner, A.J. and W.R. Holland, Intercomparison of quasigeostrophic simulations of the western North Atlantic circulation with primitive equation results, J. Phys. Oceanogr., 8, 735-754, 1978.

Veronis, G., Wind-driven ocean circulation—Part 1. Linear theory and perturbation analysis, Deep Sea Res., 13, 17-29, 1966a.

Veronis, G., Wind-driven ocean circulation—Part 2. Numerical solutions of the non-linear problem, Deep Sea Res., 13, 30-35, 1966b.

PERIODIC QUASI-GEOSTROPHIC MODELS

B. L. Hua
IFREMER
B. P. 337
29273 Brest Cedex
France

ABSTRACT. A vertical mode formulation of quasi-geostrophic equations is given followed by their spectral implementation for doubly periodic domains and zonal channels. Emphasis is put on the mean flow constraints in such formulations. Recent results concerning stratified geostrophic turbulence are given, with special attention to intermittency and three-dimensional isotropization of such flows.

1. INTRODUCTION

Periodic models differ from the box-type quasi-geostrophic models in that they seek to determine the dynamical characteristics of a local block of ocean, whose interconnection to the remaining larger scale dynamics (oceanic gyre scale) is assumed to be statistically unimportant. They were first introduced by Bretherton and Karweit [1975] in order to simulate the MODE-1 region observations [Owens and Bretherton, 1976].

The basic assumption is that the boundary conditions are periodic, either for a channel-like geometry where fluid exiting at one boundary reenters by its homologue or on a doubly periodic beta-plane where both latitudinal and longitudinal boundary conditions are supposed to be periodic. Thus a fundamental assumption of spatial horizontal homogeneity is made for either the latitudinal direction (channel) or in both directions (doubly periodic box). Like their EGCM counterparts, periodic models allow the simultaneous explicit representation of several octaves of scales from scales close to gyre scale to the ones which absorb the enstrophy cascade, encompassing the mesoscale geostrophic part of the oceanic circulation.

They need to be distinguished, however, from regional models [Haidvogel et al., 1980] in that the latter are separated from the remainder of the oceanic gyre by arbitrary boundaries, which are imposed by both the exterior surrounding fluid and the interior fluid, across which fluid is free to move in a non-periodic manner. A general recent survey of both periodic and regional models can be found in Haidvogel [1983]. Finally, another class of box-type models

J. J. O'Brien (ed.), Advanced Physical Oceanographic Numerical Modelling, 233–253.

within closed boundaries [Bretherton and Haidvogel, 1976; Haidvogel and Rhines, 1983] have also been studied, with the different goal of investigating physical processes in presence of boundaries, and they will not be considered here.

The assumptions of periodic models permit a focus on specific processes by a selection of the physical processes retained in the model, such as the study of the relative importance of the various terms of the quasi-geostrophic equations (nonlinear terms, beta-effect, stratification, mean shear effects, etc.).

2. QUASI-GEOSTROPHIC EQUATIONS

The quasi-geostrophic equations presented here are written in modal form instead of as a layer representation for the vertical structure. Following Flierl [1978], we use a normal modes expansion in the vertical, which are the solutions of the following Sturm-Liouville problem:

$$d/dz[(f^2/N^2(z))\ d/dz(F_i(z))] = -\tau_i F_i(z) \qquad (1)$$

where f is the Coriolis parameter, N(z) is the Brunt-Vaissala profile of the considered region and $\tau_i = (kz_i)^2$, kz_i being the vertical wavenumber associated with the i-th mode. Boundary conditions for the vertical structure are rigid lids at top (z=0) and bottom (z=-H). For the case where N is constant, the eigenmodes are:

$$F_0(z) = 1;\quad F_i(z) = \sqrt{2}\ \cos(i\pi z/H)$$

$$\tau_i = (\pi i f/NH)^2$$

The streamfunction field can be approximated by the truncated series:

$$\Phi(x,y,z,t) = \sum_{i=0}^{i=NV} \Phi(x,y,t)\ F_i(z) \qquad (2)$$

where we have kept the barotropic mode and NV first baroclinic modes. The potential vorticity conservation for mode i is :

$$\partial/\partial t\ q_i + \sum_{j,k} \varepsilon_{ijk}\ J(\Phi_j, q_k) + \beta\partial/\partial x\ \Phi_i + M_i =$$

$$-V1 \sum_j F_i(-H)\ F_j(-H)\ \nabla^2\Phi_i \quad -V3\nabla^6\ \Phi_i \qquad (3)$$

where

*$q_i = \int dz\ q\ F_i(z) = (\nabla^2 - \tau_i)\ \Phi_i$ is the potential vorticity of mode i since the total potential vorticity q is

$$q = \nabla^2\Phi + \partial/\partial z\ (f^2/N^2 \partial/\partial z \Phi). \tag{4}$$

* $\varepsilon_{ijk} = \int dz\ F_i(z)F_j(z)F_k(z)$ is the triple interaction coefficient between modes i, j and k;

* $\beta = df/dy$ is the planetary vorticity gradient;

* M_i = forcing term of mode i;

*V1 represents the friction coefficient through the bottom Ekman layer at z=-H. Note that it represents a coupling effect between the different vertical modes. Other types of sinks could be specified, such as a modal Rayleigh friction which has no coupling effect on the modes;
* V3 is the hyperviscosity coefficient used to absorb the enstrophy cascading to the highest horizonatal wavenumbers.

3. INTEGRAL CONSTANTS OF MOTION

Multiplying (3) by Φ_i and integrating over the domain, discarding forcing terms and viscous dissipation, one obtains the energy budget for each modal component:

$$\partial/\partial t\ 1/2 \iint dxdy\ [|\nabla\Phi_i|^2 + \tau_i\Phi_i^2] =$$

$$\iint dxdy \sum_{j,k} \varepsilon_{ijk}\ J(\Phi_i,\Phi_j)q_k \tag{5}$$

provided that Φ_i is either periodic or a function of time only ($\Phi_i = f_i(t)$) along the boundaries, since divergence terms vanish in that case [Pedlosky, 1979]. Thus (5) is verified for both doubly-periodic and channel-like geometry. The right-hand side term represents the nonlinear energy exchanges between modes, but, when summing over all modes i, the truncated series expansion still conserves total energy, since one can readily verify that the sums of the right-hand sides vanish:

$$\partial/\partial t\ 1/2 \iint dxdy \sum_{i=1}^{i=NV} [|\nabla\Phi_i|^2 + \tau_i\Phi_i^2] = 0. \tag{6}$$

The modal enstrophy budget is found by multiplying (1) by q_i and integrating over the whole domain

$$\partial/\partial t\ 1/2 \iint dxdy\ q_i^2 = \iint dxdy \sum_{j,k} \varepsilon_{ijk}\ J(q_i,q_k)\Phi_j. \tag{7}$$

We have made use of the east-west periodicity of both doubly periodic domains and zonal channel-like geometery for discarding the contribution coming from the β flux term [Pedlosky, 1979]

$$\underline{Q} = \underline{ix}\ [-\beta/2\ (\partial\Phi/\partial x)^2] \tag{8}$$

where $\underline{ix}$ is the unit vector in the latitudinal direction.

Thus total enstrophy (when summing (7) over all modes i) is conserved in the two cases we are considering, while this is not verified for domains involving solid meridional boundaries [Haidvogel and Rhines, 1983]. Again (7) involves nonlinear enstrophy exchanges between modes and should be opposed to a layered model formulation where both total enstrophy and layer-wise enstrophy are conserved.

One should point out that although the quasi-geostrophic equations are consistent with the primitive equations to leading order in a Rossby number development, auxiliary conditions [McWilliams, 1977] must be appended to insure that the model be fully determined for the case of simply connected domains. These conditions are such as to prevent energy and mass fluxes through solid boundaries and to assure correct integral budgets of circulation and mass. We shall encounter such examples of integral constraints in the next section.

4. SPECTRAL MODEL

4.1. Doubly Periodic Box

Assuming horizontal homogeneity, we also assume that the streamfunction modal components are periodic over the square domain $0 \leqslant x,y \leqslant 2\pi L$, where L is the dimensional size of the square box.

$$\Phi_i(x,y,t) = \Phi_i(x+L,y,t)$$
$$\Phi_i(x,y,t) = \Phi_i(x,y+L,t)$$

This implies that the total streamfunction field $\Phi(x,y,z,t)$ is also periodic over the whole water column.

With periodic boundary conditions, a stable spectral method based on Fourier series is accurate and efficient [Gottlieb and Orszag, 1977]. We therefore expand the streamfunction field in Fourier series:

$$\Phi_i(x,y,t) = \sum_{m=-NH/2}^{m=NH/2-1} \sum_{n=-NH/2}^{n=NH/2-1} \hat{\Phi}^i_{mn}(t) \exp i(k_m x + l_n y) \tag{9}$$

with $(k_m, l_n) = (m,n)/\hat{L}$, where $\hat{L}$ is the non-dimensional box size. In what follows, we use the box size as the length scale, thus $\hat{L}=1$. The largest wavenumber retained in the model is $k_{max}=(NH/2\hat{L})$.

The spectral procedure utilized here has a number of important advantages over finite-difference methods. First, accuracy is greater for a given number of independent degrees of freedom, since finite-differences methods converge at a given finite polynomial order depending on the chosen scheme, while spectral approximations can converge faster than any algebraic function. Second, spectral methods share the property of Arakawa's finite-difference schemes in that they preserve important integral constraints of motion (see section 3) aside from time-differencing errors and viscous dissipation. This implies that stable calculations are thus more readily achieved. Finally, the proper choice of the expansion basis allows us to readily account for the boundary conditions.

Since the streamfunction field is real, we have:

$$\hat{\Phi}^i_{mn}(t) = [\hat{\Phi}^i_{-m-n}(t)]^*$$

where []* designates the complex conjugate of the quantity within brackets. In what follows, $\hat{a}$ designates the Fourier transform of a. Subtituting (9) into (3) yields a coupled set of equations for the evolution of the spectral coefficients $\hat{\Phi}^i_{mn}(t)$.

$$-d/dt\ (K^2+\tau_i)\hat{\Phi}^i_{mn} + \hat{J}^i_{mn} + \beta ik\ \hat{\Phi}^i_{mn} + \hat{M}^i_{mn} =$$

$$[V1\ \sum_j F_i(-H)F_j(-H)K^2 + V3\ K^6]\hat{\Phi}^i_{mn} \qquad (10)$$

where $K^2=m^2+n^2$.

These equations are advanced in time using a leapfrog differencing scheme, with periodic applications of a leapfrog trapezoidal step to diminish the computational mode. The most efficient way to evaluate the nonlinear terms such as

$$\hat{J}^i_{mn} = \sum_{i,j} \varepsilon_{ijk}\ J(\Phi_j,q_k) = \sum_{i,j} \varepsilon_{ijk}\ (\Phi_{jx}\ q_{ky} - \Phi_{jy}\ q_{kx}) \qquad (11)$$

which appear in (10) (note that we have dropped the m,n subscripts for Fourier coefficients in the right hand sides of (11)), is to apply the transform method introduced by Orszag [1971]. The key idea is based on the use of Fast Fourier transforms in order to transform efficiently between spectral representations and physical-space representations. Moreover, Temperton has considerably improved the efficiency of the two-dimensional transforms algorithms for the case where the initial fields are real. Thus at each timestep, we first evaluate each factor of (11) and assemble their product in physical space before transforming it back to Fourier space. Such nonlinear terms generate waves that can be of increased wavenumbers (corresponding to the sum and differences of the wavenumbers of each factor). Problems can therefore arise from this aliasing; for instance, the growth of high wavenumbers can lead to numerical nonlinear instability, for product waves with $k > k_{max}$, since it will be misrepresented as $k^*=2k_{max}-k$ [Haidvogel, 1983]. For enstrophy conserving schemes, the likelihood of such explosive nonlinear instability is eliminated. Our spectral scheme along with (11) only preserves energy. The elimination of nonlinear instability can be obtained at the cost of evaluating (11) twice: first on the regular grid and then on a grid which is shifted by a distance of one-half grid interval and by taking the average of both evaluations. This doubling of the computational effort can, however, be decreased if one remarks that most of the computational efforts goes into eliminating aliasing interactions in the "corners" of the box $|k|$, $|l| \leqslant K_{max}$. Since they include modes that are not likely to be accurately treated anyway (because of spectral truncation effects), it is logical to restrict the computational effort to an isotropically truncated domain such as the circle: $k^2 + l^2 \leqslant k_{max}^2$, therefore reducing the computational effort by roughly 20% [Orszag, 1971].

Examination of (10) suggests that the computational effort for the nonlinear terms grows like NV^3 [Flierl, 1978], where NV is the number of vertical modes. This can be avoided if one remarks that:

$$\sum_{j,k} \varepsilon_{ijk} J(\Phi_j, q_k) = \sum_{j,k} \varepsilon_{ijk} \operatorname{div}(\underline{u}_j\, q_k). \tag{12}$$

Thus, by first evaluating $\underline{u}_j$ and q_k and then assembling and storing the products $(\underline{u}_j\, q_k)$ the computational effort can be rendered to grow like $(NV+1)^n$, where n is close to 1.3, on a CRAY-1S.

4.2. Mean Flow Representation

We have so far restricted our attention to the eddy part of the flow. Many studies have taken into account the effects of a mean flow [Bretherton and Karweit, 1975; Haidvogel and Held, 1980; Salmon, 1980].

We shall first consider the simpler case of a zonal baroclinic flow, although the discussion can be readily extended to a mean shear flow involving a meridional component. The introduction of a mean shear is performed through a polynomial subtraction to the Fourier series considered above [Vallis, 1985]. (Polynomial subtraction methods are sometimes used with Fourier series expansion in order to improve the accuracy of the convergence for spectral problems involving non-periodic boundary conditions):

$$\Phi(x,y,z,t) = -U(z,t)y + \sum_{m,n} \hat{\Phi}_{mn}(z,t) \exp i(k_m x + l_n y). \tag{13}$$

For the sake of simplifying the presentation, we assume that the mean shear projects only onto the first mode, although the following result can be generalized to any mean shear profile.

One can remark that the expansion (10) is already complete for fields satisfying (9), e.g. for the eddy part of the field only. In order to avoid the Gibbs phenomenon that could be induced by a non-periodic part of the flow, one has to verify a few consistency requirements, since the subtracted polynomial is not orthogonal to the expansion basis:

$$y = -2 \sum_{n=1}^{\infty} (\sin 2\pi n y)/n + \pi \qquad \text{for } 0<y<1. \tag{14}$$

The spectral equations for each Fourier component (k_m, l_n) are:

$$\partial/\partial t[\hat{q}^i_{mn} + \delta_{i1}\tau_1 U\hat{u}_{mn}] + \hat{J}^i_{mn} =$$

$$-\sqrt{-1}k_m \sum_j U\, \varepsilon_{ij1}[\hat{q}^j_{mn} + \tau_1{}^2\, \hat{\Phi}^j_{mn}] \tag{15}$$

where $\hat{u}_{mn} = 0$ if $m \neq 0$
$= (\sqrt{-1}/n$ if $m = 0$.

δ_{ij} is the Kronecker symbol.

In the special case of $U(z,t)=U(z)$, the mean shear is constant with respect to time, the rate of change of q^1_{mn} can be determined unambiguously from (15), i.e. (15) forms a closed set of equations (otherwise the system is underdetermined). Strictly speaking, ad hoc prognostic equations giving the rate of change of $U(z,t)$ prescribed in conjunction with (15) but neglecting the rate of change of U in (15) like in Salmon [1980] are not consistent.

Furthermore, the constancy with respect to time of U is linked to the consistency requirements of mass conservation in a quasi-geostrophic formulation [McWilliams, 1977; Bretherton and Karweit, 1975], implying that the domain-averaged vertical velocity be zero

$$\partial/\partial t \iint dxdy\, \partial/\partial z\, \Phi(x,y,z,t) = 0. \tag{16}$$

Such a mean flow representation allows, for instance, the study of turbulent flows induced by the baroclinic instability of a mean shear maintained externally constant with time. The energetics of the system are no longer closed, since an infinite external reservoir of energy is needed to maintain $U=U(z)$, a constant with respect to time.

The case of a mean barotropic flow is more subtle. In the case of a general barotropic mean flow $U=(U(t),V(t))$, the consistency relationship comes from the overall Sverdrup balance (this can be directly verified by integrating over the whole domain the barotropic mode vorticity equation (3) slightly modified to accomodate the presence of the mean shear as in (15). In the absence of both mean components of Ekman pumping and of bottom topography, this implies that

$$\beta V(t) = 0.$$

However, one can readily verify from (15) that the rate of change of the zonal component $U(t)$ can be independently prescribed without violating quasi-geostrophic constraints. In the presence of a mean bottom topography, the independently prescribable quantity would be the barotropic component of the mean flow which lies parallel to (f/h) isolines [Bretherton and Karweit, 1975]. For the special choice of $U=U0$ = constant, one can verify that the β-plane quasi-geostrophic equations are invariant in a zonal galilean transformation, so that the constant barotropic zonal mean flow case is trivial. In general, one is still left with an indeterminancy for the rate of change of the barotropic zonal component $U(t)$.

4.3. Zonal Channel

For zonal channel domains, an appropriate spectral representation [Vallis, 1985] which satisfies the channel boundary conditions:

$$\begin{aligned}
&\Phi(x,y,z,t) = \Phi(x+L,y,t)\\
&\partial\Phi/\partial x(x,0,z,t) = \partial\Phi/\partial x(x,L_y,z,t) = 0\\
&\Phi_i = \Phi_B(z,t),\ i=1,2 \text{ on the two solid boundaries: } y=0,\\
&\text{and } y=L_y
\end{aligned} \qquad (17)$$

can be chosen as (other choices of spectral representations can be made):

$$\Phi_i = \sum_{n=1}^{N-1} \bar{a}^i_n \cos l_n y + \sum_{\substack{m=N-1\\ m\neq 0}}^{N-1} \sum_{n=1}^{N-1} \hat{a}^i_{mn} \sin l_n y \exp(ik_m x) \qquad (18)$$

with $(k_m, l_n) = (m\pi/L_x, n2\pi/l_y)$, for vertical mode i.

The $\bar{a}^i_n$ designate the Fourier components of the flow which are zonally-averaged. The consistency requirement [McWilliams, 1977] for the momentum budget along the latitudinal boundaries is that:

$$\partial\bar{\Phi}_i/\partial y = 0 \text{ on } y=0 \text{ and } y=L_y. \qquad (19)$$

Both consistency requirements (16) and (19) are verified by the expansion basis in (18). This set of equations forms a closed set of equations allowing the determination of the boundary values Φ_i internally. Such a model is best fit for studies on interactions between eddies and mean zonal jets components [Klein and Pedlosky, 1985; Boville, 1982].

5. TWO-DIMENSIONAL TURBULENCE

A large fraction of studies addressed by periodic models have concerned what is known as geostrophic turbulence which is somewhat akin to the more general field known as two-dimensional tuburlence. The properties of two-dimensional turbulence are discussed much more extensively in Kraichnan and Montgomery [1980] and we shall present only a short summary of its properties here. Its governing equations correspond to the barotropic quasi-geostrophic vorticity equation, on an f-plane, with its two invariants being kinetic energy and kinetic enstrophy:

$$\partial/\partial t \ \nabla^2\Phi + J(\Phi,\nabla^2\Phi) = 0 \tag{20}$$

$$E = 1/2 \iint dxdy \ |\nabla\Phi|^2$$

$$V = 1/2 \iint dxdy \ |\nabla^2\Phi|^2. \tag{21}$$

Consistent with (20) having no anisotropic operator, flows governed by such equations exhibit isotropy and homogeneity characteristics. Furthermore, the free flow evolutions with time are such that the wavenumber peak of the energy spectrum will decrease towards higher wavenumbers while the enstrophy spectrum (and also tracer variance) is fluxing towards higher wavenumbers.

For forced flows, asymptotically with time inertial ranges presenting power law regimes will develop away from the forcing range [Kraichnan, 1967]. For small wavenumbers a $K^{-5/3}$ spectrum will develop, with a constant energy flux regime, while for wavenumbers higher than the forcing range a constant enstrophy flux (and zero energy flux) regime with a k^{-3} slope is present.

In a finite periodic domain, the transfer of energy to larger scales is arrested at the lowest available wavenumber which corresponds to the box size. Meanwhile, the inclusion of a damping mechanism (such as Newtonian viscosity) acting more selectively on the smallest scales will absorb the enstrophy cascade, preventing the piling up of enstrophy at the highest wavenumber resolved in the numerical resolution.

Two essential features of flows described by (20) are relevant for geophysical flows:

(i) its unpredictability: two initial states differing only slightly will have their solution diverge exponentially in a finite time, corresponding to the eddy turn-around time [Leith, 1971]. This behavior is linked to the nonlinear character of (20).

(ii) its intermittency: recent direct simulations have shown that besides turbulent regimes compatible with Kraichnan's phenomenology, flows obeying (20) can also exhibit coherent structures that are very robust and long lived [Basdevant et al, 1981; McWilliams, 1984].

6. GEOSTROPHIC TURBULENCE

Obviously (3) differs by many terms from the ideal case described by (20). We shall review very shortly some of these differences and their geophysical relevance, focusing more on the most recent simulations in the field. The reader is referred to Haidvogel [1983] for a more comprehensive review.

6.1. β-arrest

The presence of a gradient in the planetary vorticity supports Rossby wave solutions that can interfere with the turbulent nonlinear interactions when the wave periods are at least as short as eddy turn-around times [Rhines, 1977]. Therefore since Rossby wave periods decrease with increasing scales, turbulence will be inhibited at the largest scales and the red energy cascade will be arrested at:

$$K = \sqrt{\beta}(2E)^{-1/4}. \qquad (22)$$

Moreover, the flow will present a growing anisotropy with time with a tendency to evolve towards alternating zonal jets (Figures 1a, 2a). Associated with suppressed nonlinearity, the flow tends to exhibit an increased predictability [Basdevant et al., 1981].

Figure 1a

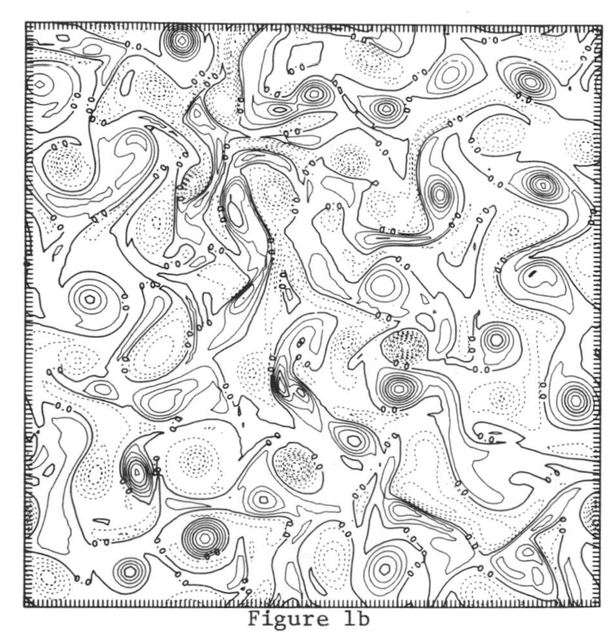

Figure 1b

Figure 1. Vorticity field at non-dimensional time t=10 for the case where (a) $\beta \neq 0.$; (b) $\tau \neq 0$. Note that for both fields the red cascades have arrested at a finite size but that (a) is characterized by a high zonal anisotropy.

6.2. τ-arrest

This term was coined by McWilliams [1983]. In analogy with the β-arrest mechanism, baroclinic modes exhibit another wave-like tendency associated with the vortex stretching term present in the potential vorticity q of mode i:

$$q_i = \nabla^2 \Phi_i - \tau_i \Phi_i.$$

The internal baroclinic mode evolution, (neglecting for the time being its interaction with other modes) would take the simplified expression:

$$\partial/\partial t \; q_i + J(\Phi_i, q_i) = 0. \tag{23}$$

For wavenumbers K smaller than $\sqrt{\tau}$, the self-advection of $J(\Phi_i, q_i) \approx 0$). This arrest of baroclinic cascades implies an enhanced predictability for the flow. Direct verification of such τ-arrest mechanism has been simulated numerically (McWilliams, personal communication) (Figures 1b, 2b). On a β-plane, (23) would be further modified by the presence of the planetary vorticity gradient: this yields the presence of another competing arrest mechanism and the arrest scale would depend on the respective scales associated with β and τ and the initial scale of the flow.

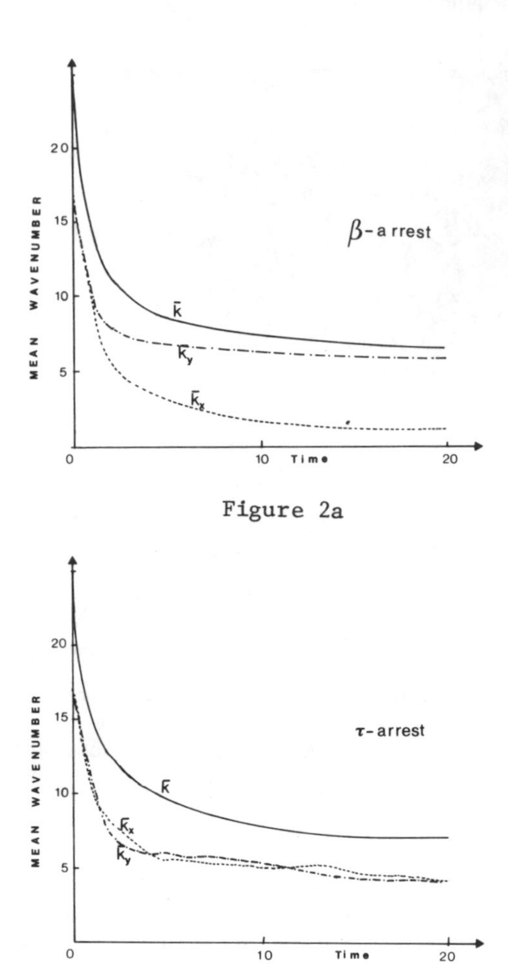

Figure 2a

Figure 2b

Figure 2. Mean wavenumber of the energy spectrum (21) evolution with time: K designates the isotropic wavenumber while K_x (resp. K_y) refers to the zonal (resp. meridional) mean wavenumber. (a) $\beta \neq 0$; (b) $\tau \neq 0$. Note the discrepancy between K_y and K_x for case (a).

6.3. Baroclinic Turbulence and Three-dimensional Isotropization

The general stratified case obeys the full (3) equations, and one has to take into account the nonlinear interactions coupling the various modes. Charney [1971] remarked that there existed in the case of constant N profile an isomorphism between the two-dimensional relative vorticity $\nabla^2\Phi$ and quasi-geostrophic potential vorticity (4), pending the transformation $\underline{K_2} = (k,l) \longrightarrow \underline{K_3} = (k,l,\sqrt{\tau})$. However, Charney's remark is not complete since (3) still involves advection terms that remain 2D even in the stratified case, while potential vorticity is a 3D operator. In pursuing Charney's isomorphism (Hua and Haidvogel, 1985) we have shown that (3) can be shown to be totally isomorphic to (20) and in particular that the three-dimensional constant-N case triadic interaction rules (Figure 3) are identical to the classical 2D triangular wavenumber interaction rules [Kraichnan, 1967]. This has been checked numerically, using equations (3) for both freely decaying turbulence and stationary turbulence forced by baroclinic instability of a constant mean shear (Figures 4a, 4b). The oceanic case of non-constant N is more complex since it involves the relaxation of homogeneity in the vertical through an explicit dependence of z in (3). We have found that although freely decaying turbulence showed tendencies for three-dimensional isotropization "à la Charney" (Figure 5a), the forced cases verified a weaker form of 3D isotropy, i.e. that modal spectra remain parallel in the inertial ranges though they do not collapse any more to a unique curve (Figure 5b).

Figure 4a shows that peaks in the dominant energetic modes, namely the barotropic and first baroclinic modes correspond to the same horizontal wavenumbers, suggesting a strong coupling between both modes at the barotropic β-arrest wavenumber for stationary stratified turbulence forced by baroclinic instability of a mean shear. This seems to be also verified from oceanic measurements [Hua et al., 1985] where the barotropic and baroclinic peak wavenumbers were found to coincide at $(75\text{Km})^{-1}$ which was significantly different from the τ-arrest scale of $\sqrt{\tau} = (45\text{Km})^{-1}$ for that region.

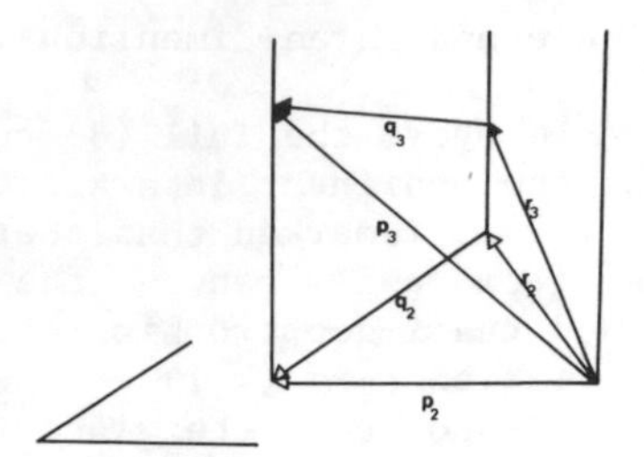

Figure 3a

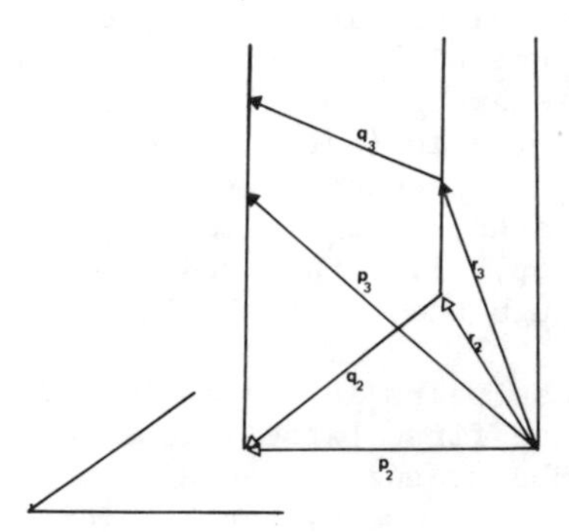

Figure 3b

Figure 3. Triadic interaction rules for N=constant and N non-constant as deduced from ε_{ijk} values. The 3D addition rules that prevail for constant-N result from the additive properties of sine vertical eigenmodes, while this result is no longer true for general vertical eigenmodes.

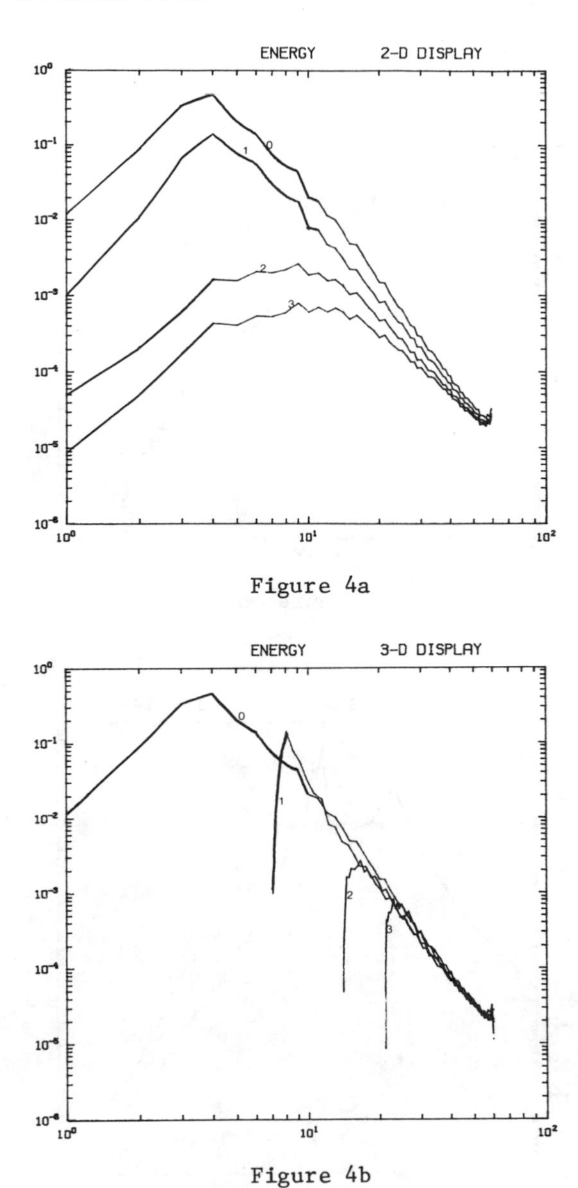

Figure 4. Modal Energy spectra from a simulation with NH=128 and NV=3 for stationary stratified (constant N) turbulence forced by baroclinic instability of a mean shear. Each curve corresponds t a given mode i. (a) spectra are displayed as a function of $\underline{K}_2$; (b) as a function of $\underline{K}_3$.

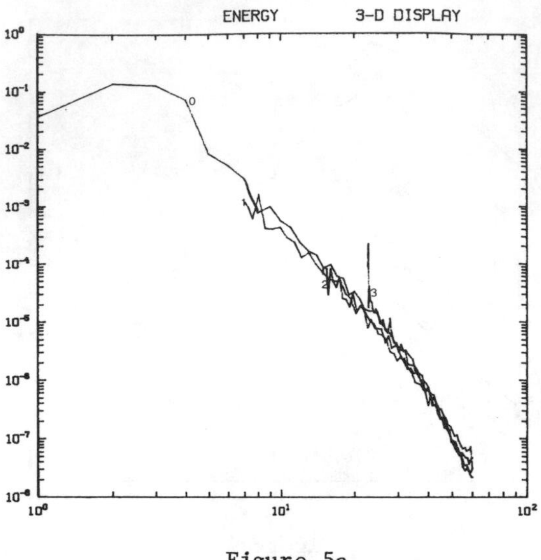

Figure 5a

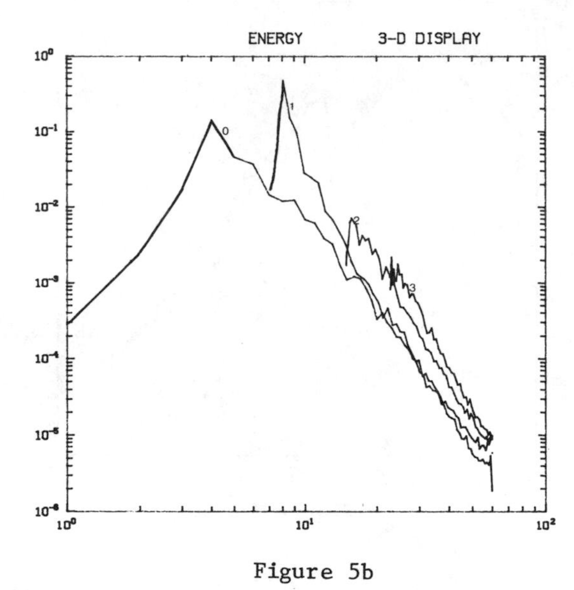

Figure 5b

Figure 5. Same as in figure 4b in the case of an exponential Brunt-vaissala profile: (a) freely decaying turbulence; (b) stationary turbulence case.

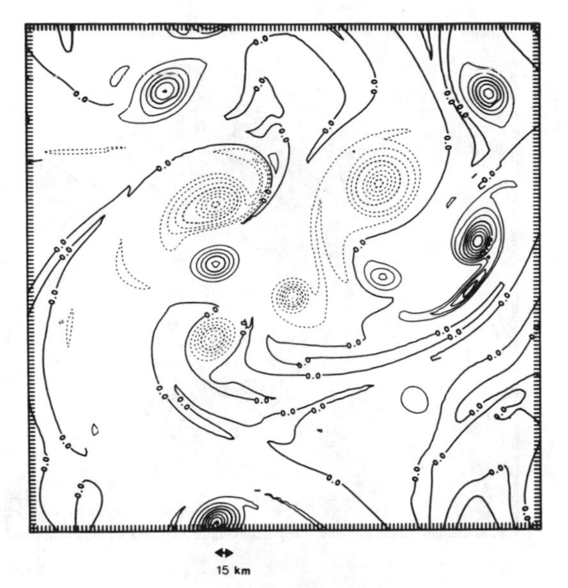

Figure 6a

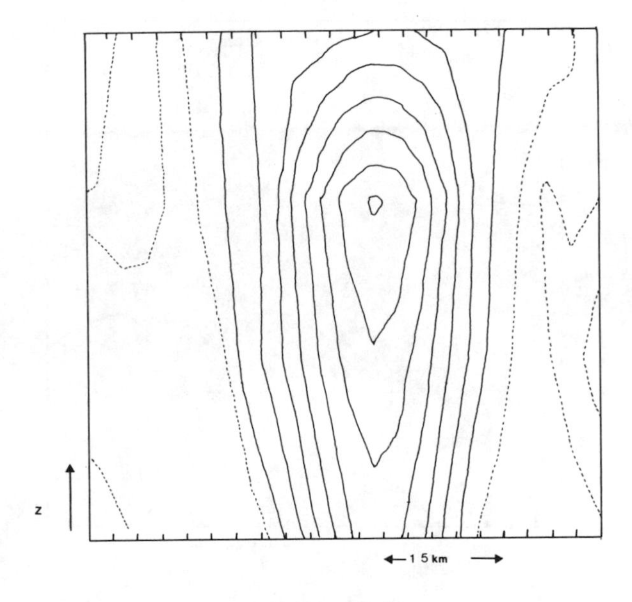

Figure 6b

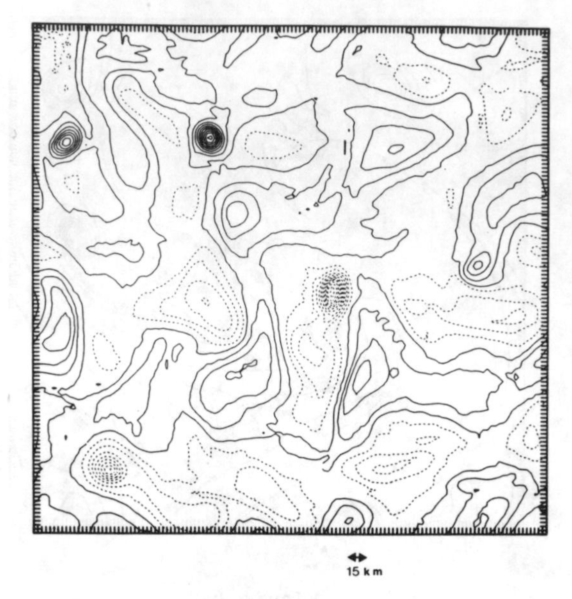

Figure 6c

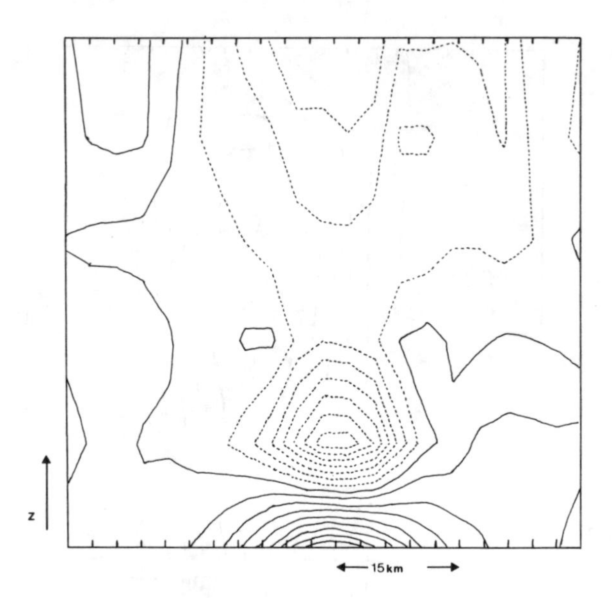

Figure 6d

Figure 6. Vorticity field for freely decaying stratified turbulence (a) and stationary stratified turbulence (c). Vertical sections through a isolated structure are given for each respective case (b) (decaying) and (d) (stationary).

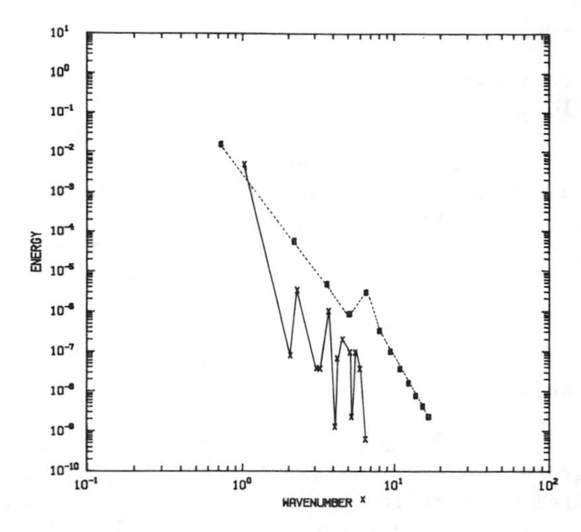

Figure 7a

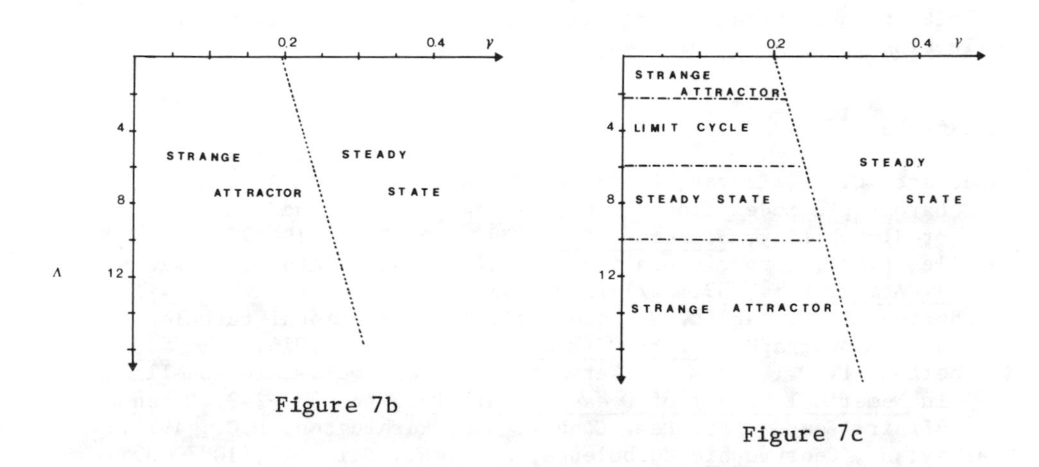

Figure 7b

Figure 7c

Figure 7. (a) spectra of 2-layer turbulence in a channel (see eq. 18). 0 symbols correspond to zonal components ($\bar{a}_n^i$) while x correspond to higher non-zonal harmonics ($\hat{a}_{mn}^i$). Note the negligible fraction of energy retained in the latter harmonics. However, (b) and (c) illustrate the differences in the chaotic behavior of equations (15) when retaining (c) or discarding the higher harmonics terms. Λ measures the supercriticality of the mean flow while $\nu = r/\sqrt{\Lambda}$, where r is the Rayleigh friction parameter.

6.4. Intermittency

Quasi-geostrophic flows also exhibit intermittency properties similar to barotropic flows, with the possibility for potential vorticity to coalesce into three-dimensionally isolated structures (Figures 6a to 6d). Like their barotropic counterpart, their three-dimensional scales are dependent on whether they are amid a freely decaying turbulent field or in a forced environment. An extensive study of the stability of such structure for freely decaying stratified flows has been done by McWilliams (personal communication). A more extensive discussion of the relevance to geophysical phenomena can be also found in McWilliams [1983], in particular for atmospheric observations.

6.5. Unpredictability

Recently there has been abundant literature on the subject. Figures 7a to 7c illustrate the very different behaviors exhibited by systems such as (18) (Klein and Pedlosky, 1985) when taking into account higher harmonics terms and neglecting them ($m \neq 1$), despite the fact that such components store very little of the total energy (Figure 7a): whole ranges of parameter space where the dynamical system can exhibit strange attractors or limit cycles behaviors disappear when neglecting the higher components contributions.

REFERENCES

Basdevant, C., B. Legras, R. Sadourny, and M. Beland, A study of barotropic model flows: intermittency, waves and predictability, J. Atmos. Sci., 38, 2305-2326, 1981.

Boville, B. A., Strongly nonlinear vacillation in baroclinic waves, J. Atmos. Sci., 39, 1227-1240, 1982.

Bretherton, F. P., and D. B. Haidvogel, Two-dimensional turbulence above topography, J. Fluid Mech., 78, 29-154, 1976.

Bretherton, F. P., and M. J. Karweit, Mid-ocean mesoscale modelling, in Numerical Models of Ocean Circulation, pp. 237-249, Ocean Affairs Board, Nat. Res. Counc., NAS, Washington, D.C., 1975.

Charney, J., Geostrophic turbulence, J. Atmos. Sci., 28, 1087-1095, 1971.

Flierl, G. R., Models of vertical structure and the calibration of two-layer models, Dyn. Atmos. Oceans, 2, 341-381, 1978.

Gottlieb, D. and S. A. Orszag, Numerical Analysis of Spectral Methods: Theory and Applications, 170pp., Society for industrial and applied mathematics, J. W. Arrowsmith, Bristol, 1977.

Haidvogel, D. B., Numerical models of the ocean circulation, in CNES summer school, July, 1983.

Haidvogel, D. B., and I. M. Held, Homogeneous quasi-geostrophic turbulence driven by a uniform temperature gradient, J. Atmos. Sci., 37, 2644-2660, 1980.

Haidvogel, D. B., and P. B. Rhines, Waves and circulation driven by oscillatory winds in an idealized ocean basin, Geophys. Astrophys. Fluid Dyn., 25, 1-63, 1983.
Hua, B. L., J. C. McWilliams, and W. B. Owens, An objective analysis of the POLYMODE Local Dynamics Experiment: 2--Streamfunction and potential vorticity fields during the intensive period, J. Phys. Oceanog., (in press), 1985.
Hua, B. L. and D. B. Haidvogel, Numerical simulations of the vertical structure of quasi-geostrophic turbulence, J. Atmos. Sciences (submitted), 1985.
Haidvogel, D. B., A. R. Robinson, and E. E. Schullman, The accuracy, efficiency and stability of three numerical models with application to the open ocean problems, J. Comp. Phys., 34, 1-53, 1980.
Klein, P. and J. Pedlosky, A numerical study of baroclinic instability of large supercretiaslity, J. Atmos. Sciences, (in press), 1985.
Kraichnan, R. H., Inertial ranges in two-dimensional turbulence, Phys. Fluids, 10, 1417-1428, 1967.
Kraichnan, R. H., and D. Montgomery, Two-dimensional turbulence, Reports in Progress in Physics, 547-619, 1980.
Leith, C. E., Atmospheric predictability and two-dimensional turbulence, J. Atmos. Sci., 28, 145-161, 1971.
McWilliams, J. C., A note on a consistent quasi-geostrophic model in a multiply-connected domain, Dyn. Atm. Oceans, 1, 427-441, 1977.
McWilliams, J. C., On the relevance of two-dimensional turbulence to geophysical fluid motions, J. de Mécanique théorique et appliquée, Numéro spécial, Turbulence bidimensionele, 83-97, 1983.
McWilliams, J. C., The emergence of isolated coherent vortices in turbulent flows, J. Fluid Mech., 42, 21-43, 1984.
Orszag, S. A., Numerical simulation of incompressible flows within simple boundries: 1. Galerkin (spectral) representations, Studies in Applied Mathematics, L, 4, 1971.
Owens, W. B., and F. P. Bretherton, A numerical study of mid-ocean mesoscale eddies, Deep Sea Res., 25(1), 1-14, 1978.
Pedlosky, J., Geophysical Fluid Dynamics, 624pp., Springer-Verlag, New York, 1979.
Rhines, P. B., Waves and turbulence on a β-plane, J. Fluid Mech., 69, 417-443, 1975.
Rhines, P. B., The dynamics of unsteady currents, in The Sea, vol. 6, edited by E. D. Goldberg, I. N. McCave, J. J. O'Brien, and J. H. Steele, Wiley & Sons, New York, 1977.
Rhines, P. B., Geostrophic turbulence, Ann. Rev. Fluid Mechanics, 11, 401-441, 1979.
Salmon, R., Baroclinic instability and geostrophic turbulence, Geophys. Astrophys. Fluid Dyn., 10, 25-52, 1980.
Vallis, G. K., On the spectral integration of the quasi-geostrophic equations for doubly-periodic and channel flow, J. Atmos. Sci., 42, 95-99, 1985.

SUPERCOMPUTERS IN OCEAN MODELLING

Mark E. Luther
The Florida State University
Mesoscale Air-Sea Interaction Group
Tallahassee, Florida 32306-3041

INTRODUCTION

The increasing availability of supercomputers to the ocean modeling community in recent years is opening vast new possibilities for oceanographic research. Many research institutions in Europe and the United States now have access to supercomputers. This access not only changes the way one approaches a problem, but enables one to think of new problems that were previously intractable. Ocean modeling lends itself particularly well to supercomputing because of the structure of the partial differential equations involved and the numerical algorithms one must use to solve them. Ocean models usually "vectorize" very easily and require the storage of a large number of variables at a particular instant, enabling one to take full advantage of the capabilities of a supercompter.

This chapter consists of three papers. The first, by Francois Ronday, outlines the general development of supercomputing and discusses the automatic vectorization of a model code on the Cray family of supercomputers. The second paper, by Mark E. Luther, gives a comparison between the Cyber 205 and the Cray X-MP supercomputers and discusses automatic vectorization on both machines and explicit vectorization on the Cyber 205. Luther illustrates the techniques of explicit vectorization in a primitive equation model of the Indian Ocean that has been optimized for the Cyber 205. The third paper, by John M. Klinck, discusses the vectorization of a quasigeostrophic model of the Drake Passage on the Cyber 205 and presents some interesting results from that model.

J. J. O'Brien (ed.), Advanced Physical Oceanographic Numerical Modelling, 255.

VECTOR COMPUTERS AND VECTOR PROGRAMMING

F. Ronday
GeoHydrodynamics and Environment Research
B5 Institut de Physique - Sart Tilman
B - 4000, Liege
Belgique

1. INTRODUCTION

The first computers specifically designed to use parallelism in order to operate efficiently on vectors or arrays of numbers appeared in 1975. The Cray-1 and Cyber 205 are the main examples of pipelined or vector computers.

Parallelism has been introducd because improvements in speed alone cannot produce the required performance. The CDC proposal was to design a machine with four high performance pipelined units.

The advent of highly parallel architectures also introduces the problem of designing new compiler and numerical algorithms that execute efficiently on them [Hockney and Jesshope, 1981; Perott, 1979].

2. HISTORY OF PARALLELISM

Figure 1 shows a ten-fold increase in computing speed every five years. This increase has been possible by combining the technological improvement with the introduction of greater parallelism at all levels of the computer architecture.

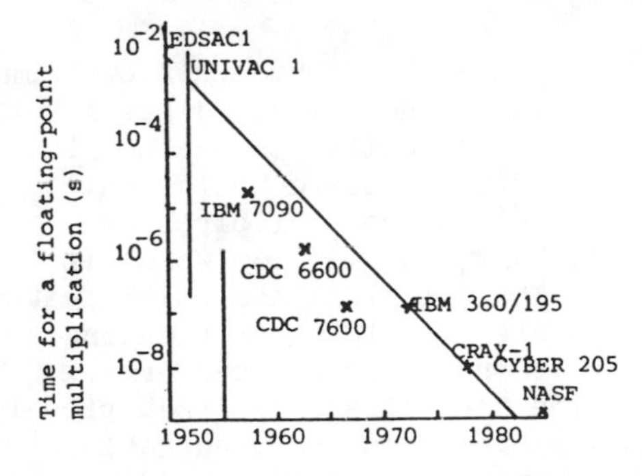

Figure 1. The computer arithmetic speed since 1950 [Hockney and Jesshope 1981]

J. J. O'Brien (ed.), Advanced Physical Oceanographic Numerical Modelling, 257–264.

The technological improvements make possible the implementation of various highly parallel architectures.

From 1950 to 1975, the basic speed of the computers has increased by a factor of 10^3, and the performance of the computers as measured by the inverse of the multiplication time has increased by a factor of 10^5. This additional speed has been made possible by architectural improvements, principally the introduction of the parallelism.

To compare computers it is useful to compare the clock period, which is a measure of the technological speed, with the number of floating operations per second (measure of performance). On the EDSAC1, the clock period (C.P.) was 2 μs. Since the Cray-X-MP has a CP of 9,5 ns the improvement in technology is 200 whereas the improvement of performance (see Figure 1) is 10^6.

The architecture of the first generation of computers is described as serial and is usually referred to as the Von Neumann organization: each operation of the computer has to be performed sequentially. Parallelism refers to the ability to overlap or perform many of these tasks simultaneously. The principle ways of introducing parallelism [Hockney and Jesshope, 1981] are:

a) pipelining: the application of assembly-line technique to improve the performance of an arithmetic or control unit;
b) functional: providing several independent units for performing different functions and allowing these to operate simultaneously on different data;
c) array: providing an array of identical processing elements under common control, all of which perform the same operation simultaneously but on different data stored in private memories;
d) multiprocessing: the provision of several processors, each obeying its own instructions and communicating via a common memory.

A real computer may combine some of all these parallel features. Up to 1980, multiprocessor designs have been concerned.

3. PIPELINED (VECTOR) COMPUTERS

Seymour Cray left Control Data in 1972 and founded his own company (Cray Research Inc.). Four years after he delivered his first computer (Cray-1) to the Los Alamos Scientific Laboratory, the Cray-1 evolved from the CDC 6600 amd CDC 7600. Its technological characteristics are 12 functional units, now all pipelined, a faster clock period of 12.5 ns and a 16 bank, one million word bipolar memory with a 50 ns cycle time. The principal and novel feature was eight vector registers, each capable of holding 64 floating points numbers and a set of 32 machine instructions for performing arithemetic on these vectors. The Cray-1 was the first pipelined vector computer. It has regularly achieved measured rates of 130 M flops (million of floating point operations per second) on real problems.

In this situation, k is wavenumber, ω is phase velocity and c is group velocity. We shall extend these concepts to more complicated problems later.

Let us consider various time differencing schemes for the oscillation equation (3).

3.1. Euler Scheme

$$u^{n+1} = u^n + i\omega\Delta t\, u^n$$

The amplification factor is

$$G = 1 + i\omega\Delta t$$

$$|G| = [1 + (\omega\Delta t)^2]^{1/2} = 1 + O(\Delta t^2) \text{ if } |\omega\Delta t| < 1$$

This is unstable in the absolute sense but stable in the von Neumann sense, since $|G| < 1 + O(\Delta t)$. However, since we do not expect any physical growth for the oscillation equation, we should not allow even $O(\Delta t)$ growth.

3.2. Backward Scheme

$$u^{n+1} = u^n + i\Delta t\omega\, u^{n+1}$$

$$G = (1 - i\Delta t\omega)^{-1}$$

$$|G| = [1 + (\omega\Delta t)^2]^{-1/2} < 1$$

The scheme is stable but dissipative, since the amplitude of the solution decreases in time ($|G|<1$).

3.3. Trapezoidal Rule

$$u^{n+1} = u^n + \frac{i\omega\Delta t}{2}(u^n + u^{n+1})$$

$$G = \frac{1 + i\omega\Delta t/2}{1 - i\omega\Delta t/2}$$

Since $|G| = 1$, the scheme is neutral.

3.4. Leapfrog Scheme

$$u^{n+1} = u^{n-1} + i2\omega\Delta t u^n$$

$$G = G^{-1} + i2\omega\Delta t$$

$$G = i\omega\Delta t \pm (-(\omega\Delta t)^2 + 1)^{1/2}$$

The two other pipelined vector computers are CDC machines (STAR-100, Cyber 203 and Cyber 205) and the TI ASC computer (Texas Instruments Advanced Scientific Compter). The STAR-100 achieved measured rates of 100 M flops on long vectors, but its magnetic-core memory (with a 1,2 μs cycle time) has been surpassed by semi-conductor memory and was introduced to the market in 1979 as the Cyber 203. This machine has been renamed the Cyber 205 and is now very competitive with the Cray-1. The Cyber 205 differs from the Cray-1 in that all vector instructions are processed to and from the main memory. There are no vector registers. The machine can have two or four high performance pipelines. The maximum asymptotic performance on the Cyber 205 is 200 M flops (addition of miltiplication) in 64 bit arithmetic on a 4 pipes machine. The Cyber 205 was first delivered to the Meteorological Office (U.K.) in 1981.

The TI ASC is based on one, two, or four identical general purpose pipelines each capable of rates of 50 M flops. Instructions could be taken from one or two instruction processing units, an example of multiprocessing. This machine suffers from a scalar unit slower than other competitive computers such as the CDC 7600. A TI ASC computer was purchased by the Geophysical Fluid Dynamics Laboratory at Princeton.

The Cray-1 was modified in 1982; the Cray-X-MP appeared with two processors and a clock period of 9.5 ns. The theoretical performance of this machine is 400 M flops.

4. PARALLELISM IN VECTOR COMPUTERS (CRAY AND CYBER)

In vector computers, the parallelism is based on the following operations:

4.1. De-coupling:

Performing different functions, such as logic, addition or multiplication, and allowing these to operate simultaneously on diferent data. Figure 2 illustrates the de-coupling of the operations.

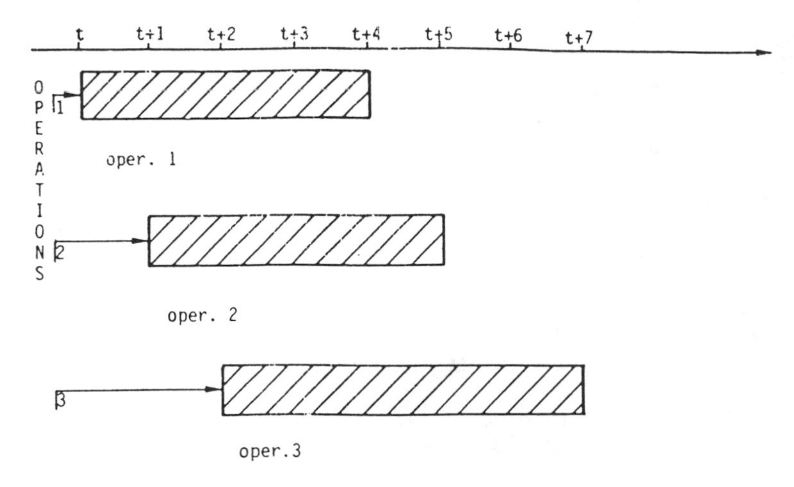

Figure 2. De-coupling: Figure 2 shows that three operations are carried out simultaneously during time (t+2) to time (t+4). The restriction on the parallelism comes from the fact that during a C.P. only one operation can start.

4.2. Segmentation

All the functional units are pipelined and may accept a new set of arguments every clock period. For example a floating point addition needs 75 ns or (6 clock periods) on the Cray-1. Each functional unit is divided (segmented) into elementary operations whose length requires exactly one clock period.

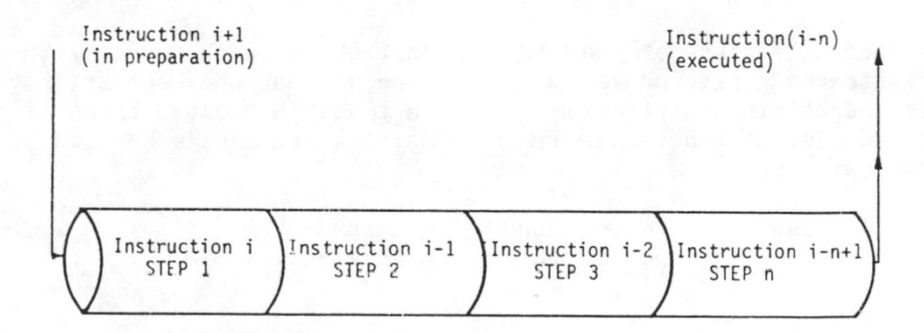

Figure 3. Segmentation of a functional unit (n steps).

In the steady state, the unit provides one result each clock period: the execution time of the unit is divided by a factor n. All the functional units of the vector computers are entirely segmented.

Table I. <u>Differences between scalar and vector operations</u>

```
      DO 10 I = 1,N
      A(I) = SIN[R(I)]
   10 B(I) = A(I) + R(I)
```

i) Scalar

```
I = 1   A(1) = SIN[R(1)]
        B(1) = A(1) + R(1)

I = 2   A(2) = SIN[R(2)]
        B(2) = A(2) + R(2)

I = N   A(N) = SIN[R(N)]
        B(N) = A(N) + R(N)
```

ii) Vector

```
Computation of A A(1) = SIN[R(1)]
                 A(2) = SIN[R(2)]
                 ================

                 A(N) = SIN[R(N)]

Computation of B B(1) = A(1) + R(1)
                 B(2) = A(2) + R(2)
                 ==================

                 B(N) = A(N) + R(N)
```

4.3. Chaining:

A special feature of the Cray-1 architecture is the ability to chain together a series of vector operations so that they operate together as one continuous pipeline. This is possible because there is a decoupling of funcitonal units. Chaining requires adherance to the following rules:

1) The use of different functional units
2) An operand cannot be used twice in a chain.

Example 1:

```
      DO 10 I = 1,N
   10 A(I) = B(I) + C(I)
```

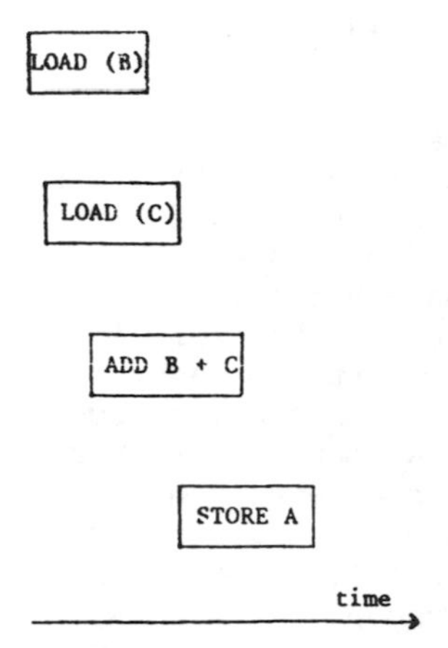

Example 2:

```
      DO 10 I = 1,N
   10 C(I) = A(I) + S * B(I)
```

L
S

LOAD B

B * S = V

LOAD A

Add V + A

STORE C

time

5. AUTOMATIC VECTOR PROGRAMMMING

Automatic vectorization is a process by which the Fortran compiler translates an interactive sequential procedure into parallel procedure without requiring alteration of the Fortran program. No special provisions are required that would affect the transportability of the program. However, the compiler does provide a utility procedure that can enhance vectorization.

The Cyber 200 Fortran allows an explicit vectorization of the programs. Unfortunately the Fortran programs are then not transportable. The Cyber 200 Fortran uses for example, the concept of descriptor (a pointer to a vector). (For explicit programming, see the CDC Cyber 200 Fortran reference manual.)

The most commonly used data structure which can be vectorized is the array or dimensioned variable (in one dimensiononly). The main program structure is the DO-loop.

Example 3:

```
      DO 1 I = 1,N
    1 C(I) = A(I) + B(I)
```

is an ideal combination to vectorization.

In practice, only the innermost DO-loop can be vectorized. The compiler analyzes this DO-loop of the Fortran program and determines whether vector processing methods can be applied to improve overall program efficiency.

DO-loops that have certain characteristics can be automatically vectorized. Some input/output statements or some IF statements, for example, are not acceptable in a DO-loop that is to be vectorized. On the Cray and Cyber 205 the control variable must be integer. The statements of a DO-loop can contain scalar variables or constants in a provided expression change by fixed increments each time around the loop. The index expressions may also involve simple arithmetic operatiors. Under these restictions vectorizaton is still possible.

6. REFERENCES

CDC, CDC CYBER 200 FORTRAN version 2 reference manual, CDC Publication, 6048500, 1981.

CRAY, CRAY-1 and CRAY X-MP computer systems, reference manual, Cray Research Inc. Publication, SR0009, 1983.

Hockney, R. W., and C. R. Jesshope, Parallel Computers, 423 pp., Adam Hilger Ltd, Bristol, 1981.

Perrott, R. H., A language for array and vector processors, ACM Trans. Prog. Lang. Syst., 1, 117-95, 1979.

OCEAN MODELLING ON SUPERCOMPUTERS

Mark E. Luther
Mesoscale Air-Sea Interaction Group
The Florida State University
Tallahassee, Florida 32306-3041

ABSTRACT. We describe the use of supercomputers in general and the necessary considerations for their efficient use in ocean modelling. A brief history of supercomputer development is given, and the relative merits of the Cray X-MP and Cyber 205 (the two leading supercomputers at present) machines are compared. Automatic vectorization is described for both machines. Explicit vectorization via the Cyber 205 Fortran vector extensions is discussed.

We next describe a specific example of a model of the Indian Ocean. The model was originally run on a Cyber 170/760 scalar computer in standard Fortran and has been rewritten in the explicit Cyber 205 Vector Fortran. Sections of model code before and after vectorization are presented to illustrate the usage of the special features of the Cyber 205. The vectorized version of the model runs approximately 32 times faster than the scalar version. Some key results of the model simulation are also described.

1. INTRODUCTION

A supercomputer may be defined as the most powerful general purpose computer available at a particular time. Today's supercomputers are the Class VI machines -- the Cray 1S, the Cray X-MP and the Cyber 205. Since the Cray X-MP and the Cyber 205 are the two leading supercomputers at the time of this writing (and the ones most familiar to the author) we will concentrate primarily on them. We will refer to the Cray X-MP as simply the X-MP and the Cyber 205 as the 205. Although the Cray X-MP is much more powerful than the Cray 1S, their basic architectures are very similar. The top of the line X-MP is roughly equivalent in performance to the most powerful 205; however, there are significant differences in their architectures that can cause them to perform quite differently on the same problem. One must therefore be wary of benchmarks that claim to show one machine out-performing the other.

There are other machines currently in use that can be classified as supercomputers. IBM and Floating Point Systems are building

J. J. O'Brien (ed.), Advanced Physical Oceanographic Numerical Modelling, 265–297.

systems with an IBM front end computer controling several FPS add-on array processors which are capable of supercomputer speeds. Several Japanese companies also build supercomputers that are very similar to the Cray machines. Most comments made here about the X-MP would apply to these machines as well.

The supercomputers of today all exploit parallelism to varying degrees to obtain their great number crunching speeds, which are on the order of hundreds of million floating-point operations per second (MFLOPS). It should be pointed out that speed alone is not what makes a supercomputer super. Large, fast central memories are just as important for ocean modelling and many other problems as are high MFLOPS rates. If one is to perform millions of floating-point operations per second, one must be able to store and readily access millions of floating-point operands. The high resolution and large basin size required in many ocean modelling applications can easily push the central memory limits of even the largest supercomputer.

In this paper we will try to make the reader aware of the considerations one must make when implementing ocean models on supercomputers. First, we will briefly describe the history of supercomputers and compare the architectures of the X-MP and the 205. We will describe the form of parallelism called "pipelining" [Kogge, 1981] and discuss its utilization via the Fortran programming language through automatic vectorization of the program code and through the Cyber 205 vector extensions to Fortran. We will not discuss the general problem of algorithm selection for a specific architecture; for this, the reader is referred to the exhaustive review by Ortega and Voigt [1985]. In the final section, we will give a specific example of the vectorization of a model of the Indian Ocean [Luther and O'Brien, 1985; Luther, O'Brien and Meng, 1985] to illustrate the concepts described in the previous sections. The model is a nonlinear reduced gravity model and is forced by observed winds. This model has been very successful in reproducing many of the observed features of the circulation in the northwest Indian Ocean, especially along the Somali and Arabian coasts. We will show how many of the special features of the Cyber 205 Fortran vector extensions are employed to optimize the model's use of the 205 hardware, and will give pre- and post-vectorization timing figures.

2. HISTORY

We will attempt to give an abbreviated history of supercomputing. For a more complete treatment of the subject, the reader is directed to Kuck [1978], Riganati and Schneck [1984], or Ortega and Voigt [1985]. Early efforts in high-speed computation relied on improving the clock rate or cycle time of the computer; however, electrons can only travel so fast and connections have a finite length. To achieve the dramatic improvements in computation rate, modern computers exploit some form of parallel processing. The idea of parallel processing dates back to at least 100 B.C. [Worlton, 1981], to a Greek computing tablet that has three positions for simultaneous calculation of different parts of a problem. Richardson [1922] suggested the use of 64,000 human computers working in parallel to perform numerical weather prediction. The very first electronic computer, the ENIAC [Eckert et al., 1945], was capable of parallel computation, but its designers decided that serial computation was preferred, due to the problems of coordinating parallel operations.

In the mid-1950's, the UNIVAC-I and IBM 7090 and 7094 computers developed parallel operation between data transfer and central processor (concurrent I/O). In the mid-1960's, the IBM 360/91 and 95 and the CDC 6600 appeared. Both the IBM and the CDC machines had multiple pipelined functional units and large interleaved memories. Pipelining is a form of parallelism analogous to an industrial assembly line (see Figure 1), in which an operation to be repeated many times is divided into several sub-units that then work like an assembly line on the data. The CDC 6600 was a Seymour Cray design and was a predecessor of the Cray 1. Vector instructions were later developed for the 6600's pipelines which could be invoked from Fortran by calling a "Q8" subroutine. These "Q8" calls are still used on the Cyber 205.

In the late 1960's and early 1970's, the CDC 7600, the Texas Instruments ASC, the CDC STAR 100 and the ILLIAC IV appeared on the computing scene. The 7600, another Seymour Cray design and predecessor to the Cray 1, was essentially a very fast scalar machine even though it did have pipeline capability. The TI-ASC was one of the first true vector processors, employing the pipeline technique on long strings or vectors of data. The STAR 100, also a vector processor, evolved into the Cyber 203 and ultimately into the Cyber 205. The ILLIAC IV was the first parallel array processor, consisting of an 8x8 array of arithmetic processing elements. This architecture was especially well suited to fluid flow problems, but led to numerous programming difficulties.

Seymour Cray left Control Data Corporation in the early 1970's to form Cray Reserch, Inc. and began work on the Cray 1. The first Cray 1 was installed at Los Alamos National Laboratory in 1976. The Cray 1 was a vector processor with one million words of central memory, and employed vector registers to handle data transfer between memory and the vector pipes [Russell, 1978]. The Cray 1 was the first supercomputer generally available to the ocean modeling commmunity through the National Center for Atmospheric Research in Boulder, Colorado.

At the time of this writing, the Cray 1 has evolved into the Cray X-MP [Chen, 1984], and the CDC STAR 100 has evolved into the Cyber 205 [Lincoln, 1982]. Presently, the X-MP is available with one, two or four central processors operating in parallel and with one to eight million 64-bit words of memory. The clock period on the X-MP is 9.5 ns, and the maximum vector speed on a four processor machine is about 800 MFLOPS. The 205 is available with one, two, four, or eight vector pipes and with up to 16 million 64-bit words of memory. The clock cycle is 20 ns, which is considerably slower than the X-MP, but because of the way the vector pipes are structured and because it can operate on 32-bit words, it also has a peak vector rate of 800 MFLOPS. Note, however, that peak vector rates are never attained in any realistic program applications, due to overhead in loading the vector pipes and in transferring the data from memory to the vector registers in the X-MP (more about this later).

Both Cray Research and Control Data have announced more powerful versions of the X-MP and the 205, to be available in the immediate future. The first Cray 2 has already been delivered to the Lawrence Livermore National Laboratory. It has a cycle time of 4 ns and is capable of 1000 MFLOPS (1 GFLOPS). Cray has announced the Cray 3, with 16 parallel processors, a cycle time of 2 ns, peak vector rates of 10 GFLOPS and a main memory of one billion 64-bit words. ETA Systems, Inc., which has taken over supercomputer devlopment from Control Data Corporation, is building the ETA 10, which is essentially eight 4-pipe, 4-million word Cyber 205's operating in parallel and sharing a 256 million 64-bit word memory. The ETA 10 is capable of peak vector rates of 10 GFLOPS. Both machines should begin beta (initial field) testing in early 1987. By the time this paper is published, however, much of this information will be outdated.

3. CRAY X-MP - CYBER 205 COMPARISON

The X-MP and 205 are very similar in many ways. Both employ vector pipeline architecture and have comparable peak vector rates and very large memories. There are significant differences that can severely affect relative performance from one machine to the other. The primary difference between the two machines is in the way they handle data in and out of the vector pipes (see Figure 2). The X-MP uses vector registers of 64 words each to transfer data to and from the vector pipes. The vector pipelines themselves are each specialized functional units for integer add, shift and, logical, and for floating point addition, multiplication, and division. The 205 uses memory to memory vector operations, wherein the data streams directly from main memory to the vector pipe and back again. The 205 has separate scalar and vector arithmetic units. It has general purpose, reconfigurable vector pipes with fully segmented functional units.

As mentioned earlier, the peak vector rates are not attained in reality because of the overhead associated with filling the vector pipeline, that is, the delay between the time that the first operands enter the pipe and the time that the first results emerge from the pipe. This overhead is frequently called the start-up time. The start-up time for a vector multiply on the X-MP is 10 cycles, while on the 205 it is 70 cycles. The effect of this start-up time on total vector performance is a function of vector length. For short vectors, the start-up time can be a substantial fraction of the total time required for the vector operation. A useful measure of the efficiency of a vector operation is the half-performance length [Hockney and Jesshope, 1981], which is the vector length needed to achieve one-half the peak vector operation rate. For the 205, the half-performance

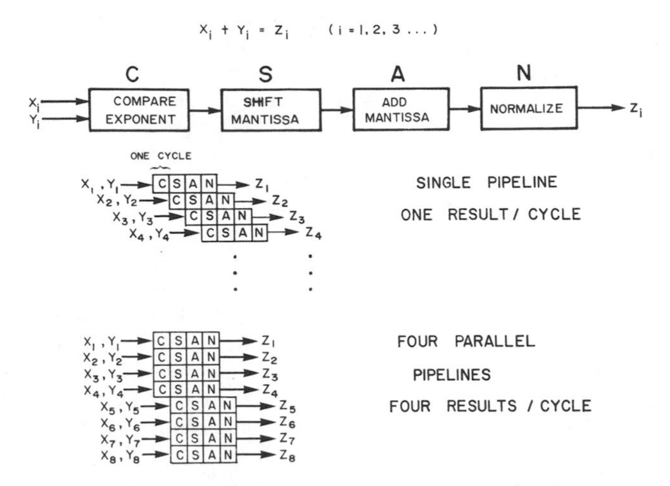

Figure 1. Simplified schematic of a pipelined floating point addition. The operation is segmented into four sub-operations: compare exponent, shift mantissa, add mantissa and normalize. These operations are performed on the vector or data string in assembly line fashion as either a single pipeline (top) or as multiple parallel pipelines (bottom). After Riganati and Schneck [1984].

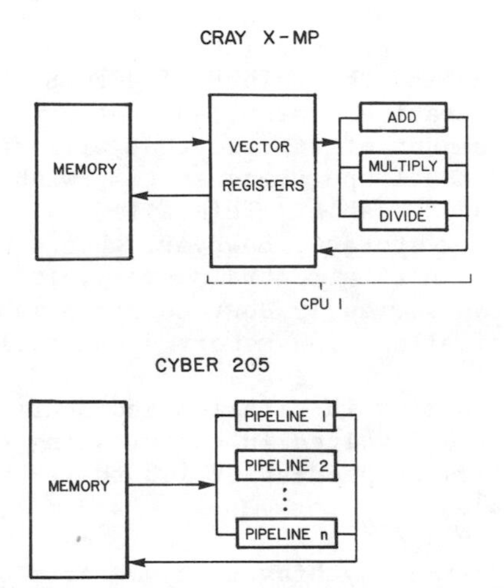

Figure 2. Schematic of the Cray X-MP (top) and the Cyber 205 (bottom). The X-MP utilizes vector registers of 64 words each to transfer data into and out of the specialized pipelined functional units in each CPU. The 205 transfers data directly from main memory to the reconfigurable pipelines and back again. The X-MP may have multiple central processing units, while the 205 has multiple parallel pipelines.

length is 100, while on the X-MP it is about 10. For short vectors, the X-MP is therefore much faster, while the 205 performs more efficiently in very long vectors. Much of the advantage gained by the low start-up time on the X-MP is offset by the time required to load the vector registers from memory. The X-MP only works on vectors of length 64, no matter how long the data stream. The 205 can work on vectors of any length up to $2^{16}-1$, because it uses 16 bits to decribe the vector length.

When programming on the X-MP, one therefore tries to keep the data in the vector registers for as long as possible. For example, whereas one might break a computation into several separate statements on a scalar machine, it is more efficient on the X-MP to combine these statements into one long statement. The compiler is then more likely to generate instructions to perform "chaining", where the results from one arithmetic unit are passed directly into another arithmetic unit without passing through a vector register. The machine will keep the operands in the vector registers until the entire computation is complete, thus minimizing the time required to load the registers and store the results.

On the other hand, when programming the 205, one tries to make the vectors as long as possible to minimize the effects of start-up time. This is easily accomplished in most ocean modelling applications because of the large model domains and high resolution typically employed. For instance, a calculation involving arrays of dimension M by N can be treated as a single vector operation on vectors of length M N, using special features of Cyber 205 vector Fortran to mask off the boundary points. One also tries to maximize the use of expressions of the form vector + scalar * vector, called a linked triad, because this computation proceeds at the speed of a single vector add, effectively doubling the number of MFLOPS attained.

The X-MP has a real memory operating system, so that an application is restricted to the amount of storage that will fit in central memory. The 205 uses a virtual operating system, with a 48-bit address space, and is bit-addressable. This allows a program to utilize up to 4.4×10^{12} words of storage; however, when a program accesses storage locations that are not in central memory, it must "page" them in from disk, which can seriously degrade performance. The 205 also has the capability of allocating memory dynamically during program execution.

The X-MP has multiple CPU's working in parallel and sharing central memory. Due to the overhead involved in coordinating the CPU's, a two processor system offers a speed-up of 1.5 to 1.9 times over a one CPU system [Chen, 1984].

Vectorization on both the X-MP and the 205 is performed automatically by the compiler. Most everything that vectorizes automatically on the X-MP will also vectorize on the 205, with a few exceptions that will be discussed below. The 205 has the added advantage that one can write explicit vector code using the Cyber 205 Fortran vector extensions. One can also invoke hardware instructions directly from Fortran, but of course, the code becomes non-portable.

4. AMDAHL'S LAW

Any realistic modelling application will involve a mixture of both scalar and vector computations. Amdahl's Law [Amdahl, 1967] states that the speed of such a mixed mode calculation will be some intermediate speed determined by the fraction performed in each mode and will usually be dominated by the slower speed, ie., that even a small percentage of scalar computation can be very costly. More specifically, let $t = t_s + t_v$ be the total time for a computation, where t_s is the time for the scalar portion of the code and t_v the time for the vector portion. If 0 is the total number of operations, then the total computation rate is $R = 0/t$, the scalar rate is $R_s = 0_s/t_s$ and the vector rate is $R_v = 0_v/t_v$.

We then have

$$t = \frac{0}{R} = \frac{0_s}{R_s} + \frac{0_v}{R_v} \tag{1}$$

Define $P_v = 0_v/0$ as the fraction of the computation that is vectorized to get

$$\frac{1}{R} = \frac{1}{R_s} + P_v \left(\frac{1}{R_v} - \frac{1}{R_s}\right). \tag{2}$$

This relationship is plotted in Figure 3 for a hypothetical machine with a scalar rate of 5 MFLOPS and a vector rate of 100 MFLOPS. As we can see, if half of the operations are vectorized, the computation speeds up by less than a factor of two. To find out how much vectorization is required to make a machine run N times faster than scalar, let $R = NR_s$ and we find

$$P_v = \frac{N-1}{N}\left(\frac{Rv}{R_v - R_s}\right) \tag{3}$$

For our hypothetical machine, if we desire a speed-up of 10 times scalar, we find that $P_v = 0.9$. It is clear that it pays to vectorize as much code as possible.

5. AUTOMATIC VECTORIZATION

If we wish to maintain portability in our models, we can code them in good, clean, standard Fortran and let the compiler do the vectorization automatically. Both the X-MP and the 205 compilers recognize DO-loop structures that can be vectorized and generate the appropriate vector hardware instructions. There are certain conditions that inhibit vectorization and that can be easily avoided, while there are other situations, such as those involving recursion, that will not vectorize under any condition. We will discuss these conditions with examples of some simple DO-loops that do and do not vectorize.

5.1. Condition 1

The DO-loop increment need not be one.

Example 1:

```
      DO 1 I=1, 100
    1 A(I) = B(I) + C(I)
```

This loop vectorizes trivially.

Example 2:

```
      DO 2 = 1, 100, 7
    2 A(I) = B(I) + C(I)
```

This loop vectorizes without any problem on the X-MP, which simply loads every 7th element of B and C into vector registers, and stores the results from a register into every 7th element of A. On the 205 the loop also vectorizes, but first the operands must be contiguous in memory. The 205 will dynamically allocate three temporary arrays in what is called dynamic stack. It will then perform a periodic gather on B and C, placing every 7th element into dynamic stack. The vector add will be performed on the two temporary arrays and the result stored in the third. A periodic scatter will then place the results into every 7th element of A. When the operation is complete, the 205 will free the dynamic stack occupied by the temporary arrays. There is thus considerable overhead associated with these data motion instructions, but this overhead is comparable to that required to load and store the vector registers on the X-MP.

Example 3:

```
      DO 3 K=1, 100
    3 V(I,J,K) = W(I,K) + C(K)
```

This loop vectorizes on both machines, and again the X-MP loads the appropriate elements of W and all of C into the vector registers and stores the results from the registers back into the appropriate elements of V, while the 205 will perform a periodic gather on W (with starting point I and period I_{max}), and a periodic scatter on V (with starting point (I,J) and period $I_{max}*J_{max}$), storing the temporary vectors in dynamic stack.

Example 4:

```
      DO 4 K=1, 100
    4 A(II(K)) = B(K)*C(II(K))
```

This loop will not vectorize automatically on the X-MP or on the 205, because the subscript is not of the proper form (this is called indirect addressing). On the 205, this loop can be vectorized by explicitly invoking a random gather and scatter through the Cyber 205 Fortran vector extensions (which are described in the next section), but, at the time of this writing, the compiler will not recognize this as a vectorizable loop.

5.2. Condition 2

Vectorization of the inner loop is a necessary but not sufficient condition for the outer loop also to vectorize.

Example 5:

```
      DIMENSION X(100,100), Y(100,100), Z(100,100)
      DO 5 J=1, 100
      DO 5 I=1, 100
    5 X(I,J) = Y(I,J) + Z(I,J)
```

On the X-MP, only the inner loop will vectorize; however, because of the way the X-MP uses the vector registers, there is no advantage to be gained by vectorizing the outer loop. The 205 automatically vectorizes both loops, creating vectors of length 10,000. This is the type of calculation where the 205 gives its best performance.

Example 6:

```
      DO 6 J=1, 100
      DO 6 I=1, 100, 2
    6 X(I,J) = Y(I,J) + Z(I,J)
```

Again, only the inner loop vectorizes on the X-MP, similarly to Example 2. The 205 will perform a periodic gather/scatter and will vectorize both loops; however, due to the overhead associated with the gather/scatter operations, it would be more efficient to write this loop in explicit vector notation and perform the computation on every element of the arrays but only store every other element into X. This is accomplished with the WHERE statement and the BIT data type, which will be discussed in the next section.

Example 7:

```
      DO 7 J=1, 100
      DO 7 I=2, 100
    7 X(I,J) = X(I-1,J) + Z(I,J)
```

This loop will not vectorize on either machine because it is recursive, ie., one pass through the loop uses results of a previous pass through the loop (a dependency in Cray terminology). The 205 does have special hardware instructions, which the compiler will generate, that will optimize this computation in the scalar processor.

5.3. Condition 3

The total iteration count must be less than 2^{16} for a nest of loops. This condition applies only to the 205, because the 205 uses a field of 16 bits to describe the length of a vector.

5.4. Condition 4

No control statement except CONTINUE can appear in the loop.

Example 8:

```
      DO 8 I=1, 100
      C(I) = B(I) + S*A(I)
      CALL SUB2 (A(I)), B(I), C(I))
    8 CONTINUE
```

This loop will not vectorize on either machine, because control is transferred out of the loop by the subroutine call. The subroutine call should be moved outside the loop and the entire arrays A, B and C should be passed to it. Loop 8 would then be a linked triad, which is the fastest vector instruction possible on the 205.

5.5. Condition 5

Relational operators must not appear in any part of the loop.

Example 9:

```
      DO 9 I=1, 100
```

```
        A(I) = B(I) + S
        IF (A(I).LT.0) A(I)=1.0
      9 CONTINUE
```

Neither machine will vectorize this loop because of the IF statement. There are special subroutine calls on the X-MP and special Fortran vector extensions on the 205 that can make this loop vectorize.

5.6. Condition 6

Only real, integer, logical or complex data elements can appear in the loop. On the 205, if complex data are present, the UNSAFE option must be specified for the compiler to have a chance to vectorize the loop.

5.7. Condition 7

No input or output can appear in the loop.

Example 10:

```
        DO 10 I=1, 100
        F(I) = A(I) + B(I) + C(I)
        PRINT 101, F(I), A(I), B(I)
     10 CONTINUE
    101 FORMAT (1H, 3(E15.7, 2X))
```

This loop cannot vectorize because of the print statement. This is inefficient Fortran even on a scalar machine, but it is sometimes done to help locate errors in a program.

5.8. Condition 8

No external function references of subroutine calls, other than intrinsic functions recognized by the compiler, can appear in the loop.

Example 11:

```
        DO 11 I=1, 100
     11 A(I) = FUNC (B(I),C(I))
```

This loop will not vectorize because of the function reference. On the 205, the function FUNC could be written as an explicit vector function using the Fortran extensions.

5.9. Condition 9

No data elements that appear in EQUIVALENCE statements can appear on the left side of an assignment statement, unless the UNSAFE option is specified. This applies only to the 205. The 205 compiler cannot check for recursions among equivalenced arrays, so it refuses to vectorize loops where they appear on the left side of an assignment statement. If the UNSAFE compiler option is specified, the compiler assumes the programmer has checked for recursions and vectorizes the loop.

Example 12:

```
      SUBROUTINE SAM (A,B,C,N)
      DIMENSION A(N), B(N), C(N), T(1)
      COMMON/TEST/AA(100)

      EQUIVALENCE (T(1), AA(10))
      DO 12 I=1,50
   12 T(I) = B(I) + C(I)
```

This loop will not vectorize on the 205 unless the UNSAFE option is specified. The loop could be possibly recursive; for instance, if AA were equivalenced in the main program to B or C, the loop would be recursive.

5.10. Condition 10

No vector assignment statements can appear in a loop. This applies to the 205 only, since the X-MP does not support explicit vector syntax.

Example 13:

```
      DIMENSION A(4,100), B(4,100), T(4,100)

      DO 13 J=1,100
   13 T(1,J;4) = B(1,J;4) + C*T(1,J;4)
```

This loop will not vectorize because the vector syntax appears within. This is extremely inefficient and would probably run slower than scalar because of the start-up time and the very short (4 element) vector length. The loop should be rewritten in explicit vector notation as

```
T(1,1;400) = B(1,1;400) + C*T(1,1;400)
```

which is equivalent to the doubly nested loop

```
      DO 13 J=1, 100
      DO 13 I=1, 4
   13 T(I,J) = B(I,J) + C*T(I,J)
```

which you will recognize as a linked triad, the fastest possible vector operation on the 205. We will discuss the explicit vector syntax in the next section.

5.11. Condition 11

Loop-dependent subscripts must have one of the forms I, I+M, I-M or I*M, where I is a control variable (or a Constant Increment Integer (CII) in Cray terminology) and M is an integer constant. In addition, on the 205 the subscript expression must be part of the subscript and not calculated as a separate index.

Example 14:

```
      DO 14 I=1, 100
      J=N*I
      K=N+I
   14 A(I) = B(K) + C(J)
```

This loop will not vectorize on the X-MP or on the 205. If the loop is rewritten as

```
      DO 14 I=1, 100
   14 A(I) = B(N+I) + C(N*I)
```

it will vectorize on both machines, not only because the subscripts are of the proper form, but also because the expressions are contained within the subscripts.

5.12. Condition 12

In references to dummy arrays, the terminal value of the loop cannot be a variable, unless the UNSAFE option is specified. This applies only to the 205, because the compiler cannot trace back from the dummy argument to see if the total loop count is less than 2^{16} (see Condition 3).

Example 15:

```
      SUBROUTINE SUB1 (A,B,C,N)
      DIMENSION A(N), B(N), C(N)
      DO 15 I=1, N
   15 A(I) = B(I) + C(I)
```

This loop will not vectorize on the 205 because the compiler cannnot determine the value of N. If the UNSAFE option is specified, the compiler assumes that the programmer knows what he is doing and vectorizes the loop. The X-MP will always vectorize this loop, since it works on only 64 elements at a time.

5.13. Condition 13

Loop independent subscripts (invariant integers) are recognized by the compiler on both machines and the loop is vectorized if otherwise possible.

Example 16:

```
      DO 16 I=1, 400
   16 W(I,J) = V(I,J,K) + C(I)
```

This loop vectorizes on the X-MP and the 205, since the values of J and K are constant within the loop.

5.14. Condition 14

The control variable cannot appear anywhere except in a subscript reference. On the X-MP, the control variable, which is itself a CII, may appear in expressions defining a CII.

Example 17:

```
      DO 17 I=1, 100
   17 D(I) = I
```

This loop vectorizes on the X-MP because I is a CII. It will not vectorize on the 205, but there is a vector function to perform this task:

```
      D(1;100) = Q8VINTL (1.,1.;D(1;100)).
```

5.15. Condition 15

When a loop appears to have recursive properties (dependencies) the compiler will not vectorize the loop.

Example 18:

```
      DO 18 I= 2, 100
   18 W(I,J) = W(I-1,J) + 2.0*A(I)
```

This loop is recursive, since the Ith pass through the loop depends on the result of the (I-1)th pass through the loop. This cannot be done in a vector pipeline, since the Ith and the (I-1)th elements are computed simultaneously. The 205 has very fast scalar hardware instructions, called StackLib calls, that will perform recursive computations very efficiently.

Good, clean, standard Fortran DO loops will usually be vectorized automatically by the compilers on the 205 and the X-MP, with the exceptions noted above. Most DO loops that do not vectorize can usually be made to vectorize with just a slight restructuring of the code. With automatic vectorization, portable code can be maintained, while still taking full advantage of the hardware features of supercomputers.

6. CYBER 205 FORTRAN VECTOR EXTENSIONS

There are some instances where automatic vectorization does not generate the most efficient hardware instructions. There are other situations where the compiler simply cannot recognize vectorizable computations that the machine's hardware can perform. There are still other situations where it is more convenient to write code in explicit vector syntax. On the Cray machines, the only choices are to use automatic vectorization or to write CAL assembly language subroutines. On the Cyber 205, however, the Fortran Vector Extensions provide another alternative, one that can produce much more efficient machine instructions as well as more efficient Fortran code. The programs are no longer portable, and are difficult to understand for those unfamiliar with the syntax, so that one must weigh the great gains in execution speed that are possible against these disadvantages.

The 205 defines a vector as a set of contiguous virtual memory locations. A vector is uniquely described by three attributes -- data type, starting address, and length. The type of the vector can be Real, Half Precision, Double Precision, Complex, Integer or Bit (the Half Precision and Bit data types are unique to the 205). The starting address is any virtual memory location from 0 to $2^{48}-1$, and the length may be from 0 to 65535 elements (0 to 32767 for double precision and complex data types). There is thus a distinction between a vector and an array. An array may be multidimensional (because our brains insist on thinking of things that way) while a vector is linear, as is computer memory. A vector may coincide with an array -- in fact, it almost always does, unless it exists in dynamic stack.

A vector is specified by a descriptor of the form

```
TYPE (STARTING ADDRESS;LENGTH).
```

A decriptor points to a vector. For example, if the array X is declared as

```
REAL X(100,100)
```

the vector that coincides with the entire array X is expressed as

```
X(1,1;10000);
```

that is, the vector of memory locations beginning at X(1,1) and extending for 10,000 elements. The type of the vector is the type of the array. This is called an explicit descriptor. Another descriptor might point to a vector that coincides with only a portion of the array X, such as

```
X(2,2;798)
```

In explicit vector syntax, the DO loop

```
      DO 19 J=1, 100
      DO 19 I=1, 100
   19 X(I,J) = Y(I,J) + Z(I,J)
```

would be written as

```
      X(1,1;10000) = Y(1,1;10000) + Z(1,1;10000).
```

The Cyber 205 Vector Fortran language supports a new data type declaration DESCRIPTOR. A DESCRIPTOR is a 64-bit word that describes a vector. The lower 16 bits of the word contain the starting address as in Figure 4. The type of the descriptor is determined as for other Fortran variables, either by the first letter or by a type declaration statement. Information is stored into the descriptor variable with the ASSIGN statement:

```
      DESCRIPTOR XD, YD, ZD
      ASSIGN XD, X(1,1;10000)
      ASSIGN YD, Y(1,1;10000)
      ASSIGN ZD, Z(1,1;10000)
```

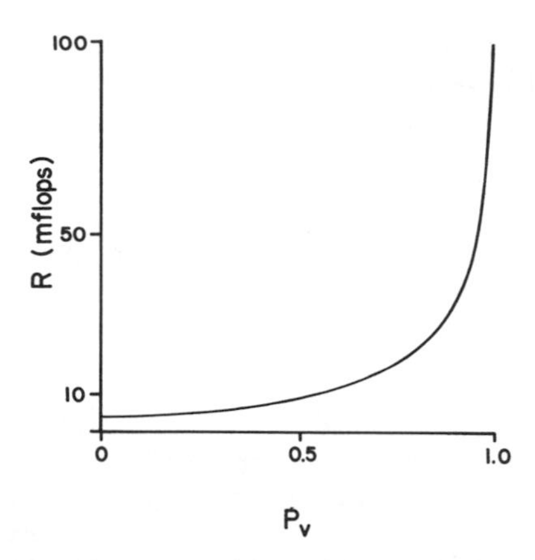

Figure 3. Amdahl's law for a hypothetical machine with a scalar rate of 5 MFLOPS and a vector rate of 100 MFLOPS. The total rate of the mixed computation, R, is plotted against the fraction of the code vectorized, P_V.

DESCRIPTORS

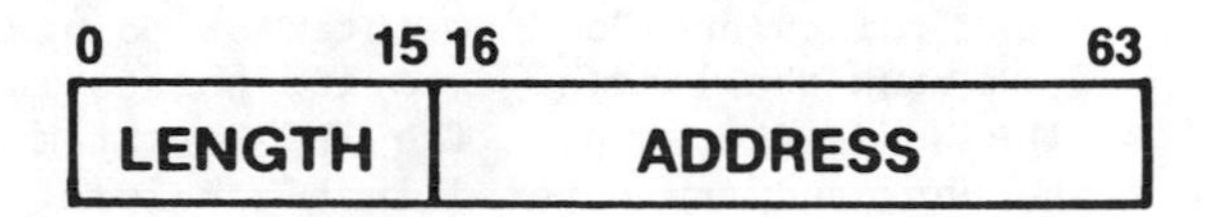

$0 \leq$ LENGTH ≤ 65535

$0 \leq$ ADDRESS $\leq 2.2 \times 10^{12}$ WORDS

Figure 4. The Cyber 205 DESCRIPTOR data type. A DESCRIPTOR points to a vector and occupies one 64-bit word in storage, with the first 16 bits containing the vector length and the last 48 bits containing the starting address in virtual storage of the vector.

XD thus points to the same vector as X(1,1;10000), etc. This is implicit vector notation. Explicit descriptors are constructed by the compiler as they are needed, whereas implicit descriptors are constructed by the programmer using the ASSIGN statement. The ASSIGN statement stores information from an explicit descriptor into a variable. Implicit descriptors must be used to obtain information about a vector, i.e., its starting address or its length. Loop 19, written in implicit vector syntax, becomes

```
XD = YD + ZD
```

which is a considerable abbreviation.

Another useful new data type supported by Cyber 205 Fortran is the BIT data type. Since the 205 is bit-addressable, one can define arrays or vectors of single bits. This is very useful for loop control, for instance, for determining which points in an ocean model are interior points or are land points.

Example 20:

```
DIMENSION U(JMAX,KMAX)

BIT UATSEA(JMAX,KMAX)


UATSEA(1,1;JMAX*KMAX) =

      U(1,1;JMAX*KMAX).NE.SPVAL
```

Here the data points U are flagged with a special value where they are outside the irregular model domain, so that the bit vector UATSEA will be "off" at those points and "on" at interior points. This bit vector can be used to mask the land or exterior points in a vector computation using the WHERE statement. The WHERE statement is similar to a vector IF statement, and can be used in block form.

Example 21:

```
WHERE (UATSEA(1,1;JMAX*KMAX))

U(1,1;JMAX*KMAX) = (Interior Equation in Vector Syntax)

END WHERE
```

This WHERE block will perform the vector computation for every element of U, but will only store the results of the computation where the bit vector UATSEA is turned "on". Since the computation proceeds at vector speeds, the extra computation involved at the exterior points is not significant, even when a large portion of the vector is masked.

Other methods of computing on portions of vectors involve the data motion vector functions EXPAND, COMPRESS, SCATTER, GATHER, MASK and MERGE. The use of GATHER and SCATTER by the compiler was described in the previous section. There are two types of GATHER/SCATTER, a periodic and a random. The periodic GATHER/SCATTER with starting location m and period n moves every nth element, beginning with the mth element, to/from a vector from/to a larger vector. The compiler will generate these instructions for DO-loops with increments other than one; however, unless the DO-loop increment is large, or unless there are a very large number of vector operations to be performed, it is often more efficient to create a periodic bit vector and use the WHERE block syntax. The random GATHER/SCATTER uses an index list, i.e., an integer vector, to point to the elements of the vector to be gathered or scattered. The MASK, MERGE, COMPRESS and EXPAND vector functions perform similar data motion functions to the random GATHER and SCATTER, except that they use bit vectors to point to the elements of the vector to be manipulated. The reader is referred to the CDC Cyber 200 Fortran Reference Manual for a more detailed description of these featrures.

7. VECTORIZING AN OCEAN MODEL - AN EXAMPLE

7.1. Motivation

We will now describe the vectorization of a model of the Indian Ocean [Luther and O'Brien; 1985, Luther et al., 1985] for the Cyber 205. This model will illustrate the implementation of the techniques described in the last section. Even though this model was coded specifically for the 205, many of the concepts involved apply also to the X-MP.

The northwest Indian Ocean is an extremely interesting area to model because of the strong seasonal signal in both the oceanic and atmospheric circulations. The seasonally reversing monsoon winds drive corresponding seasonal reversals in the upper ocean currents. During the (northern hemisphere) summer monsoon, a two-gyre circulation pattern develops of the coast of Somalia (Figure 5) in response to the southwesterly winds. Swallow and Fieux [1982] have documented this two gyre system in almost every year where sufficient historical data were available. The warm northern gyre of this system was described as the "great whirl" by Findlay [1866]. Associated with this two gyre system are wedges of cold upwelled water extending away from the coast to the north of each of the two gyres. These wedges have been observed in satellite infrared imagery by Brown et al. [1980] and by Evans and Brown [1981]. During the late stages of the summer monsoon, the two gyre system collapses abruptly, with the southern gyre propagating northward and merging with the great whirl [Swallow et al., 1983].

During the collapse of the summer monsoon system, several large eddies form along the coast of Oman [Cagle and Whritner, 1981] as in Figure 6. These eddies occur in the same locations at the same season

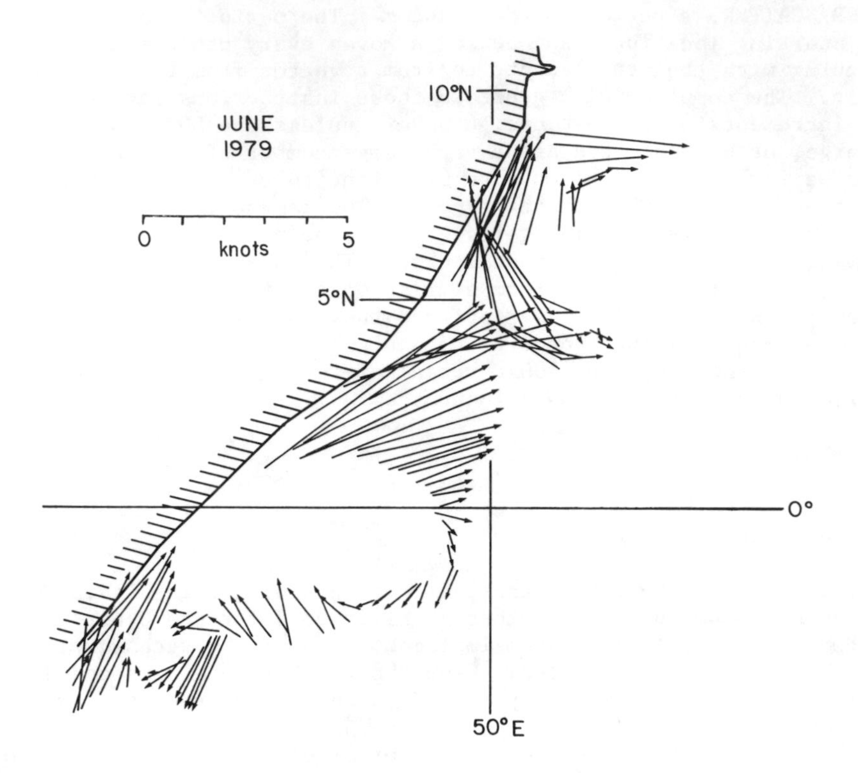

Figure 5. The two gyre current system in the summer Somali Current as observed in June 1979. From Swallow and Fieux [1982].

year after year. They persist through the winter monsoon and can still be detected as late as February.

To explain these features, we have constructed a numerical model of the northwestern Indian Ocean that is forced by observed winds [Luther and O'Brien, 1985; Luther et al., 1985]. The model equations are quite similar to those used by Lin and Hurlburt [1981] and by Anderson [personal communication]. All previous models of the Somali Current region [e.g. Cox, 1970, 1976, 1979; Hurlburt and Thompson, 1976; Lin and Hurlburt, 1981] used idealized basin geometries and wind stress forcing functions. Our model follows the approach of Busalacci and O'Brien [1980] and uses a realistic basin geometry with observed winds as forcing.

7.2. Model Formulation

The model ocean consists of a warm, thin, active upper layer of density ρ overlaying a cool, deep, motionless lower layer of density $\rho+\Delta\rho$ on a sphere of radius a. If we define the eastward (ϕ) and northward (Θ) components of the upper layer transport as $U = uH$ and $V = vH$ respectively, where (u, v) are the depth-independent (ϕ, Θ) velocity components in the upper layer and H is the thickness of the upper layer, the equations of motion are

$$\frac{\partial U}{\partial t} + \frac{1}{a\cos\Theta}\frac{\partial}{\partial\phi}\left(\frac{U^2}{H}\right) + \frac{1}{a}\frac{\partial}{\partial\Theta}\left(\frac{UV}{H}\right) - (2\Omega\sin\Theta)V$$

$$= \frac{-g'}{2a\cos\Theta}\frac{\partial H^2}{\partial\phi} + \frac{\tau(\phi)}{\rho} + A\nabla^2 U \quad (4a)$$

$$\frac{\partial V}{\partial t} + \frac{1}{a\cos\Theta}\frac{\partial}{\partial\phi}\left(\frac{UV}{H}\right) + \frac{1}{a}\frac{\partial}{\partial\Theta}\left(\frac{V^2}{H}\right) + (2\Omega\sin\Theta)U$$

$$= \frac{-g'}{2a}\frac{\partial H^2}{\partial\Theta} + \frac{\tau(\Theta)}{\rho} + A\nabla^2 V \quad (4b)$$

$$\frac{\partial H}{\partial t} + \frac{1}{a\cos\Theta}\left(\frac{\partial U}{\partial\phi} + \frac{\partial}{\partial\Theta}(V\cos\Theta)\right) = 0 \quad (4c)$$

where $g = \frac{\Delta\rho}{\rho} g$ is the reduced gravitational acceleration, Ω is the earth's rotation rate, and A is a kinematic eddy viscosity. The wind stress, $\vec{\tau} = (\tau(\phi), \tau(\Theta))$, is applied as a body force over the upper layer [Charney, 1955]. The transport form of the reduced gravity equations has the advantage that the discretization of the advective terms in (4a) and (4b) involves spatial averaging of the dependent variables, thus improving the numerical stability of the solution.

Equations (4a-c) are solved on a 1/8° latitude by 1/4° longitude staggered finite difference mesh (a C-grid) using a leap frog time integration scheme. The model basin geometry is shown in Figure 7. Along closed (land) boundaries the boundary condition is that U=V=0. Along the southern boundary and a part of the eastern boundary an open

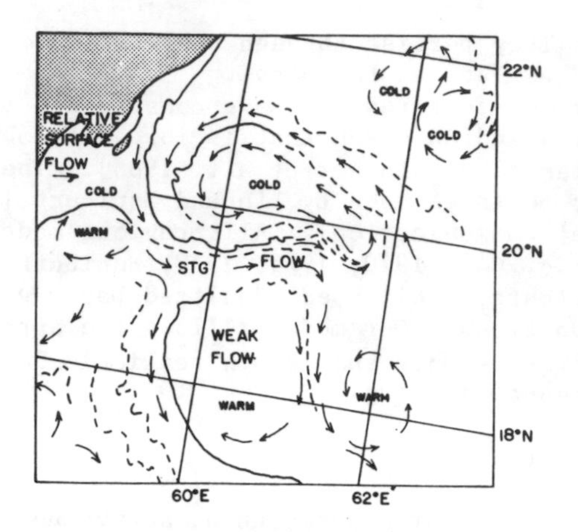

Figure 6. Interpretation of infrared image of the coastal region south of the Gulf of Oman from 26 October, 1980. Strong fronts are indicated by solid lines, weak fronts by dashed lines. A strong flow is indicated associated with a major feature which propagates from the upwelling off the south tip of the island of Masirah. A cold cyclonic eddy is indicated to the north of this tongue-like feature. A sequence of warm and cold eddies is indicated in progression down the coast of the Arabian Peninsula, and major features associated with points of land are repeated along the coast. From Cagle and Whritner [1981].

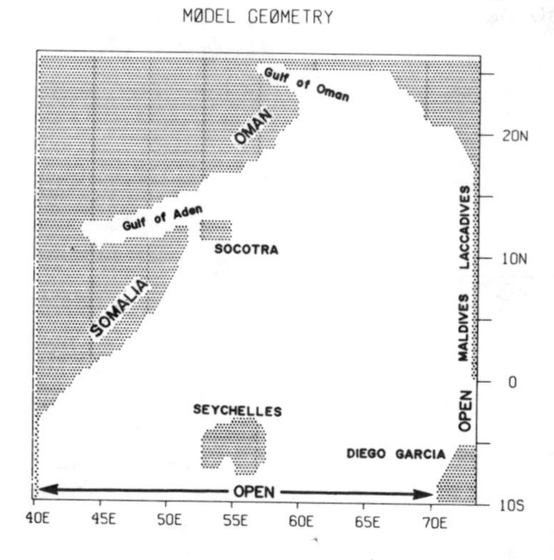

Figure 7. Model basin geometry. Shading indicates land areas. The shallow banks around Socotra, the Seychelles and the Chagos Archipelago are represented as land areas. The eastern boundary north of the equator is closed by the Maldive and Laccadive islands. Open boundary conditions are imposed along the southern boundary and along a portion of the eastern boundary.

boundary condition is applied, similar to the "Modified Orlanski" boundary condition decribed in Camerlengo and O'Brien [1980]. For a more detailed description of the model, see Luther and O'Brien [1985] and Luther et al. [1985].

The model is forced by observed winds from several different sources. The primary source of the wind data is from ship observations. The wind data sets that we have used most extensively thus far are a monthly mean climatology (MMC) of ship winds and the FGGE Level IIIb dynamically assimilated 1000 mbar winds [Luther et al., 1985]. These winds are smoothed and interpolated onto the model mesh and timestep.

The model is integrated from an initial state of no motion with the one year wind cycle being repeated for several years. After a brief (less than one year) spin-up, the model fields develop a regular seasonal cycle, where highly nonlinear eddies occur in the same place and at the same time of year during each year of model simulation. The model seasonal cycle reproduces most of the features of the observed seasonal cycle in the northwest Indian Ocean, including the reversal of the Somali Current with the onset of the summer and winter monsoons, the two gyre system and associated upwelling regions of the summer Somali Current, (Figure 8), and the eddies along the coast of the Arabian peninsula (Figure 9).

Detailed analysis of the model results reveals several new physical mechanisms at work. The Somali Current is shown to reverse before the local alongshore component of the winds in April, due to the relaxation of the pressure gradient that was maintained by the winter monsoon winds. The great whirl of the two-gyre system forms to the north of 4°N under the influence of the local wind stress curl, while the southern gyre forms near the equator due to the alongshore component of the local wind stress. The collapse of the two-gyre system appears to be remotely triggered by downwelling equatorial Rossby waves. The strong gradient in the wind stress curl associated with the jet-like nature of the winds drives a differential Ekman pumping which in turn induces a tilt in the pycnocline with a geostrophic jet along this pycnocline slope [Luther et al., 1985]. During the late stages of the summer monsoon, as the winds decrease rapidly, this jet breaks up into the eddies off the coast of the Arabian peninsula through barotropic (or horizontal shear) instability.

8. MODEL VECTORIZATION ON THE 205

The finite difference form of (4a-c) is easily vectorized because the value at the new timestep depends only on values at previous timesteps. There is, therefore, no recursion or dependency, and the calculation can proceed in any order. Two time levels of the model dependent variables U, V and H are stored as J_{max} by K_{max} by 2 arrays. In the present version of the model, $J_{max} = 135$ and $K_{max} = 74$, so that each time level may be treated as a vector of length 9990. The resulting long vector calculations are ideally suited to the 205's architecture.

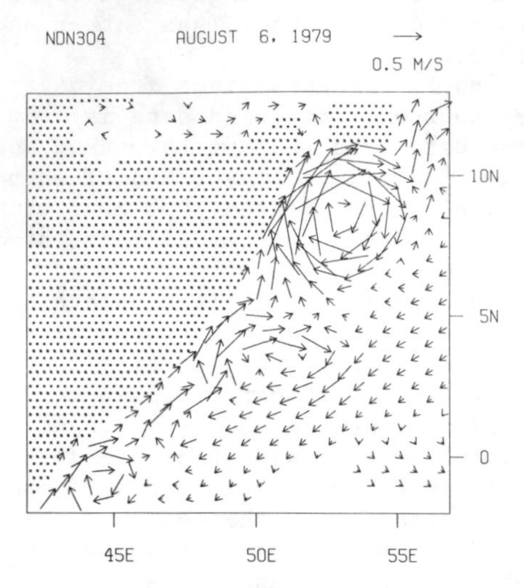

Figure 8. Two gyre current system in the summer Somali Current from the model. Only the Somali coast region is shown. The southern gyre is located south of 5°N and the great whirl is located between 6°N and 11°N. The southern gyre is moving northward at this time and will merge with the great whirl by late August. Areas of thinner upper layer (not shown) indicating upwelling are found along the coast to the north of each gyre.

UPPER LAYER THICKNESS

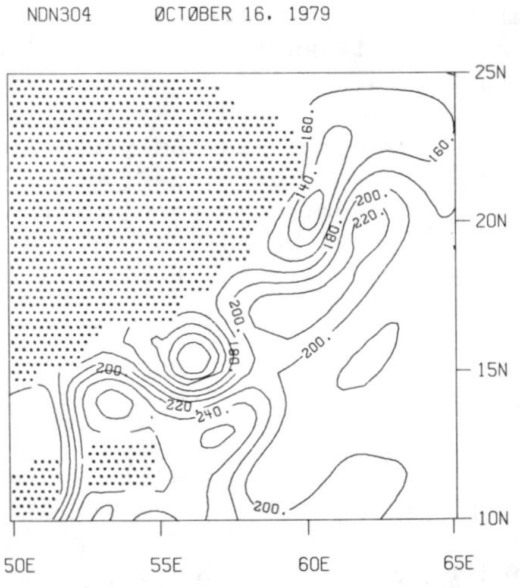

Figure 9. Eddies along the coast of the Arabian Peninsula from the model in mid-October. Only the northwest portion of the model basin is shown. These eddies are formed during September, as the wind stress and its curl decrease drastically, allowing the eddies to form through instability processes.

The scalar version of the model (Figure 10) uses IF tests within the intermost DO loop to determine whether a point is in the interior or whether it is over land. Land points are flagged with special values. After the interior computations are performed, a block of nested IF tests determines if the point is on a land boundary, and if so, performs the appropriate correction. The open boundary conditions are applied in a separate loop.

In the vectorized version, all DO loops are replaced by explicit vector syntax. Figure 11 shows the interior and boundary condition portion of the vectorized code. The irregular geometry is handled by bit control vectors. The bit vectors UATSEA, VATSEA and HATSEA are "on" for all interior mesh points and "off" for all boundary and land points. The interior vector equations are performed within a WHERE block with these bit vectors as control, so that the computation is performed at every mesh point, but the result is stored only at the interior points. Northern and southern boundary points on the U mesh and eastern and western boundary points on the V mesh are reflected using gathers and scatters. The other boundary points on U and V are zero initially and for all time, since the bit control vectors prevent them from ever being changed.

Implicit descriptors are assigned to vectors with different starting locations within the U, V and H arrays so that they point to the appropriate time level and the appropriate point in the finite difference stencil (Figure 11). The length of these vectors is J_{max} x K_{max} $-2(J_{max}+1)$ which includes just the internal points of one time level as in Figure 12. This is to accomodate the finite difference stencil, which involves the central point and its eight nearest neighbors (Figure 13). We create a 3 by 3 array of descriptors to define the stencil for each variable, as in loop 130 of Figure 11, where $L = J_{max} \times K_{max} - 2(J_{max} + 1)$ and either NT2=1 or NT2=2, depending on which level of the U, V and H arrays holds the current time level. The finite differences in space can be computed in vector form using these arrays of descriptors; for example, the descriptor UD(2,2) is the vector of all the center points of the U stencil, i.e., the U(J,K)'s, while UD(1,1) is the vector of all points in the southwest corner of the stencil, i.e., the U(J-1,K-1)'s, etc. Descriptors are also defined for the other variables, such as the wind stress field, and the entire interior computation is performed as a long vector calculation involving only implicit descriptors under bit control. Intermediate results are collected in temporary vectors to make optimum use of linked triads.

A problem arises because of the spherical coordinate system and the many coefficients that vary with latitude. To achieve the long vector lengths, these coefficients must be stored in two dimensional arrays of dimension J_{max} by K_{max}, rather than in one dimensional arrays of length K_{max} as they were in the scalar code. This results in a considerable increase in storage requirements, but it increases the length of the vector computations from J_{max} to $J_{max} \times K_{max} - 2(J_{max}-1)$, or in the present version of the model, from 135 to 9718. This long vector length enables the model code to attain vector speeds very near the theoretical maximum speed.

```
C
C  START K LOOP
C
      DO 300 K=2,KM1
C
C  START J LOOP
C
C
        DO 200 J=2,JM1
C
      IF (U(J,K,NT2).EQ.SPV.OR.V(J,K,NT2).EQ.SPV) GO TO 200
C
C  COMPUTE REPEATED FINITE DIFFERENCE TERMS ON U MESH
C
       ZJU1=U(J,K,NT2)+U(J,K-1,NT2)
       ZJU2=U(J-1,K,NT2)+U(J-1,K-1,NT2)
       ZJU3=U(J-1,K,NT2)+U(J,K,NT2)
       ZJU4=U(J,K+1,NT2)+U(J,K,NT2)
       ZJU5=U(J,K+1,NT2)+U(J,K-1,NT2)
       ZJU6=U(J+1,K,NT2)+U(J,K,NT2)
       ZJU7=U(J+1,K,NT2)+U(J-1,K,NT2)
       ZJU8=U(J,K+1,NT2)-U(J,K-1,NT2)
       ZJU9=2.*U(J,K,NT2)
C
C  ON V MESH
C
       ZJV1=V(J,K,NT2)+V(J+1,K,NT2)
       ZJV2=V(J,K+1,NT2)+V(J+1,K+1,NT2)
       ZJV3=V(J,K+1,NT2)+V(J,K-1,NT2)
       ZJV4=V(J,K,NT2)+V(J-1,K,NT2)
       ZJV5=V(J,K+1,NT2)+V(J,K,NT2)
       ZJV6=V(J,K,NT2)+V(J,K-1,NT2)
       ZJV7=V(J+1,K,NT2)+V(J-1,K,NT2)
       ZJV8=2.*V(J,K,NT2)
       ZJV9=V(J,K+1,NT2)-V(J,K-1,NT2)
C
C  AND ON H MESH
C
       ZJH1=1./H(J+1,K,NT2)
       ZJH2=1./H(J,K,NT2)
       ZJH3=1./(H(J+1,K,NT2)+H(J,K,NT2)+H(J+1,K+1,NT2)+H(J,K+1,NT2))
       ZJH4=1./(H(J+1,K-1,NT2)+H(J,K-1,NT2)+H(J+1,K,NT2)+H(J,K,NT2))
       ZJH5=H(J+1,K,NT2)*H(J+1,K,NT2)
       ZJH6=H(J,K,NT2)*H(J,K,NT2)
       ZJH7=H(J,K-1,NT2)*H(J,K-1,NT2)
       ZJH8=1./H(J,K-1,NT2)
       ZJH9=1./(H(J,K,NT2)+H(J+1,K,NT2)+H(J,K-1,NT2)+H(J+1,K-1,NT2))
       ZJH0=1./(H(J-1,K,NT2)+H(J,K,NT2)+H(J-1,K-1,NT2)+H(J,K-1,NT2))
C
C  J LOOP: U EQUATION
C
              U(J,K,NT1)=U(J,K,NT)+Z5*(-W1(K)*(ZJU6*ZJU6*ZJH1-
     +                   ZJU3*ZJU3*ZJH2)-X1*(ZJU4*ZJV2*ZJH3-ZJU1*ZJV1*
     +                   ZJH4)+W2(K)*(ZJV2+ZJV1)-W3(K)*(ZJH5-
     +                   ZJH6)*DPHIN+A*TX(J,K)+X2*(W5(K)*(ZJU7-ZJU9)+
     +                   (ZJU5-ZJU9)*X3-W6(K)*ZJU8))
C
C  J LOOP: V EQUATION
C
              V(J,K,NT1)=V(J,K,NT)+Z5*(-Y1*(ZJV5*ZJV5*ZJH2-
     +                   ZJV6*ZJV6*ZJH8)-W7(K)*(ZJU1*ZJV1*ZJH9-
     +                   ZJU2*ZJV4*ZJH0)-W8(K)*(ZJU1+ZJU2)-Y2*
     +                   (ZJH6-ZJH7)+A*TY(J,K)+X2*(W0(K)*(ZJV7-
     +                   ZJV8)+(ZJV3-ZJV8)*X3-WA(K)*ZJV9))
C
C
C   J LOOP: H EQUATION
C
        IF (H(J  ,K  ,NT2).NE.SPV.AND.
     +      H(J+1,K  ,NT2).NE.SPV.AND.
     +      H(J-1,K  ,NT2).NE.SPV.AND.
     +      H(J  ,K+1,NT2).NE.SPV.AND.
     +      H(J  ,K-1,NT2).NE.SPV.AND.
     +      H(J-1,K-1,NT2).NE.SPV.AND.H(J-1,K+1,NT2).NE.SPV.AND.
     +      H(J+1,K-1,NT2).NE.SPV.AND.H(J+1,K+1,NT2).NE.SPV)
C
     +  H(J,K,NT1)=H(J,K,NT)-Z5*WB(K)*((U(J,K,NT2)-U(J-1,K,NT2))*DPHIN
     +            +(V(J,K+1,NT2)*COST3(K)-V(J,K,NT2)*COST2(K))*DTHEN)
C
 200     CONTINUE
 300 CONTINUE
C
C
```

```
C-----------------------------------
C***** BOUNDARY CONDITIONS ******
C-----------------------------------
C
C
C   CORRECT BOUNDARY VALUES ON U AND V
C
        DO 330 K=2,KM1
        DO 330 J=2,JM1
C
        IF (U(J,K,NT2).EQ.SPV) GO TO 330
C
        IF (U(J-1,K,NT2).EQ.SPV.OR.U(J+1,K,NT2).EQ.SPV) THEN
            U(J,K,NT1)=0.0
            GO TO 330
        END IF
C
        IF (U(J,K-1,NT2).EQ.SPV) THEN
            U(J,K,NT1)=-U(J,K+1,NT1)
            GO TO 330
C
        ELSE IF (U(J,K+1,NT2).EQ.SPV) THEN
            U(J,K,NT1)=-U(J,K-1,NT1)
            GO TO 330
        END IF
C
        IF (U(J-1,K-1,NT2).EQ.SPV.OR.U(J-1,K+1,NT2).EQ.SPV.OR.
     +      U(J+1,K-1,NT2).EQ.SPV.OR.U(J+1,K+1,NT2).EQ.SPV)
C
     +      U(J,K,NT1)=0.0
C
  330 CONTINUE
C
        DO 350 K=2,KM1
        DO 350 J=2,JM1
C
        IF (V(J,K,NT2).EQ.SPV) GO TO 350
C
        IF (V(J,K-1,NT2).EQ.SPV.OR.V(J,K+1,NT2).EQ.SPV) THEN
            V(J,K,NT1)=0.0
            GO TO 350
        END IF
C
        IF (V(J-1,K,NT2).EQ.SPV) THEN
            V(J,K,NT1)=-V(J+1,K,NT1)
            GO TO 350
C
        ELSE IF (V(J+1,K,NT2).EQ.SPV) THEN
            V(J,K,NT1)=-V(J-1,K,NT1)
            GO TO 350
        END IF
C
        IF (V(J-1,K-1,NT2).EQ.SPV.OR.V(J-1,K+1,NT2).EQ.SPV.OR.
     +      V(J+1,K-1,NT2).EQ.SPV.OR.V(J+1,K+1,NT2).EQ.SPV)
C
     +      V(J,K,NT1)=0.0
C
  350 CONTINUE
```

Figure 10. Interior and boundary condition section of model code before vectorization. Any finite difference terms used more than once are computed and stored in the ZJ-temporary variables. Special values are imbedded in the U, V and H arrays to indicate land points. IF tests then determine whether a point is an interior point, a boundary point or a land point.

```
C
C                   FLIP TIME INDICES TO POINT TO FIRST OR SECOND
C                   PLANES OF U, V AND H
      NSAVE=NT1
      NT1=NT2
      NT2=NSAVE
C
C                   FOR START-UP AND FOR EVERY NSTEP-TH TIME STEP
C                   USE FORWARD DIFFERENCES
C
      IF (MOD(N,NSTEP).EQ.1) THEN
      NT=NT2
      Z5=0.5S0*DELT
C
      ELSE
C
C                   USE CENTERED TIME DIFFERENCES
C
      NT=NT1
      Z5=DELT
C
      END IF
C
      Z0=-2.0S0*ARAD*DTHE/Z5
      Z5A=1.0S0/Z5
C
C                    LHS OF EQUATIONS ON PLANE NT1
C
         ASSIGN UNT1D,U(2,2,NT1;L)
         ASSIGN VNT1D,V(2,2,NT1;L)
         ASSIGN HNT1D,H(2,2,NT1;L)
C
C                    AND RHS TERMS MOSTLY ON PLANE NT2
C
         DO 130 J=1,3
          DO 130 I=1,3
           ASSIGN UD(I,J),U(I,J,NT2;L)
           ASSIGN VD(I,J),V(I,J,NT2;L)
           ASSIGN HD(I,J),H(I,J,NT2;L)
  130     CONTINUE
C
C                   CHOOSE  CENTERED OR FORWARD DIFFERENCES
C
          ASSIGN UNTD,U(2,2,NT;L)
          ASSIGN VNTD,V(2,2,NT;L)
          ASSIGN HNTD,H(2,2,NT;L)
C
C                   REFLECT BOUNDARY POINTS AT PLANE NT1
C
      ASSIGN UNOD,U(2,1,NT1;L)
      ASSIGN USOD,U(2,3,NT1;L)
      ASSIGN VEAD,V(1,2,NT1;L)
      ASSIGN VWED,V(3,2,NT1;L)
C                  U EQUATION
C
      WHERE (UATSEA(1;L))
         T1D=(UD(3,2)+UD(2,2))*(UD(3,2)+UD(2,2))/HD(3,2)
         T2D=(UD(1,2)+UD(2,2))*(UD(1,2)+UD(2,2))/HD(2,2)
         T2D=W1D*(T1D-T2D)
         T1D=A*TXD-T2D
         T2D=(UD(2,3)+UD(2,2))*(VD(2,3)+VD(3,3))/(HD(3,2)+
     1        HD(2,2)+HD(3,3)+HD(2,3))
         T3D=(UD(2,2)+UD(2,1))*(VD(2,2)+VD(3,2))/(HD(3,1)+
     1        HD(2,1)+HD(3,2)+HD(2,2))
         T1D=T1D-X1*(T2D-T3D)
         T1D=T1D+W2D*(VD(2,3)+VD(3,3)+VD(2,2)+VD(3,2))
         T1D=T1D-W3D*DPHIN*(HD(3,2)*HD(3,2)-HD(2,2)*HD(2,2))
         T2D=W5D*(UD(3,2)+UD(1,2)-2.S0*UD(2,2))+X3*(UD(2,3)+
     1        UD(2,1)-2.S0*UD(2,2))
         T2D=T2D-W6D*(UD(2,3)-UD(2,1))
         T1D=T1D+X2*T2D
         UNT1D=UNTD+Z5*T1D
      ENDWHERE
C
C                   REFLECT NORTHERN BOUNDARY POINTS
C
      ASSIGN TEMP1D,.DYN.LNORTH
C
      TEMP1D=Q8VGATHR(UNOD,INOD;TEMP1D)
      UNT1D=-Q8VSCATR(TEMP1D,INOD;UNT1D)
C
      FREE
C
C                   REFLECT SOUTHERN BOUNDARY POINTS
C
      ASSIGN TEMP1D,.DYN.LSOUTH
C
      TEMP1D=Q8VGATHR(USOD,ISOD;TEMP1D)
      UNT1D=-Q8VSCATR(TEMP1D,ISOD;UNT1D)
C
      FREE
```

```
C
C                       V EQUATION
C
C
      WHERE (VATSEA(1;L))
         T1D=(VD(2,3)+VD(2,2))*(VD(2,3)+VD(2,2))/HD(2,2)
         T2D=(VD(2,2)+VD(2,1))*(VD(2,2)+VD(2,1))/HD(2,1)
C
         T1D=A*TYD-Y1*(T1D-T2D)
C
         T2D=(UD(2,2)+UD(2,1))*(VD(2,2)+VD(3,2))/(HD(2,2)+
     1        HD(3,2)+HD(2,1)+HD(3,1))
C
         T3D=(UD(1,2)+UD(1,1))*(VD(2,2)+VD(1,2))/(HD(1,2)+
     1        HD(2,2)+HD(1,1)+HD(2,1))
C
         T1D=T1D-W7D*(T2D-T3D)
         T1D=T1D-W8D*(UD(2,2)+UD(2,1)+UD(1,2)+UD(1,1))
         T1D=T1D-Y2*(HD(2,2)*HD(2,2)-HD(2,1)*HD(2,1))
C
         T2D=W0D*(VD(3,2)+VD(1,2)-2.S0*VD(2,2))+X3*(VD(2,3)+
     1        VD(2,1)-2.S0*VD(2,2))
C
         T1D=T1D+X2*(T2D-WAD*(VD(2,3)-VD(2,1)))
C
         VNT1D=VNTD+Z5*T1D
C
      ENDWHERE
C
C
C
C                       REFLECT EASTERN BOUNDARY POINTS
C
      ASSIGN TEMP1D,.DYN.LEAST
C
      TEMP1D=Q8VGATHR(VEAD,IEAD;TEMP1D)
      VNT1D=-Q8VSCATR(TEMP1D,IEAD;VNT1D)
C
      FREE
C
C                       REFLECT WESTERN BOUNDARY POINTS
C
      ASSIGN TEMP1D,.DYN.LWEST
C
      TEMP1D=Q8VGATHR(VWED,IWED;TEMP1D)
      VNT1D=-Q8VSCATR(TEMP1D,IWED;VNT1D)
C
      FREE
C
C
C                       H EQUATION
C
C
      WHERE (HATSEA(1;L))
         T1D=WBD*(DPHIN*(UD(2,2)-UD(1,2))+DTHEN*(VD(2,3)*HC1D-
     2        VD(2,2)*HC2D))
         HNT1D=HNTD-Z5*T1D
      ENDWHERE
C
```

Figure 11. Interior and boundary condition code from vectorized version of model. The variables NT1 and NT2 are flipped at each timestep so that level NT1 of each array holds both the n+1 and the n-1 timestep, while level NT2 holds the n timestep. This is the same in both the scalar and vector versions. Occassional forward time differences are used to surpress the computational mode. Descriptors are ASSIGNed to vectors within the U, V and H arrays to determine the finite difference mesh points to be used in the interior equations and the reflected boundary conditions. The U, V and H vector computations are performed in WHERE blocks under control of the BIT vectors UATSEA, VATSEA and HATSEA. These BIT vectors are "true" only at interior points. Intermediate results are stored in temporary vectors to make optimum use of linked triads. The reflected boundary conditions are computed by first gathering the appropriate mesh points into a temporary vector and then scattering these values, with a sign change, into their proper locations in the U and V arrays, using the vector functions Q8VGATHR and Q8VSCATR.

Another large savings in execution time is gained by using all half-precision variables. On our two-pipe Cyber 205, this results in four results per clock period from the vector pipes, rather than two results per clock period as with full precision variables. For the linked triads in the interior computation, this gives theoretical vector speeds of 800 MFLOPS. The use of half-precision variables also doubles the available working storage.

In one version of the vector code, the GATHER and SCATTER function calls in the reflected boundary condition are replaced by a WHERE block under control of a bit vector that is turned "on" only at the appropriate boundary points. The resulting code requires the same amount of execution time as the version with the GATHER and SCATTER calls, even though the WHERE blocks discard more than 97% of the calculations at each boundry. Since the calculation performed is a simple replacement with a sign change, it is as expensive to perform the GATHER and SCATTER as it is to perform the unnecessary computations. Had the computation been more complicated, or had the boundary points been an even smaller percentage of the vector, the GATHER/SCATTER would have been more efficient.

The end result of recoding the model in explicit vector Fortran is a substantial savings in execution time. In the original code, no major loops would vectorize automatically. The entire computation was performed in scalar mode. On FSU's Cyber 170/760 computer, which is a fast scalar machine, a one year model integration requires 7078 seconds of execution time. The vectorized version in half-precision requires 223 seconds for a one year integration on our two-pipe Cyber 205, which makes it 32 times faster than the Cyber 170/760 version. A large fraction of that time is taken by input and output, so that the computations themselves actually speed up by a much larger factor.

We are now expanding the model basin to include the entire Indian Ocean north of 25°S. At a resolution of 1/8° in both directions, this will require a 350 by 250 by 2 array for each dependent variable and for each component of the wind stress. This results in a vector length of 86,798, which is larger than 2^{16}, the maximum vector length on the 205. We must employ the technique of "strip-mining," wherein the vector computations are performed in pieces, each piece having a vector length less than 2^{16}. This can be done by first ASSIGNing all the descriptors in the interior equations to the starting locations corresponding to the point (2,2) in each level of each array as before, with a vector length of $2^{16}-1$, performing the vector computations, then re-ASSIGNing the descriptors to starting locations corresponding to their previous starting locations plus $2^{16}-1$ with length $(J_{max} * K_{max} -2(J_{max} -1) -(2^{16} -1))$ and performing the vector computation again. This procedure retains the very long vector length, but still has the problem with the extra storage required for the spherical coefficients.

Our next step in the level of model sophistication is to add vertical resolution, so that we may include the observed vertical structure of the currents in this region. With our present four million word Cyber 205, we can include up to eight layers in the vertical. This will require 3.5 million words of storage and will also require the development of a vectorized multi-grid elliptic solver that can handle an irregular computational domain.

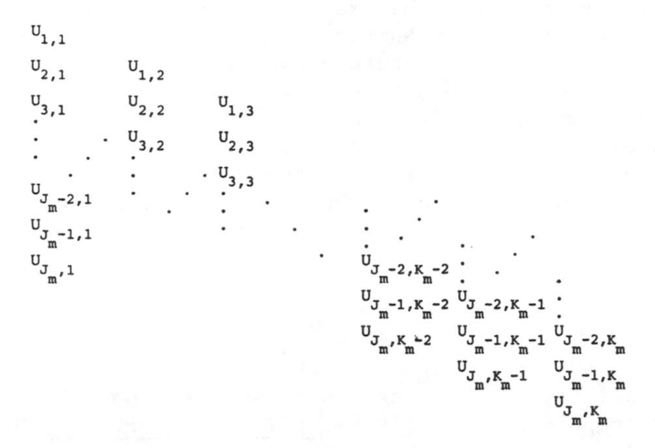

Figure 12. Arrangement of model mesh points in computer storage. Descriptors are ASSIGNed to different vectors within the same array to point to the various mesh points involved in the finite difference computation. For instance, the descriptor UD(2,2) points to the vector beginning at the mesh point U(2,2,NT2) and ending at the point U(J_{max}-1,K_{max}-1,NT2) where NT2 indicates the appropriate time level of the U array. UD(2,2) is thus the vector of all the center points of the finite difference stencil at time level NT2.

	$H_{j,k+1}$	$U_{j,k+1}$	$H_{j+1,k+1}$	
	$V_{j,k+1}$		$V_{j+1,k+1}$	
$U_{j-1,k}$	$H_{j,k}$	$U_{j,k}$	$H_{j+1,k}$	$U_{j+1,k}$
	$V_{j,k}$		$V_{j+1,k}$	
	$H_{j,k-1}$	$U_{j,k-1}$	$H_{j+1,k-1}$	

Figure 13. Finite difference stencil for U on the staggered mesh.

7. ACKNOWLEDGEMENTS

This work is supported by The Office of Naval Research and the Control Data Corporation. Figure 3 and the examples in section 4 were taken from lecture notes by Dr. Robert Numrich of Cray Research, Inc. and the examples in section 5 from lecture notes by Dr. Larry Rudinsky of Control Data Corporation. Much of the vector syntax in the Indian Ocean model was written by Dennis Lehane of Control Data Corporation. The author would also like to thank Dr. Ruth Anne Manning and Dr. Gary Beckwith of Control Data Corporation for their assistance in vectorization of the model and the members of the Mesoscale Air-Sea Interaction Group for their many helpful comments and suggestions.

8. REFERENCES

Amdahl, G., The validity of the single processor approach to achieving large scale computing capabilities, AFIPS Conf. Proc., 30, 483-485, 1967.

Brown, O. B., J. G. Bruce, and R. H. Evans, Evolution of sea surface temperature in the Somali Basin during the southwest monsoon of 1979, Science, 209, 595-597, 1980.

Busalacchi, A. J., and J. J. O'Brien, The seasonal variability in a model of the tropical Pacific, Journal of Physical Oceanography, 10, 1929-1951, 1980.

Cagle, B. J., and R. Whritner, Arabian Sea project of 1980 - Composites of Infrared Images, Technical Report, 61 pp, Office of Naval Research, Western Regional Office, Pasadena, CA, 1981.

Camerlengo, A. L., and J. J. O'Brien, Open boundary conditions in rotating fluids, Journal of Computational Physics, 35, 12-35, 1980.

Charney, J. G., The generation of ocean currents by the wind, Journal of Marine Research, 14, 477-498, 1955.

Chen, S., Large-scale and high-speed multiprocessor system for scientific applications: Cray X-MP-2 series, in Proceedings of the NATO Workshop on High Speed Computations, West Germany, edited by J. Kowalik, NATO ASI Series, Vol. F-7, Springer-Verlag, Berlin, 59-67, 1984.

Cox, M. D., A mathematical model of the Indian Ocean, Deep Sea Research, 17, 47-75, 1970.

Cox, M. D., Equatorially trapped waves and the generation of the Somali Current, Deep-Sea Research, 23, 1139-1152, 1976.

Cox, M. D., A numerical study of Somali Current eddies, Journal of Physical Oceanography, 9, 311-326, 1979.

Eckert, J. Jr., J. Mauchly, H. Goldstein, and J. Brainerd, Description of the ENIAC and comments on electronic digital computing machines, Applied Mathematics Panel Report No. 171.2R, Univ. Pennsylvania, Philadelphia, 1945.

Evans, R. H., and O. B. Brown, Propagation of thermal fronts in the Somali Current system, Deep-Sea Research, 28, 521-527, 1981.

Findlay, A. G., A directory for the navigation of the Indian Ocean, 1062 pp., Richard Holmes Laurie, London, 1866.
Hockney, R., and C. Jesshope, Parallel Computers: Architecture, Programming and Algorithms, Adam Hilger, Bristol, 1981.
Hurlburt, H. E., and J. D. Thompson, A numerical model of the Somali Current, Journal of Physical Oceanography, 6, 646-664, 1976.
Kogge, P., The Architectrure of Pipelined Computers, McGraw-Hill, New York, 1981.
Kuck, D., The Structure of Computers and Computation, John Wiley, New York, 1979.
Lin, L. B., and H. E. Hurlburt, Maximum simplification of nonlinear Somali Current dynamics,in Monsoon Dynamics, edited by M. J. Lighthill and R. P. Pearce, Cambridge University Press, 1981.
Lincoln, N., Technology and design tradeoffs in the creation of a modern supercomputer, IEEE Trans. Computers, C-31, 349-362, 1982.
Luther, M. E., J. J. O'Brien, and A. H. Meng, Morphology of the Somali Current System during the southwest monsoon, Coupled Ocean-Atmosphere Models, edited by J.C.S. Nihoul, Elsevier, Amsterdam, 405-437, 1985.
Luther, M. E., and J. J. O'Brien, A model of the seasonal circulation in the Arabian Sea forced by observed winds, Progress in Oceanography, 14, 353-385, 1985.
Ortega, J. M., and R. G. Voigt, Solution of partial differential equations on vector and parallel computers, SIAM Review, 27, 149-240, 1985.
Richardson, L. S., Weather Prediction by Numerical Process, Cambridge University Press, London, 1922.
Riganati, J. P., and P. B. Schneck, Supercomputing, Computer, 17, 97-113, 1984.
Russel, R., The Cray-1 computer system, Comm. ACM, 21, 63-72, 1978.
Swallow, J. C., and M. Fieux, Historical evidence for two gyres in the Somali Current, Journal of Marine Research, 40, supplement, 747-755, 1982.
Swallow, J. C., R. L. Molinari, J. G. Bruce, O. B. Brown, and R. H. Evans,) Development of near-surface flow pattern and water mass distribution in the Somali Basin, in response to the southwest monsoon of 1979, Journal of Physical Oceanography, 13, 1398-1415, 1983.
Worlton, J., A Philosophy of Supercomputing, Tech. report LA-8849-MS-UC-32, Los Alamos National Lab., Los Alamos, NM, 1981.

CHANNEL DYNAMICS AND ITS APPLICATION TO THE ANTARCTIC CIRCUMPOLAR CURRENT

J. M. Klinck
Department of Oceanography
Texas A&M University
College Station, TX 77843
USA

ABSTRACT. A one-layer, quasi-geostrophic, nonlinear numerical model is used to analyze wind-driven flow in a zonal channel with partial meridional boundaries. The model is chosen to correspond to the Antarctic Circumpolar Current, especially that region in the vicinity of Drake Passage. Numerical techniques are chosen to take advantage of vectorized computer operations. The leap frog integration of the vorticity equation vectorizes directly. The streamfunction is obtained from the vorticity field through a Red-Black SOR iterative scheme. The geometry of solid walls in the zonal channel is specified through BIT control arrays. This feature of the program allows the geometry to be changed easily and allows vectorized calculations over entire two-dimensional arrays. Performance of this model improves by a factor of 50 or better relative to a similar calculation on a scalar computer. Model results show that meridional boundaries strongly reduce the total transport in a zonal channel. Further, there is a qualitative change in the flow depending on whether the meridional walls overlap or not. If the walls overlap (all latitude lines blocked), the flow is Sverdrup-like and has the appearance of modified closed basin flow. If walls do not overlap, there is a strong, laterally sheared current in the zonal band that is not blocked. Bottom topography also provides a lesser drag on the flow, further reducing the total transport.

1. INTRODUCTION

The Antarctic Circumpolar Current (ACC) has a number of features that make it unique among ocean currents. The most obvious is that it extends around the globe in a continuous band (Figure 1), thereby providing an exchange pathway from one ocean basin to another. A second, perhaps less obvious, feature is that the ACC flows through that part of the ocean where considerable deep and bottom water production occurs. In fact, some of the water involved in the production process comes from the ACC; thus, it affects the characteristics of the newly produced water masses. The continuous

J. J. O'Brien (ed.), Advanced Physical Oceanographic Numerical Modelling, 299–328.

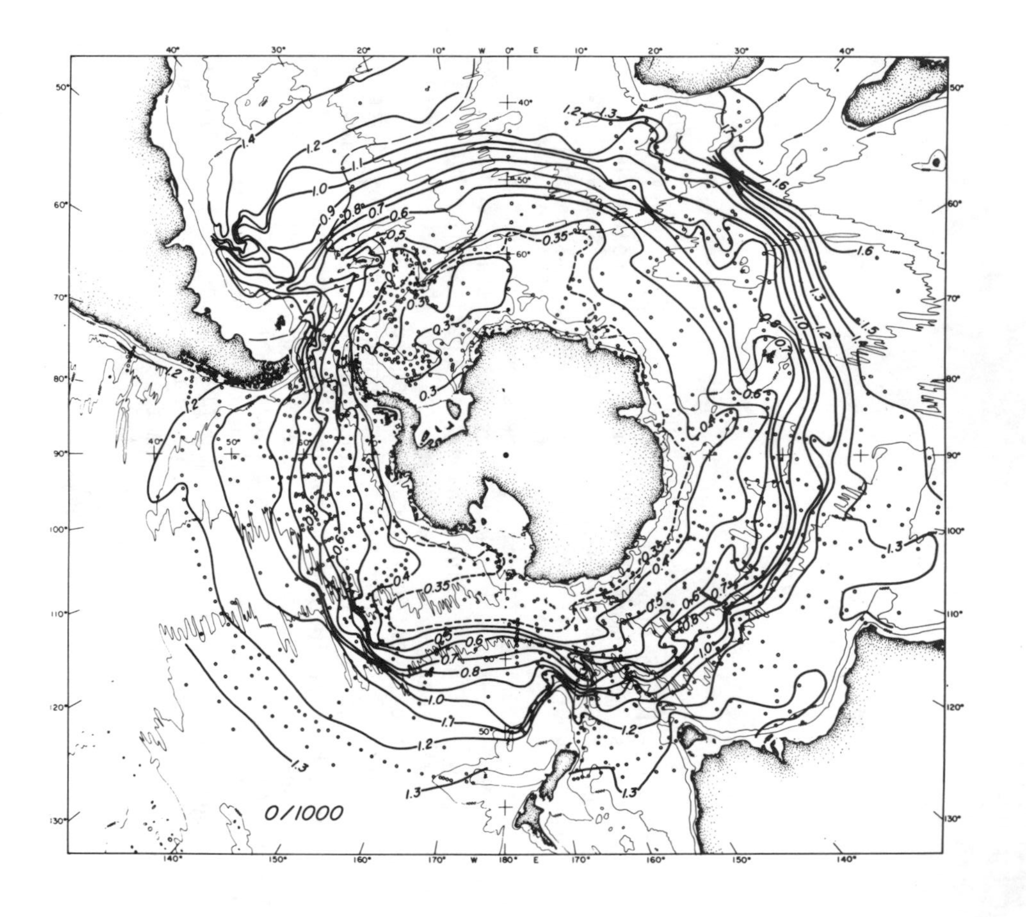

Figure 1. The Southern Ocean and dynamic topography of the sea surface relative to the 1000dbar constructed from historical hydrographic data at locations shown by dots (Gordon et al., 1978).

nature of this current and its involvement in water formation and interbasin exchanges, both of which affect heat exchange and ocean climate, make it necessary that the dynamics of this part of the global ocean be understood.

Historical hydrographic data from the Southern Ocean is sparse and does not have the coverage found in other, more accessible, oceans. The first systematic coverage of the Southern Ocean was obtained in the 1920's and 1930's by the British Discovery Committee. These expeditions investigated the broad scale (100's of kilometers) horizontal character of the ACC and provided a rough understanding of the water formation mechanisms in this area. A comprehensive study by the United States Research Vessel Eltanin (1962 to 1972), and its continuation as the Islas Orcadas (1976-1978) represents the first sampling of the ACC with modern oceanographic instruments. In contrast to earlier studies, water samples were obtained to the bottom, and measurements of nutrients, in addition to the usual physical properties, were made.

Large field experiments, International Southern Ocean Studies (ISOS) and First GARP Global Experiment (FGGE) among others, during the last decade have improved the description of the ACC and the processes involved with its maintenance. The ISOS program provided a measure of the transport of water through Drake Passage [Whitworth, 1983]. This transport is a key observation against which numerical models of this region must be tested. The FGGE drifters provide a description of the surface currents in regions of the Southern Ocean that had historically been undersampled.

Even though comprehensive measurements of the oceanography of the Southern Ocean have been recent, theoretical models of the ACC have existed since the beginning of modern physical oceanography (beginning with the work in the late 1940's of Sverdrup, Stommel, Hidaka, Munk and others who considered the dynamics of large-scale ocean flow).

The work presented here is a study of the effect of lateral and bottom topography on wind-driven flow in a zonal channel, which is assumed to be a simple representation of the ACC. Previous work has indicated the importance of both of these effects in balancing the surface stress, although few have included both together or included inertial effects. A secondary purpose is to demonstrate that modern vector computers can solve this problem with sufficient rapidity that the model can be used in an "experimental" mode. That is, simulations can be run to consider cases that may not correspond to the real ocean in an attempt to understand geophysical fluid dynamics.

The following section presents a short overview of the dynamics of the ACC and enumerates some shortcomings in our understanding of the dynamics of this current system. Section 3 describes the model used to analyze dynamical balances within the ACC. Section 4 presents the numerical techniques by which the model governing equations are solved. Some of the details of the numerical procedures that are associated with vector computation are included. Section 5 presents the dynamical studies that are considered and the numerical results pertinent to these studies. Section 6 gives conclusions derived from the dynamical experiments and considers the performance increase of vector computing in numerical studies of large-scale ocean dynamics using high resolution numerical models.

2. OVERVIEW OF WIND-DRIVEN FLOW IN A ZONAL CHANNEL

The basic dynamical balance for large-scale, wind-driven circulation in a closed basin involves wind forcing, which transports water meridionally and a viscous western boundary layer, which dissipates the energy due to the wind. This basic balance ignores, of course, various details of the circulation associated with inertia effects, time dependence and instability. However, it does provide a basic balance, the minimum required dynamics, for a wind-driven ocean in a closed basin.

Hidaka and Tsuchiya [1953] applied the above dynamics to a model of the circulation of the Southern Ocean. They represented the Southern Ocean by an annular channel of constant width and depth in which the water is forced by surface wind stress and slowed by lateral eddy friction. The results obtained from this model presented a dilemma. If the lateral friction coefficient was given a reasonable value, the transport of the ACC was an order of magnitude too large. If friction was increased sufficiently to give a reasonable transport, the friction coefficient was then larger than could be justified. The turbulence required would have homogenized the ACC -- a condition that is not observed. Consequently, the assumed dynamics were not sufficient to explain the circulation of the ACC.

This simple calculation illustrates the problem of elucidating the dynamics of the ACC. There must be some mechanism to remove the energy due to the wind stress on the ocean surface. Several possibilities have been suggested, but there is no real consensus as to the basic dynamical balance of the ACC. Two hypotheses are considered below.

Munk and Palmen [1951] advanced the idea that drag due to topography (form drag) provides sufficient retarding force on the flow to balance the surface wind forcing. They show in a scaling calculation that the four major north-south ridges in the Southern Ocean provide sufficient drag to balance the surface wind stress. For this mechanism to be effective, the deep currents must move at several cm/sec; such speeds are observed in the ACC. Moreover, the smaller scale (100 km wide) seamounts may be even more effective in slowing the flow in the ACC through topographic Rossby lee wave production.

A second drag mechanism, proposed by Stommel[1957], occurs because of continental land masses which allow a net pressure gradient (along a line of constant latitude) which can balance the wind stress -- a kind of lateral form drag. Latitude lines that do not intersect land masses, as at the latitude of Drake Passage, must balance the wind stress by some other means, say friction. Stommel observed that even though some latitude lines are not blocked, the ocean is less than 1000m deep somewhere along every line. The conclusion is that the dynamics for flow in a closed basin apply over most of the Southern Ocean. However, the dynamics of flow through Drake Passage are "different" and need further consideraton.

Gill [1968] considered an elaboration of the dynamics suggested by Stommel [1957] with a linear, viscous model for a zonal channel with partial meridional boundaries. This model shows that reasonable flow was obtained with lateral viscous coefficients that were somewhat large, though not as large as Hidaka and Tsuchiya. Additionally, solutions reveal that a viscous western boundary current develops on the partial meridional boundaries and that much of the recirculating flow through "Drake Passage" enters this dissipative current. The conclusion is that like closed basins, much of the dissipation occurs in western boundary layers and that these play a role in the dynamics of the ACC.

McWilliams, et al. [1978] describe solutions from a numerical model of wind-driven flow in a two-layer zonal channel with partial meridional barriers. The horizontal grid resolution of the model is small enough to allow dynamical flow instabilities that create eddies. This extra dynamical freedom in this model allows turbulent exchanges to be more representative of exchanges in the ocean compared to eddy viscosity formulation used in the previous models. Energy budgets from these solutions reveal that eddies are created by instabilities, transfer energy downward, and are dissipated by bottom friction. Bottom topography increases energy dissipation through form drag and reduces the total transport in the channel from that occurring in a flat bottom situation.

The problem to be considered, then, is how do meridional boundaries affect flow in a wind-driven zonal channel and, furthermore, how do variations in ocean depth contribute to the dissipation of forcing in such a channel. The model presented here addresses aspects of these questions.

3. ONE-LAYER, QUASI-GEOSTROPHIC MODEL

A one-layer, quasi-geostrophic numerical model is used to analyze the effect of lateral and bottom topography on wind-driven flow in a zonal channel. The one-layer model considers the vertically-integrated transport in the ACC. The Southern Ocean has a small stratification when compared to the subtropical gyres, so a homogenous fluid is appropriate. The quasi-geostrophic assumption implies that the time scale of the flow is small relative to the Coriolis period and that the flow is in geostrophic balance with the pressure field and is nearly horizonally non-divergent. For a detailed analysis of the approximations involved in these assumptions, the reader is refered to Pedlosky [1979, section 5.2].

The non-dimensional governing equation for the one-layer model is

$$q_t + \psi_x q_y - \psi_y q_x = c_0 \, \mathrm{curl}\left(\frac{\vec{\tau}}{\rho_0}\right) - r\nabla^2\psi + \frac{\nabla^4\psi}{Re} \qquad (1)$$

where q is the potential vorticity, defined as

$$q = \nabla^2\psi + \beta y + \eta_b - F\,\psi \quad , \qquad (2)$$

and the streamfunction is defined as

$$u = -\psi_y \quad \text{and} \quad v = \psi_x \; . \tag{3}$$

The variables are nondimensionalized as

$$(x^*, y^*) = L\,(x, y) \qquad t^* = T\,t$$
$$(u^*, v^*) = U\,(u, v) \qquad \eta^* = N_0 \eta$$

where the symbols have the usual oceanographic meaning [Pedlosky, 1979]. The definition of nondimensional parameters in (1) and (2) are given in Table I.

Table I. Nondimensional parameters

$\beta = \frac{\beta_0 L^2}{U}$	$\varepsilon = \frac{U}{fL}$	$Re = \frac{U\,L}{A_H}$
$r = \frac{L}{DU} \left\{\frac{fA}{2}\right\}^{1/2}$	$F = \frac{f^2 L^2}{gD}$	$N = \frac{f_0 UL}{g}$
$T = \frac{L}{U}$	$c = \frac{L}{DU} \max[\, \mathrm{curl}(\frac{\vec{\tau}}{\rho})\,]$	
$\eta_b = \frac{h(x,y)}{\varepsilon\, D}$		

Inspection of (1) shows the dynamics included in this model. The left hand side of (1) is the time change of potential vorticity from local changes and from advection of relative and planetary vorticity by the geostrophic flow. The terms on the right hand side represent, in order, creation of vorticity by the surface wind (through surface Ekman suction), relative vorticity dissipation by bottom friction (through bottom Ekman suction), and lateral diffusion of relative vorticity (representing sub-grid scale dissipation). The dynamics occur on a beta plane so steady western boundary layers, Rossby wave transients, and Rossby lee waves are possible.

The dynamics included in this model are taken from models of the time dependent flow in an enclosed basin forced by a surface wind stress. However, one factor makes this model different from the classical basin models; the meridional boundaries are not complete, water can exit one side of the basin and reappear at the other side. Thus, the flow is in a doubly connected domain, and the transport of water through this recirculating gap must be calculated as a part of the model solution.

The circulation in this model is forced by the curl of the surface wind stress. The annual average wind stress curl for the Southern Hemisphere [Figure 2, Han and Lee, 1981] has a maximum (negative) along approximately 55°S. The zero curl line along 45°S is the general boundary between the ACC from the subtropical gyre. This boundary is

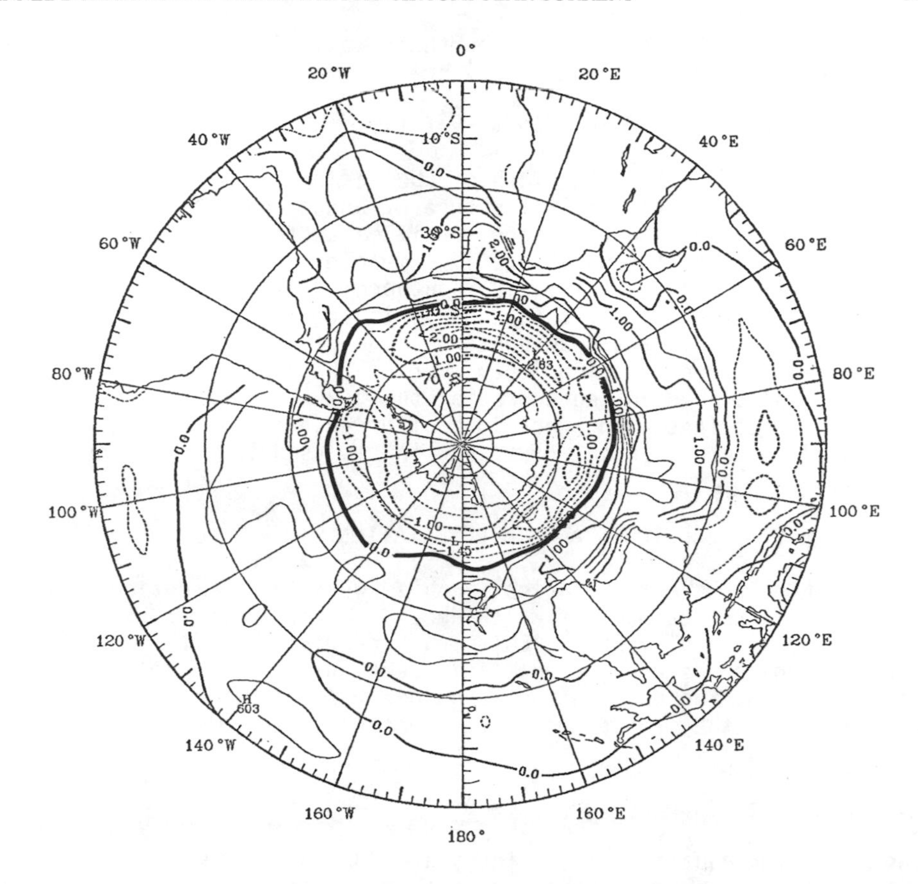

Annual Mean
Vertical Component of Wind Stress Curl
$\times 10^{-7}(kg/m^2 s^2)$
Han and Lee (1981)

Figure 2. Annual average wind stress curl in the Southern Hemisphere. A band of large (negative) curl exists over the ACC from 45°S to 70°S [Han and Lee, 1981].

represented in the model as the northern wall of the channel. The wind stress curl used in the model is

$$\mathrm{curl}\left(\frac{\vec{\tau}}{\rho} \right) = c_0 \sin (\pi y / L)$$

where L is the north-south width of the basin. This structure approximates the meridional structure of annual average wind curl.

The boundary conditions are specified as free slip on all walls (Figure 3). Since the walls are impermeable, the streamfunction has a constant value along a wall. The southern wall of the basin (representing the Antarctic Continent) is chosen, without loss of generality, to have a streamfunction value of zero. The value for the northern wall (representing South America and the bounding subtropical gyre) is calculated in the course of the time integration of the governing equations. The technique of Bryan and Cox [1972], which is discussed in detail in the next section, is used in this calculation.

4. NUMERICAL TECHNIQUES AND IMPLEMENTATION

The solution of the mathematical problem outlined in section 3 involves a number of choices many of which depend on the computer being used. The introduction of vector architecture requires that the reasons for various choices be considered. This section gives some of the considerations for chosing the techniques used in this calculation, especially as they relate to vector versus scalar computer architecture.

4.1 Numerical Integration Procedure

The governing equation (1) is integrated using centered finite difference representations for spacial derivatives and leap frog differences for time derivatives. For numerical stability, the advective terms are evaluated at the middle (n) time level (where time is discretized as $n\Delta t$) and the friction terms are evaluated at the oldest (n-1) time level (lagged friction). The Jacobian (advective) term is evaluated with the Arakawa [1966] procedure so that both energy and enstrophy are conserved in the calculation. The wind forcing is applied at the middle time level. A forward timestep is performed every 100 steps to avoid time splitting. The calculation of the streamfunction is diagnostic, that is, does not involve time derivatives, and is done after the new vorticity value is calculated from evaluation of (1).

The stability of the model integration depends on a number of factors. Since this is an explicit integration scheme, the CFL stability condition applies. The speed that must be resolved is the larger of the advective speed or the Rossby wave phase speed. Other stability conditions are obtained from each of the frictional terms. The bottom friction term requires that the timestep be shorter than 1/r (the bottom frictional spinup time). For lateral frictional effects, the stability condition is that $(Re \, \Delta x^2/\Delta t)$ must be less than π^2.

These estimates of the maximum timestep are only approximate because it is the stability of the entire scheme that is important and not that of the component parts. In general, the scheme should be stable for a timestep somewhat smaller than the most restrictive of these conditions.

There are also spatial resolution considerations which must be addressed to properly choose the grid spacing, Δx. The dynamics represented by (1) allow several types of boundary layers which must be resolved if the model results are to be realistic. The inertia boundary layer is a balance of advective and planetary vorticity terms and has a width given by $1/\beta^{1/2}$. The bottom friction layer balances bottom friction and planetary vorticity advection (this is the Stommel western boundary layer) and has a width of r/β. Finally, the Munk western boundary layer balances planetary vorticity against lateral friction and has an associated width of $(Re\beta)^{-1/3}$. These restrictions affect the choice of parameters by requiring that the inertia layer be at least Δx wide and that either the Munk layer is $2\Delta x$ wide or the Stommel layer is Δx wide, whichever is wider. For the parameters chosen in the following simulations, the lateral frictional layer is resolved, while the bottom frictional layer is narrower than the grid spacing.

4.2 Initial And Boundary Conditions

All of the simulations discussed here are started from a condition of no flow and no forcing. The forcing is turned on after two timesteps and is increased linearly over 40 steps to the final steady forcing amplitude.

The boundary condition on the flow is that the walls are slippery, which implies that there is zero relative vorticity at the walls. The potential vorticity along the walls is required for the calculation of the new interior potential vorticity. Therefore, the potential vorticity on the walls is

$$q = \beta y + \eta_b - F\psi \qquad \text{along the walls.} \tag{4}$$

These choices of techniques and boundary conditions were made to take advantage of (or at least not to impede) vectorized computer operations. In spite of the complexity of the finite difference analog of (1), there is no recursion in the resulting computer operations, that is, no value of q at the new time (n+1) depends on any other value of q at the new time. Therefore, all of the calculations can take advantage of either parallel or pipelined computer architecture.

4.3 Model Geometry

The dynamical problem under consideration requires that the model geometry (Figure 3) change. In many computer models, geometry is specified by the use of elaborate indexing schemes. This model uses a special feature of the CYBER 205 FORTRAN, a BIT control array, to specify the model geometry, thus adding flexibility to the computer

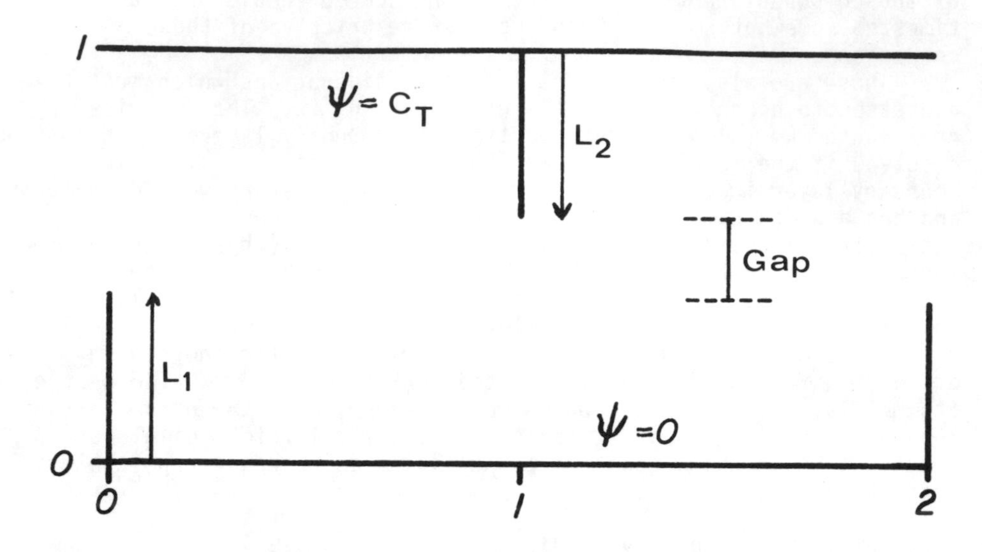

Figure 3. Geometry of model basin. As a test channel for analyzing the effects of lateral topography on zonal channel flow. The wall lengths (L_1 and L_2) are given several values as discussed in the text.

code and allowing the geometry to be changed easily. A BIT control array can prevent the processor from storing a newly calculated array element. Essentially, the calculation at a particular grid point is never done, and the array element never changes. Therefore, a set of BIT arrays can be used to define "active" points in the arrays, which are interior points, versus "inactive" points which are boundary and land points. The geometry of the model can be changed simply by changing a few arrays. All of the programs for the integration of (1) are controlled by BIT arrays.

The added benefit of this type of geometry control is that the total length of the "vector" in a given computer operation is the total number of elements in the two-dimensional array. Vector lengths of 1000 to 10,000 are easily obtained and such lengths allow the best performance increase of the vector computer over a sequential computer since the overhead of "filling the pipeline" is spread over a large number of operations.

4.4 Transport Calculation

The calculation of the total transport at each timestep uses the technique of Bryan and Cox [1972]. The streamfunction at any time is obtained by solving a Helmholtz equation forced by the potential vorticity for the same time

$$\nabla^2\psi^{n+1} - F\psi^{n+1} = q^{n+1} - \beta y + \eta_b \quad . \tag{5}$$

The boundary conditions are $\psi = 0$ and $\psi = C_T$ on the southern and northern boundaries, respectively. The geometry of the two boundaries is given by BIT arrays as discussed above. The value of C_T is calculated from the line integral of the momentum equation around the northern boundary

$$\oint \left[\frac{\partial \vec{u}}{\partial t} + \vec{u}\cdot\nabla\vec{u} - \frac{1}{Re}\nabla^2\vec{u} + r\vec{u} - \vec{\tau}\right]\cdot d\vec{l} = 0 \tag{6}$$

where dl is the line element along the northern boundary. The Coriolis and pressure gradient terms do not contribute to the line integral and are not included in (6).

The streamfunction solution is obtained in two parts: a forced solution with zero value on both boundaries and a free solution (no forcing) with a zero value on the southern boundary and unity on the northern boundary. The governing equation for the forced part is

$$\begin{gathered}\nabla^2\psi_F - \psi_F = q - \beta y + \eta_b \\ \psi_F\,(\text{north}) = \psi_F(\text{south}) = 0,\end{gathered} \tag{7}$$

and for the free problem

$$\nabla^2\psi B - F\psi_B = 0 \qquad \psi_B\,(\text{north})=1 \quad \psi_B(\text{south})=0. \tag{8}$$

The total solution is the sum

$$\psi = C_T \psi_B + \psi_F \quad . \tag{9}$$

Let $\vec{u}_B$ and $\vec{u}_F$ be the velocity field associated with ψ_B and ψ_F, respectively. Then (6) becomes

$$C_T \oint \vec{u}_B \cdot d\vec{l} + \oint \vec{u}_F \cdot d\vec{l} - \oint \vec{u}^{n-1} \cdot d\vec{l} + + \Delta t \oint \left[\vec{u} \cdot \nabla \vec{u} - \frac{1}{Re} \nabla^2 \vec{u} + r\vec{u} - \vec{\tau}\right]^n \cdot d\vec{l} = 0 \tag{10}$$

This equation can easily be solved to obtain the value of C_T. Notice the line integral for the free solution is independent of time and need only be calculated once.

4.5 Elliptic Solver

The Helmholtz equations (7 and 8) are solved with a red-black successive-over-relaxation (SOR) procedure, chosen for several reasons. Direct solvers and alternating-direction implicit solvers require that the domain of the equation be known. In situations where the geometry changes from one simulation to the next, these methods are not easily implemented. They also suffer from difficulties in code vectorization. SOR is an efficient, iterative technique [Roache, 1982] which is not difficult to implement even in complicated geometry. The optimal relaxation parameter is not strongly sensitive to geometry changes as long as the basic geometry (dimensions of the array) does not change, and in any case, the optimal parameter can be determined for each geometry with little effort. SOR is, however, recursive (the current calculation depends on the answer to the immediately previous one), which prevents vectorized calculations. Recursion is eliminated by performing the relaxation on every other array element in a pattern like that of the red and black squares of a checker board (or black and white squares of a chess board) (Figure 4). Two sweeps of the array are required for a single relaxation. The performance of the red-black SOR is the same as the standard SOR in terms of iterations required to converge; however, the performance improvement, measured as CPU time to converge, on a vector machine is substantial.

A further problem with this streamfunction calculation is that the flow is periodic over that part of the domain associated with the recirculating gap. This periodicity requires that the relaxation be performed at the left and right boundaries of the array and that these calculations use values from the other side of the domain. One approach to calculating a periodic boundary is to code explicitly the boundary calculations and do the normal relaxation over the interior. This is not a good approach because the geometry will be "hardwired" into the code and the vectorized calculations will be on rather short vectors (one line across the array).

A better way to handle the calculations in a periodic domain, which allows for vectorization is as follows. The red-black relaxation can be performed over the entire array under control of the BIT maps that specify geometry. However, the calculations at grid points in the gap will be incorrect because the wrong value from the other side of the array is used; the value used is one grid point above or below the correct point (see Figure 4). The calculation is corrected quite simply by subtracting the incorrect term and adding the correct one. This correction is performed after every sweep and involves only those grid points which are in the gap. These extra few computations do not negate the benefit of the vectorized calculation over the whole array.

4.6 Vector Versus Scalar Computers

The benefit of a vectorized calculation is made clear by a direct comparisons of CPU time required for model simulations on a scalar computer (Amdahl V/6B) versus the CYBER 205 (Table II). Such comparisons are not of identical code because of the BIT map specification of geometry and the slight differences in the elliptic solver (SOR versus red-black SOR). Also, the simulation on the CYBER used HALF PRECISION (4 byte) variables while the Amdahl code used DOUBLE PRECISION (8 byte). The use of full precision (8 bytes) on the CYBER will double the required CPU time. However, it is clear from the CPU time required by each machine for a similar simulation that the vectorized code runs about fifty times faster, making it worthwhile to take the time to convert the code to vectorized format.

Table II. Comparison of execution speed for vector versus scalar computers. Array size is 81 by 41.

		CPU seconds		
case	steps	Amdahl v/6B	CYBER 205	ratio
open gap	300	989.95	14.23	1/63
closed basin	600	1589.22	11.60	1/137

5. SIMULATIONS AND RESULTS

Three sets of experiments are discussed. The first set considers the effect of lateral geometry on the circulation and transport of the ACC. These simulations are an extension of the analysis of Gill [1968] in that both lateral and bottom friction effects are considered, inertia terms are included, and different geometries are specified. The second set of experiments considers the effect of transverse submarine ridges on the transport in the ACC. The third set of experiments demonstrates

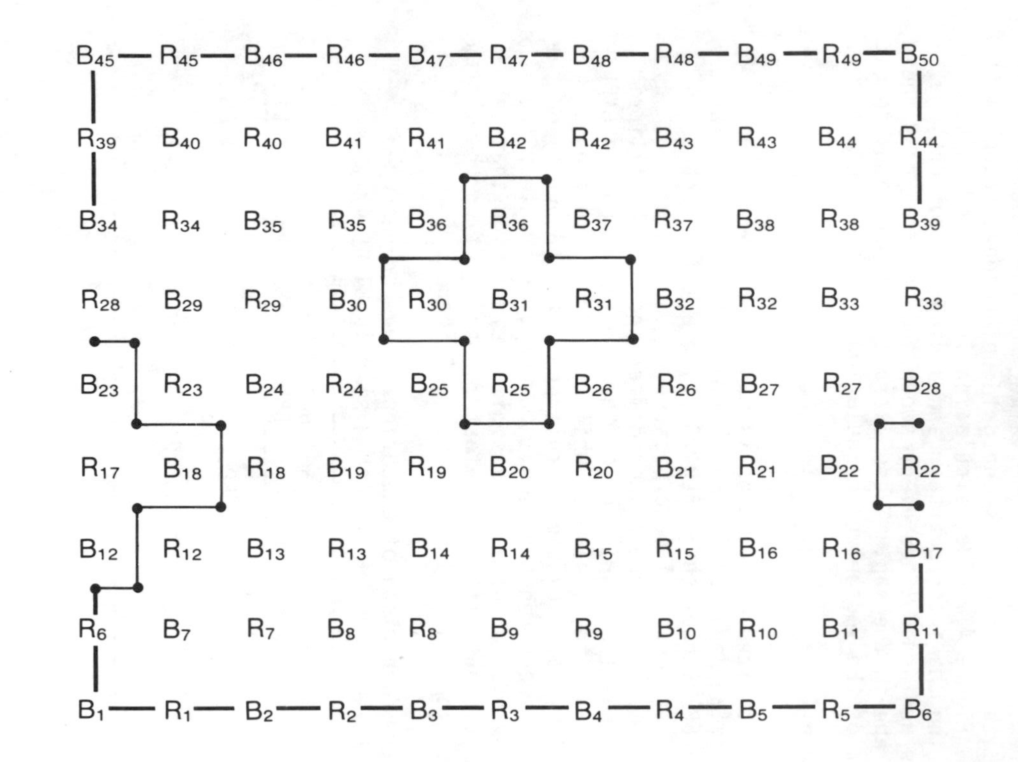

Figure 4. Stencil for Red-Black SOR and indexing for correcting SOR in the periodic domain. The grid points are labeled R and B for "red" and "black" respectively. Subscripts denote FORTRAN indexing as a one-dimensional array. The interior relaxation stencil is indicated by the cross centered at B-31. The relaxation at the periodic gap is shown for point R-17. For a relaxation sweep to the right and up, the interior stencil at R-17 uses point B-28 from the right boundary instead of the correct point, R-22. Such errors are easily corrected after a relaxation sweep.

the feasibility of performing fine-resolution calculations for a model of the ACC with low friction and reasonable lateral geometry. The computer code is sufficiently efficient on the CYBER 205 that multi-year simulations (using about 10 minutes of CPU per year) are possible allowing the investigation of the effects of seasonal wind variations on the transport and current structure and consideration of several values for the less well known parameters.

5.1 Geometric Effects On Transport

A set of numerical experiments is considered which tests the influence of overlapping meridional boundaries on flow in a zonal channel. Essentially the model represents a zonal channel forced by a negative wind stress curl that has a sinusoidal meridional structure and is uniform in the zonal direction. The wind curl is a maximum in the center of the channel and zero at the northern and southern boundaries. Two meridional boundaries are placed in the channel, one perpendicular to each of the walls of the channel (Figure 3). The lengths of these walls are varied to test the effect of lateral boundaries on the flow and on the total transport. The basic parameter values for this set of experiments (Table III) are chosen to correspond to the ACC. Both bottom and lateral friction operate in the model and the friction coefficients are in the mid to low end of the acceptable range for oceanic flow. The friction coefficients are low compared to the values chosen for most model studies of the ACC.

Table III. Parameter values for single gyre experiments.

Dimensional parameters		
$L = 1000$ km	$H = 5$ km	$\Delta x = 25$ km
$T = 210$ days	$u = 5$ cm/s	$\Delta t = .32$ days
$f_0 = 1.2 \times 10^{-4}\ s^{-1}$	$\beta_0 = 1.1 \times 10^{-13}\ cm^{-1}\ s^{-1}$	
$A_V = .78\ cm^2/s$	$A_H = 1.4 \times 10^7\ cm^2/s$	
$\tau_0 = .6\ dynes/cm^2$	(single gyre sinusoid)	
Nondimensional parameters		
$\beta = 200$	$r = .25$	
$F_x = .3$	$Re = 40$	
$\tau^x = -\cos(\pi y) + .6$		

The first case (Figure 5) considers a case where the gaps are quite small (20%) compared to the basin width. This case can be compared with more familiar closed basin calculations. Several features of the circulation deserve comment. The overall structure of the circulation in the eastern basin looks rather like the closed basin, wind-driven models of Stommel [1948] and Munk [1950]. A slight effect of nonlinearity is evident in the oscillation of the streamlines

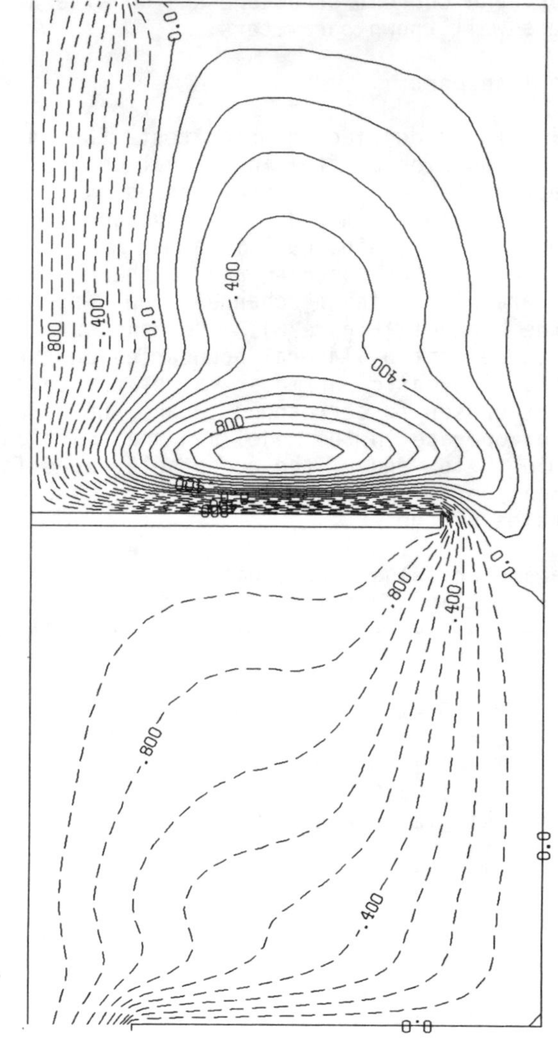

Figure 5. Steady stream function for wind-driven flow in a rectangular basin with offset walls. Each wall is 800 km long. The stream function has been normalized by the value along the northern wall. The total transport is 5.7 Sv. The parameters for the simulation are given in Table III.

in the northwestern corner of the gyre. Such oscillations were identified by Moore [1963] as standing Rossby waves and serve as additional dissipation for the flow.

There is a net transport though the basin of 5.7 Sv. along the northern side of the channel, north of the zero streamline (1 Sv. = 10^6 m^3/s). In this simulation, all of the recirculating flow passes through the dissipative western boundary layer. The flow in the remainder of the gap is quite slow by comparison and is bi-directional, with the southern part of the gap having westward flow and the northern part having eastward. This structure implies that part of the sub-polar gyre penetrates a short distance west of the gap.

A series of three cases are now considered to see the effect of partial barriers. The wall lengths are chosen so that 1) each wall extends to midchannel; 2) the walls overlap by 200km; and 3) there is a 200km gap in midchannel. These simulations test the importance of the meridional boundaries on flow in a zonal channel. According to Stommel [1957], overlapping walls should lead to a more Sverdrup-like flow while open gap situations should lead to frictionally slowed flow like that found by Hidaka and Tsuchiya [1953].

The parameters for all of the simulations are given in Table III; the only change for the different cases is the length of the walls. All of the simulations were run for 600 time steps (about 200 days) which was sufficient for steady state as determined by mean kinetic and potential energy (not shown).

The steady streamfunctions for these three cases are presented in Figures 6, 7 and 8. Figure 6 is the steady solution for the case where the walls cover all latitude lines without overlapping. There is a western boundary current on each of the walls though the flow along the northern wall is considerably stronger than that along the southern wall. The net transport occurs along the northern side of the channel (north of the zero streamline), and a considerable fraction of the flow seems to enter the western boundary layer. The slight nonlinearity of the flow is again evident in the wavy streamlines in the region where the northern end of the western boundary current turns to rejoin the interior. The total transport for this case is 50 Sv. which is a little small compared to the known transport value of 125 Sv. [Whitworth, 1983]. Since the geometry of the model is rather different than the Southern Ocean, a direct comparison may not be appropriate.

As the walls overlap, the transport should decrease because the dissipative boundary layers are longer and more of the flow is Sverdrup-like. The solution for the case in which the walls overlap by 200km (Figure 7) does show decreased transport (18 Sv.) and the character of the flow is somewhat different from that in Figure 6. The gyre in the eastern basin is more evident. Almost all of the net transport flow through the western boundary layer on the central wall. The flow in this case eastern basin looks rather like the frictionally dominated flow west of a partial barrier [de Ruijter, 1982]. The flow as a whole is like the flow around a partial barrier, like the tip of Africa.

If the walls are such that there is a 200 km gap in mid-channel, (Figure 8) the steady flow is much stronger (129 Sv.). There is some

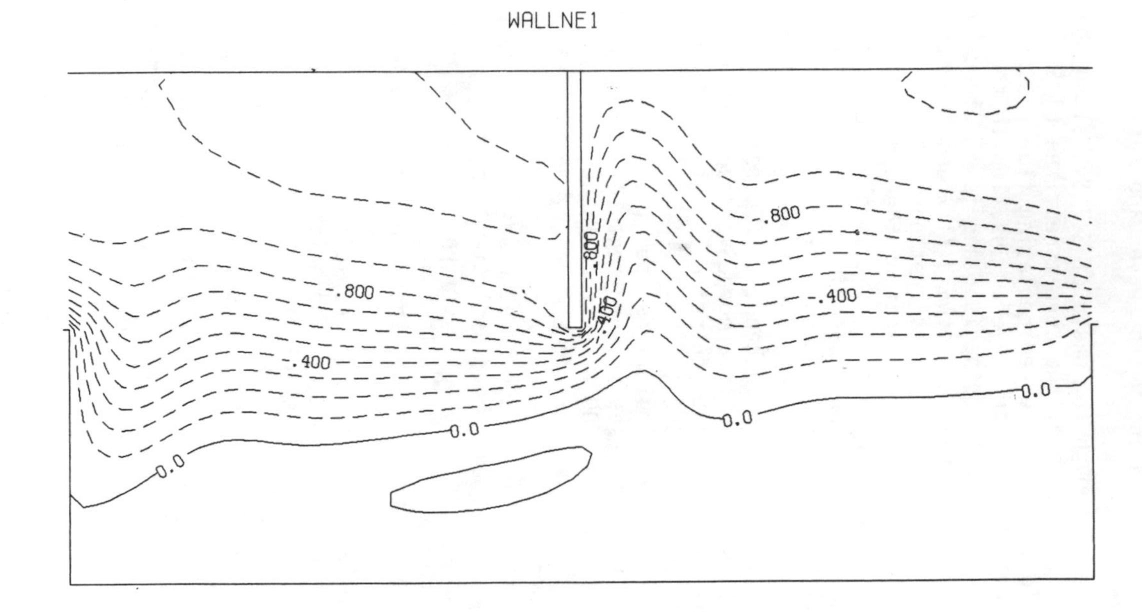

Figure 6. Steady streamfunction for wind-driven flow in a rectangular basin with offset walls. Each wall is 500 km long; the ends are at the same latitude. The streamfunction has been normalized by the value along the northern wall. The total transport is 50sv. The parameters for the simulation are given in Table III.

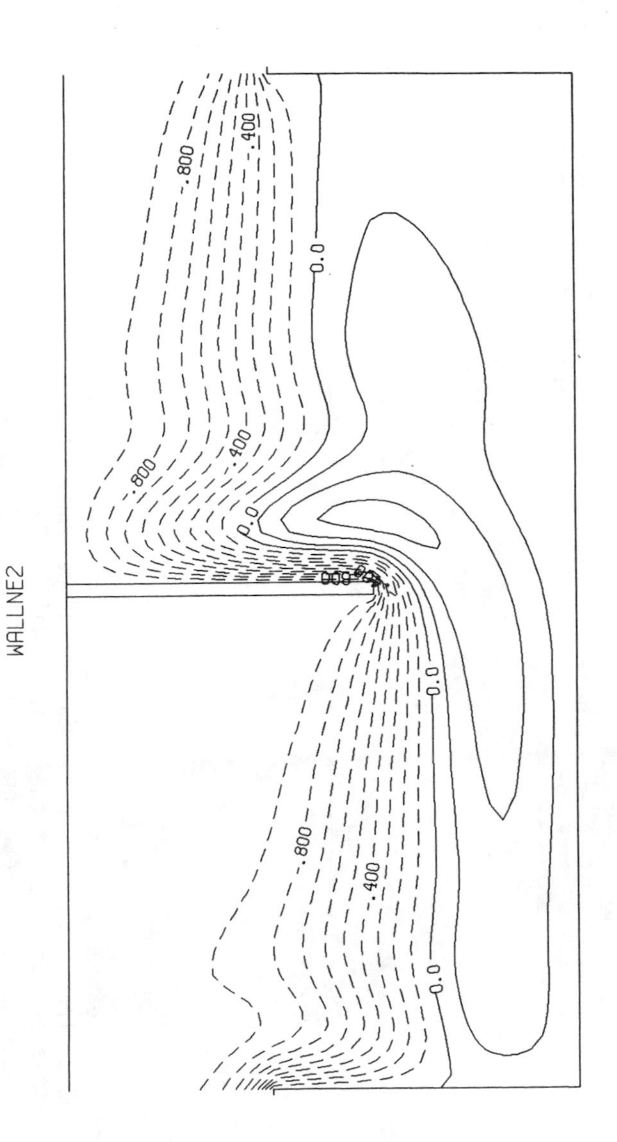

Figure 7. Steady streamfunction for wind-driven flow in a rectangular basin with offset walls. Each wall is 600 km long; there is a 200 km overlap in the center of the channel. The streamfunction has been normalized by the value along the northern wall. The total transport is 18 Sv. The parameters for the simulation are given in Table III.

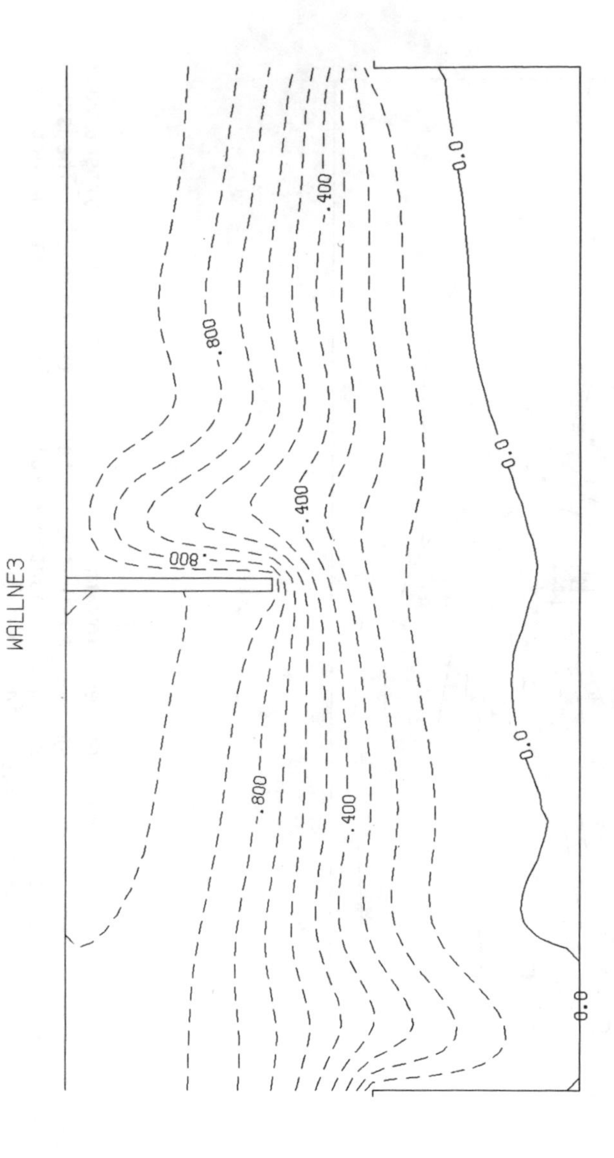

Figure 8. Steady streamfunction for wind-driven flow in a rectangular basin with offset walls. Each wall is 400 km long; there is a 200 km gap in the center of the channel. The streamfunction has been normalized by the value at the northern wall. The total transport is 129 Sv. The parameters for the simulation are given in Table III.

deflection of the flow around the barrier but the flow looks more like zonal flow in a channel rather than gyre-like circulation. The flow is stronger since more of the dissipation occurs as lateral and bottom friction in the center of the basin rather than in western boundaries, where the gradients are large and, hence, dissipation is more effective.

These three solutions illustrate how the flow in a zonal channel changes as the length of the walls change. The net transport as a function of wall length for this size basin and wind stress is shown in Figure 9; simulations not shown here are used to construct this figure. Two asymptotic points are also shown on the plot. If each wall extends all the way across the channel, there will be zero transport. If there are no meridional walls then the steady solution (a function of y only) can easily be found giving a transport of 2860 Sv., clearly an enormous increase in the transport compared to the other cases. The behavior of the graph is as expected; as the length of the walls increase, the transport decreases. This decrease in transport indicates that the presence of the walls allows dissipative boundary currents to develop and supports pressure differences, both effects balancing the forcing effect of the wind. There is a strong increase in the transport with decreasing wall length when the walls no longer overlap. The implication of this fact is that the dynamics of the flow change if there are zonal bands that are not blocked by barriers. In these bands, the flow is mainly zonal and the wind stress is balanced by lateral frictional dissipation. Elsewhere in the basin, the flow is more like Sverdrup flow that is seen in closed basins.

5.2 Bottom Topography Effects

Several simulations were considered in a square basin with a small gap and having one or more meridional ridges on the bottom. One set of experiments (not shown) had a ridge in the center of the basin about 200 km wide with heights ranging from 50 m to 800 m. The circulation was strongly controlled by the presence of the ridge, and streamlines tended to follow isobaths. The higher ridges reduced the transport by 30%. Cases were also considered with two and three ridges, each of height 100m. The two ridges had the same effect on transport as one ridge twice as high. Three ridges, however, did not have three times the effect of one. These simulations indicate the importance of bottom topography, but its effect on a one-layer model is much too strong and there is some question as to how to scale the topography so that it might have a reasonable effect [Pedlosky, 1979]. Topographic influences are better considered with multi-layer models which allow vertical stratification to play a role.

5.3 Fine Resolution ACC Simulations

The previous simulations show that the lateral geometry of a zonal channel plays an important role in determining the wind-driven transport. Therefore, to obtain reasonable results for models of the ACC, the lateral geometry of the Southern Ocean needs to be considered.

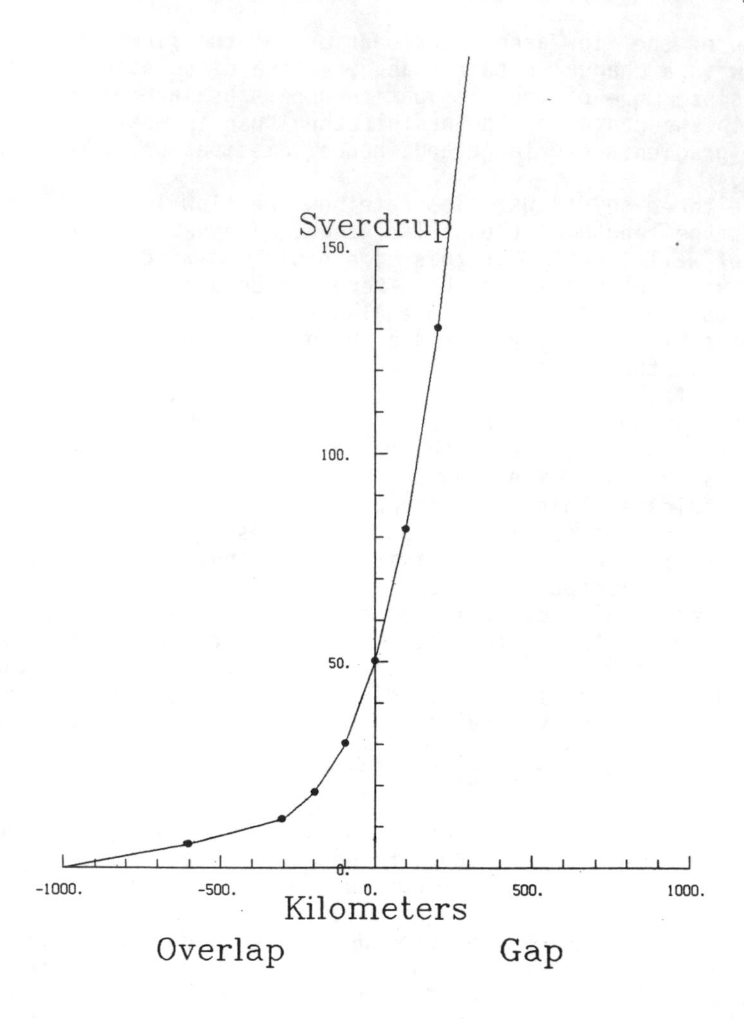

Figure 9. Transport as a function of wall overlap for the rectangular basin as determined by the simulations shown in Figures 6-8 and other simulations not shown. Two asymptotic points are added. There is no transport if the walls span the channel. If there are no walls, the streamfunction can easily be solved giving a transport of 2860 Sv. (using values given in Table III).

The following two simulations are demonstrations of the feasability of modeling the ACC with relatively small friction coefficients if lateral geometry is specified. The effect of bottom topography is also considered.

These fine resolution simulations represent the south Atlantic and south Indian Ocean regions of the ACC, an area 6000 km by 4000 km, with a grid resolution of 20 km. The principal boundaries in this domain are South America, Africa and the Antarctic Penninsula, which are represented as straight walls in the model. The detailed topography of these boundaries is not so important at this stage in model development since the mere presence of rigid walls allows appropriate dynamics to be active.

The parameters for the simulations are given in Table IV. Note that the lateral and bottom friction parameters are at the low range of reasonable values for large-scale ocean flow. The wind curl is specified so as to produce a double gyre with the southern gyre about twice as large as the northern gyre. This wind curl structure is a crude representation of the conditions in this area [Han and Lee, 1981]. Two oceanic gyres should develop from this wind forcing: a subtropical and a subpolar gyre. The trade winds are underrepresented by this simple structure of wind curl, and the subtropical gyre is unrealistically weak. The simulations are run for 2400 time steps which is about one model year.

Table IV. Parameter values for the fine resolution ACC simulations.

Dimensional parameters		
$L = 1000$ km	$H = 5$ km	$\Delta x = 20$ km
$T = 231$ days	$u = 5$ cm/s	$\Delta t = .17$ days
$f = 1.2\text{x}10^{-4}\ s^{-1}$	$\beta = 1.1\text{x}10^{-13}\ cm^{-1}\ s^{-1}$	
$A = 1.6\ cm^2/s$	$A = 1.0\text{x}10^{7}\ cm^2/s$	
$\tau_0 = .5\ dynes/cm^2$	(modified double gyre)	
Nondimensional parameters		
$\beta = 228$	$r = 2.$	
$F = .3$	$Re = 50$	

The flow field for a flat bottom channel after 1 year of forcing is shown in Figure 10. This is not really a steady state as the energy of the flow and the total transport are still increasing at a slow rate even though the structure of the current developed after about 1000 steps. Because of the low friction, there is insufficient dissipation to balance exactly the wind. With more integration, the model flow will increase until dissipation balances the wind, either through sufficient lateral shear or through development of barotropic instabilities which act as a sink for kinetic energy.

The streamfunction is dominated by the circumpolar Current which

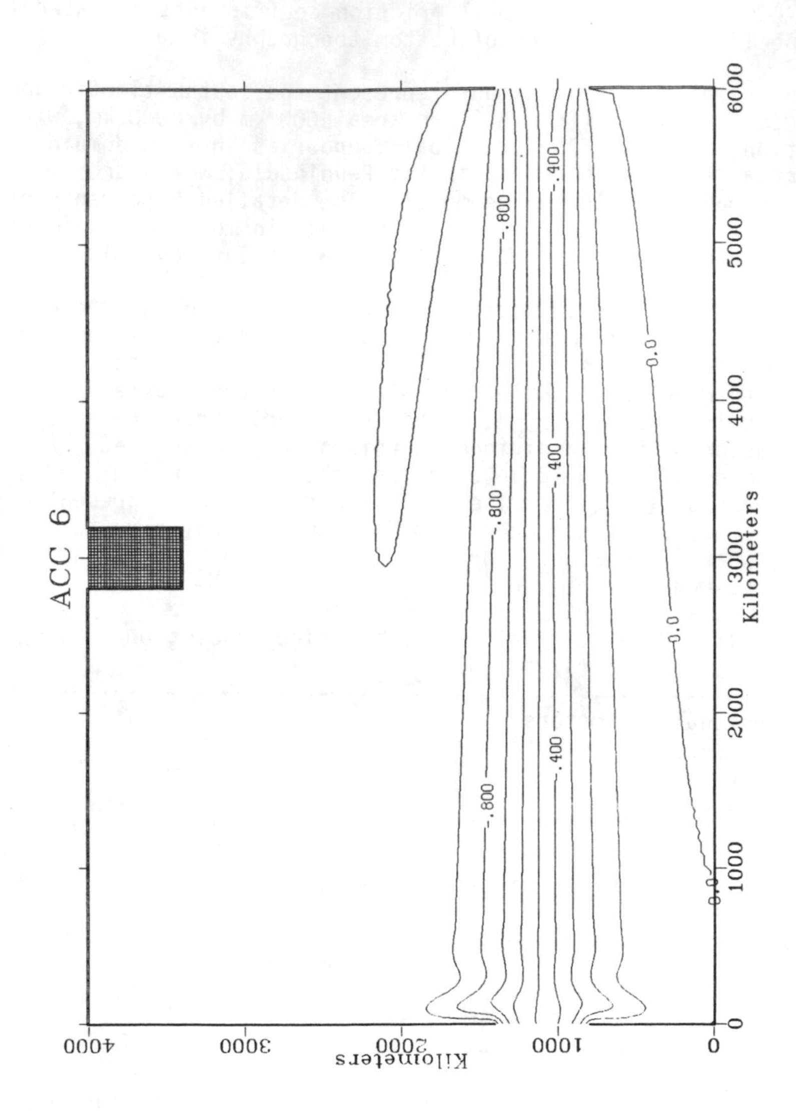

Figure 10. Transport streamfunction for fine-resolution simulation of the ACC in a flat bottom domain (parameter values given in Table IV). A double gyre wind with the polar gyre having twice the curl of the subtropical gyre is used. Streamfunction is normalized by the value on the northern boundary. Net transport in this channel of 191 Sv.

has a width of about 100 km and transport of 191 Sv. There is a subtropial gyre but its strength is so small compared to the ACC that it does not appear in the figure. Both the Weddell and subtropical gyres are only slightly represented and have been pushed to the west by the strong flow through the basin. There is a strong boundary current downstream of "Drake Passage" which is similar to the northward flowing Falkland current. A similar, though unobserved, southward boundary current is also indicated. As discussed by Gill (1968) and de Ruijter (1982), the exact nature of the flow will depend on the location of the maximum wind relative to the ends of the partial boundaries. With more realistic winds and continental outlines, the current patterns will be more realistic.

One major problem with this simulation is that the transport is still growing and has reached values of over 190 Sv. Clearly, some dissipative mechanism has been omitted from the model.

Munk and Palmen [1951] calculate that drag on the ACC due to bottom topography should be sufficient to balance the surface wind stress. This idea can be tested with this model by adding bottom topography. In order to keep the simulation simple, topography is specified only in the gap, representing Drake Passage, the Falkland Plateau, and the island chains (Scotia and South Sandwich) east of Drake Passage. The topography (Figure 11) is taken to be a flat plateau 100 m high. The actual topography is reduced by a factor of about ten because of the strong effect of bottom variations on one-layer models [Pedlosky, 1979].

The steady solution after 2400 timesteps is shown in Figure 12. The general character of the flow is not substantially different from the flat bottom simulation except for the deflection of flow by bottom topography. Because of the reduced flow in the ACC, the subtropical gyre is more visible. The principal difference is that the transport has stabilized at a value of 94 Sv. which is about 80% of the observed transport. A second important difference is that the flow south of the ACC is considerably reduced over the values obtained in the flat bottom case.

6. CONCLUSIONS

The conclusions derived from this study are of two types, those related to vector computing and those related to zonal channel dynamics.

One of the purposes of this work is to consider the benefit to numerical modellers of geophysical fluid dynamics afforded by the new generation of vector computers. This work deals with the CDC CYBER 205 computer, but the comments can apply to other vector computers. The advantage of vector computers over sequential scalar computers in processing speed is considerable, approaching a factor of 50 or better (for the simulations considered here). There is a price associated with this increase, however; the programmer must learn a new programming language (it looks like FORTRAN, but there are many new and different features to the language). Furthermore, the programmer must consider the calculation to be attempted and tailor the algorithms to

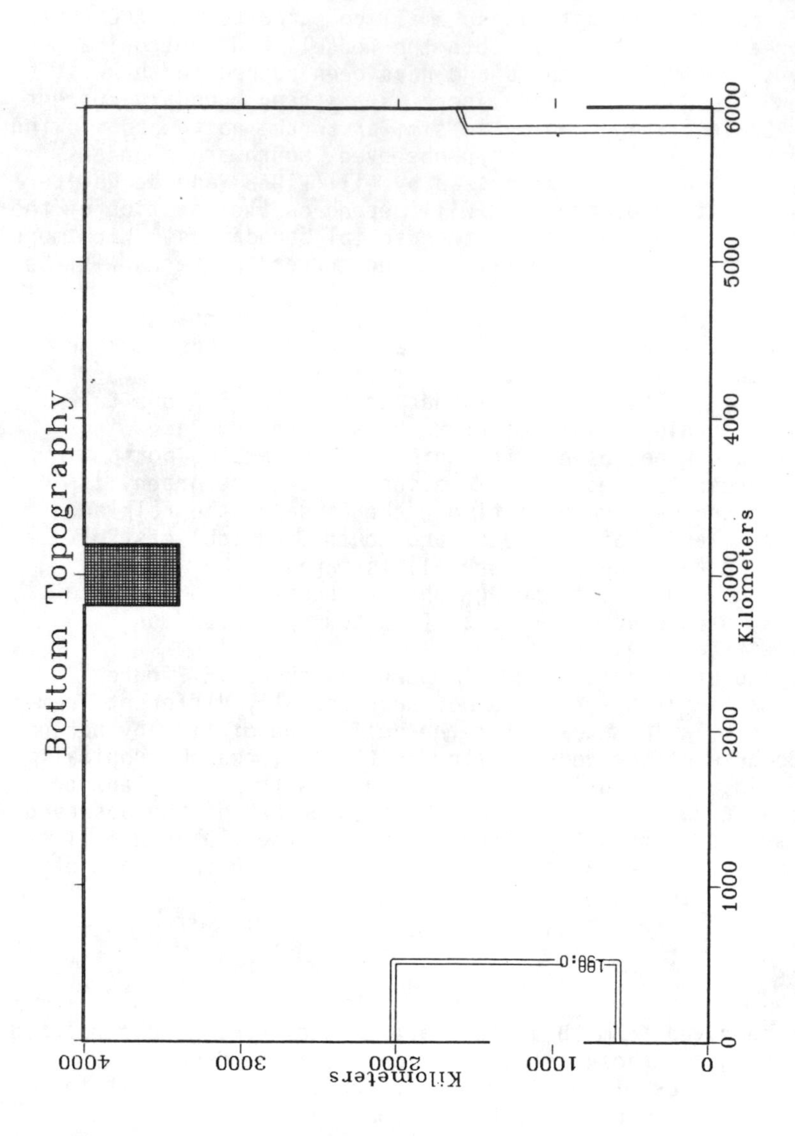

Figure 11. Bottom topography for ACC fine resolutions simulation. Contour lines show departures of the topography from the mean depth. Contour lines are 50 and 100 m.

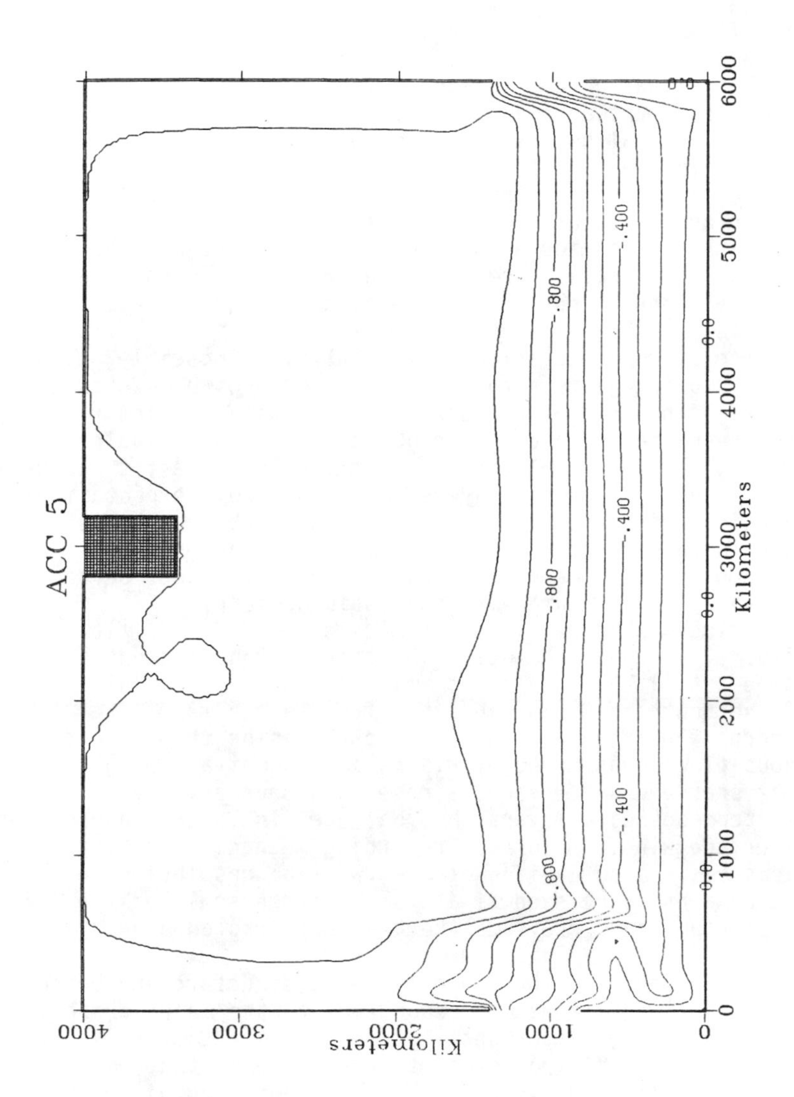

Figure 12. Transport streamfunction for fine resolution simulation of the ACC. Bottom topography is shown in Figure 11. Streamfunction is normalized by the value on the northern boundary. Net transport in this channel is 94 Sv.

the vector computer. "Conventional wisdom" compiled over the years on scalar computers may not transfer, unmodified, to vector machines. Sometimes, older methods that are too slow on scalar computers have a special advantage on the vector computer.

A benefit of the new architecture peculiar to the CYBER 205 is the ability to use BIT control vectors to "deactivate" grid points in a calculation. This allows general solvers of various sorts to be coded for general geometry, further reducing the programming required for a given problem. It also allows geometry to be changed with ease, a particular benefit in oceanic flow modeling.

As a final comment on modeling calculations, it is sometimes beneficial to perform a calculation over a large vector even if a few incorrect calculations are made and subsequently corrected. The advantage is that large vectors are processed with sufficient speed that the few corrections do not cancel the advantage gained by the larger calculation.

The fine resolution calculation shows that it is possible, with vector computers, to integrate a zonal channel model with mesoscale resolution over a large part of the ocean. The model dynamics presented here are rather simple given our current understanding of ocean circulation, but the simulations are sufficiently fast that one could extend the model to several layers with reasonable expectation that the simulations could be run on currently available machines. One might also be able to consider a one-layer model of the whole ACC (20,000 km by 4000 km) with mesoscale resolution to consider seasonal wind forcing effects in a domain with realistic geometry.

Several conclusions bear on the dynamics of wind-driven flow of a homogeneous fluid in a zonal channel. The calculations presented here consider in some detail the idea of Stommel [1957] that the lateral boundaries of the Southern Ocean cause the flow to act more like inviscid Sverdrup flow that is found in closed basins than like the strongly viscous flows considered by Hidaka and Tsuchiya [1953]. It is the presence of pressure differences across land boundaries and dissipative western boundary layers that balance, in large measure, the force of the surface wind. A further retarding mechanism is bottom topography, as seen in a comparison of the two fine resolution solutions. It remains to be seen if it is the large scale or the small scale ocean topography that provides the greatest drag on ocean circulation.

Despite the crudeness of the geometry, wind structure and bottom topography, the model gives some correspondence to the major currents in the Southern Ocean. One important feature of the ACC that is not represented is the system of fronts and strong currents that have been observed [Nowlin and Clifford, 1982]. These currents contain over half of the total transport of the ACC in Drake Passage and it is expected that they play a similar role over the remainder of the ACC. To date, no model has had a horizontal grid spacing sufficiently small to resolve these narrow features. One extention to the current model would be the inclusion of smaller resolution and vertical stratification (and density advection) to see the dynamics of fronts and the role that they play in the ACC.

7. ACKNOWLEDGMENTS

I would like to thank Control Data Corporation for a grant of two hours of CPU time on the CYBER 205 computer at Colorado State University and the Texas Engineering Experiment Station for providing access to the CSU computer. Richard Goodrum of Control Data Corporation provided me with invaluable help on the technical aspects of using the CYBER 205.

Support was provided by the National Science Foundation, Division of Ocean Sciences (grant OCE-8308959) during the time when this research was conducted.

8. REFERENCES

Arakawa, A., Computational design for long-term numerical integration of the equations of fluid motion: two-dimensional incompressible flow, Part I, J. Comp. Phys., 1, 119-143, 1966.

Bryan, K., and M.D. Cox, The circulation of the World Ocean: A numerical study. Part I, a homogeneous model. J. Phys. Oceanogr., 2 , 319-335, 1972.

de Ruijter, W., Asymptotic analysis of the Agulhas and Brazil current systems, J. Phys. Oceanogr., 10, 361-373, 1982.

Gill, A. E., A linear model of the ACC, J. FL. Mech., 32, 465-488, 1968.

Gordon, A. L., E. Molinelli, and T. Baker, Large-scale relative dynamic topography of the Southern Ocean, J. Geophys. Res., 83, 3023-3032, 1978.

Han, Y.-J., and S.-W., Lee, A new analysis of monthly mean wind stress over the global ocean, Report No. 26, Climate Research Institute, Oregon State University, 1981.

Hidaka, K., and M. Tsuchiya, On the Antarctic Circumpolar Current, J. Mar. Res., 12, 214-222, 1953.

McWilliams, J.C., W.R. Holland, and J.H.S. Chow, A description of numerical Antarctic Circumpolar Currents, Dyn. Atm. O, 2 , 213-291, 1978.

Moore, D. W., Rossby waves in ocean circulation, Deep Sea Res., 10, 735-747, 1963.

Munk, W., On the wind driven ocean circulation, J. Met., 7, 79-93, 1950.

Munk, W. H., and E. Palmen, Note on the dynamics of the Antarctic Circumpolar Current, Tellus, 3, 53-55, 1951.

Nowlin, W. D. Jr., and M. Clifford, The kinematic and thermohaline zonation of the Antarctic Circumpolar Current at Drake Passage, J. Mar. Res., 40(Suppl.), 481-507, 1982.

Pedlosky, J., Geophysical Fluid Dynamics, 624 pp. Springer-Verlag, New York, 1979.

Roache, P. J., Computational Fluid Dynamics, 446 pp., Hermosa Pub., Albuquerque, New Mexico, 1982.

Stommel, H., The westward intensification of wind driven ocean currents, Trans. AGU, 29, 202-204, 1948.

Stommel, H., A survey of ocean current theory, Deep Sea Res., 4, 149-184, 1957.

Whitworth, T., III, Monitoring the transport of the Antarctic Circumpolar Current at Drake Passage, J. Phys. Oceanogr., 13, 2045-2057, 1983.

WORLDWIDE OCEAN TIDE MODELLING

Ernst W. Schwiderski
Naval Surface Weapons Center
Dahlgren, Virginia 22448-5000

ABSTRACT. Area-time averaged Navier-Stokes equations of turbulent fluid motions are simplified by single-layer approximations to model tidal elevations and currents in the real world oceans. The resulting ocean tidal equations (OTE's) are solved by a finite difference scheme, which allows for a realistic integration of empirical tide data into the computed tidal field. The unique hydrodynamical-interpolation technique takes advantage of the massive empirical values, to search by systematic variation for the a-priori unknown turbulent momentum exchange or mixing and friction parameters occurring in the OTE's. The characteristic features of the new interpolation scheme are pointwise demonstrated for the constructed M_2-tide model by tidal charts displaying side-by-side empirical and computed data for direct evaluation. The global quality of the model is shown by the computed realistically balanced budgets of angular momentum and energy. The latter results reveal the true nature of eddy dissipation as a momentum exchange: while the former is completely negligible, the latter is of major significance.

1. INTRODUCTION

The global phenomenon of ocean tides has always captivated the awareness and imagination of mankind. A fascinating account of the early history of tidal research was written by Harris [1897]. Two interesting events may be recalled from this account. In one of his writings, Galileo (about 1600) was concerned with the effect of uncertainties in tidal observations upon Aristotle (about 322 B.C.) and wrote: ". . . for which diversities, and their causes incomprehensible to Aristotle, some say, that after he had a long time observed it upon some cliffes of Negropont, being brought to desperation, he threw himself into the adjoining Euripus, and voluntarily drowned himself". This account may be right or wrong, but it did foreshadow a tidelike history of tidal studies with all the ingredients of frustration and fruition inherent in scientific research. Indeed, through the ages tidal research encountered stunning successes and failures, puzzling myths and paradoxes, and inspired international contests and

J. J. O'Brien (ed.), Advanced Physical Oceanographic Numerical Modelling, 329–372.

working groups (Schwiderski [1980a]). According to Cartwright's [1977] count of tidal publications since Newton [1687] to 1969, tidal investigations increased exponentially and by extrapolation there should be over 5000 papers available today.

Another major event mentioned by Harris [1897] refers to Pliny the Elder (about 77 A.D.) who recognized that the tides are caused by the sun and moon by some swelling of the waters and that spring tides occur some time after full or new moon, i.e., after the driving forces are at their maximum. The retardation interval called "age of the tide" by Whewell [1833] was attributed by Pliny the Elder to "the effect of what is going on in the heavens being felt after a short interval; as we observe with respect to lightning, thunder and thunderbolts". Later, Young [1823] correctly explained the phenomenon by analogy to similar effects shared by forced and damped linear oscillators (compare first and second tidal paradoxes below and in section 4.1).

According to communications from my Norwegian colleagues, the Vikings realized the need to know high and low tidal water levels on their far-reaching voyages. Later tidal observations in coastal waters were collected into tide tables such as the famous Liverpool Tide Tables which were published by a clergyman. Nowadays, newspapers, radio and television broadcasts, and special bulletins publish daily, monthly, and longer-range tide predictions of interest to many users in seashore areas. Such predictions are based on massive tidal records which have been collected around the world, for example, by the U.S. National Ocean Service [1985], the British Admiralty [1984] and the International Hydrographic Bureau [1978]. Starting with Eyris [1968], Snodgrass [1968] and Filloux [1969], sensitive pressure gauges have been invented which can be placed into the deep oceans to record hourly tidal levels of the open oceans for up to one year. The collected data have been published by Cartwright et al. [1979]. Also, precision gravitimeters (Pratt [1960] and Thiel et al. [1960]) and satellite altimeters are presently being used to measure the geocentric tidal signal from which the ocean tidal level may be extracted by indirect analyses.

Of course, the mathematical theory of ocean tides was initiated by Newton [1687], who introduced the equilibrium theory of tides assuming an instantaneously responding solid ocean. It successfully explained the mainly semidiurnal nature of ocean tides and has attracted the attention of mathematicians and scientists ever since. However, the equilibrium theory predicted falsely a high tide under the moon or sun and lost its meaning as an ocean tide model. But it retained its meaning as an earth tide model and as a model of the tide generating forces of the moon and sun.

Although Newton recognized the hydrodynamical response

of the oceans, it was Laplace [1775] who derived the first ocean tidal equations (LTE's; section 3.2) from the Euler-Lagrange equations of inviscid fluid motions. The empirical and theoretical analyses of ocean tides were significantly aided by Lord Kelvin (Thomson [1868]) who followed a suggestion by Laplace and introduced the method of harmonic analysis into tidal research (section 3.2) which was later improved by Darwin [1883], Doodson [1921], Cartwright and Tayler [1971] and Cartwright and Edden [1973]. Numerous renowned mathematicians and scientists sought realistic solutions of the LTE's for idealized ocean basins, estuaries and channels (Lamb [1932] and Schwiderski [1980a]). While the constructed solutions revealed some important characteristic features of ocean tides, they lacked significantly in realistic detail, leading to heated controversies.

For example, elementary solutions (Schwiderski [1980a]) were found which showed the ocean's tidal response exactly 180 degrees out of phase with the forcing Newtonian equilibrium tide. This stunning reversal of Newton's result is the first tidal paradox (Lamb [1932]), which is common to all forced and undamped linear oscillators. It explains partly the above mentioned age of tides observed by Pliny the Elder. The same elementary solutions also displayed a resonance catastrophe of Laplacian ideal tides at certain real ocean depths. Of course, realistic ocean tides display all possible phase shifts relative to the forcing equilibrium tide (section 4.1), and they are nowhere in catastrophic resonance. Nonlinear interactions and dissipation neglected in the LTE's control the phase shift of the oceanic response and prevent any catastrophic resonance

With the advancement of science and technology, practical interest in open ocean tides became an urgent problem in a large variety of applications, such as in offshore industry and navigation, in commerce, management, and exploration of oceanic biological and chemical resources, in marine geodesy and geography, in ocean acoustics and general oceanography, in geophysics, seismology, and tectonics, in aeronomy, geomagnetism, and meteorology, and in astronomy and space technology. Some of the applications require tide predictions anywhere in the oceans of specified accuracy of 10cm and even 3cm. Other applications require precision modelling of energy dissipation and angular momentum exchanges in the ocean. For instance, modern empirical estimates of the retardations of the earth's rotation and moon's revolution can be used to check the quality of the modelled hydrodynamical parameters such as bottom friction and turbulent momentum exchange. Accordingly, as in other oceanic circulation problems, modelling of ocean tidal motions requires proper attention to four overlapping details: (I) modelling of internal hydrodynamical properties, (II) specifying effective driving forces, (III) discretization of continuous equations of motion, and (IV) empirical evaluations and verification of results. Ocean tidal motions are distinguished from other oceanic circulations by their relative simplicity such as known periodicity and depth-constant driving forces and by the availability of massive empirical tide data from all over the oceans. The latter knowledge can be used to improve directly the quality of the generally unknown turbulent hydrodynamical parameters of the model by hydrodynamical interpolation (section 3.3).

Following the invention of large-scale computers, modern numerical methods were introduced by various researchers starting with Hansen [1948, 1966], who also recognized the turbulent nature of tidal motions. In spite of major contributions by Pekeris and Accad [1969], Zahel [1970, 1977, 1978], Hendershott [1972], Estes [1977, 1980], Accad and Pekeris [1978], and others, progress was slow and the constructed tidal models failed to agree with empirical tide data over large areas of the oceans even when most of them were used as boundary data (section 3.2). The computed tidal maps fell far behind empirical maps, which were hand-drawn by simple rules of thumb from available observations by, for instance, Dietrich [1944a,b], Luther and Wunsch [1975], and Cartwright et al. [1980]. Somewhat more systematic tidal charts were produced by least-squares interpolation of empirical tide data using special interpolating basis functions, for example, by Munk et al. [1970], Irish et al. [1971], Jachens and Kuo [1973], Parke and Hendershott [1980] and Parke [1982]. According to private communications by Cartwright and LeProvost, satellite altimeter measurements of the geocentric sea level used by Mazzega indicate a possible improvement of such empirical methods. Nevertheless, as the computations by Parke and Hendershott [1980] and Parke [1982] indicate, empirically modeled tidal elevations do not necessarily allow a computation of the corresponding flow field and turbulent friction parameters needed in many applications.

Considering the successes and difficulties experienced with hydrodynamical techniques with and without empirical boundary data and with purely empirical interpolation methods, it seemed promising to seek a certain combination of both approaches. In view of the intense turbulence of tidal motions (section 3.1), it appeared also physically well justified to use the empirical knowledge to seek systematically for realistic hydrodynamical parameters specifying the otherwise undetermined mean turbulent flow (sections 2.1 and 2.2). This idea was realized by a novel hydrodynamical interpolation technique (Schwiderski [1978a, 1979, 1980a,b,c, 1981b]), which is described and evaluated in the following sections. The computed tidal charts (Schwiderski [1979, 1981a, 1982, 1983]) achieved the desired 10 cm prediction accuracy uniformly over all open oceans and proved their accuracy in numerous applications.

2. BASIC HYDRODYNAMICAL NOTIONS

2.1 Laminar Fluid Motions

In a Cartesian rectangular coordinate frame (x,y,z) and time t, the Navier-Stokes equations (NSE's) of laminar motions of isotropic viscous Newtonian fluids can be written conveniently in matrix notation (for detailed derivations see, e.g., Eliassen and Kleinschmidt [1957], Schlichting [1968], Whitaker [1968]):
Conservation of mass

$$\frac{\partial \rho}{\partial t} + \nabla'(\rho \mathbf{q}) = 0 \qquad (1)$$

and conservation of (linear) momentum

$$\rho[\frac{\partial \mathbf{q}'}{\partial t} + \nabla'(\mathbf{q}\mathbf{q}') - \mathbf{q}'(\nabla'\mathbf{q})] = \mathbf{f}' - \nabla' p + \nabla'\mathbf{T}', \qquad (2)$$

where bold letters denote matrices (or column vectors) and primes their transpose.
The notations are:

(p, ρ) = pressure and density scalors,
$\mathbf{q}'$ = (u, v, w) = velocity vector,
$\mathbf{f}'$ = (f^x, f^y, f^z) = body forces/volume,
∇' = $(\partial_x, \partial_y, \partial_z)$ differential operator,
∇p = $\mathbf{grad}(p)$, $\nabla'\mathbf{q} = \mathrm{div}(\mathbf{q})$,

$$\mathbf{Rm} = \rho \mathbf{q}\mathbf{q}' = \rho \begin{bmatrix} uu & uv & uw \\ vu & vv & vw \\ wu & wv & ww \end{bmatrix} \qquad (3)$$

= Reynolds tensor of (momentum flux)/(area·time)

$$\mathbf{T} = \begin{bmatrix} T^{xx} & T^{xy} & T^{xz} \\ T^{yx} & T^{yy} & T^{yz} \\ T^{zx} & T^{zy} & T^{zz} \end{bmatrix} \qquad (4)$$

= deviatoric stress tensor.

By invoking Stokes extension of Newton's law of friction, one arrives at the linear constitutive equation

$$\mathbf{T} = 2\,\nu\,\rho\,[\mathbf{S} - \tfrac{1}{3}\,\mathbf{I}\,(\nabla'\mathbf{q})] \qquad (5)$$

with the rate of strain tensor

$$\mathbf{S} = \frac{1}{2}[(\nabla \mathbf{q}') + (\nabla \mathbf{q}')'] \tag{6}$$

and the kinematic viscosity ν. At this point it is important to realize that the temperature-dependent viscosity ν is generally a specific fluid parameter. However, if the fluid motion is divergence free and irrotational, i.e., potential, then the viscosity is undetermined between zero and infinity. The particles of the fluid move relative to each other to appear frictionless. This explains the insensitivity of mean turbulent motions to relatively large variations of the analogous momentum exchange coefficient (eddy viscosity) often experienced in numerical modelling of tidal and other oceanic motions (sections 2.2 and 3.2).

For incompressible fluid motions (ρ = constant), the NSE's (1) and (2) reduce to:

$$\nabla' \mathbf{q} = 0 \tag{7}$$

$$\rho[\frac{\partial q'}{\partial t} + \nabla'(\mathbf{q}\mathbf{q}')] = \mathbf{f}' - \nabla' p + \rho \nabla' \{\nu[(\nabla \mathbf{q}') + (\nabla \mathbf{q}')']\} \tag{8}$$

with the laminar constitutive equation

$$\mathbf{T} = 2\nu\rho\ \mathbf{S} = \nu\rho[(\nabla \mathbf{q}') + (\nabla \mathbf{q}')'] \quad . \tag{9}$$

In order to specify a possible unique solution of the partial differential equations (7) and (8) with given ρ, ν, and $\mathbf{f}$, additional boundary conditions must be imposed on the velocity. For example, on solid boundary surfaces Σ it is common practice to specify the physically justified no-slip condition

$$\mathbf{q}(x,y,z;t)\Big|_{\text{on } \Sigma} = 0 \qquad \text{for all } t > t_o \quad . \tag{10}$$

In agreement with experimental observations, mathematical theory (e.g., Ladyzhenskaya [1969]) has established existence, uniqueness, and stability of an ordinary laminar solution of the NSE's (7) and (8) under the no-slip (or equivalent) boundary condition (10), provided the boundary surface is sufficiently smooth and the characteristic Reynolds number Rn satisfies the non-critical condition

$$Rn = L^c|\mathbf{q}^c|/\nu < \overset{*}{Rn} \quad . \tag{11}$$

Here, L^c and $|\mathbf{q}^c|$ denote the characteristic length and velocity of the flow and $\overset{*}{Rn}$ is the instability-critical Reynolds number at which stability ceases.

The ordinary solution is representable by ordinary functions, which describe a laminar flow in which the fluid particles move along smooth and almost parallel path lines. The solution is stable, i.e., any temporary or any small time-dependent disturbance of the equations of motion and/or boundary conditions decay or remain small, respectively. Hence, the solution can be found by starting from any initial state assumed at some time $t = t_o$. Obviously, the method of perturbation-scale analysis used so extensively to simplify the NSE's and/or boundary data (Mark Cane's article in this book) rests solidly on the stability of the unique solution. As the various models considered in this book show, it is almost incredible how much the NSE's can be cut down and still yield reasonable solutions.

Again, in agreement with experimental observations, it has been rigorously established (e.g., Joseph [1976]) that if the Reynolds number exceeds the stability condition (11), then the ordinary solution loses its stability against any kind of perturbations, and bifurcation occurs into various different flow patterns (e.g., Busse and Whitehead [1971]; Schwiderski [1972]). Some of the bifurcating flow patterns are laminar and relatively stable against small disturbances, as long as the Reynolds number remains below the turbulence-critical Reynolds number Rn^{**} at which turbulence takes hold. None of the critical flow patterns is determined by the NSE's and laminar boundary conditions such as the no-slip condition (10). Neither experiments nor theory indicate any clues to principles or conditions by which a unique flow pattern could be selected. The motions appear to be governed by hysteresis and pure chance, yet any catastrophic flow development is prevented by the quadratic Reynolds tensor of momentum flux (3). The critical motions become completely unpredictable above the turbulent-critical Reynolds number Rn^{**}, which is of typically conservative order $5 \cdot 10^4$. Needless to say, the method of perturbation-scale analysis loses its validity in the unstable flow regime where small perturbations may cause large effects. While such a situation seems hopeless from the mathematical point of view, rich experience over some hundred years, reinforced by several authors in this book, has convincingly proven that the scale analysis may still be justified, if, borrowing the apt words from Mark Cane's article, "proper care" (section 2.2), is being taken.

2.2. Turbulent Fluid Motions

In spite of a century old history of intensive research devoted to the phenomenon of turbulence, it is still far from being fully understood particularly in the microscopic particle domain. Nevertheless, significant advances have been realized in the macroscopic, i.e., statistically averaged regime. Through ingenious modelling techniques based on an experienced combination of art and science, numerous well-defined mean turbulent flow models have been developed, which proved their realistic values in many real-world problems.

In order to treat a turbulent flow with "proper care", it is important to remember that fluid particles in turbulent motion move in an unpredictably unstable and random fashion. In tidal and other oceanic motions, the time-dependent turbulent variation may be triggered by a large variety of mechanisms such as earth tides (section 4.3), bottom topography, inhomogenities of density, and/or atmospheric disturbances. The excited fluctuating or eddying commotions are known to claim the entire spectrum of short waves and fall into the range of internal or baroclinic waves even without any density variations.

Realizing the random nature of turbulent motions, it seems natural to seek certain statistical averages ($\overline{\mathbf{q}},\overline{p},\overline{\rho}$) of the corresponding turbulent velocity, pressure, and density, which, hopefully, remove the instability and indeterminancy in such a way to allow realistic flow predictions of practical interest, at least for some time interval. In practice one may simply assume, a priori, that the modelled mean motions exist, are well-defined, and are sufficiently realistic. Obviously such an uncertain procedure necessitates an, a posteriori, experimental verification; otherwise, even a constructed unique and stable solution may be useless. For example, beginning with Reynolds [1894] simple models of mean turbulent incompressible motions have been introduced in which the mean variables $\overline{\mathbf{q}}$ and $\overline{p}$ have been defined as averages of $\mathbf{q}$ and p, over "sufficiently long" time intervals (e.g., Schlichting [1968]; Whitaker [1968]). While such time mean models proved useful if the existence of reasonably steady mean turbulent motions could be assumed, they are generally not viable in time-dependent mean flow problems. Clearly, if one averages the periodic tidal motions over a long period, then the mean flow solution is null, which is unique and stable but not of interest. The time filters were improved (e.g., Eliassen and Kleinschmidt [1957]) by averaging $\mathbf{q},p,\rho$) over "some chosen" volume-time space ($\Delta x,\Delta y,\Delta z;\Delta t$).

In practice, the necessary averaging procedure is intimately tied (usually without mentioning) to the integration technique selected to solve the NSE's under laminar boundary conditions, although no unique and stable solution is specified. For example, if "the solution" is assumed to be representable in terms of a finite or infinite Fourier series, then uniqueness and stability may be enforced by imposing a harmonic wave spectrum which consists only of multiples of some basic wave lengths (Δx, Δy, Δz) and/or period Δt. Though such a forced exclusion of the possible full non-harmonic wave spectrum seems somewhat artificial from the mathematical point of view, it is physically irrelevant if the modelled solution is well-defined and of interest. Similarly, in numerical modelling of turbulent motions uniqueness and stability enforcing averaging procedures are automatically specified by choosing an appropriate grid system and interpolating basis functions for the definition of finite differences or finite elements. In any case, all known averaging techniques share the need to choose some appropriate length and time scales upon which the definition and usefulness of the desired solution depends. The significance of such filter scales was first recognized by Prandtl [1925], who introduced the well-known mixing lengths in his celebrated mixing theory of turbulent motions.

The important dependence of a specified mean turbulent motion of an incompressible fluid on the chosen filter scales may be illustrated in connection with the conceptually simple finite difference techniques, which are still mostly used in modelling of large-scale tidal and other oceanic motions. Consider (Schwiderski [1978a, 1980a,b,c]) a flow parcel contained in a rectangular test or grid cell of volume $(\Delta x \Delta y \Delta z)$ around the point (x,y,z) at time t during a time interval or integration step Δt. The mass fluxes (M^x,M^y,M^z) through the respective surface areas $(\Delta y \Delta z)$, $(\Delta x \Delta z)$ and $(\Delta x \Delta y)$ are always measurable quantities regardless of the laminar or turbulent nature of the motion. Hence, an area-time mean velocity $\overline{\mathbf{q}}$ at (x,y,z,t) may be defined by the divided fluxes

$$\overline{\mathbf{q}}' = (\overline{u},\overline{v},\overline{w}) = \left(\frac{M^x}{\Delta y \Delta z}, \frac{M^y}{\Delta x \Delta z}, \frac{M^z}{\Delta x \Delta y}\right)/(\rho \Delta t) \qquad (12)$$

without any knowledge of the particle or point velocity $\mathbf{q}$. An analogous area-time mean pressure p may be defined as the divided momentum acting on a surface element during Δt. Clearly, any finite-difference analog of the NSE's is in fact written in terms of area-time mean values rather than point values. For instance, the discretized equation of conservation of mass (9) assumes the obvious form

$$\Delta M^x + \Delta M^y + \Delta M^z = 0 \qquad (13)$$

in terms of mass fluxes (compare Eq. 56).

If the flow is laminar and sufficiently regular, then the area-time mean values $(\overline{\mathbf{q}},\overline{p})$ are ε-approximations of their existing point limit values $(\mathbf{q},p)$ for $(\Delta x,\Delta u,\Delta z;\Delta t)\rightarrow(0,0,0;0)$ However, if the flow is turbulent, then the existence of the limits is by no means obvious since point velocities and pressure at (x,y,z,t) are not uniquely determined. Hence, if the area-time scales are chosen too small, then the defined averages may be as unpredictable as their limits. On the other side, if the filter scales are chosen larger and larger, then more and more as well as wider and wider fluctuating commotions are filtered out and remain unaccounted for in the mean values $(\overline{\mathbf{q}},\overline{p})$. Thus it is important to remember that the mean turbulent motions depend significantly on the chosen filter scales; they should be neither too large nor too small. In practice it requires experience, intuition, and trial-and-error computations to seek optimum scales in order to retain relevant flow phenomena as realistically as possible without encountering instability and indeterminancy.

Following Reynolds [1894], it is useful to decompose the turbulent variables ($\mathbf{q}$,p) into their corresponding means ($\overline{\mathbf{q}}$,$\overline{p}$) and fluctuating or eddying parts ($\tilde{\mathbf{q}}$,$\tilde{p}$) such that

$$\mathbf{q} = \overline{\mathbf{q}} + \tilde{\mathbf{q}} \quad , \text{ with } \quad \overline{\tilde{q}}=o, \tag{14}$$

$$p = \overline{p} + \tilde{p} \quad , \text{ with } \quad \overline{\tilde{p}}=o. \tag{15}$$

As is well-known, the decomposed variables lead to the area-time mean of the quadratic Reynolds tensor of (momentum flux)/(area·time) (Eq. 3)

$$\overline{\mathbf{Rm}} = \rho(\overline{\mathbf{q}\mathbf{q}'}) = \rho(\overline{\mathbf{q}}\overline{\mathbf{q}}') + \rho(\overline{\tilde{\mathbf{q}}\tilde{\mathbf{q}}'}). \tag{16}$$

While the first part of $\mathbf{Rm}$ depends only on the mean velocity $\overline{\mathbf{q}}$, the second term contains only the undetermined velocity $\tilde{\mathbf{q}}$ of the fluctuating commotions and its negative value

$$\tilde{\mathbf{R}}\mathbf{s} = -\rho(\overline{\tilde{\mathbf{q}}\tilde{\mathbf{q}}'}) = -\rho \begin{bmatrix} (\overline{\tilde{u}\tilde{u}}) & (\overline{\tilde{u}\tilde{v}}) & (\overline{\tilde{u}\tilde{w}}) \\ (\overline{\tilde{v}\tilde{u}}) & (\overline{\tilde{v}\tilde{v}}) & (\overline{\tilde{v}\tilde{w}}) \\ (\overline{\tilde{w}\tilde{u}}) & (\overline{\tilde{w}\tilde{v}}) & (\overline{\tilde{w}\tilde{w}}) \end{bmatrix} \tag{17}$$

is known as Reynolds tensor of "apparent turbulent stresses". The notion "stresses" is formally justified (e.g., Schlichting [1968] p. 527), because it has the dimension of stresses. Of course, it has also the dimension of energy per volume (e.g., Eliassen and Kleinschmidt [1957]). But by definition ($-\tilde{\mathbf{R}}\mathbf{s}$) is a tensor of (momentum flux)/(area·time) just as the first term of Eq. (16). As will be seen in section (4.3), this important distinction explains a plausible property of tidal motions.

Using Eqs. (14) to (17), one finds the NSE's of mean turbulent motions of incompressible fluids

$$\nabla'\overline{\mathbf{q}} = o \tag{18}$$

$$\rho[\frac{\partial\overline{\mathbf{q}}'}{\partial t} + \nabla'(\overline{\mathbf{q}}\overline{\mathbf{q}}')] = \mathbf{f}' - \nabla'\overline{p} + \\ + \rho\nabla'\{\nu[(\nabla\overline{\mathbf{q}}') + (\nabla\overline{\mathbf{q}}')']-(\overline{\tilde{\mathbf{q}}\tilde{\mathbf{q}}'})\}, \tag{19}$$

which contain the undetermined Reynolds stress tensor. Following Boussinesq [1896b] it is common practice to replace the unknown Reynolds stress tensor by the symmetric rate of strain tensor (Eq. 6) such that

$$\tilde{\mathbf{R}}\mathbf{s} = 2\rho\mathbf{AS} = \rho\mathbf{A}[(\nabla\mathbf{q}') + (\nabla q')'] \quad , \tag{20}$$

which resembles closely the substitution (9) for the viscous stress tensor **T**. Only the kinematic viscosity is replaced by the diagonal matrix

$$\mathbf{A} = \mathbf{Diag}(A^x, A^y, A^z) \tag{21}$$

of unknown eddy viscosities [Boussinesq 1896b] or more appropriately (section 4.3) of momentum austausch (= exchange) or mixing coefficients as noted by Prandtl [1925]. It may be mentioned that the symmetric rate of strain tensor **S** in the substitution (20) has been augmented by a skew-symmetric tensor (e.g., Bagriantsev [1983]), which appeared to improve results in some models of oceanic motions.

After inserting Eq. (20) into Eq. (19), one arrives at the NSE's of mean turbulent motions of incompressible fluids

$$\nabla'\mathbf{q} = 0 \tag{22}$$

$$\rho[\frac{\partial \mathbf{q}'}{\partial t} + \nabla'(qq')] = \mathbf{f}' - \nabla'p + \rho\nabla'\{\mathbf{A}[(\nabla q') + (\nabla q')']\} \tag{23}$$

where here and henceforth all averaging bars over **q** and p have been omitted for simplicity. As is customary, it has been assumed that the diagonal matrix of momentum exchange coefficients **A** absorbs the usually negligible kinematic viscosity matrix (ν**I**). Evidently, the NSE's of mean turbulent motions (22) and (23) differ formally from the NSE's of laminar incompressible flows (7) and (8) only in the replacement of (ν**I**) by **A**. Nevertheless, the mean NSE's are based on an assumed area-time scale, which must be chosen along with the mixing coefficients **A**, such that a unique, stable, and realistic solution is determined under some chosen boundary conditions. In mean turbulent motions, the no-cross-flow and free-slip conditions

$$\mathbf{q}\Big|_{\text{normal to }\Sigma} = 0 \quad \text{and} \quad \mathbf{T}\Big|_{\text{tangential to }\Sigma} = 0 \tag{24}$$

appear most useful on a solid boundary surface Σ, since the boundary layers are very thin relative to the chosen length scales. Still, the laminar no-flow conditions (10) may be equally useful and is, in fact, needed if flow separations are to be modelled as is pointed out by Bill Holland in this book. Computer experiments with both boundary sets (Schwiderski [1978a, 1980a,b c]) appeared to indicate only an inconclusive preference for (24) for global tidal motions. In any case, if a unique and stable solution of Eqs. (22) and (23) is specified, then the method of pertubation-scale analysis is again justified.

Obviously, the indeterminancy of turbulent motions has been essentially reduced to three unknown parameters A^x, A^y, and A^z, which act as scale factors for the momentum exchanges of the fluctuating motions across the corresponding surface elements $(\Delta y \Delta z)$, $(\Delta x \Delta z)$, and $(\Delta x \Delta y)$. Considering the above mentioned dependence of the mean turbulent velocities on those surface areas, it is physically plausible to assume

$$A^x = a^x \Delta y \Delta z, \quad A^y = a^y \Delta x \Delta z, \quad A^z = a^z \Delta x \Delta y \qquad (25)$$

with a^x, a^y, and a^z as reduced mixing coefficients. For example, a larger averaging area $(\Delta y \Delta z)$ results in a smaller mean velocity $u=\overline{u}$ and a larger fluctuating velocity $\tilde{u}$. Hence, the effect of $\tilde{u}$ on u should be enlarged by the averaging area. A similar argument could be made for the averaging time interval Δt. Since Δt is usually constant as an integration timestep, its explicit entrance in Eq. (25) is not necessary.

Some dependence of the mixing coefficients on the mesh size of discrete ocean circulation models has been noticed by Cox [1970], Friedrich [1970], Holland and Hirschmann [1972] and others. In fact, in flow models with uniform mesh sizes and almost uniform flow dimensions, it is appropriate to work with only one coefficient of mixing $A=A^x=A^y=A^z$. In oceanic circulations with small depth scales relative to almost equal lateral scales, it is more appropriate to assume two coefficients A^h and A^v for horizontal and vertical mixing, respectively. The most commonly quoted values are in the ranges $A^h = 10^6$ to 10^{11} cm^2/sec and $A^v = 10^0$ to 10^1 cm^2/sec with $\nu \doteq 10^{-2}$ cm^2/sec. The necessity of the scaling law (25) was determined by numerous computer experiments with global ocean tide models by Schwiderski [1978a, 1980a,b,c; sections 3.1 to 3.3; 4.3]. The same law is also suggested by the plausible decay and dispersion features of the discrete local waves generating the forced global tides (section 3.4).

The momentum exchange coefficients provide the researcher with a great flexibility of choice to model realistic mean turbulent motions. As was recognized by Prandtl [1925], these coefficients are characteristic parameters of the particular motion considered and not just fluid constants as the molecular viscosity. Hence, they must be chosen either by experienced guesses or by a systematic search using experimental information (Schwiderski [1978a, 1980a,b,c]; section 3.3, also Alan Davies in this book). The law of turbulent momentum exchange (20) is called linear or nonlinear, if, respectively, its coefficient matrix **A** is chosen independent of or dependent on the absolute flow velocity. The linear law was introduced by Boussinesq [1896] and is effectively used in a wide range of applications where relatively slow motions with negligible quadratic terms are being considered. Since the Reynolds stress tensor (17) is a quadratic function of the fluctuating velocity $\tilde{\mathbf{q}}$, Prandtl [1925] introduced the quadratic law and derived the first momentum austausch coefficient A which depended linearly on the velocity. Prandtl's mixing theory was extended by Smagorinsky [1963], Leith [1968], Crowley [1968,1970], Kraav [1969], O'Brien [1971], Heaps [1972], Davies [1977], Nihoul [1977], Davies and Furness [1980], and others.

3. OCEAN TIDAL EQUATIONS

3.1. Basic Continuous Ocean Tidal Equations

In order to develop hydrodynamical models of tidal motions in the real world oceans, it is necessary to estimate their Reynolds numbers to determine their laminar or turbulent nature. While the kinematic molecular viscosity of water is essentially fixed at $\nu = 0.01\ cm^2/sec$, the empirically known characteristic velocities vary considerably over the ocean with typically low values of $|\mathbf{q}^c| = 0.1$ cm/sec and $|\mathbf{q}^c| = 10$cm/sec for deep and shallow oceans, respectively. With the corresponding low horizontal characteristic lengths (grid sizes) $L^c = \Delta x = \Delta y = 100$ km or 1 km and low vertical characteristic lengths (depths) $L^c = H^c = \Delta z = 3$ km or 30m, one computes, respectively, the horizontal and vertical Reynolds numbers (Eq. 11)

$$\overset{h}{Rn} = 0.1 \cdot 10^7/0.01 = 10 \cdot 10^5/0.01 = 10^8$$

$$\overset{v}{Rn} = 0.1 \cdot 3 \cdot 10^5/0.01 = 10 \cdot 3 \cdot 10^3/0.01 = 3 \cdot 10^6 .$$

Clearly, even these low estimates of the horizontal and vertical mesh Reynolds numbers in open and coastal waters exceed the well-established conservative turbulence-critical Reynolds number $\overset{**}{Rn} = 5 \cdot 10^4$ (section 2.1) by a significant margin. Thus, tidal motions in real ocean basins must be considered as supercritically turbulent and can only be modelled in the mean as outlined in section 2.2.

To derive sufficiently general ocean tidal equations, the NSE's (22) and (23) of mean turbulent incompressible motions may be expressed in a rotating Cartesian frame (x,y,z) with z increasing upwards and z=0 specifying the geoidal sea surface at hydrostatic conditions. If $H = H(x,y)$ denotes the hydrostatic ocean depth, then a well-defined mean tidal motion may be assumed to exist under the following hydrodynamical boundary conditions:

(B1) $z = \zeta^s = \zeta^s(x,y,t)$ = sea surface = geocentric surface tide,

(B2) $z = \zeta^b - H(x,y)$ = perturbed sea bottom relief,

(B3) $\zeta^b = \zeta^b(x,y,t) = \zeta^e - \zeta^{eo}$ = geocentric bottom tide,

(B4) $\zeta^e = \zeta^e(x,y,t)$ = earth tide (section 3.2),

(B5) $\zeta^{eo} = \zeta^{eo}(x,y,t)$ = earth ocean-load tide (section 3.2),

(B6) $\zeta^o = \zeta = \zeta(x,y,t)$ = ocean tide (to be modelled),

(B7) $p^s = P$ = constant surface pressure,

(B8) $(\mathbf{q}^s)' \nabla(\zeta^s - z) \approx -w^s = -\zeta^s_t$, no cross-flow over sea surface,

(B9) $\mathbf{T}^s = (T^{sx}, T^{sy}, T^{sz})' = \mathbf{T}' \nabla(\zeta^s - z) = 0$, free surface slip,

(B10) $(\mathbf{q}^b)' \nabla(z + H - \zeta^b) \approx w^b = \zeta^b_t$, no cross-flow through bottom

(B11) $\mathbf{T}^b = (T^{bx}, T^{by}, T^{bz})' = \mathbf{T}' \nabla(z + H - \zeta^b)$

$$\approx \rho A^z [(u^b_z + w^b_x), (v^b_z + w^b_y), (w^b_z + w^b_z)]', \text{ bottom stress,}$$

(B12) Some lateral boundary conditions (section 3.2).

All approximations of the boundary conditions (B8) to (B11) are based on the strong inequalities

$$H_x << 1, \quad \zeta^s_x << 1, \quad \zeta^b_x << 1$$
$$H_y << 1, \quad \zeta^s_y << 1, \quad \zeta^b_y << 1,$$

such that

$$\nabla(\zeta^s - z) \approx (0,0,-1)' \text{ and } \nabla(z + H - \zeta^b) \approx (0,0,1)'.$$

These and the following major "single-ocean-layer assumptions (A1) through (A4) are well justified by perturbation-scale analysis, provided a unique and stable mean turbulent motion exists (section 2.2):

(A1) Depth-constant body forces/volume, i.e.,

$$\mathbf{f} = \rho[\nabla\psi - \mathbf{c}] = \rho[\psi_x - c^x), (\psi_y - c^y), (\psi_z - c^z)]',$$
with
$$\psi(x,y,z,t) \approx G[\eta^*(x,y,t) - z] \text{ and } \mathbf{c} \approx (c^x, c^y, c^z = 0),$$

where G is the gravitational and c^x and c^y the well-known horizontal Coriolis (section 3.2) accelerations of the earth. The functions ψ and η^* denote, respectively, the total geopotential due to the centrifugal and all gravitating forces and the corresponding tide-generating equilibrium tide.

(A2) Depth-constant mean horizontal velocities, i.e.,

$$u(x,y,z,t) \approx u(x,y,t)$$

$$v(x,y,z,t) \approx v(x,y,t)$$

(A3) Negligible vertical velocity, i.e.,

$$w(x,y,z,t) \approx 0$$

(A4) Hydrostatic pressure, i.e.,

$$p_z(x,y,z,t) \approx -G$$

or with (B7)

$$p \approx P + \int_z^{\zeta^s} G\rho dz = P + G\rho[\zeta^s(x,y,t)-z].$$

It may be noted that the assumption (A1) concerning the geopotential ψ is well justified since the ocean depth is very small relative to the radius of the earth. Evidently, it is this property of the driving tidal forces that justifies all other single-ocean-layer assumptions (A2) to (A4). The consequential reduction of the three-dimensional problem to a two-dimensional one is usually not possible in other ocean circulations, which are driven by surface and/or depth-dependent density variations. Also, these assumptions are consistent with the area-time averaging procedure assumed to define a mean turbulent tidal motion, provided the depth scale is chosen as the instantaneous ocean depth $(H+\zeta)$. By integrating (or averaging) the NSE's of mean turbulent motions (22) and (23) with the boundary conditions (B1) to (B11) and the single-ocean-layer assumptions (A1) to (A4) over the ocean depth, one finds such terms as

$$\int_{\zeta^b \, -H}^{\zeta^s} u \, dz \approx u(\zeta^s-\zeta^b+H) = u(H+\zeta), \tag{26}$$

$$\int_{\zeta^b \, -H}^{\zeta^s} (A^z u_z)_z \, dz = A^z u_z^s - A^z u_z^b \approx -A^z u^b, \tag{27}$$

$$\int_{\zeta^b \, -H}^{\zeta^s} W_z \, dz = w^s - w^b \approx \zeta_t^s - \zeta_t^b = \zeta_t \; ; \tag{28}$$

and v replacing u.

Based on the arguments presented in section 2.2, it may be assumed that the horizontal averaging scales Δx and Δy are almost equal, then with $\Delta z = H+\zeta$ one has the horizontal and vertical momentum exchange coefficients

$$A^x = A^y = A = a\,H(\Delta x+\Delta y)/2, \quad A^z = A^v = a^v \Delta x \Delta y, \tag{29}$$

where the reduced coefficients a and a^v may be constants or functions of velocity at one's disposal.

The approximate bottom stress (B11) is further simplified by the bottom friction law (including A2 and A3)

$$\mathbf{T}^b \approx \rho A^z[u_z^b, v_z^b, o]' \approx \rho B[u, v, o]' \quad , \tag{30}$$

which may be linear or quadratic depending on whether the bottom friction coefficient is chosen as

$$B \begin{cases} = \beta_o & = \beta_1 \Delta x \Delta y \quad \text{or} \quad (31) \\ = \alpha_o (u^2+v^2)^{\frac{1}{2}} & = \alpha_1 \Delta x \Delta y (u^2+v^2)^{\frac{1}{2}} . \quad (32) \end{cases}$$

The latter scale-dependent formulations appear more appropriate (Schwiderski [1978a, 1980a,b,c, 1984; section 4.2]) in view of the relationship between A^v and B (Eqs. 29 and 30). In any case, the reduced friction parameters α_0, α_1, β_0, or β_1 are at one's disposal just as the reduced lateral mixing parameter a in Eq. (29).

As was argued for the momentum exchange coefficients (section 2.2), the linear law of friction (Eq. 31) should be applicable if the motion is sufficiently slow such that all quadratic terms become negligible. As was recognized by Boussinesq [1896a], the quadratic law of friction (Eq. 32) should be more realistic in fast motions. It was successfully applied by Taylor [1919] in his often quoted paper on tidal motions in the Irish Sea (section 4.2). Grace [1930b] applied both laws to tidal motions in the Gulf of Suez and experienced a slight preference for the linear law. Inconclusive results were also experienced by other researchers such as Mofjeld and Lavelle [1983] in the Bering Sea. Other friction laws were considered by Johns [1966], Kagan [1972], McGregor [1972], and others.

After carrying out all approximations and simplifications, one arrives at the basic continuous ocean tidal equations (COTE's)

$$\zeta_t + [u(H+\zeta)]_x + [v(H+\zeta)]_y = 0, \tag{33}$$

$$u_t+(uu)_x+(uv)_y+c^x=G(\eta^*-\zeta^s)_x - \frac{Bu}{H+\zeta} + \tag{34}$$

$$+ 2(Au_x)_x+[A(u_y+v_x)]_y,$$

$$v_t+(vu)_x+(vv)_y+c^y=G(\eta^*-\zeta^s)_y - \frac{Bv}{H+\zeta} + \qquad (35)$$
$$+ [A(v_x+u_y)]_x+2(Av_y)_y,$$

which must be integrated under lateral boundary conditions realistic for mean turbulent tidal motions in global and/or coastal ocean basins. With suitably chosen horizontal ($\Delta x \approx \Delta y$) and time ($\Delta t$) scales as well as corresponding lateral momentum exchange and bottom friction coefficients A and B (Eqs 29 and 31 or 32), these COTE's are probably flexible enough to model most tidal motions with satisfactory results. In fact, all successful models of global and coastal ocean tides have been constructed by numerical integration of these equations with additional simplifications. For solutions in global oceans see section 3.2 and in coastal waters see, e.g., the papers by Alan Davies, Bruno Jamart, and Christian LeProvost in this book. However, if significant interactions with other ocean circulations are of interest, then the single-ocean-layer assumptions (A1) to (A4) are not justified and more layers need to be considered (Alan Davies article in this book).

For further simplifications of the basic COTE's (33) to (35), it is necessary to recognize their nonlinear and depth-dependent properties. If the tidal currents are sufficiently strong as in shallow coastal waters, then it may be necessary to retain the quadratic inertial terms and use the nonlinear laws of bottom friction and possibly lateral mixing in the momentum equations (34) and (35). Perhaps more important are the nonlinearities in the continuity equation and in the denominator of the bottom friction terms if the absolute tidal height ζ rivals or even exceeds the shallow static water depth. Indeed, the nonlinearly generated harmonic frequencies may also change the mean sea level. Considering the depth-dependent bottom friction term and the assumed depth-dependent mixing coefficient A (Eq. 29) in the momentum equations (34) and (35), one realizes that lateral mixing is significant in deep oceans while bottom friction is dominant in shallow waters. In fact, most coastal tide models neglect lateral mixing with realistic results. However, in global ocean tide models, both lateral mixing and bottom friction in proper proportions are necessary to achieve realistic results (Schwiderski [1978a, 1980a,b,c]). Finally, it may be mentioned that in coastal models it is adequate to assume $\eta^*=0$ and $\zeta^s=\zeta$, i.e. the coastal tides are mainly driven by the deep ocean tides, which must be known along the open-ocean boundaries by velocities and/or elevations.

3.2. Global Continuous Ocean Tidal Equations

In modeling of mean turbulent tidal motions in the global

oceans of the earth of radius R and angular velocity Ω, it is natural to rewrite the basic COTE's (33) to (35) in spherical coordinates(λ,θ) denoting longitude east and colatitude south, respectively. If (u,v) denote the corresponding east and north (opposite θ) velocities, then

$$(c^{\lambda},c^{\theta}) = 2\Omega(-v,u)\cos\theta \tag{36}$$

is the horizontal Coriolis acceleration. If one assumes the horizontal length scales $(\Delta\lambda,\lambda\theta)$ of the mean turbulent motions to be specified by a one-degree (or larger) grid system (section 3.3), then the basic COTE's (33) to (35) can be fully linearized in all open oceans (section 3.3) by neglecting ζ against H, deleting all quadratic inertial terms, and invoking the linear laws of lateral mixing and bottom friction. The corresponding lateral momentum exchange and bottom friction coefficients (Eq's 29 and 32) assume the form

$$A = aH(1+\mu\sin\theta)/2 \quad \text{and} \quad B = b\mu\sin\theta, \tag{37}$$

where $\mu(=1,2,4,8)$ designates a useful grading parameter of the spherical grid system (section 3.3). The unknown reduced parameters a and b have been determined for a one-degree graded grid system by Schwiderski [1978a, 1980a,b, and c] using a novel hydrodynamical interpolation technique (section 3.3). The computed values are

$$a \approx 200 \text{ m/sec} \quad \text{and} \quad b = 1.0 \text{ cm/sec}, \tag{38}$$

where b appeared sharply fixed and a allowed a variation of 25% and more without significant changes of the computed data (section 2.1). For a depth range of $10 \leq H \leq 7000$ m the lateral mixing coefficient A (Eq. 37) varied in the realistic range $1.3 \cdot 10^{7}$ to $1.3 \cdot 10^{10}$ cm^2/sec. For a discussion of the value of b see section 4.2.

It may be mentioned that the author conducted numerous computer experiments to test the feasibility of lateral mixing and friction coefficients of the form

$$A \sim H^{\gamma} \quad \text{and} \quad B \sim H^{-\varepsilon} \tag{39}$$

with γ=0, 1/2, and 3/2 and ε=1/2 and 1, but the computed results clearly rejected such parameters. For instance, for γ=0 (used by Hansen [1966], Zahel [1970], 1977, 1978], and Estes [1977, 1980]) lateral mixing appropriate for deep oceans of 3000m depth or more was found far too strong in shallow waters of depths 10 to 100m (see sections 3.4 and 4.3). Similarly, for ε=1 (used by Pekeris and Accad [1969]) bottom friction adequate in shallow waters at depths 10m to 30m was found completely ineffective by enforcing assumption at any depth of 100m and more (see also sections 4.2 and 4.3).

As was mentioned in section 1, the primary astronomical equilibrium tides $\hat{\eta}$ (or potentials $G\hat{\eta}$) of the moon and sun can be decomposed into a series of spherically harmonic partial tides ζ of the form (Thomson-Lord Kelvin [1868])

$$\hat{\eta} = \Sigma\eta = \Sigma K_{\nu}(\theta)\cos(\sigma t+\chi+\nu\lambda) \tag{40}$$

with a nonharmonically clustered frequency spectrum. Here, $K_{\nu}(\theta)$, ν, σ, and χ are well-known amplitude functions, species numbers, frequencies, and astronomical constants, respectively. In order to achieve a total tide prediction accuracy of about 10cm, it is necessary to consider the eleven leading harmonic equilibrium tidal modes listed in Table I. Since the global COTE's have been assumed to be linear in all open ocean regions, the oceans respond without interactions to each forcing equilibrium constituent of (40) with an harmonic ocean tide ($\chi=0$)

$$\zeta = \xi(\lambda,\theta)\cos[\sigma t-\delta(\lambda,\theta)] = \sum_{n=0}^{\infty} \zeta_n(\lambda,\theta,t) \tag{41}$$

of equal frequency σ but different location-dependent amplitudes ξ and phases δ. Let (e.g., Schwiderski [1985])

$$\zeta_n = \sum_{m=0}^{n} P^m(\cos\theta)[C^{\pm}_{mn} \cos(\sigma t\pm m\lambda-\delta^{\pm}_{mn}]$$

denote the spherical surface harmonics of the ocean tide ζ, then

$$\eta^o = \sum_{n=0}^{\infty} \alpha_n \zeta_n \quad \text{and} \quad \zeta^{eo} = -\sum_{n=0}^{\infty} \alpha_n h'_n \zeta_n \tag{42}$$

are its gravitational and terrestrial effects with

$$\eta^{eo} = -\sum_{n=0}^{\infty}\alpha_n k'_n \zeta_n \tag{43}$$

as the gravitational effect of ζ^{eo}, where (h'_n k'_n) designate

TABLE I: LEADING EQUILIBRIUM TIDES

TIDAL MODE	$\sigma \cdot 10^{-4}(\sec^{-1})$	K(cm)
$\nu=2$: **Semidiurnal Species**, $\eta = K \sin^2 \theta \cos(\sigma t + 2\lambda)$		
M_2: Lunar Principal	1.40519	24.2334
S_2: Solar Principal	1.45444	11.2841
N_2: Lunar Elliptical	1.37880	4.6398
K_2: Luni-Solar Declination	1.45842	3.0704
$\nu=1$: **Diurnal Species**, $\eta = K \sin 2\theta \cos(\sigma t + \lambda)$		
K_1: Luni-Solar Declination	0.7921	14.1565
O_1: Lunar Principal	0.67598	10.0574
P_1: Solar Principal	0.72523	4.6843
Q_1: Lunar Elliptical	0.64959	1.9256
$\nu=0$: **Long-Period Species**, $\eta = (1-3\cos^2\theta)\cos\sigma t$		
Mf: Lunar Fortnightly	0.053234	2.0871
Mm: Lunar Monthly	0.026392	1.1013
Ssa: Solar Semiannual	0.003892	0.9723

Data taken from Bartels [1957], Neumann and Pierson [1966], Cartwright and Tayler [1971], and Cartwright and Edden [1973].

Love numbers and $\alpha_n = 3\rho/\rho^e(2n+1)$ with the mean density of the earth ρ^e. The corresponding earth tide ζ^e and its gravitational effect η^e are

$$\zeta^e = h_2\eta \quad \text{and} \quad \eta^e = k_2\eta \tag{44}$$

with (h_2, k_2) as Love numbers.

The importance of these effects on ocean tides has been pointed out by Proudman [1928], Grace [1930a], Farrell [1972] and Hendershott [1972]. Although the needed Love numbers have been derived for simple earth models (e.g., by Farrell [1972]), the use of Eqs. (42) and (43) leads to complicated integro-differential equations. In analogy to

$$\eta + \eta^e - \zeta^e = \alpha\eta \quad , \quad \alpha = 0.69 \tag{45}$$

Schwiderski [1978a, 1980a,b, and c] used the following Takahasi [1929] and Accad and Pekeris [1978] approximations

$$\zeta - (\eta^o + \zeta^{eo} - \eta^{eo}) \approx \beta\zeta \quad , \quad \beta = 0.90 \tag{46}$$

such that in the COTE's (34) and (35)

$$\eta^* - \zeta^s = (\eta+\eta^e+\eta^o-\eta^{eo}) - (\zeta+\zeta^e-\zeta^{eo}) \approx \alpha\eta-\beta\zeta . \tag{47}$$

The approximation (46) is well-justified since the terms in parenthesis are about 10% of ζ and display a similar horizontal cotidal pattern as ζ (section 4.1).

The special simplifications of the basic COTE's (33) and (35) for worldwide oceans yield the global COTE's

$$U_t = \frac{GH}{R\sin\theta}(\alpha\eta-\beta\zeta)_\lambda + 2\Omega V\cos\theta - b\frac{U}{H}\mu\sin\theta -$$

$$- a\,H\,L^\lambda\,(U,U_\lambda,U_\theta,U_{\lambda\lambda},U_{\theta\theta};\,V,V_\lambda), \tag{48}$$

$$V_t = \frac{GH}{R}(\beta\zeta-\alpha\eta)_\theta - 2\Omega U\cos\theta - b\frac{V}{H}\mu\sin\theta -$$

$$- a\,H\,L^\theta\,(V,V_\theta,V_{\lambda\lambda},V_{\theta\theta};\,U,U_\lambda,U_\theta), \tag{49}$$

$$\zeta_t = -\frac{1}{R\sin\theta}[U_\lambda - (V\sin\theta)_\theta] \tag{50}$$

with the depth-integrated mean velocities

$$(U,V) = (u,v)H \tag{51}$$

and with (L^λ, L^θ) known linear functions of the listed arguments (see Schwiderski 1978a for more details). These equations must be integrated, for instance, under the lateral no-cross and free-slip boundary conditions (24) modified by hydrodynamical interpolation of empirical data (section 3.3). For a=0, b=0, α=1, and β=1, the global COTE's (48) to (50) reduce to the considerably simpler Laplace [1775] tidal equations, which can be derived from the Euler-Lagrange equations of ideal fluid motions which neglect all viscous and turbulent momentum exchange terms of the NSE's. Any energy dissipation is then assumed to occur at the shoreline or beyond by some unspecified mechanism.

3.3. Global Discrete Ocean Tidal Equations

In order to convert the global COTE's (48) to (50) to a discrete analog by finite differences, the world oceans are covered by a grid system of spherically graded "one-degree" mesh sizes $\Delta\lambda=\mu^\circ$ and $\Delta\theta=1^\circ$, which also determine the horizontal length scales of the mean turbulent tidal motions. The grading parameter μ(=1,2,4, and 8) assures a more uniform resolution and enhances the accuracy and stability of the numerical procedure (Schwiderski [1978a, 1980b,c]). The bathymetric data of Smith et al. [1966] are modified by hydrodynamical considerations to achieve a more realistic resolution of strong tidal distortions and retardations by shallow shelves and island ridges (Schwiderski [1978a,b]). The mathematical boundary of the ocean basin is zigzagging along gridlines and encloses all connected cells of mean depths H in the range 10 to 7000 m.

Following Hansen [1966], Zahel [1970], [1977, 1978] and Estes [1977, 1980], central finite differences are used on staggered (U,V,ζ) points (Richardson [1922] and Bert Semtner's article in this book) such that for m=μ,2μ,...,360 and n=1,2,...,168 (excluding Antarctica south of θ=168°)

$$\lambda_m^u = (m-\mu)^0 \quad , \quad \lambda_m^v = (m-\mu/2)^0 = \lambda_m^\zeta$$
$$\theta^u = (n-1/2)^0 = \theta_n^\zeta, \quad \theta_n^v = n^0. \tag{52}$$

To enhance the stability of the discrete analog and to facilitate the following hydrodynamical interpolation of empirical tide data, a mixed integration rule in time of the form

$$\int_{t_j}^{t_{i+1}} F(t)dt \approx \Delta t[\kappa F^{j+1} + (1-\kappa)F^j], \quad t_j = \Delta t(j-1) \tag{53}$$

is used. The parameter $\kappa=1$ is chosen at all points of the continuity equation (56), while $\kappa=1$ and $\kappa=0$ are chosen at the center and off-center points of the momentum equations (54) and (55).

The resulting global discrete ocean tidal equations (DOTE's) are

$$(1+A^4)U^{j+1}_{m,n}=U^j_{m,n}+A^1\sin\tilde{t}_j+A^2\cos\tilde{t}_j+A^3(\zeta^j_{m-\mu,n}-\zeta^j_{m,n})+$$
$$+A^5U^j_{m+\mu,n}+A^6U^j_{m-\mu,n}+A^7U^j_{m,n+1}+A^8U^j_{m,n-1}+$$
$$+A^9(V^j_{m-\mu,n-1}+V^j_{m-\mu,n})+A^{10}(V^j_{m,n}+V^j_{m,n-1}),\quad(54)$$

$$(1+B^4)V^{j+1}_{m,n}=V^j_{m,n}+B^1\cos t_j+B^2\sin t_j+B^3(\zeta^j_{m,n+1}-\zeta^j_{m,n})+$$
$$+B^5(V^j_{m+\mu,n}+V^j_{m-\mu,n})+B^6V^j_{m,n+1}+B^7V^j_{m,n-1}+$$
$$+B^8U^j_{m+\mu,n+1}+B^9U^j_{m,n}+B^{10}U^j_{m+\mu,n}+B^{11}U^j_{m,n+1},\quad(55)$$

$$\zeta^{j+1}_{m,n}=\zeta^j_{m,n}+C^1(U^{j+1}_{m,n}-U^{j+1}_{m+\mu,n})+C^2V^{j+1}_{m,n}+C^3V^{j+1}_{m,n+1}\quad(56)$$

with $\tilde{t}_j=\sigma\Delta t\cdot(j-1/2)$ and where the coefficients A^i and B^i depend on m and n and C^i on n alone (Schwiderski [1978a]). The explicit global DOTE's are supplemented by the simple no-cross-flow and free-slip boundary conditions

$$U^{j+1}_{m,n}=0 \text{ and } V^{j+1}_{m-\mu,n}=V^{j+1}_{m,n} \text{ for u-points}\quad(57)$$

$$V^{j+1}_{m,n}=0 \text{ and } U_{m+\mu,n}=U^{j+1}_{m,n} \text{ for v-points.}\quad(58)$$

As is well-known (e.g., Marchuk and Kagan [1984]) the global DOTE's and boundary conditions (54) to (58) yield a unique and stable solution, provided the timestep Δt and the reduced lateral mixing and bottom friction coefficients a and b (Eq. 38) satisfy certain stability conditions (section 3.4). Furthermore, these, a priori, unknown parameters may be chosen within their respective stability ranges to yield realistic results, i.e., to fit empirical tide data around the oceans as closely as possible. With a timestep fixed at $\Delta t \approx 3$ min (60 points for a quarter semi-diurnal period) the search for realistic mixing and friction parameters a and b has been accomplished by Schwiderski [1978a, 1980a,b,c] by the following hydrodynamical interpolation technique:

Suppose that in the grid cell (m,n) the tidal amplitude $\tilde{\xi}_{m,n}$ and phase $\tilde{\delta}_{m,n}$ are empirically known, then the empirical tide at the time t_{j+1} is determined by (Eq. 41)

$$\tilde{\zeta}^{j+1} = \tilde{\xi}_{m,n} \cos(\sigma j \Delta t - \tilde{\delta}_{m,n}). \tag{59}$$

The continuity gap

$$\Delta\zeta^{j+1} = \tilde{\zeta}^{j+1}_{m,n} - \zeta^{j+1}_{m,n} \tag{60}$$

between the empirical and modelled values (Eqs. 56 and 59) may be reduced in three overlapping steps:

STEP 1: By systematic trial-and-error variation of the parameters a and b uniformly over all oceans.

STEP 2: By controlled variation of the bottom friction coefficient b in boundary and/or shallow grid cells, where the assumed simplifications of the basic COTE's to the global COTE's (section 3.2) are not fully justified. Due to the mixed time integration, this step can be easily implemented, since the parameter b is only contained in the coefficients A^4 and B^4 on the left side of the Eqs. (54) and (55). Also, due to the assumed depth dependence of the mixing coefficient A (Eq. 37), it is negligible in the values of A^4 and B^4.

STEP 3: In order to further narrow any remaining continuity gap, limited cross flows are allowed over the mathematically defined zigzagging ocean boundary.

The described integration of the global DOTE's has been carried out by Schwiderski [1978a, 1979, 1981, 1982, 1983] for all eleven partial tides listed in Table I. For each component, the time integration was terminated when the amplitudes and phases reached a steady state uniformly over all oceans. A pointwise and global evaluation of the computed models is given in sections 4.1 to 4.3.

3.4. Stability Analysis of the Global DOTE's

Following standard procedures (Zahel [1970]; Schwiderski [1978a, 1980a,b,c]) the stability and more informatively, the decay and dispersion characteristics of the global DOTE's (54) to (56) with "locally constant" coefficients (H= constant, B=0, Ω =0) can be analyzed by seeking wave-type eigensolutions corresponding to the grid points (52)

$$U^j_{m,n} = U_o \tau^j \exp i[\omega_1(m-\mu)\Delta\lambda + \omega_2(n-1/2)\Delta\theta],$$

$$V^j_{m,n} = V_o \tau^j \exp i[\omega_1(m-\mu/2)\Delta\lambda + \omega_2 n\Delta\theta], \qquad (61)$$

$$\zeta^j_{m,n} = \zeta_o \tau^j \exp i[\omega_1(m-\mu/2)\Delta\lambda + \omega_2(n-1/2)\Delta\theta]$$

with arbitrary wave numbers (ω_1,ω_2) and nonzero amplitudes (U_o,V_o,ζ_o). After substituting Eqs. (51) into the DOTE's (54) to (56) one finds a cubic characteristic equation for the three eigenvalues $\tau = \tau_o$, τ_1, and τ_2, which depend on the chosen wave numbers (ω_1,ω_2), on the averaging space-time scales $\Delta\lambda$, $\Delta\theta$, H, and Δt, and on the momentum exchange coefficient $A \sim aH$. As is shown in detail in Schwiderski [1978a], for any wave numbers the scale parameters and the mixing coefficient of the modelled mean turbulent motion can be chosen such that τ_o is real, τ_1, and τ_2 are conjugate complex, and

$$|\tau_k| = 1-aH\Gamma \leq 1 \text{ for } k=0,1,2 \qquad (62)$$

where Γ depends essentially only on the scale parameters. Under this physically plausible condition, the desired solution will be stable and locally composed of standing eigenwaves and pairs of progressing eigenwaves with the same dispersion rates and all with the same decay rates. This conclusion is based on the fact that the eigensolutions (62) form a complete set of the localized DOTE's (54) to (56).

With the fixed one-degree mesh sizes and an assumed timestep $\Delta t \sim 3$ min, the stability condition (62) is fulfilled in the realistic depth range $10m \leq H \leq 7000m$, provided the reduced momentum exchange coefficient a is chosen in the range

$$0 \leq a \leq a_r \approx 730 \text{ m/sec}. \qquad (63)$$

Thus, the stability condition of the DOTE's places a surprising but physically plausible upper limit and no lower limit on a. The value a = 200 m/sec used in the model (Eq. 38) was exclusively determined by hydrodynamical interpolation STEP 1 in section 3.3 and not forced by any stability requirements. The stability condition (62) shows that without bottom friction waves decay faster in deep oceans than in shallow waters. This physically obvious feature can be easily traced to the assumed depth-dependence of the lateral momentum exchange coefficient A (Eq. 37). If one assumes a scale-independent A, then the decay is the same for deep and shallow oceans, which leads to unrealistic results (sections 2.2 and 4.3).

4. VALIDATION OF OCEAN TIDE MODEL

4.1. Pointwise Evaluation of Modelled Tides

As has been emphasized and illustrated by example in section 2.2., in the turbulent regime any modelled mean motion requires verification by empirical knowledge no matter how well-posed (unique and stable) the flow model may be defined. In the present tide model, the necessary evaluation is extensively incorporated into the integration of the DOTE's (54) to (56) by the hydrodynamical interpolation technique, which checks and adjusts the model at each timestep to join the computed open ocean tides as smoothly as possible to hydrodynamically compatible empirical tide data known around the oceans (section 3.3.). Nevertheless, no evaluation is complete without visual pointwise inspection of the end product. This is particularly important in the present case, since the novel hydrodynamical interpolation is fundamentally different from other classical methods of integrating or interpolating empirical tide data and, hence, requires an analysis of its characteristic features.

In order to display the characteristic properties of ocean tides, the amplitude and phase data (Eq. 41) are plotted in corange (i.e., equiamplitude) and cotidal (i.e., equiphase) maps as shown for the semidiurnal and diurnal lunar principal tides M_2 and O_1 in Figures 1 and 2. While the corange lines depict the maximum tidal elevations, the cotidal lines depict the arrival times of the crests of the tidal waves relative to the cresting time of the forcing equilibrium tide over Greenwich. For instance, when the M_2 or O_1 equilibrium tides are at their maximum over Greenwich, the responding ocean tides crest along the $0°=360°$ lines. About one hour later the M_2 wave advances to the 30° line and the O_1 wave to the 15° line. Thus, one can see the ocean tidal waves rotating around amphidromic points with zero amplitudes. This long-known phenomenon is due to the Coriolis force and to the presence of the continents.

It is interesting to note that the distribution of the amphidromic systems is quite similar particularly among the harmonic tides of the same species. For example, in Figures 3 and 4 one can see the lunar and solar semidiurnal principal tidal waves sweeping around the North Atlantic in just about 12 hours and in about the same modulation. Yet, a close look reveals the second tidal paradox: slower rotating tidal waves arrive earlier than faster progressing waves. In fact, if the moon and sun pass simultaneously (new moon) or oppositely (full moon) over Greenwich meridian, then the slower M_2 wave (period 12.42 hours) arrives at the coast of Northwest Ireland about five hours later (cotidal line 150°). The faster S_2 wave (period 12 hours) arrives at about the same point about six hours later (cotidal line 180°); i.e., one

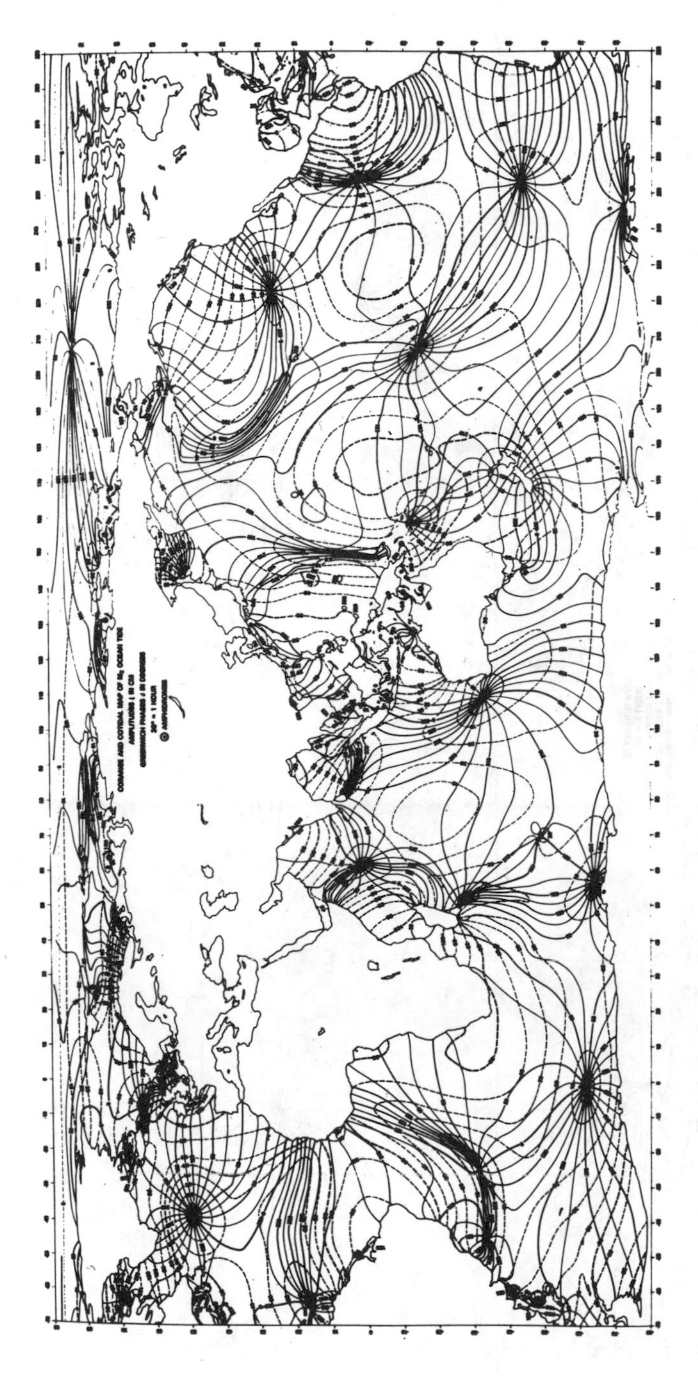

Figure 1. Cotidal (solid) and Corange (dashed) Map of Global M_2 Ocean Tide

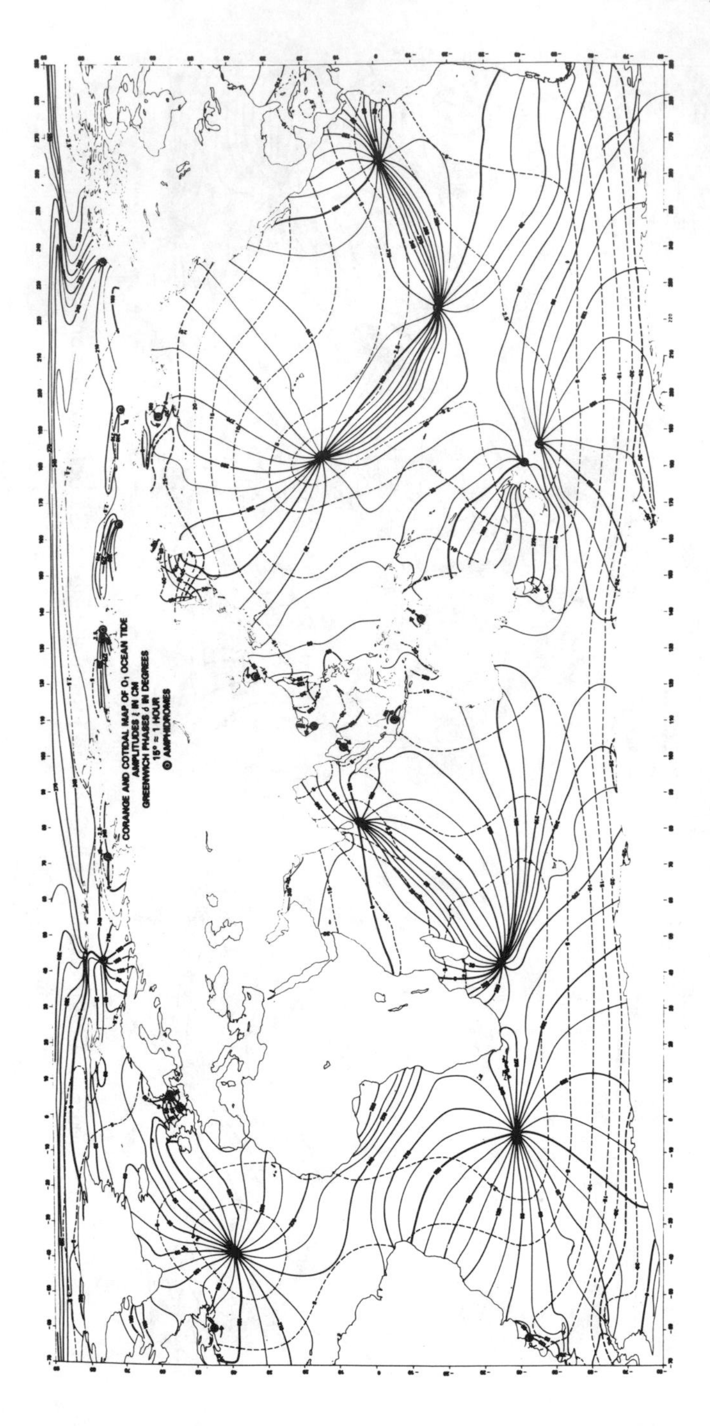

Figure 2. Cotidal (solid) and Corange (dashed) Map of Global O_1 Ocean Tide

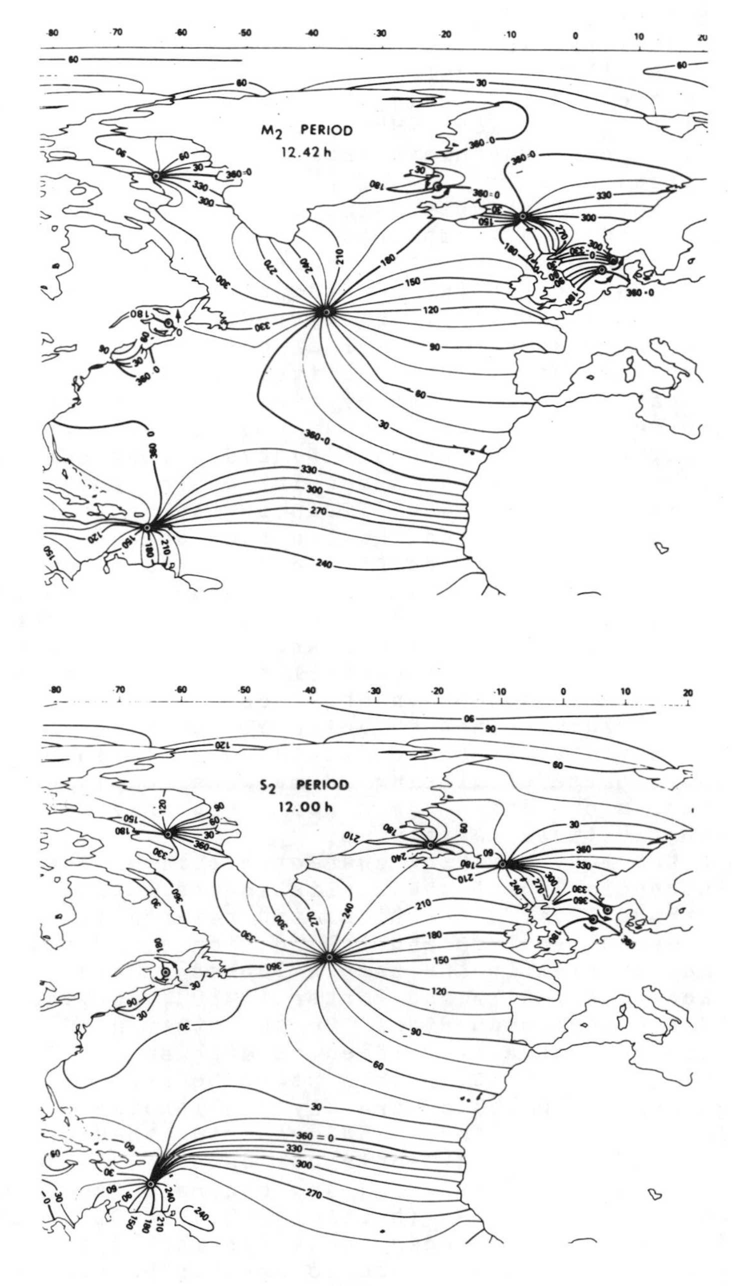

Figures 3 and 4. M_2 and S_2 Amphidromic Rotations

hour after the slower M_2 wave. Evidently, this interesting paradox, which is shared by all forced and damped linear oscillators (Young [1823]) explains the observation of Pliny the Elder (section 1) concerning the belated occurrence of spring tides or age of tides (Whewell [1833]). Clearly, at new or full moon over Northwest Ireland, there is almost maximum low tide.

To facilitate a reliable pointwise evaluation of the modelled ocean tides, all amplitude and phase data are gridwise computer printed in map-like charts (Schwiderski [1979, 1982, 1983]) as shown for the M_2 tide of the Northwest Atlantic (Tables II and III). All hydrodynamically interpolated coastal and nearshore empirical data are visibly marked by subbars and subbrackets, respectively. Offshore deep-sea tide guage locations are marked by tides below the respective computed tide data. They may be compared to the corresponding empirical data, which did not justify hydrodynamical interpolation (section 3.3), listed in special tables such as Table IV supplementing Tables II and III.These tables enable the interested viewer to scan the charted tide data pointwise and evaluate the smoothness of the fitted (integrated and excluded) empirical tide data within the adjacent computed tidal field. Since the realistic fit displayed by the table sample holds uniformly over all ocean areas and for all computed tidal modes with sufficient empirical data, it is estimated that the modelled eleven-component tides allow a tide prediction of 10cm accuracy or better uniformly over all open oceans. This estimate may be somewhat too optimistic in coastal areas because of missing or marginal empirical tide data or because a one-degree grid scale can only resolve averages of subgrid tidal variations.

Perhaps the most striking characteristic of the hydrodynamical interpolation is the realistic resolution of the well-known strong tidal distortions and retardations over shallow continental shelves and island ridges. This important capability may be seen in the sample Tables II and III along the U.S. Coast (e.g., Gill and Porter [1980]; Schwiderski [1983]), and the Caribbean Ridge. On the other hand, various empirical tide data were identified as marginal or false. It may be pointed out that it is this power to properly resolve tidal irregularities which apparently distinguishes the new hydrodynamical interpolation technique from other methods of integrating or interpolating empirical data. For example, Hansen [1948], Hendershott [1972] and others solved the simpler Laplace tidal equations (section 3.2.) as Dirichlet problems with empirical boundary data but failed to resolve tidal irregularities and achieved no results better than those researchers who used no empirical data. Although direct interpolation of empirical data by superposed special basis functions (e.g., Parke and Hendershott [1980]; Parke [1982]) may yield somewhat better results in open oceans, they too

TABLE II. M_2 Ocean Tide Amplitude Chart

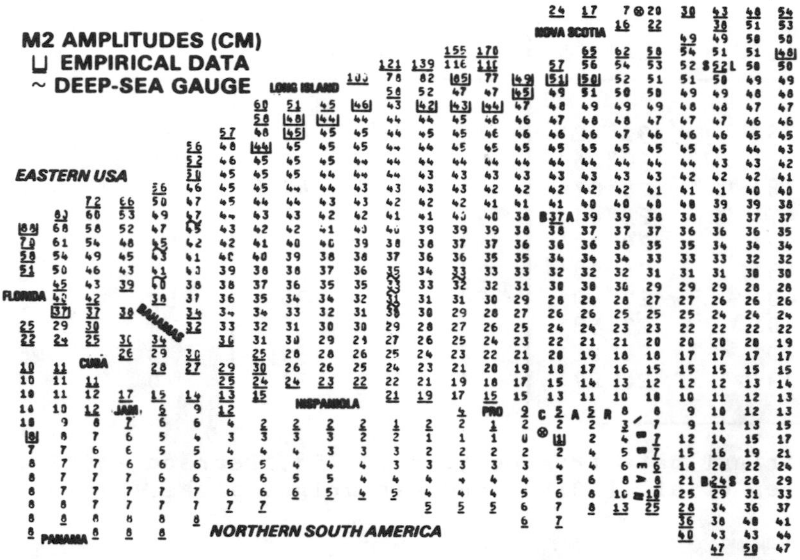

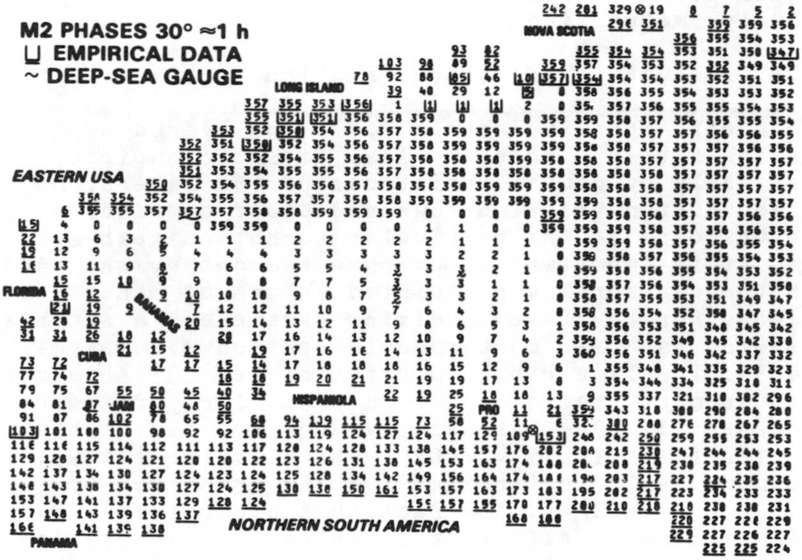

M2 AMPLITUDES (CM)
⊔ EMPIRICAL DATA
~ DEEP-SEA GAUGE

24 17 7 ⊗ 20 30 43 48 54
NOVA SCOTIA 16 22 38 51 53
49 49 50 50
155 170 65 62 58 54 51 51 48
121 139 116 116 57 56 54 53 52 $52L 50 50
LONG ISLAND 103 78 82 85 77 49 51 50 52 51 51 50 49 49
58 52 47 47 45 49 51 50 50 49 49 48 48
60 51 45 46 43 42 43 44 47 48 49 49 49 48 48 47 47
58 48 44 44 44 44 45 46 46 47 48 48 47 47 47 46 46
57 48 45 45 45 44 45 45 46 46 46 46 47 46 46 46 45 45
56 48 44 45 45 45 44 44 45 45 45 45 45 45 45 45 45 44 43
52 46 45 45 45 44 44 44 44 44 44 44 44 44 44 44 43 43 42
EASTERN USA 50 45 45 45 44 44 44 43 43 43 43 43 43 43 43 42 42 42 41
56 46 45 45 44 44 43 43 43 43 42 42 42 42 42 41 41 41 40 40
72 66 50 47 45 44 44 43 43 42 42 42 41 41 41 40 40 40 40 39 39 38
83 60 53 49 47 44 43 43 42 42 41 41 40 40 38 B37A 39 39 38 38 38 37 37
88 68 58 52 47 45 43 42 42 41 40 40 39 39 39 38 38 37 37 37 36 36 36 35
70 61 54 48 45 42 42 41 40 40 39 38 38 37 37 36 36 36 36 35 35 34 34 34
58 54 49 45 43 41 40 40 39 38 38 37 36 36 35 35 34 34 34 33 33 33 32 32
51 50 46 43 41 40 39 38 38 37 36 35 34 33 33 33 32 32 32 31 31 31 30 30
45 43 39 40 38 38 37 36 35 35 33 33 32 32 31 30 30 30 29 29 29 28 28
FLORIDA 40 42 38 37 36 35 34 34 32 31 31 31 30 29 28 28 28 27 27 26 26 26
37 37 38 BAHAMAS 34 34 34 33 32 31 30 30 29 28 27 26 26 25 25 25 24 24 24
25 29 30 32 33 32 31 30 30 29 28 27 26 25 24 24 23 23 22 22 22 22
22 24 25 30 34 36 31 30 29 29 27 26 25 24 23 22 21 21 20 20 20 20 19
26 29 30 25 28 28 26 25 24 23 22 21 20 19 18 18 17 17 17 17
10 11 CUBA 28 27 29 30 26 26 25 24 23 21 20 19 18 17 16 15 15 15 15 15
10 11 11 25 24 24 23 22 22 21 19 18 17 15 14 13 12 12 12 13 14
10 11 12 17 15 14 13 15 HISPANIOLA 21 19 17 15 15 13 11 10 10 10 11 12 13
10 10 12 JAM 6 9 12 4 PRO 9 C 5 A 5 R 8 8 9 10 12 13
10 9 8 7 6 6 4 2 2 2 2 1 2 2 1 2 ⊗ 2 2 3 7 9 11 13 15
10 8 7 6 5 4 3 3 3 3 2 2 1 1 1 0 11 2 4 7 12 14 15 17
7 7 6 6 5 5 4 4 4 3 3 3 2 2 2 2 2 4 5 7 15 16 19 21
8 7 7 6 6 6 5 5 4 4 4 3 3 3 3 3 4 4 6 6 18 20 22 24
8 7 7 7 6 6 6 6 5 5 4 4 4 4 4 4 5 6 8 8 21 B24S 26 29
8 7 7 7 7 7 6 6 6 5 4 5 4 4 4 5 6 8 10 10 25 29 31 33
8 8 7 7 7 7 7 7 5 5 5 6 7 8 13 25 28 34 36 37
8 8 8 8 8 8 6 7 36 38 40 41
8 PANAMA 8 8 8 NORTHERN SOUTH AMERICA 40 43 43 44
47 50 47

M2 PHASES 30° ≈1 h
⊔ EMPIRICAL DATA
~ DEEP-SEA GAUGE

242 201 329 ⊗ 19 0 7 5 2
NOVA SCOTIA 296 351 359 359 356
356 355 354 353
93 82 355 354 354 353 351 350 347
103 98 89 52 359 357 354 353 352 352 349 349
78 92 88 85 46 10 357 354 354 354 353 352 351 351
LONG ISLAND 39 40 29 12 5 0 358 356 355 354 353 353 352
357 355 353 356 1 1 1 1 2 0 35. 357 356 355 355 354 353
355 351 351 356 358 359 0 0 0 359 359 358 357 356 355 355 354
353 352 350 354 356 357 358 359 359 359 359 358 358 357 357 356 356 355
352 351 350 352 354 356 357 358 359 359 359 359 358 358 357 357 357 356 356
352 352 352 354 355 356 357 358 358 358 359 358 358 358 357 357 357 357 357
351 353 354 355 355 356 357 358 358 358 358 358 358 358 358 357 357 357 357
EASTERN USA 350 352 354 355 356 356 357 358 358 358 359 359 359 358 358 358 357 357 357 357
358 354 352 354 355 356 357 357 358 358 359 359 359 359 359 359 358 358 357 357 357 357
6 355 355 357 357 357 358 358 359 359 359 0 0 0 0 359 359 358 358 357 357 356 356
15 4 0 0 0 0 359 359 0 0 0 0 1 1 0 0 359 359 359 358 357 356 356 355
22 13 6 3 2 1 1 1 2 2 2 2 2 1 1 1 0 359 359 358 357 356 355 354
19 12 9 6 5 4 4 4 3 3 3 3 3 2 2 1 0 358 358 357 356 355 354 353
16 13 11 9 8 7 6 6 5 5 4 3 3 3 2 1 0 359 358 356 355 354 353 352
15 15 10 9 9 8 8 7 7 5 3 3 3 1 1 0 358 357 356 354 353 351 350
FLORIDA 16 12 9 10 10 10 9 8 7 5 3 3 2 1 0 358 357 355 353 351 349 347
21 19 9 BAHAMAS 7 12 12 11 10 9 7 6 4 3 2 0 358 356 354 352 350 347 345
42 28 19 20 15 14 13 12 11 9 8 6 5 3 1 358 356 353 351 348 345 342
31 31 26 10 12 20 17 16 14 13 12 10 9 7 4 2 359 356 352 349 345 342 338
21 15 12 19 17 16 16 14 13 11 9 6 3 360 356 351 346 342 337 332
73 72 CUBA 17 17 15 14 17 18 18 18 16 15 12 9 6 1 355 348 341 335 329 323
77 74 72 18 18 19 20 21 21 19 19 17 13 8 3 354 344 334 325 318 311
79 75 67 55 50 45 40 34 HISPANIOLA 22 19 25 18 18 13 9 355 337 321 310 302 296
84 81 87 JAM 80 48 50 25 PRO 11 21 359 343 318 300 290 284 280
91 87 86 102 78 65 55 60 94 139 115 115 73 58 52 11 6 32. 300 288 276 270 267 265
103 101 100 100 98 92 92 106 113 119 124 127 124 117 129 109 ⊗ 153 240 242 250 259 255 253 253
116 116 115 114 112 111 113 117 120 124 128 133 138 145 157 176 202 208 215 230 247 244 244 245
129 128 127 124 121 120 120 122 123 126 131 138 145 153 163 174 188 20. 208 219 230 235 238 239
142 137 134 130 127 124 123 124 125 128 134 142 149 156 164 174 18. 19. 203 217 227 224 235 236
148 143 138 134 130 127 124 125 130 138 150 161 153 157 163 173 183 195 202 217 223 234 233 233
153 147 141 137 133 129 128 124 155 157 155 170 177 200 210 218 218 230 230 231
157 140 143 139 136 137 NORTHERN SOUTH AMERICA 168 188 220 227 228 229
166 141 139 138 229 227 226 227
PANAMA 225 225 224

TABLE III. M_2 Ocean Tide Greenwich Phase Chart

TABLE IV: Deep-Sea Empirical and Modeled M_2 Tides (Supplement to Tables II and III)

LONG. W	LAT. N	AMPLITUDES (cm) EMP.	MOD.	PHASES (deg) EMP.	MOD.
75°38'	32°42'	48	47	356	357
76°25'	30°26'	44	45	358	2
76°48'	28°27'	41	41	2	8
76°47'	28°01'	40	41	9	8
67°32'	28°14'	34	33	359	3
69°45'	28°08'	35	35	1	3
69°40'	27°59'	34	36	359	3
69°40'	27°58'	35	36	1	3
69°20'	26°28'	32	31	3	5
69°19'	26°27'	31	31	0	5

Empirical data from Mofjeld [1975], Pearson [1975], Zetler et al. [1975], and Cartwright et al. [1979].

fail to join deep ocean tides realistically with the cooscillating shallow water tides.

4.2. Global Evaluation of Bottom Friction

As has been explained in section 3.3., the hydrodynamical interpolation fits the modeled tidal field to empirical tide data essentially by global variations of the lateral momentum exchange and bottom friction laws and their coefficients, which are unknowns of the particular turbulent mean motions. Therefore, it is important to compare the determined parameters with those generally postulated values of other ocean circulations. For the lateral mixing parameter, a satisfactory agreement has been mentioned in section 3.2 (see also section 4.3). The linear bottom friction law (37) has been determined as most realistic with a sensitively fixed value of b=1.0cm/sec (Eq. 38). The widely accepted law is the quadratic law (32) with $\alpha_o \approx 0.0026$, which Taylor [1919], (see also Proudman [1952]) determined for the Irish Sea by balancing its energy budget. The fact that other investigations yielded significantly different α_o values is usually ignored. For instance, as is mentioned by Proudman [1952] similar studies in the Bristol and English Channels yielded α_o=0.0014 and α_o=0.0213, which differ from Taylor's value by (-46)% and (+720)%, respectively.

Proudman [1952, p.310] also showed that the quadratic law (32) can be linearized into an equivalent linear law

(37) by the conversion formula

$$b \sin\theta = \frac{8}{3\pi} \alpha_o |\mathbf{q}^c|, \tag{64}$$

where $|\mathbf{q}^c|$ is the characteristic amplitude of the periodic velocity. It allows a comparison of the determined linear global ocean b-value with common quadratic α_o-values. Table V displays such equivalent friction coefficients α_o and b for different characteristic tidal velocities in the open oceans, the shelf seas, and the Irish Sea with a global ocean θ-average of 50°. Clearly, the determined global ocean b-value fits very well into the usual range of bottom friction coefficients.

TABLE V: Equivalent Quadratic and Linear Bottom Friction Coefficients

Sea Area	$\lvert\mathbf{q}^c\rvert$ (cm/sec)	$\alpha_o=0.0014$	Irish Sea $\alpha_o=\mathbf{0.0026}$	$\alpha_o=0.0213$	Ocean b=1.0cm/sec
Ocean	2	b=0.0031	b=0.0058	b=0.0472	$\alpha_o=\mathbf{0.4512}$
Shelf	10	b=0.0155	b=0.0288	b=0.2360	$\alpha_o=0.0902$
Irish	114	b=0.1768	b=0.3284	b=2.6906	$\alpha_o=0.0079$

It is important to note that if one compares Taylor's Irish Sea $\alpha_o=0.0026$ to the open ocean $\alpha_o=0.4512$ equivalent to b=1.0cm/sec, one finds the simple relationship

$$\frac{\alpha_{o,\mathrm{Irish}}}{\alpha_{o,\mathrm{Ocean}}} = \frac{0.0026}{0.4512} \approx \frac{(7.5\mathrm{km})^2}{(100\mathrm{km})^2} = \frac{\text{Area Scale Irish}}{\text{Area Scale Ocean}} \tag{65}$$

Apparently, if one transfers the quadratic bottom friction law from the Irish Sea (with a resolution scale of about $(7.5\mathrm{km})^2$) to the open oceans (with a scale of $(100\mathrm{km})^2$), the different resolution scales must be observed as is physically explained in section 3.1 by the reduced parameter α_1 in Eq. (32).

With the determined bottom friction coefficient B (Eqs. 37 and 38) and the computed velocity field, one can compute the mean rate of bottom friction work (mean friction power) on the ocean floors by the integral

$$\dot{W}_f = \rho \iint \langle B|\mathbf{q}|^2 \rangle \, ds, \qquad (66)$$

where <x> denotes the periodic average of x and the integral is to be taken over the entire or some portion of the ocean area. For the modeled M_2 ocean tide, the power integral has been computed (Schwiderski [1984, 1985]) for the shelf, slope, deep, and all ocean areas with the results summarized in Table VI. As is physically plausible, the largest bottom

TABLE VI: H-Depth Distribution of M_2 Tide Bottom Friction Power

Bottom Friction Power	Shelf Area $0 \le H < 1$km	Slope Area $1 \le H < 4$km	Deep Area $4 \le H \le 7$km	All Areas $0 \le H \le 7$km
$\dot{W}_f$(TW)	1.273 (67.82%)	0.405 21.58%	0.199 20.60%	1.877 100%

1 TW (Tera Watt) = 10^{12} Watts

friction work occurs over the shallow shelves with smaller amounts over the slopes and in the deep seas. The power over the slopes is well substantiated by the observation of strong internal tidal currents in those areas (e.g., Cartwright et al. [1980]). Also, the computed 0.41TW are very close to Munk's [1966] 0.5TW estimate of the power required to drive the internal currents. Munk's estimate falls well into Schwiderski's [1983] roughly estimated power range of 0.41 to 0.6TW, which is also safely below the upper limit of 0.7TW determined by Wunsch [1975].

The total bottom friction power of the modelled M_2 Tide may be compared in Table VII to Zahel's [1970, 1977] modelled results and to empirical estimates by Jeffreys [1920], Heiskanen [1921], and Miller [1966]. As can be seen, the empirical estimates validate, the present estimate and hence the global velocity field and bottom friction coefficient. The low estimates by Zahel [1970, 1977] can be traced to the applied quadratic bottom friction law with Taylor's coefficient for the Irish Sea, which suppresses any effective bottom friction particularly in deep oceans and on slopes by assumption. In this connection it may be mentioned that it is the unchecked application of the Taylor bottom friction law of the Irish Sea, which has led many researchers to restrict bottom friction to shallow border seas or even outside the modelled ocean basin. As the investigations, e.g., by Munk [1966], Wunsch [1975], Cartwright et al. [1980], and the

TABLE VII: Comparison of M_2 Tide Bottom Friction Power Estimates

SOURCE	$\dot{W}_f$(TW)
Zahel [1970, 1977], modelled	0.19*)
Zahel [1977], modelled	0.71*)
Schwiderski [1978, 1985] modelled	1.88
Jeffreys [1920], empirical	1.1
Heiskanen [1921], empirical	1.9
Miller [1966], empirical	1.9+)

*) Models adjusted for earth tide effects.
+) Adjusted to global and pure bottom friction power (Schwiderski [1983, 1984])

present model conclusively indicate, such an assumption is not substantiated by any facts.

4.3. Global Evaluation of Angular Momentum and Energy Budgets

In order to further check the global reality of the modelled mean tidal motions, it is important to compute their complete angular momentum and energy budgets. After multiplying the east-west component of the linear momentum equation (48) with the water density and the distance of the tidally deformed ocean floor from the earth's axis and subsequently averaging over a tidal period and integrating over all oceans (Schwiderski [1984, 1985]), one arrives at the secular axial angular momentum equation

$$\tau_g = \tau_p + \tau_f + \tau_m + (\tau_r = \tau_u + \tau_v) \quad (67)$$

$$-50.53=-23.25+0.00+23.53+(-50.34=-7.23-43.11)\mathrm{PJ}. \quad (68)$$

Here, τ_g, τ_p, τ_f, τ_m, and τ_r denote, respectively, the total gravity, bottom pressure, bottom friction, lateral mixing, and fictitious rotation torques, where the latter consists of the two components τ_u and τ_v. The corresponding numerical values in PJ(= Peta Joules)=10^{15} Joules have been computed for the present M_2 ocean tide model. Similarly, by multiplying both components of linear momentum by the water density and the velocity components u and v, respectively, one arrives (e.g., Hendershott [1972], Schwiderski [1983, 1984, 1985]) at the secular energy equation

$$\dot{W}_g = \dot{W}_p + \dot{W}_f + \dot{W}_m \tag{69}$$

$$3.449 = 1.668 + 1.877 + 0.004 \text{ TW}, \tag{70}$$

where $\dot{W}_g$, $\dot{W}_p$, $\dot{W}_f$, and $\dot{W}_m$ denote the mean rates of total gravity, bottom pressure, bottom friction, and lateral mixing work, respectively. Again, the corresponding numerical values have been computed for the modeled M_2 tide.

Since the numerical equations (68) and (70) follow from the COTE's (48) and (49) by identical operations, the expressed balanced angular momentum and energy budgets are a natural consequence of the applied physical laws. However, the question must be asked: are the balances achieved in realistic terms? Indeed, both budget equations reveal the following physically plausible and important properties.

In the angular momentum equation (68), the total gravity torque τ_g is apparently almost completely balanced by the fictitious rotation torque τ_r, i.e., the gravity forces acting on the ocean water exert almost no direct torque on the rotating earth. Also, in agreement with earlier conclusions by Suendermann and Brosche [1978] bottom friction holds no significant torque τ_f on the earth. This surprising result can be shown (Schwiderski [1985]) to depend only on the simple earth tide model (Eq. 44); it is valid for any kind of ocean tide model regardless of the applied linear or quadratic bottom friction law. The bottom pressure provides the only essential mechanism by which the ocean tides directly brake the earth rotation. While the gravity forces pull the tidal waves into amphidromic rotation (section 4.1), the oceans gear into the tidal deformations of the bottom and brake the earth rotation by their tidal pressure torque τ_p. In fact, detailed computations with all eleven tidal modes listed in Table I produced a deceleration constant in good agreement with reliable astronomical estimates (Schwiderski [1985]). There exists also a close balance between the bottom pressure torque τ_p and the opposite lateral mixing torque τ_m. This remarkable result seems to indicate that turbulent mixing or momentum exchange is caused by bottom pressure against the moving ocean bottom and, hence, plays as experienced an important role in global tidal modelling.

The energy equation (70) shows that the oceans lose the energy, which is supplied by all gravity forces at the rate $\dot{W}_g$, essentially by bottom pressure and friction work at the combined rates $\dot{W}_p+\dot{W}_f$. A simple relationship between the bottom pressure torque τ_p and power $\dot{W}_p$ (Schwiderski [1985]) proves that $\dot{W}_p$ is essentially mechanical energy which equals the rotational energy loss of the earth. On the other hand, since the bottom friction torque τ_p is negligible, $\dot{W}_f$ constitutes a heat loss. Its dominating amount (53% of $\dot{W}_g$) explains the experienced significant part it played in modelling tidal motions. As is physically expected, the energy loss by lateral mixing ($\dot{W}_m$) is negligible. Thus, together with the found significance of angular momentum exchange or mixing torque (τ_m) one has a clarifying demonstration that the Reynolds tensor of apparent turbulent stresses is, in fact, a momentum austaush (exchange or mixing) tensor rather than a dissipation term (Prandtl [1925]; section 2.2.). Again, this result may be compared to Zahel's [1970, 1977] models which yielded (after earth-tide adjustments) $\dot{W}_m$=3.12TW and 3.07TW, respectively. These large values can be attributed to the application of the widely used constant mixing coefficient and Taylor's quadratic bottom friction term (section 4.2.). In fact, the author conducted many computer experiments with the classical laws before the present laws were realized.

5. SUMMARY

Following a brief review of historical ocean tide research, a combined empirical-hydrodynamical method is developed to model realistic tidal elevations, currents, and dissipation in the worldwide oceans. Since such motions are supercritically turbulent, modern hydrodynamical notions and results are recalled in order to derive a physically sound procedure to remove the instability and nonuniqueness of turbulent motions by some averaging process while retaining experimentally verifiable flow properties of interest. In the resulting NSE's of mean turbulent motions, the unpredictability of turbulence manifests itself in averaging area-time scales and in momentum exchange parameters, which must be chosen to achieve the desired well posed and realistic results. Assuming that a unique and stable mean motion has been defined, the averaged NSE's are simplified to the basic COTE's by the usual single-ocean-layer assumptions on the basis of perturbation-scale analysis. For the world oceans, these equations are further simplified to the global COTE's by complete linearization and by specifying the tide-generating astronomical forces and their modifications due to earth and earth ocean-load tides. Finally, the COTE's are converted to the global DOTE's by finite differencing in space and time such that the space-time differencing scales consistently specify the corresponding averaging scales. The remaining unknown lateral momentum exchange and bottom friction laws are determined by a novel hydrodynamical interpolation technique, which integrates empirical tide data from shallow waters around the oceans into the computed tidal field.

The developed empirical-hydrodynamical method has been applied to compute the eleven major harmonic tidal constituents. The constructed models have been validated by the hydrodynamical interpolation technique, which compares the computed data with empirical data at each integration step.

Moreover, a computerized printout of all computed and empirical tide data in map-like charts has been published for direct visual evaluation. Based on a complete pointwise evaluation, it is estimated that the superposed eleven-mode tidal model allows a tide prediction with an accuracy of 10cm uniformly over all open oceans and with somewhat lesser precision in some coastal areas. Also, the global lateral momentum exchange and bottom friction coefficients obtained by hydrodynamical interpolation have been found in good agreement with other ocean circulation models. For the dominant M_2 tide, the global angular momentum and energy budgets have been found to be perfectly balanced in realistic terms. Finally, it may be mentioned that the models proved their usefulness in various applications such as marine geodesy, precision gravimetry, earth's rotation, and moon's revolution. They have been selected as a "working standard" by resolutions of the International Permanent Commission for Earth Tides and the International Association for Geodesy.

6. ACKNOWLEDGEMENTS

It is the author's most pleasant obligation to thank Dr. James J. O'Brien for his invitation to attend the NATO ASI and to prepare this lecture article. This project was supported by special funds of the Naval Surface Weapons Center with the generous sponsorship of Dr. Thomas A. Clare, Head of the Strategic Systems Department, and Mr. Carlton W. Duke, Head of the Space and Surface Systems Division.

7. REFERENCES

Accad, Y., and C.L. Pekeris, Solution of the Tidal Equations for the M_2 and S_2 Tides in the World Oceans from a Knowledge of the Tidal Potential Alone Phil. Trans. Roy. Soc., London, A, 290, p. 235, 1978.

Bagriantsev, N.V., A.I. Danilov, and V.O. Ivchenko, Orientational Effects in Geophysical Fluid Dynamics, Int. J. Eng. Sci., 21, p. 725, 1983.

Bartels, J., Gezeitenkraefte, In Handbuch der Physik XLVIII, Geophysik II, edited by S. Fluegge and J. Bartels, Springer, Berlin, 1957.

Boussinesq, J., Expression du Frottement Extérieur dans L'écoulement Tumultueux d'une Fluide, Comptes Rend. Acad. Sci., 122, p.1445, 1896a.

Boussinesq, J., Formules du Coefficient des Frottements Intérieurs dans L'écoulement Tumultueux Graduellement Varié des Liquides Comptes Rend. Acad. Sci., 122, p. 1517, 1896b.

British Admiralty, Tide Tables, Vols 1,2, and 3, 1984.

Busse, F.H., and J.A. Whitehead, Instabilities of Convection

Rolls in a High Prandtl Number Fluid, J. Fluid Mech., 47, p. 305, 1971.
Cartwright, D.E., Ocean Tides, Rep. Progr. Phys., 40, p. 665, 1977.
Cartwright, D.E., and A.C. Edden, Corrected Tables of Tidal Harmonics, Geophys. J. Roy. Astr. Soc., 33, p. 253, 1973.
Cartwright, D.E., A.C. Edden, R. Spencer, and J.M. Vassie, The Tides of the Northern Atlantic Ocean, Phil. Trans., Roy. Soc., 298, London, p.87, 1980.
Cartwright, D.E., and R.J. Tayler, New Computations of the Tide-Generating Potential, Geophys. J. Roy. Astr. Soc., 23, p. 45, 1971.
Cartwright, D.E., B.D. Zetler, and B.V. Hamon, Pelagic Tidal Constants, IAPSO Publication Scientifique No., 30, 1979.
Cox, M.D., A Mathematical Model of the Indian Ocean, Deep-Sea Res., 17, p. 45, 1970.
Crowley, W.P., A Global Numerical Model: Part 1, J. Comp. Phys., 3, p. 111, 1968.
Crowley, W.P., A Numerical Model for Viscous, Free-Surface, Barotropic Wind-Driven Ocean Circulations, J. Comp. Phys., 5, p. 139, 1970.
Darwin, G. H., Report on the Harmonic Analysis of Tidal Observations, Brit. Ass. for adv. Sci. Rep., 1883, see also Sci. Pap. 1, Cambridge, 1907.
Davies, A.M., The Numerical Solution of the Three-Dimensional Hydrodynamic Equations Using a B-Spline Representation of the Vertical Current Profile; Three-Dimensional Model with Depth-Varying Eddy Viscosity, In Bottom Turbulence, Proceedings of the 8th Liege Colloquium on Ocean Hydrodynamics, 19, edited by J.C.J. Nihoul, Elsevier Oceanography Series, p.1, 1977.
Davies, A.M., and G. K. Furness, Observed and Computed M_2 Tidal Currents in the North Sea, J. Phys. Oceanogr., 10, p. 237, 1980.
Dietrich, G., Die Gezeiten des Weltmeeres als Geographische Erscheinung, Zeitschr. d. Gessellsch. f. Erdkunde, 3/4, 1944a.
Dietrich, G., Die Schwingungssyteme der Halb-und Eintaegigen Tiden in den Ozeanen, Veroeff. Inst. Meereskunde, Univ. Berlin, N.F.A. No. 41, 1944b.
Doodson, A.T., The harmonic Development of the Tide-Generating Potential, Proc. Roy. Soc., London, A. 100, p. 305, 1921.
Eliassen, A., and E. Kleinschmidt, Dynamic Meteorology, In Handbuch der Physik XLVII, Geophysik II, edited by S. Fluegge and J. Bartels, Springer, Berlin, 1957.
Estes, R.H., A Computer Software System for the Generation of Global Ocean Tides Including Self-Gravitation and Crustal Loading Effects, NASA, TR-X-920-77-82, Goddard Space Flight Center, 1977.
Estes, R.H., A Simulation of Global Ocean Tide Recovery

Using Altimeter Data with Systematic Orbit Error, Marine Geodesy, 3, p. 75, 1980.
Eyris, M., Maregraphs de Grandes Profoundeurs, Cahiers Oceanographiques, 20, p. 355, 1968.
Farrell, W.E., Deformation of the Earth by Surface Loads, Rev. Geophys. Space Phys., 10, p. 261, 1972.
Filloux, J.H., Bourdon Tube Deep Sea Tide Gauges, In Tsunamis in the Pacific Ocean, edited by W.M. Adams, East-West Center Press, Honolulu, 1969.
Friedrich, H.J., Preliminary Results from a Numerical Multilayer Model for the Circulation in the North Atlantic, Deutsche Hydr. Zeitsch., 23, p. 145, 1970.
Gill, S.K., and D.L. Porter, Theoretical Offshore Tide Range Derived from a Simple Defant Tidal Model Compared with Observed Offshore Tides, Int. Hydrogr. Review LVII, Monaco, p. 155, 1980.
Grace, S.F., The Semidiurnal Lunar Tidal Motion of the Red Sea, Mon. Not. Roy. Astr. Soc., Geophys. Supp.I, 2, p. 273, 1930a.
Grace, S.F., The Influence of the Friction on the Tidal Motion of the Gulf of Suez, Mon. Not. Roy. Astr. Soc., Geophys. Suppl., 2, p. 316, 1930b.
Greenberg, D.A., Modeling of the Mean Barotropic, Circulation in the Bay of Fundy and Gulf of Maine, J. Phys. Oceanogr., 13, 1983.
Hansen, W., Die Ermittlung der Gezeiten Beliebig Gestalteter Meeresgebiete mit Hilfe des Randwertverfahrens, Deutsche Hydr. Zeitsch., 1, p. 157, 1948.
Hansen, W., Die Reproduktion der Bewegungsvorgaenge im Meere mit Hilfe Hydrodynamisch-Numerischer Verfahren, Mitt. des Inst. f. Meereskunde der Univ. Hamburg, V, 1966.
Harris, R.A., Manual of Tides Parts I and II, U.S. Coast and Geodetic Survey, 1897.
Heaps, N.S., On the Numerical Solution of the Three-Dimensional Hydrodynamical Equations for Tides and Storm Surges, Mem. Soc. Roy. Sci. Liege, Ser. 6, 2, p. 143, 1972.
Heiskanen, W., Ueber den Einfluss der Gezeiten auf die Saekulare Acceleration des Mondes, Ann. Acad. Sci., Fennicae A, 18, p. 1, 1921.
Hendershott, M.C., The Effects of Solid-Earth Deformation on Global Ocean Tides, Geophys. J. Roy. Astr. Soc., 29, p. 389, 1972.
Holland, W. R., and A. D. Hirschman, A Numerical Calculation in the North Atlantic Ocean, J. Phys. Oceanogr., 2, p. 336, 1972.
International Hydrographic Bureau, Tides, Harmonic Constants, Computer Tape, Monaco, 1978.
Irish, J.D., W.H. Munk, and F.E. Snodgrass, M_2 Amphidrome in the Northeast Pacific, Geophys. Fluid Dyn., 2, p. 355, 1971.
Jachens, R.C., and J.T. Kuo, The O_1 Tide in the North Atlan-

tic Ocean as Derived from Land-based Tidal Gravity Measurements, Proceedings of the Seventh Symposium on Earth Tides, Sopron, Hungary Akad. Kiado Budapest, 1973.
Jeffreys, H., Tidal Friction in Shallow Seas, Phil. Trans. Roy. Soc. A., 221, p. 239, 1920.
Johns, B., Vertical Structure of Tidal Flows in River Estuaries, Geophys. J. Res., Astr. Soc., 12, p. 103, 1966.
Joseph, D.D., Stability of Fluid Motions I, Springer, Berlin, 1976.
Kagan, B.A., Resistance Law of Tidal Flow, Izv. Acad. Sci. USSR, Atm. and Oce. Phys., 5, p. 302, 1972.
Kraav, V.K., Computation of the Semidiurnal Tide and Turbulence Parameters in the North Sea, Oceanology, 9, p. 332, 1969.
Ladyzhenskaya, O.A., The Mathematical Theory of Viscous Incompressible Flow, Gordon and Breach, New York, 1969.
Lamb, H., Hydrodynamics, Dover Publications, New York, 1932.
Laplace, P.S., Recherches sur Quelques Points de Systeme du Monde, Mem. Acad., Roy. Sci., 88, 1775.
Leith, C.E., Two-Dimensional Eddy Viscosity Coefficients, Proc. WMO/IUGG Symp. on Numerical Weather Prediction, Tokyo, p. 140, 1968.
Luther, D.S., and C. Wunsch, Tidal Charts of the Central Pacific Ocean, J. Phys. Oce., 5, p. 227, 1975.
Marchuk, G.I., and B.A. Kagan, OCEAN Tides, Mathematical Models and Numerical Experiments, Pergamon Press, Oxford, 1984.
McGregor, R.C., The Influence of Eddy Viscosity on the Vertical Distribution of Velocity in the Tidal Estuary, Geophys J. Roy. Astr. Soc., 29, p. 103, 1972.
Miller, G.R., The Flux of Tidal Energy out of the Deep Oceans, J. Geophys. Res., 71, p.2485, 1966.
Mofjeld, H.O., Empirical Model for Tides in the Western North Atlantic Ocean, Nat. Oceanic and Atmos. Admin. Rep. TR ERL 340-AOML 19, 1975.
Mofjeld, H.O., and J. W. Lavelle, Bottom Boundary Layer Studies in Tidally Dominated Regimes, paper presented at the XVIII General Assembly of the International Union of Geodesy and Geophysics Symposium on Coastal and Near Shore Zone Processes, Hamburg, August 15-27, 1983.
Munk, W.H., Abyssal Recipes, Deep-Sea Res., 13, p. 707, 1966.
Munk, W.H., F. Snodgrass, and M. Wimbush, Tides Offshore: Transition from California Coastal to Deep-Sea Waters, Geophys. Fluid Dyn., 1, p. 161, 1970.
Neumann, G., and W.J. Pierson, Jr., Principles of Physical Oceanography, Prentice-Hall, Inc., Englewood Cliffs, New Jersey, 1966.
Newton, I., Philosophiae Naturalis Principia Mathematica, London, 1687.
Nihoul, J.C.J., Three-Dimensional Model of Tides and Storm

Surges in a Shallow Well-Mixed Continental Sea, Dyn. Atm. Oceans, 2, p. 29, 1977.
OBrien, J.J., A Two-Dimensional Model of the Wind-Driven North Pacific, Investig. Pesquera, 35, p.331, 1971.
Parke, M.E., O_1, P_1, N_2 Models of the Global Ocean Tide on an Elastic Earth Plus Surface Potential and Spherical Harmonic Decompositions for M_2, S_2 and K_1, Marine Geodesy, 6, p. 35, 1982.
Parke, M.E., and M.C. Hendershott, M_2, S_2, K_1 Models of the Global Ocean Tide on an Elastic Earth, Marine Geodesy, 3, p. 379, 1980.
Pekeris, C.L., and Y. Accad, Solution of Laplaces Equations for the M2 Tide in the World Oceans, Phil. Trans. Roy. Soc., London, A. 265, p. 413, 1969.
Pearson, C.A., Deep-Sea Tide Observations off the South-Eastern United States, Tech. Memo., NOS 17, Nat. Oceanic and Atmos. Admin., Rockville, Md., 1975.
Prandtl, L., Ueber die Ausgebildete Turbulenz, ZAMM, 5, p. 136, 1925.
Pratt, J.G.D., Tides at Shackleton, Weddel Sea, Trans-Ant. Exp., 1955-58, Sci. Rep., 4, London, 1960.
Proudman, J., Deformation of Earth-Tides by Means of Water-Tides in Narrow Seas, Bull No. 11, Sect. Oceanogr., Cons. de Recherches, Venedig, 1928.
Proudman, J., Dynamical Oceanography, Dover Publications, New York, 1952.
Reynolds, O., On the Dynamical Theory of Incompressible Viscous Fluids and the Determination of the Criterion, Phil. Trans., Roy. Soc., 186, London A, p. 123, 1894.
Richardson, L.F., Weather Prediction by Numerical Methods, Cambridge University Press, New York, 1922.
Schlichting, H., Boundary-Layer Theory, McGraw-Hill Book Co., New York, 1968.
Schwiderski, E.W., Bifurcation of Convection in Internally Heated Fluid Layers, Phys. Fluids, 15, p. 1882, 1972.
Schwiderski, E.W., Global Ocean Tides, Part I: A Detailed Hydrodynamical Interpolation Model, NSWC/DL-TR 3866, 1978a.
Schwiderski, E.W., Hydrodynamically Defined Ocean Bathymetry, NSWC/DL-TR 3888, 1978b.
Schwiderski, E.W., Global Ocean Tides, Part II: The Semidiurnal Principal Lunar Tide (M_2), Atlas of Tidal Charts and Maps, NSWC TR 79-414, 1979.
Schwiderski, E.W., On Charting Global Ocean Tides, Reviews of Geophys. and Sp. Phys., 18, 243, 1980a.
Schwiderski, E.W., Ocean Tides, Part I: Global Tidal Equations, Marine Geodesy, 3, p. 161, 1980b.
Schwiderski, E.W., Ocean Tides, Part II: A Hydrodynamical Interpolation Model, Marine Geodesy, 3, p. 219, 1980c.
Schwiderski, E.W., Global Ocean Tides, Parts III-IX: S_2, K_1, O_1, N_2, P_1, K_2, Q_1, NSWC TRs 81-122, -124, -144, -218, -220, -222, -224, 1981a.

Schwiderski, E.W., Exact Expansions of Arctic Ocean Tides, NSWC TR 81-494, 1981b.
Schwiderski, E.W., Global Ocean Tides, Parts X-XII: Mf, Mm, Ssa, NSWC TRs 82-151, -147, 149, 1982.
Schwiderski, E.W., Atlas of Ocean Tidal Charts and Maps, Part I: Semidiurnal Principal Lunar Tide M_2, Marine Geodesy, 6, p. 219, 1983.
Schwiderski, E.W., Combined Hydrodynamical and Empirical Modeling of Ocean Tides, Mar. Geophys. Res., 7, p. 215, 1984.
Schwiderski, E.W., On Tidal Friction and the Decelerations of the Earths Rotation and Moons Revolution, Marine Geodesy, 9, p. 417, 1985.
Smagorinsky, J., General Circulation Experiments With the Primitive Equations. I. The Basic Experiment, Mon. Weather Rev., 91, p. 99, 1963.
Smith, S.M., H.M. Menard, and G. Sharman, Worldwide Ocean Depths and Continental Elevations Averaged for Areas Approximating One-Degree Squares of Latitude and Longitude, Scripps Inst. of Oceanography, Ref. 65-8, 1966.
Snodgrass, F.E., Deep-Sea Instrument Capsule, Science, 162, p. 78, 1968.
Suendermann, J., and P. Brosche, Numerical Computation of Tidal Friction for Present and Ancient Oceans, In Tidal Friction and the Earths Rotation, edited by P. Brosche and J. Suendermann, Springer, Berlin, 1978.
Takahasi, R., Tilting Motion of the Earth Crust Caused by Tidal Loading, Bull. Earthquake Res. Inst., 6, p. 85, 1929.
Taylor, G.I., Tidal Friction in the Irish Sea, Phil. Trans. Roy. Soc., London, A, 220, p. 1, 1919.
Thiel, E., A.P. Crary, R.A. Haubrich, and J.C. Behrendt, Gravimetric Determination of Ocean Tide, Weddel and Ross Seas, Antartica, J. Geophys. Res., 65, p. 629, 1960.
Thomson, W. (Lord Kelvin), Report of Committee for the Purpose of Harmonic Analysis of Tidal Observations, Brit. Ass. Adv. Sci. Rep., London, 1868.
U.S. National Ocean Service, Tide Tables, 1985.
Whewell, W., Essay Towards a First Approximation to a Map of Co-tidal Lines, Phil. Trans., Roy. Soc., London, 1, 147, 1833.
Whitaker, S., Introduction to Fluid Mechanics, Prentice-Hall, Inc., Englewood Cliffs, New Jersey, 1968.
Wunsch, C., Internal Tides in the Ocean, Rev. of Geophys. and Sp. Phys., 13, p. 167, 1975.
Young, T., Tides, in Encyclopedia Britanica, 8th Ed. Vol. 21, Little and Brown, Boston, 1823.
Zahel, W., Die Reproduktion Gezeitenbedingter Bewegungsvorgaenge im Weltozean Mittels des Hydrodynamisch-Numerischen Verfahrens, Mitt. des Inst. f. Meereskunde der Univ., Hamburg, XVII, 1970.
Zahel, W., A Global Hydrodynamical-Numerical 1_o-Model of the

Ocean Tides; the Oscillation System of the M_2-Tide and its Distribution of Energy Dissipation, Ann. Geophys. t., 33, p. 31, 1977.

Zahel, W., The Influence of Solid Earth Deformations on Semi-diurnal and Diurnal Oceanic Tides, In Tidal Friction and the Earths Rotation, edited by P. Brosche and J. Suender-mann, Springer, Berlin, 1978.

Zetler, B.W., H. Munk, H. Mofjeld, W. Brown, and F. Dormer, MODE Tides, J. Phys. Oceanogr., 5, p. 430, 1975.

MATHEMATICAL FORMULATION OF A SPECTRAL TIDAL MODEL

A. M. Davies
Institute of Oceanographic Sciences
Bidston Observatory
Birkenhead
Merseyside, L43 7RA
England

ABSTRACT. In this paper the stages in formulating a Galerkin-Spectral model are illustrated by developing a spectral model of the vertical profile of oscillatory flow. Such a simple model is chosen so that the steps in the method can be clearly illustrated. References to the literature are given for general background information; the reader is directed to the second section of this chapter [Davies, 1986b] for the extension to three dimensions. Some results from a three-dimensional tidal model are presented to illustrate oceanographic applications. The objective is to introduce (by references to the literature and a simple example) the Galerkin method to someone new to the topic and to illustrate its applications.

1. INTRODUCTION

1.1 Previous tidal models

Over the last twenty years, two-dimensional tidal models having a range of geographical extents from: global, e.g., Schwiderski [1983]; continental shelf, e.g., Flather [1976]; shallow seas, e.g., Davies [1976], Mathisen and Johansen [1983]; and localized higher resolution models, e.g., Proctor [1981] have been developed. Since these models are primarily based upon the vertically integrated equations, they cannot give any information on tidal current profile.

With the advent of increasing computer power, three-dimensional tidal models aimed at examining variations in tidal currents from sea surface to sea bed have evolved. The majority of these models use various finite difference grids in the vertical. (A review of these various grids is given in Cheng et al. [1976]).

An alternative to using a grid box discretization through the vertical is to represent the vertical variation of current using the Galerkin method with functions from sea surface to sea bed (a Galerkin-Spectral model).

J. J. O'Brien (ed.), Advanced Physical Oceanographic Numerical Modelling, 373–390.

1.2 Background to the Galerkin method

Several general monographs and text books on the method of weighted residuals, of which the Galerkin method and spectral models are special cases, have been written in recent years. The interested reader is referred to Gottlieb and Orszag [1977] for a review of spectral methods in general and to Finlayson [1972] for the method of weighted residuals and general baokground information. Some details of the mathematical development of the technique are given in Strang and Fix [1973].

In this paper we shall be concerned with the application of spectral methods through the vertical (i.e., from sea surface to sea bed). However the Galerkin approach, in the form of the finite element method in the horizontal space domain, has been used by a number of authors, e.g., Grotkop [1973] to formulate two-dimensional tidal models. A review of various two-dimensional finite element models and their associated numerical advantages and difficulties is given by Gray [1982].

Three-dimensional mixed finite-difference spectral models have been used by a number of authors to study tidal propagation problems. These models use a uniform finite difference grid in the horizontal (although a non-uniform grid could be applied) to represent the lateral variation of tidal elevations and currents. The vertical variation of current is represented by a functional expansion. Models of this type have been used to study the M_2 tide on the continental shelf [Davies and Furnes, 1980] and subsequently extended to model its higher harmonics produced by nonlinear interactions [Davies and James, 1983]. The method has also proved successful in modelling tidal propagation in estuaries [Owen, 1980] and in shallow sea regions [Heaps and Jones, 1981; Proctor, 1981; Wolf, 1984].

In this paper we illustrate the application of the Galerkin method to the formulation of such a model by initially considering the problem of determining the vertical variation of the current associated with an oscillatory flow. In part b of this chapter, the method is extended to the solution of the linear three-dimensional equations. Cartesian coordinates are used throughout for ease of presentation. Also, the nonlinear terms are omitted from the equations so that the basic details can be clearly illustrated, although their inclusion does not present any mathematical or numerical problems [Davies, 1980]. In many tidal problems, it is necessary to use the fully nonlinear three-dimensional equations in polar coordinates. The interested reader is referred to Davies and James [1983] for the spectral form of these equations.

By illustrating the formulation of a tidal spectral model in this manner, the main points in the development can be emphasized and introduced to someone new to the field.

2. OSCILLATORY FLOW: A SPECTRAL MODEL

2.1 Previous models and observations

The problem of determining the current profile produced by oscillatory flow (in particular the vertical variation of tidal currents) has been considered by a number of authors, e.g., Prandle [1982] and Fang and Ichiye [1983]. In these papers, eddy viscosity was assumed constant or linearly increasing with height above the sea bed.

Observational evidence [Wolf, 1980; Soulsby, 1983] suggests that a more complex vertical variation of eddy viscosity exists in nature, with viscosity reaching a maximum at about mid-depth and then rapidly decreasing [Wolf, 1980]. In this section we show that by using the Galerkin method through the vertical, it is possible to determine current profile for an arbitrary specified profile of eddy viscosity.

This method, in common with analytical solutions, yields a continuous current profile from sea surface to sea bed but has the major advantage of being able to include arbitrary, physically realistic profiles of viscosity. When a modal expansion is used in the vertical, additional insight into the vertical structure of current profiles is possible.

2.2 Spectral form of the equations

Using Cartesian coordinates, the linear hydrodynamic equations of motion can be written as

$$\frac{\partial U}{\partial t} - \gamma V = -g \frac{\partial \zeta}{\partial x} + \frac{\partial}{\partial z}\left(\mu \frac{\partial U}{\partial z}\right) \tag{1}$$

$$\frac{\partial V}{\partial t} + \gamma U = -g \frac{\partial \zeta}{\partial y} + \frac{\partial}{\partial z}\left(\mu \frac{\partial V}{\partial z}\right). \tag{2}$$

where x, y, z are Cartesian Coordinates with z the depth below the undisturbed surface and ζ elevation above that surface. Time is denoted by t with U and V the x and y components of current at depth z. Other parameters are: γ the Coriolis parameter, g acceleration due to gravity, and μ vertical eddy viscosity.

For tidal problems, a zero surface stress condition is specified, namely,

$$-\rho\left(\mu\frac{\partial U}{\partial z}\right)_0 = 0 \qquad -\rho\left(\mu\frac{\partial V}{\partial z}\right)_0 = 0. \tag{3}$$

At the sea bed, a no-slip bottom boundary condition can be formulated, namely,

$$U_h = V_h = 0 \tag{4}$$

or in many cases [Davies and Furnes, 1980], a quadratic law of bottom friction is applied, giving

$$-(\mu\frac{\partial U}{\partial z})_h = KU_h(U_h^2 + V_h^2)^{1/2} \quad (5a)$$

$$-(\mu\frac{\partial V}{\partial z})_h = KV_h(U_h^2 + V_h^2)^{1/2} \quad (5b)$$

with K a coefficient of bottom friction.

Defining a complex velocity vector

$$Q = U + iV, \quad (6)$$

and a complex slope

$$S = g\frac{\partial\zeta}{\partial x} + ig\frac{\partial\zeta}{\partial y}, \quad (7)$$

together with a non-dimensional depth

$$\sigma = z/h, \quad (8)$$

and then multiplying equation (2) by $i = \sqrt{-1}$, transforming, using (8), and adding (1) and (2), gives the single equation

$$\frac{\partial Q}{\partial t} + i\gamma Q = \frac{1}{h^2}\frac{\partial}{\partial\sigma}(\mu\frac{\partial Q}{\partial\sigma}) - S. \quad (9)$$

For oscillatory flow at a frequency ω, velocity and elevation gradients can be divided into rotary components of the form

$$Q = R_+e^{i\omega t} + R_-e^{-i\omega t} \quad (10)$$

$$S = S_+e^{i\omega t} + S_-e^{-i\omega t}. \quad (11)$$

In (10) and (11), $|R_+|$ gives the amplitude of the anti-clockwise rotating vector and $|R_-|$ the amplitude of the clockwise rotating vector. The terms S_+ and S_- are defined in an analogous manner. Details of this rotary decomposition can be found in Soulsby [1983] and Fang and Ichiye [1983].

Substituting (10) and (11) into (9) and writing the resulting equation in terms of anti-clockwise and clockwise components gives

$$i(\gamma+\omega)R_+ = \frac{1}{h^2}\frac{\partial}{\partial\sigma}(\mu\frac{\partial R_+}{\partial\sigma}) - S_+ \quad (12)$$

and

$$i(\gamma-\omega)R_- = \frac{1}{h^2}\frac{\partial}{\partial\sigma}(\mu\frac{\partial R_-}{\partial\sigma}) - S_-. \quad (13)$$

The two rotary components R_+ and R_- are expanded in terms of m complex coefficients A_r, B_r and real basis functions $f_r(\sigma)$ through the vertical, giving

$$R_+ = \sum_{r=1}^{m} A_r f_r(\sigma), \quad R_- = \sum_{r=1}^{m} B_r f_r(\sigma). \tag{14}$$

The choice of basis functions f_r in (14) is arbitrary, and a number of functions, for example, Chebyshev polynomials, Legendre polynomials, trigonometric functions, or piecewise polynomials, in particular B-splines, have been successfully applied [Davies and Owen, 1979; Heaps and Jones, 1981; Davies 1980, 1983; Owen, 1980].

It is evident from (14) that once the coefficients A_r, B_r have been determined, then the terms R_+ and R_- can be computed at any depth σ. From these terms, the amplitude and phase of the U and V components of current, together with tidal ellipse parameters, can be computed using standard expressions [Soulsby, 1983].

We now consider the solution of (12) and (13) using the Galerkin method through the vertical. Applying the Galerkin method to (12), this equation is multiplied by each basis function f_k and integrated through the vertical, giving

$$i(\gamma+\omega) \int_0^1 R_+ f_k d\sigma = \frac{1}{h^2} \int_0^1 \frac{\partial}{\partial\sigma} (\mu \frac{\partial R_+}{\partial\sigma}) f_k d\sigma - S_+ \int_0^1 f_k d\sigma. \tag{15}$$

where k=1,2,...,m.

By integrating the term involving μ by parts, we obtain

$$i(\gamma+\omega)\int_0^1 R_+ f_k d\sigma = \frac{1}{h^2} \left[\mu \frac{\partial R_+}{\partial\sigma} \Big|_1 f_k(1) - \mu \frac{\partial R_+}{\partial\sigma} \Big|_0 f_k(0) - \int_0^1 \mu \frac{\partial R_+}{\partial\sigma} \frac{df_k}{d\sigma} d\sigma\right] - S_+ \int_0^1 f_k d\sigma. \tag{16}$$

Considering surface and sea bed conditions, the zero surface stress condition (3) requires that

$$\mu \frac{\partial R_+}{\partial\sigma}\Big|_0 = \mu \frac{\partial R_-}{\partial\sigma}\Big|_0 = 0, \tag{17}$$

while the no-slip condition at the sea bed (4) requires that

$$R_+(1) = R_-(1) = 0 \tag{18}$$

In order for this boundary condition (termed an essential boundary condition [Connor and Brebbia 1976; Strang and Fix 1973]) to be satisfied for all coefficients A_r and B_r, the choice of basis functions f_r in (14) is no longer arbitrary, and it is essential to choose functions such that

$$f_r(1) = 0 \tag{19}$$

in order to satisfy condition (18).

Substituting boundary conditions (17) and (19) into (16), and using (14) to expand R_+ in terms of coefficients A_r, we obtain

$$i(\gamma+\omega) \sum_{r=1}^{m} A_r \int_0^1 f_r f_k d\sigma = -\frac{\alpha}{h^2} \sum_{r=1}^{m} A_r \int_0^1 \psi \frac{df_r}{d\sigma} \frac{df_k}{d\sigma} d\sigma$$

$$-S_+ \int_0^1 f_k d\sigma \quad (20)$$

where k=1,2,...,m.

In deriving (20), the eddy viscosity μ has been written as

$$\mu = \alpha\psi(\sigma) \quad (21)$$

with α representing the depth averaged value of μ and $\psi(\sigma)$ the vertical profile of μ. A similar equation to (20) can be derived from (13). For general basis functions f_k, the m equations in (20) are coupled together and have to be solved by matrix methods. However, these equations, can be further simplified by choosing the basis functions f_k to be eigenfunctions of an eigenvalue problem involving ψ.

2.3 Modal equations

In order to uncouple the m equations in (20), it is necessary to determine basis functions f_r, which are eigenfunctions f of an eigenvalue problem

$$\frac{d}{d\sigma}\left[\psi \frac{df}{d\sigma}\right] = -\varepsilon f, \quad (22)$$

with associated eigenvalues ε subject to boundary conditions

$$\frac{df}{d\sigma}\Big|_0 = 0, \quad f(1) = 0. \quad (23)$$

These conditions are essential in order to ensure the physical boundary conditions of zero surface stress (equation (3)) and no flow at the sea bed (equation (4)).

The eigenfunctions and eigenvalues of (22) can also be solved by using the Galerkin method [Davies, 1983]. Thus applying the Galerkin method to (22), the r'th eigen-equation is multiplied by f_k and integrated over the region 0 to 1. By integrating the term involving ψ by parts, we obtain

$$\psi\frac{df_r}{d\sigma}f_k\Big|_1 - \psi\frac{df_r}{d\sigma}f_k\Big|_0 - \int_0^1 \psi\frac{df_r}{d\sigma}\frac{df_k}{d\sigma}d\sigma = -\varepsilon_r \int_0^1 f_r f_k d\sigma \quad (24)$$

where k=1,2,...,m.

Substituting condition (23) into (24), gives

$$\int_0^1 \psi \frac{df_r}{d\sigma} \frac{df_k}{d\sigma}\, d\sigma = \varepsilon_r \int_0^1 f_r f_k d\sigma. \tag{25}$$

Substituting (25) into (20) and simplifying the resulting equations by using the orthogonality property of eigenfunctions, namely,

$$\int_0^1 f_r f_k d\sigma = \begin{cases} 0 & r \neq k \\ \neq 0 & r = k \end{cases} \tag{26}$$

we obtain from (20),

$$i(\gamma+\omega) \sum_{r=1}^{m} A_r \int_0^1 f_r f_k d\sigma = - \frac{\alpha}{h^2} A_r \varepsilon_r \int_0^1 f_r f_r d\sigma - S_+ \int_0^1 f_r d\sigma \tag{27}$$

A similar equation to (27) can be obtained for the negative rotary component of current.

In practice, it is convenient to write expansions (14) in the form

$$R_+ = \sum_{r=1}^{m} a_r \phi_r f_r, \quad R_- = \sum_{r=1}^{m} b_r \phi_r f_r \tag{28}$$

with a_r, b_r, complex coefficients, related to A_r, B_r, through $a_r \phi_r = A_r$, $b_r \phi_r = B_r$, with

$$\phi_r = 1 \,/ \int_0^1 f_r f_r d\sigma \tag{29}$$

Using expansions (28), we obtain from (27) and the corresponding equation for R_-, the working equations

$$[i(\gamma+\omega) + \frac{\alpha}{h^2} \varepsilon_r] a_r = -S_+ F_r \tag{30}$$

$$[i(\gamma-\omega) + \frac{\alpha}{h^2} \varepsilon_r] b_r = -S_- F_r \tag{31}$$

where $F_r = \int_0^1 f_r d\sigma$.

Equations (30) and (31) are an uncoupled set of m modal equations from which the coefficients a_r, b_r, r=1,2,...,m can be determined. Having computed these coefficients, the values of R_+ and R_- at any depth may be determined from equations (28). Various tidal current ellipse parameters may then be readily obtained from R_+ and R_- (see Soulsby [1983]).

It is evident from (30) and (31) that the coefficients a_r, b_r are proportional to the forcing terms S_+ and S_-. However, these terms are weighted by the term F_r which varies from mode to mode (see Table I). The importance of each mode f_r in expansion (28) is determined by the products $a_r \phi_r$ and $b_r \phi_r$, which again vary with mode number r due to changes in both coefficients a_r, b_r and the ϕ_r term (Table I).

3. TIDAL CURRENT PROFILES AND VERTICAL MODES

3.1 Tidal profiles for constant eddy viscosity

An analytical solution [Fang and Ichiye, 1983] is available in the literature for the case of constant eddy viscosity (Figure 1, Profile a) and is valuable for comparing the accuracy of various numerical solutions.

The vertical variations of the real and imaginary parts denoted by $\overline{R}_+$, $\overline{R}_-$ (Real) and $\overline{I}_+$, $\overline{I}_-$ (Imaginary) of the positive and negative components given by (28), normalized by dividing the geostrophic flow (see Fang and Ichiye [1983]), are shown in Figure 2, computed with Profile (a), $\mu = 1\ cm^2\ s^{-1}$ and Profile (b), $\mu = 1000\ cm^2\ s^{-1}$. In these profiles, ω was taken as the frequency of the M_2 tide, γ=0.00012s^{-1}, and water depth h as 100m. It is evident from Figure 2 that for low $\mu(\mu = 1\ cm^2 s^{-1})$ a high shear near bed layer occurs, with little vertical variation of flow above this region. As μ is increased, shear in this layer diminishes, and a more uniform profile from sea bed to sea surface is evident.

Velocity profiles having high shear in a lateral boundary layer can be difficult to model using the Galerkin method with continuous functions through the vertical. Chebyshev polynomials [Davies and Owen, 1979] do, however, exhibit high shear at their boundaries and can sometimes be suitable for solutions in regions where high shear boundary layers exist.

An alternative approach is to use piecewise polynomials, for example, B-splines (see Davies [1983] for details). The piecewise nature of these polynomials enables resolution to be enhanced locally by increasing the number of spline functions. For example, for the profiles shown in Figure 2, by using a high resolution of spline functions in the near bed layer, Davies [1985] was able to accurately resolve this high shear region with a minimum of computational effort.

The high shear boundary layer shown in Figure 2, is produced by the bottom stress which retards the flow. Shear boundary layers also occur at the sea surface during wind events, and some examples of the relative merits of various basis functions in resolving these layers, which are equally applicable to the bottom boundary layer, are given in the next section of this chapter.

3.2 Vertical modes

The use of a no-slip condition with eddy viscosity constant through the vertical is not supported by observational evidence [(Wolf, 1980; Bowden et al. 1959)], and the viscosity profile shown in Figure 1, Profile b, may be more appropriate. In this profile μ_0 and μ_2 are significantly less than μ_1, with h_1 and h_2 the order of 0.1h. An alternative vertical variation used by a number of authors, for which

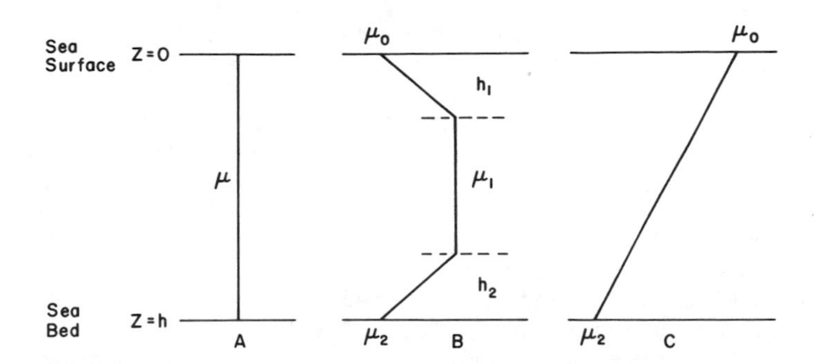

Figure 1. Various profiles of eddy viscosity used in the calculations.

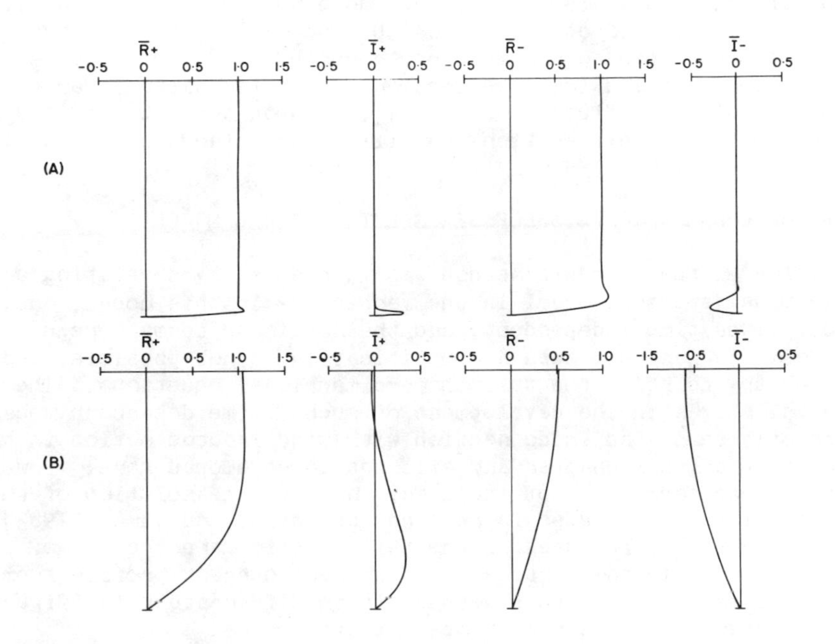

Figure 2. Profiles from sea surface to sea bed of the normalized rotary components of the tide, computed with (a) $\mu = 1\ cm^2s^{-1}$, (b) $\mu = 1000\ cm^2s^{-1}$.

an analytical solution is available [Smith, 1977], is shown in Figure 1, Profile c.

It is evident from equation (22), that the eigenvalues ε_r and eigenfunctions (modes) f_r are independent of the value of eddy viscosity which is taken into account in the parameter α and, hence, only depend upon the viscosity profile.

The first five modes computed using viscosity profiles (a), (b) and (c) in a depth h=100m are shown in Figure 3. As mode number increases, the vertical structure of the modes increases. Also, as near bed eddy viscosity is decreased relative to the mean value, so near bed shear increases significantly, particularly in the higher modes (compare modal distributions (A) and (C)).

Values of $\alpha\varepsilon_r$, ϕ_r, and F_r computed using viscosity profiles (b) and (c) (Figure 1) are shown in Table I for a range of μ values. It is apparent from this Table that ε_r increases significantly with mode number. Values of ϕ_r, which together with a_r, b_r, determine the contribution of each mode to the rotary components R_+, R_- (28), however, vary from mode to mode depending upon the viscosity profile. The term F_r, which weights the rotary components of the gradients of sea surface elevation S_+, S_- and therefore modifies these forcing terms, also varies from mode to mode (see Table I).

A similar set of equations to (30) and (31), although involving time dependency, can be derived for the problem of time dependent wind induced profiles (see the Section b of this chapter); a discussion of the influence of $\alpha\varepsilon_r$, ϕ_r, F_r upon current profiles is also given in the next section and will not be discussed here.

4. A THREE-DIMENSIONAL SHELF SEA SPECTRAL TIDAL MODEL

In section 2, the Galerkin method was introduced by developing a simple tidal spectral model in the vertical. In this model, eddy viscosity was time independent, and the nonlinear terms were not included. In general, such a simplification is not possible, and it is necessary to solve the full three-dimensional equations. The essential stages in the development of such a time dependent model are briefly described in connection with wind induced motion in the second part of this chapter and will not be developed here. However, the nonlinear terms are not included, and, for the solution of the full equations, the reader is referred to Davies and James [1983].

The essential features of the full solution are the use of the Galerkin method in the vertical to represent current profile from sea surface to sea bed, together with a finite difference grid (Figure 4) in the horizontal to represent spatial differenes.

Figure 4 shows the finite difference grid of a three-dimensional shelf model that has been used to reproduce the M_2, S_2, and M_4 tides on the continential shelf. In this model a slip condition is used at the sea bed with a quadratic law of bottom friction (equations (5a)).

Along the open boundaries of the model, a three-dimensional radiation condition in modal form [Davies and Furnes, 1980] is specified to allow for tidal forcing from the open ocean and permit

TABLE 1

Viscosity (cm s)	$\alpha\varepsilon_1$	$\alpha\varepsilon_2$	$\alpha\varepsilon_3$	$\alpha\varepsilon_4$	ϕ_1	ϕ_2	ϕ_3	ϕ_4	F_1	F_2	F_3	F_4
$\mu_0=50$, $\mu_1=1000$, $\mu_2=50$	0.17	1.6	4.8	9.8	1.7	2.1	2.7	3.7	0.7	-0.2	0.1	-0.03
$\mu_0=5000$, $\mu_2=50$	0.15	3.0	9.0	17.9	1.3	1.1	1.1	1.1	0.9	-0.1	0.1	-0.05

Table 1. Values of $\alpha\varepsilon_r$, ϕ_r and F_r computed using the viscosity profiles shown in Figure 1b, c, with a range of μ_0, μ_1, μ_2 values.

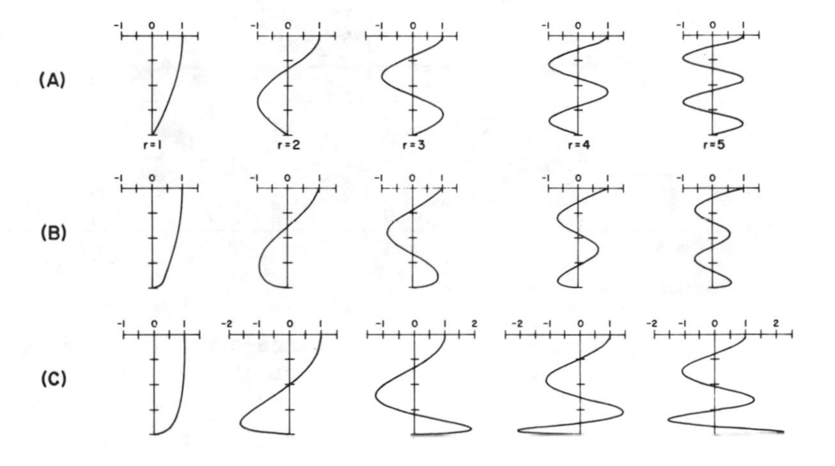

Figure 3. Vertical variation of the first five modes in a water depth h=100m, with $h_1=h_2=10$m, computed with (A) the viscosity profile given in Figure 1a, $\mu=1000$ cm^2s^{-1}, (B) profile in Figure 1b, $\mu_0=50$, $\mu_1=1000$, $\mu_2=50$ cm^2s^{-1}, (C), profile in Figure 1c, with $\mu_0=5000$, $\mu_2=50$ cm^2s^{-1}.

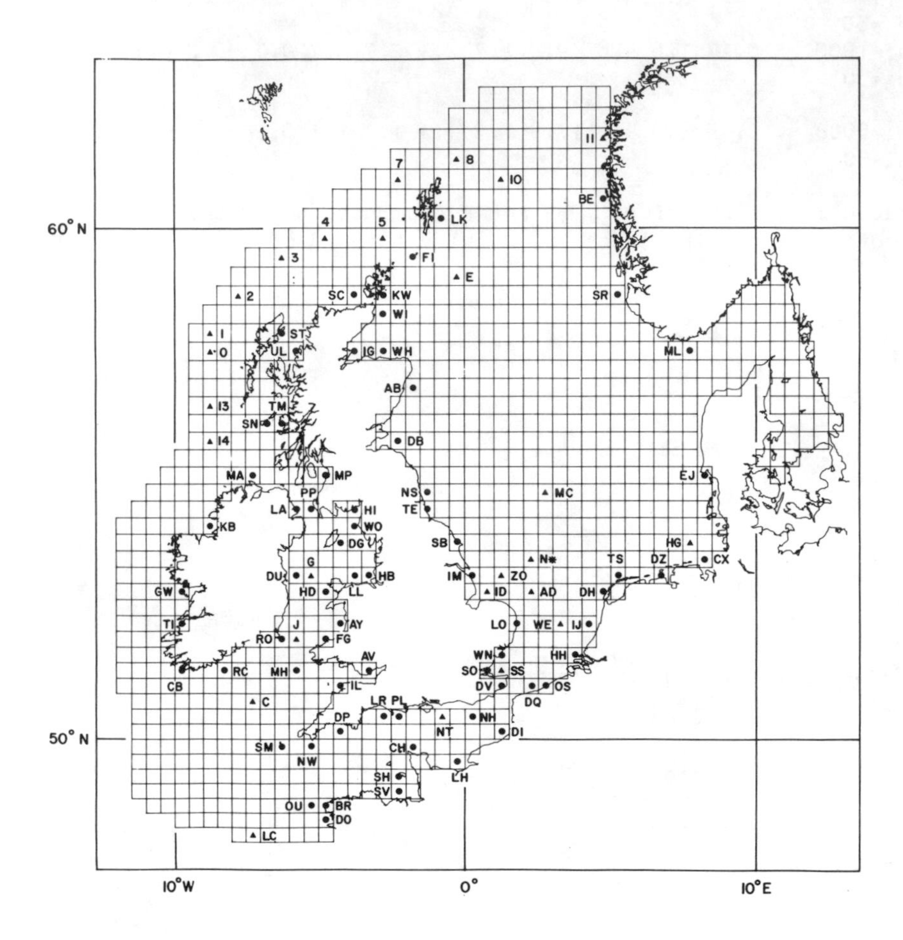

Figure 4. Finite difference grid of the three-dimensional continental shelf model, showing positions of tide gauges.

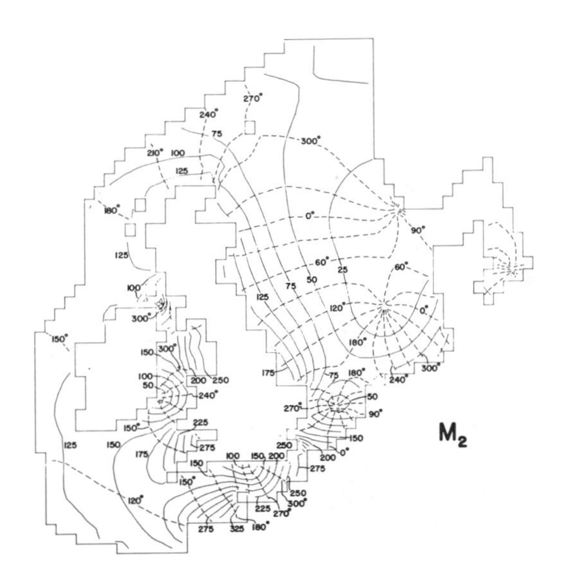

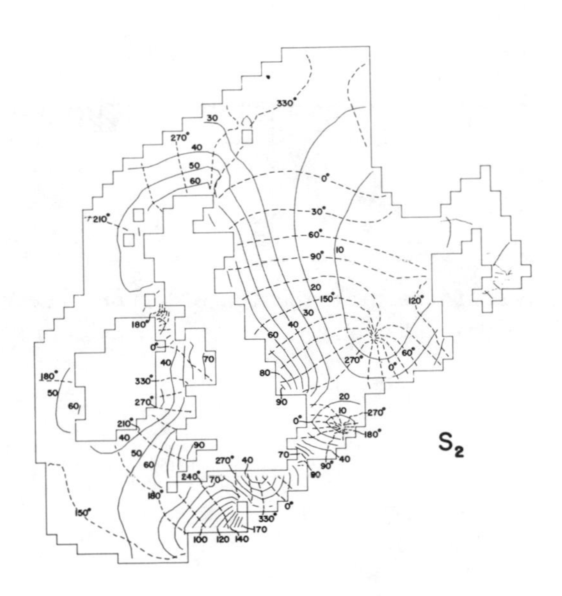

Figure 5. M_2 and S_2 co-tidal charts showing amplitude in cm (———) and phase in degrees (- - - -).

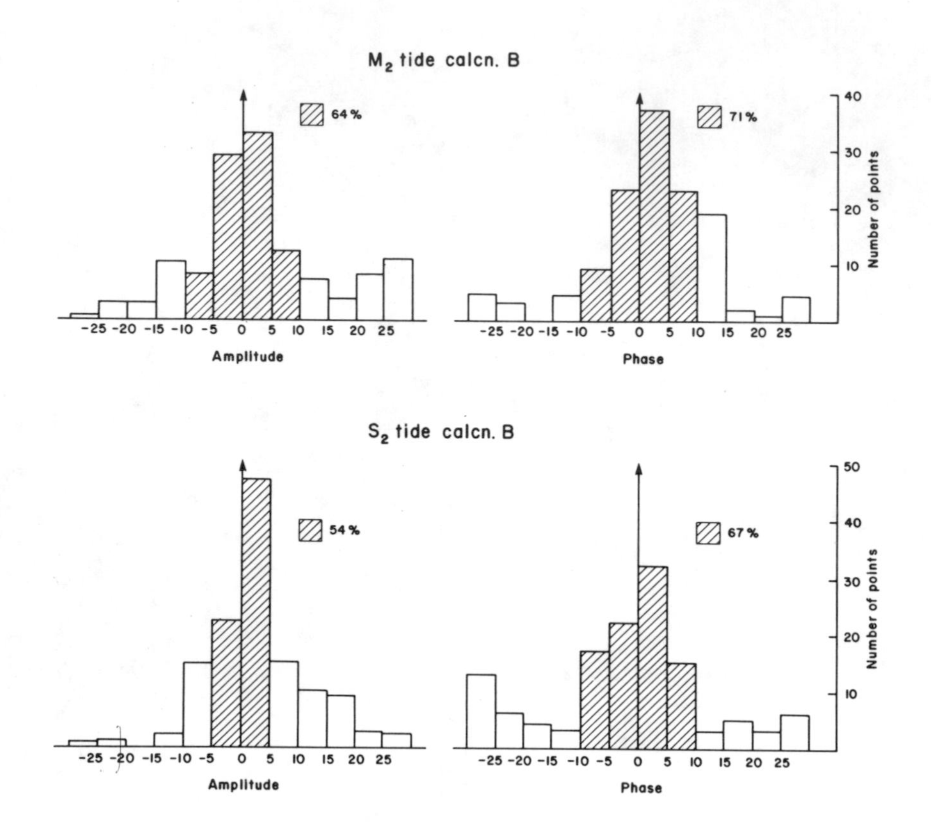

Figure 6. Histogram of the distribution of errors for the M_2 and S_2 components of the tide.

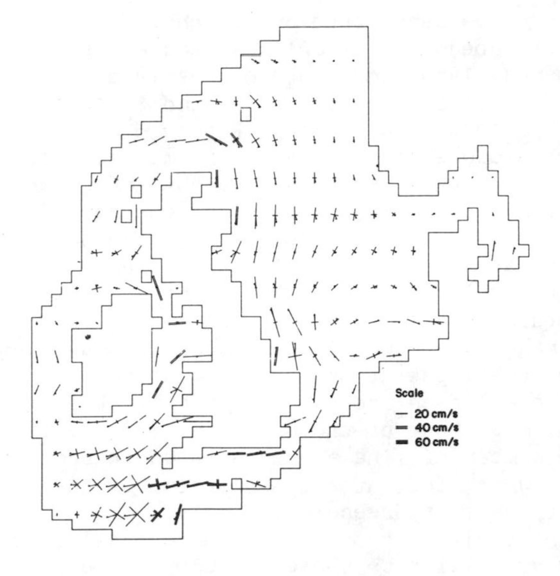

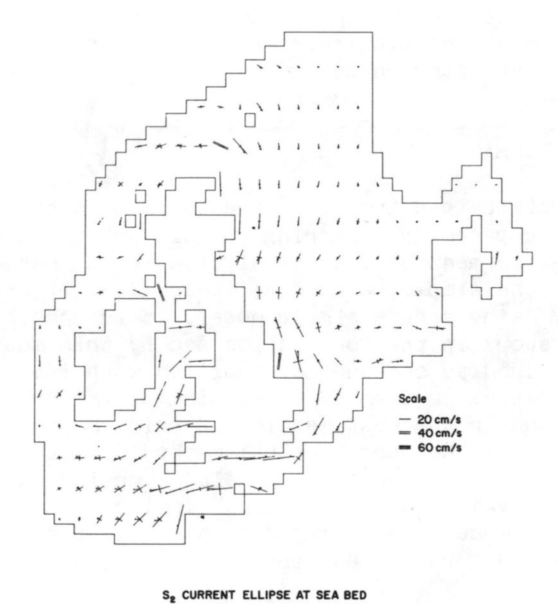

Figure 7. Computed distributions of the S_2 current ellipse at sea surface and sea bed.

reflected waves to be tansmitted out of the model. Eddy viscosity within the model depends in a nonlinear manner upon the current [Davies and Furnes, 1980] and therefore evolves with the flow field.

Computed cotidal charts for the M_2 and S_2 components of the tide computed with the model are shown in Figure 5. Error histograms based upon comparisions of observed and computed tidal elevation and phases at the tide gauges shown in Figure 4 are given in Figure 6. It is evident that, on average, the model can reproduce the tide to within an acceptable degree of accuracy (in the main determined by grid resolution) at the majority of tide gauges. To improve model accuracy, a finer grid [Davies, 1976; Mathisen et al., 1983] would probably be required.

Computed magnitudes and directions of the major and minor axes of the S_2 current ellipse at sea surface and sea bed are given in Figure 7. These ellipses are plotted at every third grid point of the model. The magnitude of the major and minor axis is denoted by the length and number of line segments shown in the diagram. Also shown is a line which indicates the position of the major axis at the time of lunar transit at Greenwich.

The spatial variations of the S_2 current ellipses shown in the Figure 5 are very similar to those found by Davies and Furnes [1980] for the M_2 component of the tide. Current magnitudes are, in general, a half to a third less than those of the M_2 tide. As expected, the magnitude of the major axis of the S_2 current ellipse diminishes with depth due to frictional effects. A full comparison of observed and computed tidal currents is presently in progress, and results will be presented in due course.

5. CONCLUDING SUMMARY

A brief review of the background of the Galerkin method has been presented in this paper by referring to relevant text books. The mathematical development of a spectral tidal model has been illustrated for the simple case of a linear model of oscillatory flow at a point. By using such a simple model, it is possible to clearly illustrate the steps in the formulation and by this means to introduce the method to someone not familiar with it.

The more complex problem of developing a time dependent spectral model can be found in the next section of this chapter. However, in order to demonstrate that the spectral method is not restricted to the formulation of idealized tidal models, some indication of its wider applicability is given by showing tidal results for a three-dimensional shelf model. A fuller discussion of the formulation of such a model can be found in Davies and Furnes [1980].

6. ACKNOWLEDGEMENTS

The author is indebted to Mr. R. A. Smith for preparing diagrams and Mrs. L. Parry for typing the text.

7. REFERENCES

Bowden, K. F., L. A. Fairbairn, and P. Hughes, The distribution of shearing stresses in a tidal current, Geophys. J. R. Astr. Soc., 2, 288-305, 1959.

Cheng, R. T., T. M. Powell, and T. M. Dillon, Numerical models of wind-driven circulation in lakes, Applied Mathematical Modelling, 1, 141-158, 1976.

Connor, J. J., and C. A. Brebbia, Finite element techniques for fluid flow, 307 pp., London, Newnes - Butterworths, 1976.

Davies, A. M., A numerical model of the North Sea and its use in choosing locations for the deployment of offshore tide gauges in the JONSDAP '76 oceanographic experiment, Dtsch. Hydrogr. Z., 29, 11-24, 1976.

Davies, A. M., Application of the Galerkin method to the formulation of a three-dimensional non-linear hydrodynamic numerical sea model, Applied Mathematical Modelling, 4, 245-256, 1980.

Davies, A. M., Formulation of a linear three-dimensional hydrodynamic sea model using a Galerkin-Eigenfunction method, Int. J. Num. Meth. in Fluids, 3, 33-60, 1983.

Davies, A. M., On determining current profiles in oscillatory flows, in press, Applied Mathematical Modelling, 1985.

Davies, A. M., Mathematical formulation of a spectral circulation model, this volume, 1986b.

Davies, A. M., and G. K. Furnes, Observed and computed M_2 tidal current in the North Sea, J. Phys. Oceanogra., 10, 237-257, 1980.

Davies, A. M., and I. D. James, Three-dimensional Galerkin-Spectral sea models of the North Sea and German Bight, in North Sea Dynamics, edited by J. Sundermann and W. Lenz, pp. 85-95, Springer-Verlag, 1983.

Davies, A. M., and A. Owen, Three-dimensional numerical sea model using the Galerkin method with a polynomial basis set, Appl. Math. Modelling, 3, 421-428, 1979.

Fang, G., and T. Ichiye, On the vertical structure of tidal currents in a homogeneous sea, Geophys. J. R. Astr. Soc., 73, 65-82, 1983.

Flather, R. A., A tidal model of the northwest European continental shelf, Mem. Soc. Roy. Sci. Liege, 10, 141-164, 1976.

Finlayson, B. A., The method of weighted residuals and variational principles, Academic Press, New York, 1972.

Gottlieb, D., and S. Orszag, A numerical analysis of spectral methods, NSF-CBMS Monograph No. 26, Soc. Ind. and Appl. Math., Philadelphia, 1977.

Gray, W. G., Some inadequacies of finite element models as simulators of two-dimensional circulation, Adv. Water Resources, 5, 171-177, 1982.

Grotkop, G., Finite element analysis of long-period water waves, Comput. Meth. Appl. Mech. Eng., 2, 133-146, 1973.

Heaps, N. S., and J. E. Jones, Three-dimensional model for tides and surges with vertical eddy viscosity prescribed in two layers. II. Irish Sea with bed friction layer, Geophys. J. R. Astr. Soc., 64, 303-320, 1981.
Mathisen, J. P., and O. Johansen, A numerical tidal and storm surge model of the North Sea, Marine Geodesy, 6, 267-291, 1983.
Owen, A., A three-dimentional model of the Bristol Channel, J. Phys. Oceanography, 10, 1290-1302, 1980.
Prandle, D., The Vertical structure of tidal currents and other oscillatory flows, Continental Shelf Research, 1, 191-207, 1982.
Proctor, R., Tides and residual circulation in the Irish Sea: a numerical modelling approach. Ph.D. thesis, Liverpool University, 1981.
Schwiderski, E. W., Atlas of ocean tidal charts and maps, Part 1: the semidiurnal principal lunar tide M_2, Marine Geodesy, 6, 219-265, 1983.
Smith, J. D., Modelling of sediment transport on continental shelves, in The Sea, 6, edited by E. D. Goldberg et al., pp. 539-578, Wiley-Interscience, 1977.
Soulsby, R. L., The bottom boundary layer of shelf seas, in Physical Oceanography of Coastal and Shelf Seas, edited by B. Johns, pp. 189-266, Elsevier Oceanography Series, No. 35, 1983.
Strang, G., and G. J. Fix, An analysis of the finite element method, Prentice Hall Inc., Englewood Cliffs, 1973.
Wolf, J., Estimation of shearing stresses in a tidal current with application to the Irish Sea, in Marine Turbulence, Proceedings of the 11th Liege Colloquium on Ocean Hydrodynamics edited by J.C.J. Nihoul, pp. 319-344, Elsevier Oceanography, Series No. 28, 1980.
Wolf, J., The variability of currents in shallow seas, Ph.D. thesis, Liverpool University, 1984.

MATHEMATICAL FORMULATION OF A SPECTRAL CIRCULATION MODEL

A. M. Davies
Institute of Oceanographic Sciences
Bidston Observatory
Berkenhead
Merseyside L43 7RA
England

ABSTRACT. This section of the chapter extends the spectral method developed earlier to deal with the problem of formulating a full three-dimensional spectral model for wind induced circulation. Some simple examples are considered to illustrate the accuracy of the method and to show how the approach can be used to gain some insight into the factors controlling wind induced current profiles. In the final section references to the literature are used to indicate the range of models which have employed spectral methods in shallow sea problems. The aim is to produce a self-contained text which will introduce the topic to a student or research scientist who is not familiar with the spectral approach.

1. INTRODUCTION

In the first section of this chapter, references to relevant background literature were presented and the major features involved in developing a one-dimensional, essentially time independant spectral model for oscillatory flow were developed. In this paper, a three-dimensional time dependent spectral model is formulated and applied to the problem of determining wind induced current profiles and characteristic time scales for the dispersion of pollutants in the shelf seas.

The problem of determining wind induced current profiles stems from the classic work of Ekman. However, with enhanced activity in off-shore oil exploration, it has become increasingly important to determine current profiles during major wind events [R. L. Gordon, 1982].

Wind induced surface currents and turbulence in the surface boundary layer are also of particular interest in air-sea interaction problems and the dispersion of pollutants, in particular oil spills. The problem of surface currents and near surface turbulence is presently being actively studied using a range of observational methods [Kullenberg, 1976; Thorpe 1984].

In section 2 of this paper, the major steps in formulating a

J. J. O'Brien (ed.), Advanced Physical Oceanographic Numerical Modelling, 391–409.

three-dimensional spectral model are given in detail. In the following section, the problem of determining a suitable formulation of sub-grid scale turbulence is briefly considered. The fourth section of the paper illustrates the application of a spectral model to the problem of wind induced current profiles, while the final part of the paper briefly indicates the application of spectral models to pollution problems.

The object of the paper is to give sufficient mathematical detail to introduce the topic to a student or scientist new to the area, rather than to present new scientific results which are more appropriate in scientific journals. Two review papers dealing with spectral models will appear in the next year [Davies 1985a,b] giving substantially more information than can be presented here; they may be of interest of anyone wishing to pursue the topic.

2. FORMULATION OF A SPECTRAL (MODAL) MODEL FOR WIND INDUCED CURRENTS

2.1. Basic linear hydrodynamic equations

In this section, we use the linear hydrodynamic equations in Cartesian Coordinates to illustrate the basic steps in developing the modal equations for wind induced currents.

Using a normalized depth coordinate (sigma coordinate), the linear hydrodynamic equations are given by

$$\frac{\partial \zeta}{\partial t} + \frac{\partial}{\partial x}\left(h\int_0^1 U d\sigma\right) + \frac{\partial}{\partial y}\left(h\int_0^1 V d\sigma\right) = 0 \tag{1}$$

$$\frac{\partial U}{\partial t} - \gamma V = -g\frac{\partial \zeta}{\partial x} + \frac{1}{h^2}\frac{\partial}{\partial \sigma}\left(\mu\frac{\partial U}{\partial \sigma}\right) \tag{2}$$

$$\frac{\partial V}{\partial t} + \gamma U = -g\frac{\partial \zeta}{\partial y} + \frac{1}{h^2}\frac{\partial}{\partial \sigma}\left(\mu\frac{\partial V}{\partial \sigma}\right) \quad . \tag{3}$$

Standard notation is used in these equations and is defined in the first part of this chapter.

For wind induced motion, unlike tidal motion, surface stress is non-zero and must equal the externally specified stress, thus,

$$-\frac{\rho}{h}\left(\mu\frac{\partial U}{\partial \sigma}\right)_0 = \tau_x^s \quad , \quad -\frac{\rho}{h}\left(\mu\frac{\partial V}{\partial \sigma}\right)_0 = \tau_y^s \tag{4}$$

with τ_x^s , τ_y^s the components of wind stress, ρ density taken as constant, and h water depth.

Similarly, at the sea bed, a stress condition of the form

$$-\frac{\rho}{h}\left(\mu\frac{\partial U}{\partial \sigma}\right)_1 = \tau_x^h \quad , \quad -\frac{\rho}{h}\left(\mu\frac{\partial V}{\partial \sigma}\right)_1 = \tau_y^h \tag{5}$$

can be applied with τ_x^h , τ_y^h the x and y components of bottom stress.

An alternative bottom boundary condition is one of no-slip, given by

$$U_h = V_h = 0 \quad . \tag{6}$$

2.2 Development of the spectral model

In the case of oscillatory motion at a single frequency (e.g., tidal motion at one point), time dependancy could be removed from the equations by using rotary decomposition. Here, by way of contrast, we wish to consider the problem of generalized wind forcing over a sea area and must therefore retain the time domain within our equations. Also, rather than using complex notation as previously, the U and V components of current are individually expanded in terms of m time and horizontally spatially dependent coefficients $A_r(x,y,t)$, $B_r(x,y,t)$ and functions (the basis set) $f_r(\sigma)$ through the vertical.

Thus,

$$U = \sum_{r=1}^{m} A_r(x,y,t)\ f_r(\sigma) \quad , \quad V = \sum_{r=1}^{m} B_r(x,y,t)\ f_r(\sigma). \tag{7}$$

Substituting (7) into equation (1) gives

$$\frac{\partial \zeta}{\partial t} = - \sum_{r=1}^{m} \left[\left(\frac{\partial}{\partial x} (A_r h) + \frac{\partial}{\partial y} (B_r h) \right) \cdot \int_0^1 f_r d\sigma \right]. \tag{8}$$

Applying the Galerkin method to equation (2), this equation is multiplied by each of the basis functions, f_k, k=1,2...,m and integrated through the vertical, giving

$$\int_0^1 \frac{\partial U}{\partial t} f_k d\sigma = \gamma \int_0^1 V f_k d\sigma - g \frac{\partial \zeta}{\partial x} \int_0^1 f_k d\sigma + \frac{1}{h^2} \int_0^1 (\mu \frac{\partial U}{\partial \sigma}) f_k d\sigma \tag{9}$$

where k=1,2...m.

By integrating the term involving μ by parts in an analogous manner to that used earlier for tidal motion, we obtain

$$\int_0^1 \frac{\partial U}{\partial t} f_k d\sigma = \gamma \int_0^1 V f_k d\sigma - g \frac{\partial \zeta}{\partial x} \int_0^1 f_k d\sigma + \frac{1}{h^2} \mu \frac{\partial U}{\partial \sigma} f_k \Big|_1 - \frac{1}{h^2} \mu \frac{\partial U}{\partial \sigma} f_k \Big|_0 - \frac{1}{h^2} \int_0^1 \mu \frac{\partial U}{\partial \sigma} \frac{\partial f_k}{\partial \sigma} d\sigma \tag{10}$$

where k=1,2...m.

Surface and sea-bed stress conditions can now be included into (10), using (4) and (5), giving

$$\int_0^1 \frac{\partial U}{\partial t} f_k d\sigma = \gamma \int_0^1 V f_k d\sigma - g \frac{\partial \zeta}{\partial x} \int_0^1 f_k d\sigma - \frac{\tau_x^h}{\rho h} f_k(1) + \frac{\tau_x^s}{\rho h} f_k(0) - \frac{1}{h^2} \int_0^1 \mu \frac{\partial U}{\partial \sigma} \frac{\partial f_k}{\partial \sigma} d\sigma \qquad (11)$$

where k=1,2...m.

It is interesting to note that by using the Galerkin method in the vertical space domain, surface and sea bed boundary conditions are combined with the partial differential equation (2) to give a set of couple equations (11) which include these conditions.

Substituting expansions (7) into (11) gives

$$\sum_{r=1}^{m} \frac{\partial A_r}{\partial t} \int_0^1 f_r f_k d\sigma = \gamma \sum_{r=1}^{m} B_r \int_0^1 f_r f_k d\sigma - g \frac{\partial \zeta}{\partial x} \int_0^1 f_k d\sigma - \frac{\tau_x^h}{\rho h} f_k(1) + \frac{\tau_x^s}{\rho h} f_k(0) - \frac{1}{h^2} \sum_{r=1}^{m} A_r \int_0^1 \mu \frac{df_r}{d\sigma} \frac{df_k}{d\sigma} d\sigma \qquad (12)$$

where k=1,2...m.

A similar equation to (12), giving the time variation of the coefficients B_r, can be derived from equation (3).

For a general basis set of functions, the m equations in (12), are coupled together. Even with an orthogonal basis set of functions, coupling still occurs through the integrals involving the vertical eddy viscosity.

Although this coupling does not present any numerical problems, it does increase the computational requirements of the method [Davies and Stephens, 1983]. As was demonstrated in the first section of the chapter for the case of tidal flow, if we restrict the form of eddy viscosity such that

$$\mu = \alpha(x,y,\tau)\psi(\sigma) \quad , \qquad (13)$$

then some uncoupling of the equations is possible. In (13) the coefficient α allows for spatial and temporal variation of μ, and $\psi(\sigma)$ restricts μ to a fixed vertical profile.

2.3 The Eigenvalue Problem

As was shown earlier for tidal problems, uncoupling can be achieved by making the basis functions f_r eigenfunctions (modes) of an eigenvalue problem, involving the vertical eddy viscosity μ. When this eigenvalue problem is solved using the Galerkin method, we obtain (see the previous section of this chapter for detail).

$$\psi \frac{df_r}{d\sigma} f_k\Big|_1 - \psi \frac{df_r}{d\sigma} f_k\Big|_0 - \int_0^1 \psi \frac{df_r}{d\sigma} \frac{df_k}{d\sigma} d\sigma = -\varepsilon_r \int_0^1 f_r f_k d\sigma. \quad (14)$$

In the case of tidal motion described earlier, the no-slip sea bed condition gave $f_k(1)=0$, and the surface boundary condition of zero stress gave $df_r/d\sigma=0$, and, hence, the first two terms in (14) were zero.

However, for wind induced motion with a slip bottom boundary condition and a sea surface shear condition, the first two terms in (14) need no longer be zero.

The eigenfunctions and eigenvalues of (14) will be real [Davies, 1982a], provided

$$\frac{df_r}{d\sigma}\Big|_0 = C_1 f_1(0) \quad , \quad \frac{df_r}{d\sigma}\Big|_1 = C_2 f_1(1) \quad (15)$$

with C_1 and C_2 arbitrary constants. In the case of $C_1 = C_2 = 0$, the first two terms in (14) are zero, and the basis functions (modes) are eigenfunctions satisfying boundary conditions

$$\frac{df_r}{d\sigma}\Big|_0 = \frac{df_r}{d\sigma}\Big|_1 = 0. \quad (16)$$

Since C_1 and C_2 are arbitrary constants, it is possible to chose them to increase the near surface shear in each mode and by this means to reduce the number of terms in expansions (7) required to compute an accurate solution. A discussion of this topic is beyond the scope of the present paper, although the interested reader is referred to Davies [1982a], for details.

A consequence of (16) and the symmetric nature of (14) is that

$$\varepsilon_1 = 0 \quad , \quad f_1 = \text{constant} \quad (17)$$

The orthogonality property of eigenfunctions,

$$\int_0^1 f_r f_k d\sigma \begin{array}{ll} = 0 & r \neq k \\ \neq 0 & r = k \end{array} \quad (18)$$

implies that for f_1 = constant, we have

$$\int_0^1 f_r d\sigma = 0 \quad , r=2,3\ldots m. \tag{19}$$

Also, it is possible to normalize the eigenfunctions such that

$$f_r(0) = 1 \quad r=1,2\ldots m. \tag{20}$$

Substituting (14) into (12), using boundary conditions (16) together with (18), (19), and (20), and writing expansions (7) as in the first section

$$U = \sum_{r=1}^{m} a_r \phi_r f_r(\sigma) \quad , \quad V = \sum_{r=1}^{m} b_r \phi_r f_r(\sigma) \tag{21}$$

with $\phi_r = 1 \Big/ \int_0^1 f_r^2 d\sigma$ (22)

we obtain the set of equations

$$\frac{\partial a_1}{\partial \tau} = \gamma b_1 - g \frac{\partial \zeta}{\partial x} - \frac{\tau_x^h}{\rho h} + \frac{\tau_x^s}{\rho h} \tag{23}$$

$$\frac{\partial a_k}{\partial \tau} = \gamma b_k - \frac{\tau_x^h}{\rho h} f_k(1) + \frac{\tau_x^h}{\rho h} - \frac{\alpha \varepsilon_k}{h^2} a_k \tag{24}$$

where k=2,3...m.

In deriving (23), we have used the fact that $f_1(0)=1$.

In an analogous manner from the v-equation of motion (3) we obtain

$$\frac{\partial b_1}{\partial \tau} = -\gamma a_1 - g \frac{\partial \zeta}{\partial y} - \frac{\tau_y^h}{\rho h} + \frac{\tau_y^s}{\rho h} \tag{25}$$

$$\frac{\partial b_k}{\partial \tau} = \gamma a_k - \frac{\tau_y^h}{\rho h} f_k(1) + \frac{\tau_y^s}{\rho h} - \frac{\alpha \varepsilon_k}{h^2} b_k \tag{26}$$

where k=2,3...m.

The modal form of the continuity equation is

$$\frac{\partial \zeta}{\partial \tau} + \frac{\partial}{\partial x}(a_1 h \phi_1 F_1) + \frac{\partial}{\partial y}(b_1 h \phi_1 F_1) = 0 \tag{27}$$

where

$$F_1 = \int_0^1 f_1 d\sigma. \tag{28}$$

Although it might appear that the coefficients a_r, b_r, $r=2,\ldots,m$ in equations (24) and (26) are uncoupled except for coupling due to the Coriolis parameter, in a shallow sea they are coupled together through the bottom stress terms, which can be related to the bottom current by

$$\tau_x^h = K\rho U_h \quad , \quad \tau_y^h = K\rho V_h \tag{29}$$

In (29), K is a coefficient of linear bottom friction.

Since U_h and V_h depend upon all the coefficients through expansion (21), then all the modal equations are coupled, unless the coefficient of bottom friction K is set to zero, or the water is sufficiently deep that U_h and V_h are zero.

It is interesting to note that because of condition (19) which arises due to boundary conditons (16), only the first mode contributes to changes in sea surface elevation. Also, changes in sea surface elevation do not contribute to changes in the coefficients of the higher modes.

3. PARAMETERIZATION OF SUB-GRID SCALE MOTION

In a continental shelf model having a grid resolution of the order of kilometers (see Figure 4), it is necessary to parameterize all scales of motion which cannot be resolved on this grid and are responsible for transferring the wind's momentum to depth. Of particular importance in this context are surface waves and the secondary circulations associated with them. Since these waves cannot be resolved, any transfer of the wind's momentum to depth produced by them has to be parameterized in the form of an effective eddy viscosity μ_W, taking into account wind and wave effects.

Based upon dye diffusion experiments (Ichiye, 1967; Huang, 1979), an effective near surface eddy viscosity μ_W incorporating wind and wave effects, of the form

$$\mu_W = 0.028 \; H^2/T \tag{30}$$

with H and T significant wave height and period, has been determined.

In the case of a fully developed sea state, using equations due to Carter [1982] to relate H and T to wind speed W gives

$$\mu_W(m^2s^{-1}) = 0.3043 \times 10^{-4}W^3 \text{ (W in m s}^{-1}). \tag{31}$$

Other dye diffusion experiments have been performed by Kullenberg [1976], who suggests a relationship of the form

$$\mu_W = \frac{1}{\gamma}\left(\frac{\rho_a}{\rho_W}\frac{C_D}{k'}\right)^2 W^2 \tag{32}$$

with ρ_a/ρ_W ratio of air to water density = 1.23×10^{-3} and $k' = 1.8 \times 10^{-2}$. In (32) C_D is the drag coefficient.

An empirical formulation proposed by Neumann and Pierson [1964], gives

$$\mu_W(cm^{-1}gr\ s^{-1}) = 0.1825 \times 10^{-4}W^{5/2} \text{(W in cm s}^{-1}). \tag{33}$$

It is evident from (31), (32), and (33) that μ_W increases rapidly with increasing wind speed.

The vertical variation of μ_W from sea surface to sea bed is also difficult to determine. By analogy, with the atmospheric boundary layer over land or the sea bed layer, a linear decrease in μ_W in the vicinity of the sea surface might be expected, with μ_W reducing to a value μ_o, given by

$$\mu_o = k_o U_* Z_o \tag{34}$$

In (34), Von Karman's constant, k_o=0.4, $U=(\tau/\rho)^{1/2}$, is frictional velocity, τ wind stress, and Z_o roughness length.

A reduction in eddy viscosity near the sea surface has been proposed by a number of authors (e.g., Dyke [1977]; Weber [1981]). However, it is possible to argue that since the sea surface is continually in motion, it does not behave as a rigid boundary. During a major wind event, the sea surface is a source of turbulence generated by the wind, and hence eddy viscosity at the sea surface will be maximum. An increase in eddy viscosity near the surface has been assumed by Dobroklonskiy [1969] and Lai and Rao [1976]. The alternative simpler approach [Heaps 1972; Davies 1983a] is to assume eddy viscosity constant through the vertical.

These various idealized profiles are shown in Figure 1. In profile A, eddy viscosity is constant at a value μ_W from sea surface to bed, while in profile B, it decreases to a value μ_o at the sea surface, and in profile C, a linear increase is assumed.

The influence upon accuracy and current profile of these various assumptions concerning eddy viscosity variation are considered in the next section.

4. CALCULATION OF THE PROFILE OF WIND INDUCED CURRENTS

4.1. Accuracy of the Galerkin Method

In theory, the choice of basis functions f_r in expansions (7) is arbitrary; however, in practice their mathematical form can increase the rate of convergence of the series. This can be illustrated by considering Ekman's problem of a uniform wind blowing over a sea area of infinite extent and depth. In this particular problem, no gradients of sea surface elevation are produced, and consequently a point model through the vertical is adequate. An infinite water depth can be simulated by ensuring that the depth h is larger than the depth to which the wind's momentum can penetrate.

In the case of a low value of μ_W (for example μ_W=1cm^2 s^{-1}), the wind's momentum cannot diffuse to depth, and high near surface shear layer results, with strong surface currents decreasing to zero very rapidly below the surface. With increased eddy viscosity, the wind's momentum can penetrate to greater depths and surface current decreases, with current at depth being enhanced, giving rise to decreased surface shear.

In order to illustrate the influence of the functional form of the basis set upon the computed current profile, it is interesting to compute steady state surface currents using two different types of basis functions. Here, for illustrative purposes, we compare (Table I) the rate of convergence of surface currents computed with an increasing number of Legendre polynomials (continous functions) with those computed using various distributions of β-splines (discontinous functions, details of which can be found in Davies [1983b], for the case of μ_W constant, but having a range of values (Table I). In this example, surface currents were produced by a northerly wind stress of 10 dyn cm^{-2}.

Referring to Table I, it is evident that with μ_W = 100cm^2 s^{-1}, surface currents computed with spline distributions A or B (based upon 10 and 20 uniformly distributed splines in the vertical) can be accurately resolved. However, in the case of lower values of μ_W, the resulting high shear surface layer could not be adequately resolved. Increasing in a uniform manner the number of splines through the vertical does improve accuracy. However, because the current very rapidly reaches a near zero value below the surface shear layer, optimum convergence can be obtained by taking advantage of the piecewise nature of these splines and increasing resolution locally in the near surface layer, using spline distribution C, having a high spline resolution close to the sea surface.

Since Legendre polynomials are continuous functions, it is not possible to locally enhance their resolution within a boundary layer, and 45 polynomials are required to accurately reproduce the current profile computed with μ = 1cm^2 s^{-1}. However, in the absence of a high shear layer, the convergence of the Legendre polynomials can exceed those of the spline function (e.g., the case of μ = 10cm^2 s^{-1}).

4.2. Modal solution of the time dependent problem

Here we again consider the case of a uniform wind suddenly applied to a sea area of infinite extent. However, in this case the wind stress has a magnitude τ_y^s = 15 dynes cm, τ_x^s = 0.0, and is only applied for a period of 3 1/2 hours. Zero bottom stress is achieved by setting K = 0.0. Before considering the solution of this problem using the modal model, it is instructive to examine the vertical variation of modes computed with: (a) μ_o = 50cm^2 s^{-1}, μ_W = 500cm^2 s^{-1}; and (b) μ_o = 4000cm^2 s^{-1}, μ_W = 500cm^2 s^{-1} (see Figure 2) in a water depth h = 260m with h_o=228m.

It is evident from Figure 2 that the first mode represents a uniform current from sea surface to sea bed. As the mode number increases, so the vertical structure of the mode increases, although, as was demonstrated in section 2, these higher modes have a zero integral through the vertical and, hence, zero transport. It is apparent from Figure 2 that for a given mode number, near surface shear is enhanced as μ_o is reduced. Also, it is important to note that for a given μ_o and μ_W, near surface shear increases with increasing mode number.

Current profiles computed with the modes shown in Figure 2(A), using expansions of five and twenty-five functions at times one-hour and five hours after the onset of the wind field, are shown in Figure 3. Also given are accurate profiles computed with an expansion of β-splines having high resolution in the near surface layer.

It is apparent from this figure that in order to compute the high shear surface layer produced by the wind field, it is necessary to use a basis set of twenty five modes. This is the case because the initial response of the sea to a suddenly imposed wind stress is the production of a high shear surface layer. This is reproduced in the spectral model by the initial excitation of all vertical modes. With a small number of modes (m=5), the near surface shear layer is not resolved; also, physically unrealistic "ripples" which are reduced as the number of modes are increased occur on the current profile.

By considering in some detail equations (23), (24) and (25), (26), it is possible to appreciate the factors controlling the profile of wind induced currents and the parameters which determine the number of modes required to accurately reproduce the current's profile. For the case described previously of a suddenly applied uniform northerly wind stress with zero bottom stress and zero gradients of sea surface elevation, then the inital response is the excitation of all coefficients $(b_1, b_2, \ldots, b_m)$.

Coupling between the sets of equations (23), (24) and (25), (26) is then responsible for transferring some of this energy into the set of coefficients $(a_1, a_2, \ldots, a_m)$. It is evident from these equations that in the absence of bottom friction, the first mode is not damped. However, the higher modes are damped by the term $\alpha\varepsilon_k/h^2$. Values of α and ε_k, together with values of ϕ_k are given in Table II. It is apparent from this table that ε_1=0; hence, the first mode is only damped by bottom friction, while for the other modes

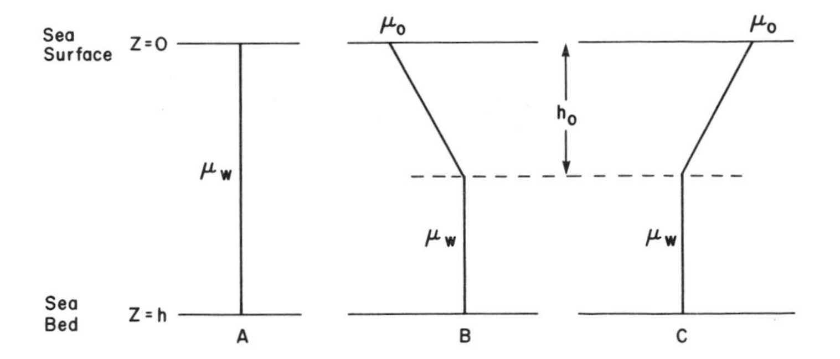

Figure 1. Idealized profiles of vertical eddy viscosity.

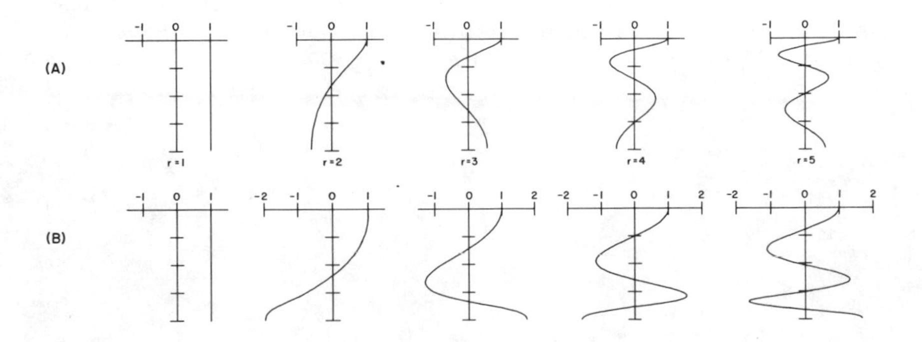

Figure 2. Vertical variation of the first five vertical modes computed with a slip bottom boundary condition using viscosity profiles (b) and (c) (Figure 1), with (A) $\mu_o = 50$, $\mu_W = 500 cm^2 s^{-1}$, (B) $\mu_o = 4000$, $\mu_W = 500\ cm^2 s^{-1}$, and $h_o = 222.8m$, $h = 260m$.

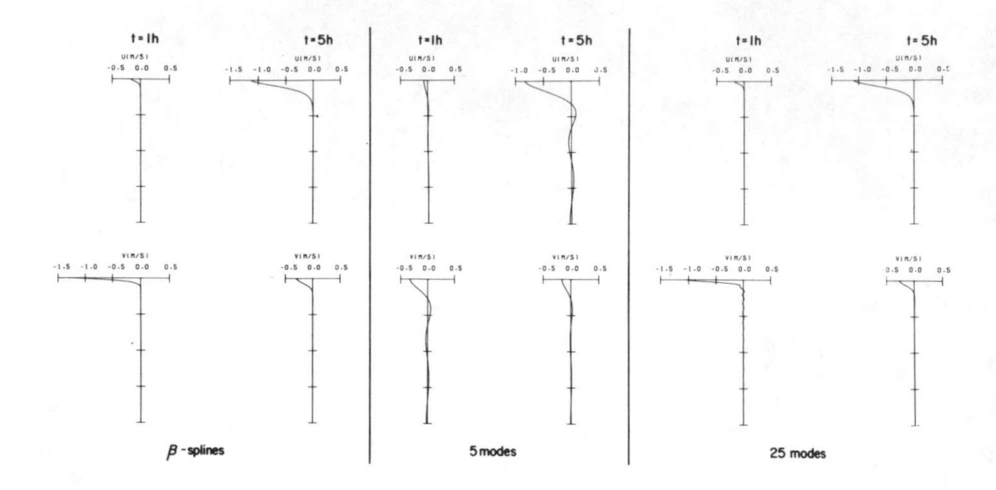

Figure 3. Profiles of the U and V components of current at times t=1 hour and 5 hours, induced by a wind impulse of 15 dynes/cm^2 and duration 3.5 hours, computed with a basis set of 14 β-splines; 5 modes and 25 modes using $\mu_o=50cm^2s^{-1}$, $\mu_T=500cm^2s^{-1}$, $h_o=222.8m$ and h=260m.

ε_r increases rapidly with mode number.

A consequence of this increase in ε_r with mode number is that although all modes are initially excited by the suddenly applied wind field, the higher modes are damped more rapidly than lower modes. This damping does, however, depend importantly on water depth. In shallow water (h of order 100m), then the contribution of the higher modes decreases rapidly, and an accurate profile can be obtained with 10 modes. In deep water (h of order 1000m), because of the h^{-2} dependence of the damping, the higher modes are not so rapidly damped, and a larger number, of order 30 modes, is required.

The term ϕ_r is also important since it acts as a weighting term, determining together with a_r, b_r the contribution of each mode to the current profile. It is evident from Table II that, in the case of $\mu=50cm^2s^{-1}$, the ϕ_r values of the higher modes are larger than those computed with $\mu=4000cm^2s^{-1}$, and consequently the higher modes will be more important in this case.

The value of depth mean eddy viscosity α is also important in determining the time variation of the modes. In the case of wind induced currents, as the wind speed increases, α might be expected to increase, and hence the rate at which the modes other than the first are damped will rise.

It is evident that the modal equations give some insight into the excitation and decay of the various modes and, hence, changes in the profile of wind induced current. The terms ϕ_r, ε_r, and depth h are particularly important in determining modal damping and the number of modes required for an accurate solution. Bottom friction also has an important role to play in coupling the modes, and details of this are given in Davies [1985a].

5. MODAL MODEL OF SHELF SEAS

Although the primary aim of this text is to present details of the formulation of a spectral model illustrated with idealized examples, the picture would not be complete without some brief examples of physically realistic calculations.

The Galerkin method has been used by a number of authors (e.g., Cooper and Pearce [1982]) to study wind induced currents in various sea areas. One example which is quite interesting, in that it used to advantage the ability of the Galerkin method to produce a continuous profile from sea surface to sea bed, is given in Davies [1982b]. In that work, turn-over times (a parameter which is important in determining the time spent by a pollutant in a given sea area) for various North Sea areas marked on Figure 4 were computed using a numerical model of the continental shelf.

Wind induced circulation on the shelf, computed using both seasonal and annual wind field, was determined. Figures 5 and 6 show the spatial distribution of surface and sea bed currents computed with the mode. In these figures, the grid point of the model is denoted by a dot, and flow is away from this point with magnitude and direction denoted by the vector. It is evident from these figures

TABLE I.

μ_W ($cm^2 s^{-1}$)	Number of Legendre Polynomials m: 10 U	10 V	20 U	20 V	45 U	45 V	Analytical U	Analytical V
100	-62	-64	-63	-63	-63	-63	-63	-63
10	-258	-104	-198	-198	-199	-199	-199	-199
1	-312	-133	-851	-504	-630	-630	-630	-630

B-spline Distribution

μ_W ($cm^2 s^{-1}$)	A m=10		B m=20		C m=30	
100	-63	-64	-63	-63	-63	-63
10	-261	-114	-214	-210	-199	-199
1	-325	-15	-626	-115	-630	-630

Table I. U and V components of surface current ($cm\ s^{-1}$) computed with an increasing number of Legendre polynomials and various spline distributions.

TABLE 2.

Viscosity	ε_1	ε_2	ε_3	ε_4	ε_5	ϕ_1	ϕ_2	ϕ_3	ϕ_4	ϕ_5
$\mu_o = 50 cm^2 s^{-1}$, $\mu_W = 500 cm^2 s^{-1}$	0.0	8.5	31.9	70.5	124.5	1.0	3.7	4.1	4.3	4.3
$\mu_o = 400 cm^2 s^{-1}$, $\mu_W = 500 cm^2 s^{-1}$	0.0	7.3	27.4	63.4	113.9	1.0	1.1	1.2	1.2	1.2

Table II. Values of ε_k, ϕ_k for the first five eigenfunctions computed with h_o=222.8, h=260m for a range of μ_o and μ_W values.

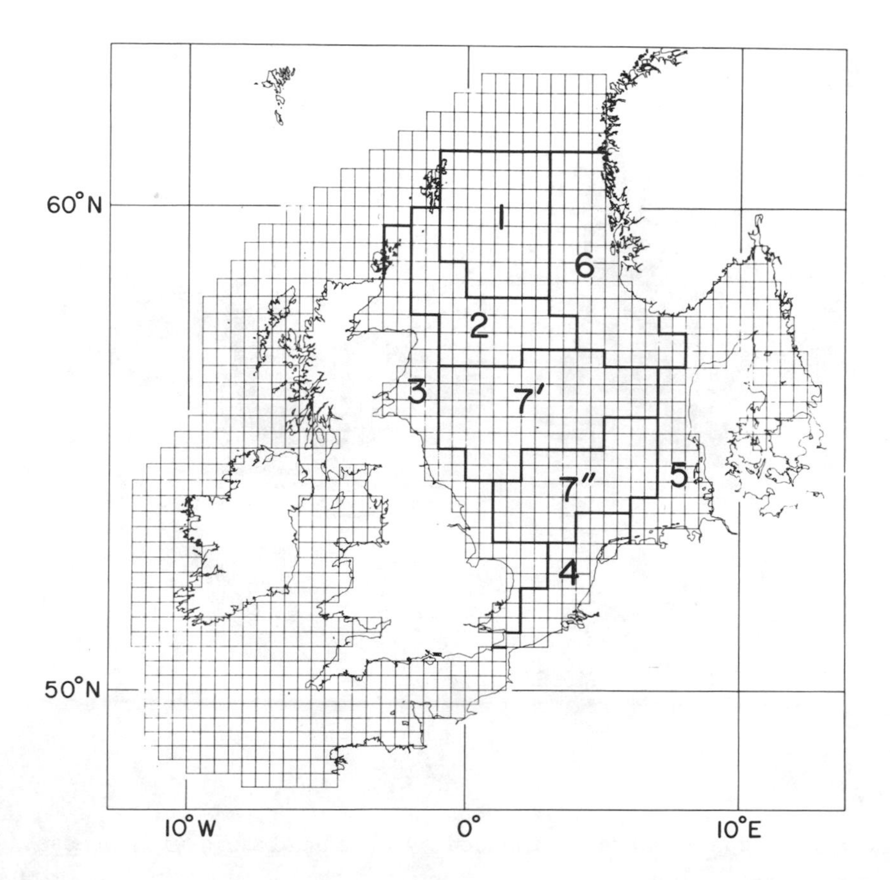

Figure 4. Finite difference grid of the model and sea areas used to compute turn-over times.

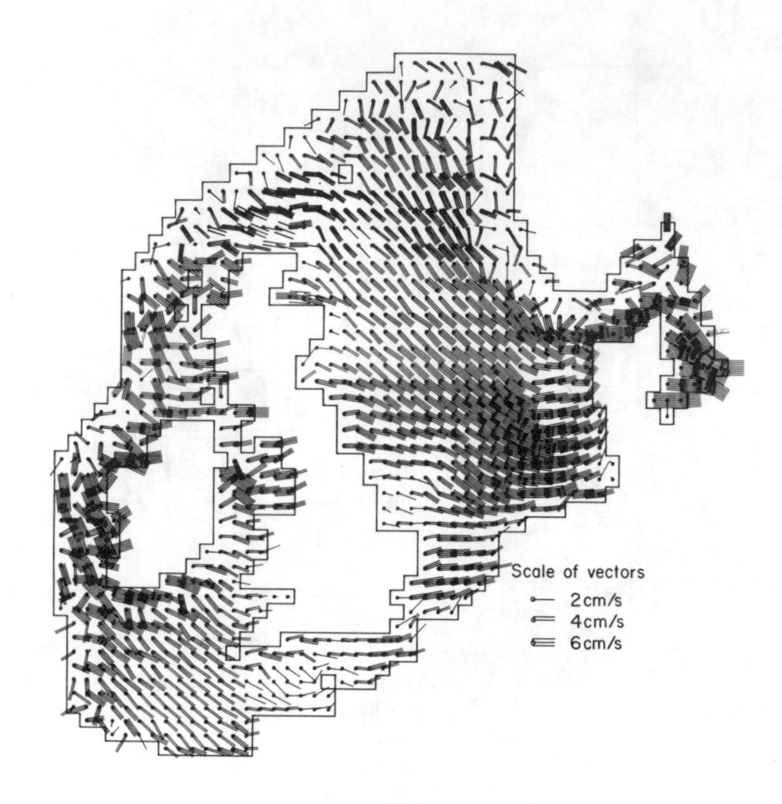

Figure 5. Surface currents induced by an annual mean wind stress.

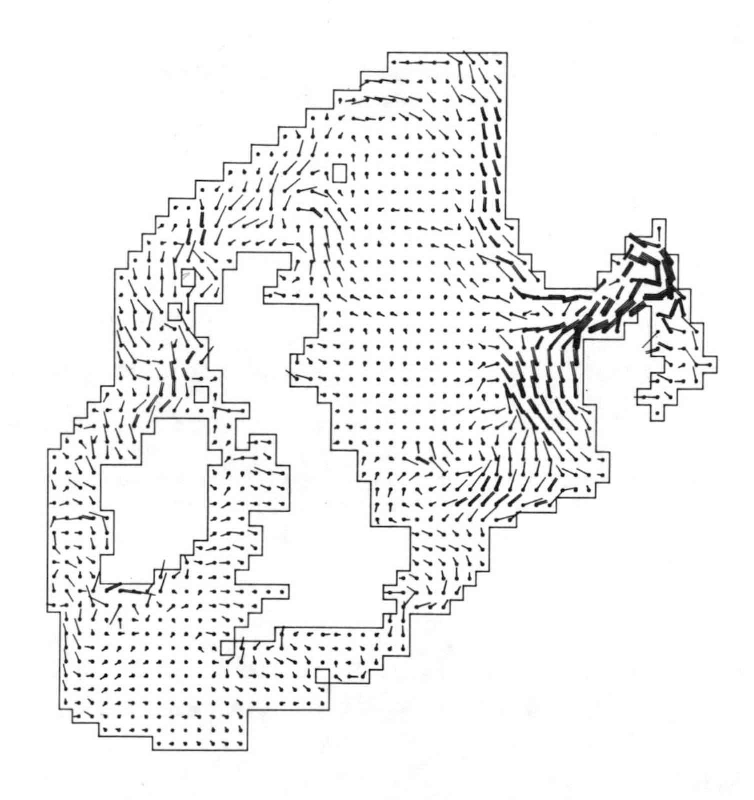

Figure 6. Sea bed currents induced by an annual mean wind stress (for scale of vectors, see Figure 5).

that, in many areas, current direction changes significantly from sea surfaces to sea bed. In order to compute the turn-over time for a given geographical area, it is important to have a continuous profile through the vertical in order to determine the depth at which the flow reverses (see Davies [1982a], for detail). For this reason, and also in order to be able to compute turn-over times in surface and sea bed layers [Davies, 1982a], the Galerkin approach has special advantages.

6. CONCLUDING SUMMARY

In this text the formulation of a three-dimensional spectral model of wind induced flow has been developed in detail. Some simple examples have been presented to illustrate the factors influencing the choice of the spectral approach. By this means it is hoped to introduce the topic to the student or established research scientist who wishes to learn about the spectral method.

Further details of the method can be found in two articles which are due to be published in the next year, [Davies 1985a,b].

Although the method has been developed here for a homogeneous sea region, it has been applied in stratified conditions (see, for example, Gordon, [1982] and Davies [1983c].

7. ACKNOWLEDGEMENTS

The author is indebted to Mr. R. A. Smith for preparing diagrams and Mrs. P. Lynch for typing the text.

8. REFERENCES

Carter, D. J. T., Prediction of wave height and period for a constant wind velocity using the JONSWAP results, Ocean Engng, 9, 17-33, 1982.

Cooper, C. K., and B. Pearce, Numerical Simulations of hurrican-generated currents, J. of Phys. Oceano., 12, 1071-1091, 1982.

Davies, A. M., Application of a Galerkin-Eigenfunction method to computing currents in homogeneous and stratified seas, in Numerical Methods for Fluid Dyn., edited by K. W. Morton and M. J. Baines, Academic Press, 1982a.

Davies, A. M., Meteorologically-induced circulation on the north west European continental shelf: from a three dimensional numerical mode, Oceanologica Acta, 5, 269-280, 1982b.

Davies, A. M., Comparison of computed and observed residual currents during JONSDAP'76, in Physical Oceanography of Coastal and Shelf Seas, edited by B. Johns, Elsevier Oceanography Series, 35, 1983a.

Davies, A. M., Formulation of a linear three-dimensional hydrodynamic sea model using a Galerkin-Eigenfunction method, Int. J. Num. Meth. in Fluids, 3, 33-60, 1983b.

Davies, A. M., Numerical modelling of stratified flow: a spectral approach, Continental Shelf Research, 2, 275-300, 1983c.

Davies, A. M., A three dimensional modal model of wind induced flow in a sea region, in press, Progress in Oceanography, 1985a.

Davies, A. M., Spectral models in Continental Shelf Sea Oceanography, in press, AGU Coastal and Estuarine Regimes Monograph Series, edited by N.S. Heaps, 1985b.

Davies, A. M., Mathematical formulation of a spectral tidal model, this volume, 1986a.

Davies, A. M., and C. V. Stephens, Comparison of the finite difference and Galerkin methods as applied to the solution of the hydrodynamic equations, Applied Mathematical Modelling, 7, 226-240, 1983.

Dobroklonskiy, S. V., Drift current in the sea with an exponentially decaying eddy viscosity coefficient, Oceanology, 9, 19-25, 1969.

Dyke, P. P. G., A simple ocean surface layer model, Rivista Italiana di Geofisica, 4, 31-34, 1977.

Gordon, R. B., Wind Driven circulation in Narragansett Bay, Ph.D. Thesis, University of Rhode Island, 1982.

Gordon, R. L., Coastal ocean current response to storm winds, J. Geophys. Res., 87, 1939-1951, 1982.

Heaps, N. S., On the numerical solution of the three-dimensional hydrodynamical equations for tides and storm surges, Mem. Soc. r. Sci Liege Ser 6, 2, 143-180, 1972.

Huang, N. E., On surface drift currents in the ocean, J. Fluid Mech., 91, 191-208, 1979.

Ichiye, T., Upper ocean boundary-layer flow determined by dye diffusion, Phys. Fluids Suppl., 10, 270-277, 1967.

Kullenberg, G. E. G., On vertical mixing and the energy transfer from the wind to the water, Tellus, 28, 159-165, 1976.

Lai, R. Y. S. and D. B. Rao, Wind drift currents in the deep sea with variable eddy viscosity, Arch. Met. Geophys. Bioklim, A, 25, 131-140, 1976.

Neumann, G. and W. J. Pierson, Principles of Physical Oceanography, Prentice Hall, 1964.

Thorpe, S. A., On the determination of K_v in the near-surface ocean from acoustic measurements of bubbles, Journal of Physical Oceanography, 14, 855-863, 1984.

Weber, J. E., Ekman currents and mixing due to surface gravity waves, Journal of Physical Oceanography, 1, 1431-1435, 1981.

OPEN BOUNDARY CONDITIONS IN NUMERICAL OCEAN MODELS

L. P. Røed
Det norske Veritas
Industry and Offshore Division
P.O. Box 300
N-1322 Høvik
Norway

and

C. K. Cooper
Conoco Inc.
Petroleum and Research Division
Ponca City, Oklahoma 74602
U.S.A.

ABSTRACT. Conditions imposed on a class of computational boundaries sometimes referred to as "open" boundaries are reviewed. A possible definition of the term "open boundary" is suggested in order to limit the discussion. Some popular and recently suggested conditions are reviewed and discussed. Emphasis is on conditions based on the Sommerfeld radiation condition including the Orlanski type and modifications thereof, but also sponges are considered. In particular the modifications suggested to handle (i) oblique incidence and (ii) wind forcing at the open boundary are considered. Some sample experiments with a barotropic ocean model are shown. These experiments serve to illustrate the high sensitivity of the interior response to the implementation of different conditions at the model's open sea boundaries.

1. INTRODUCTION

The term "limited-area integration" was first introduced by Charney et al. [1950] in their early attempt to integrate the barotropic vorticity equations numerically. The integration area had to be confined mostly due to the limited computer resources available. They were probably also the first numerical modelers who seriously had to consider what kind of conditions that should be imposed at the boundaries of the limited area. Invariably throughout the years the numerical ocean modelers have been met with the problem of limited computer resources. This usually entails some part of the

J. J. O'Brien (ed.), Advanced Physical Oceanographic Numerical Modelling, 411–436.

numerical grid ending in the open sea. Thus, in order to be able to integrate the governing equations the modeler has to invent some useful condition to be used at the model's open sea boundaries.

The area over which the integration of the governing equations of a numerical model is performed will be called the computational domain. The boundaries which limits the computational domain will be called the computational boundaries. These boundaries may further be subdivided in certain classes according to their different nature. Figure 1 gives a possible hierarchy of boundary classes.

The two main types of computational boundaries are the natural boundaries and the artificial boundaries. A natural boundary is a boundary where the fluid motion is restricted by the physics involved (e.g. impermeable coastlines) while an artificial boundary, which is commonly located in the open sea, does not restrict the fluid motion. Thus, natural boundaries are required by the existence of physical boundaries within the computational domain while the location of artificial boundaries may be arbitrarily chosen by the modeler. The condition to be used at natural boundaries are given by the physical processes active at the boundary while the condition at artificial boundaries may be chosen arbitrarily by the modeler. The prime example of a natural boundary is the closed boundary (no volume flux).

Artificial boundaries may further be divided in two classes, the open boundary (OB) and the specified boundary. At specified boundaries one (or several) of the model's dependent variables are a priori prescribed by the modeler independent of the evolution of the their interior counterparts and as such may have a significant influence on the model's interior response. In certain models it may actually dominate the interior solution. An example is the limited-area tidal model which is forced by specifying the tidal constituents at the boundary [Davies and Furnes, 1980]. Another recent example is the Harvard Open Ocean Model [e.g. Robinson and Walstad, 1985; Robinson and Leslie, 1985].

At the other extreme is the open boundary, at which conditions should be constructed so as to minimize the influence on the interior response. A useful definition of open boundaries, which will be used below, is as follows:

> An open boundary is a computational boundary at which disturbances originating in the interior of the computational domain are allowed to leave it without disturbing or deteriorating the interior solution.

Note that "deteriorating the interior solution" is included in this definition so as to rule out events like a slow but steady artificial (nonphysical) increase or decrease in the initial volume of water. Often such an event happens because the mathematical problem of solving the the underlying governing equations subject to the condition imposed at the open boundary is ill-posed.

Some boundaries utilized by modelers such as a sponge boundary or a telescoped boundary are a mixture of artificial and natural boundaries. For instance in the case of a sponge this is usually implemented by extending the integration area so as to include an area outside the computational domain (or the area of interest to the modeler). In this area some frictional parameter is increased so as to dampen out travelling disturbances. Sometimes such an area may be natural (i.e. waves hitting a sandy beach, internal waves hitting the head of a fjord, etc.) or it may be artificial and used as an open boundary just to minimize reflections, etc.. The telescoped boundaries work the same way. The integration area is telescoped (i.e extended) so as to include an area in which either the physics or the numerics becomes gradually simpler (i.e smoother topography and external forcing or larger grid spacing, etc.). Again the choice may be based on the natural physics of the extended model area or be purely artificial.

The need for artificial boundaries in general and open boundaries in particular is given not only by the limited resources of the present day computers, but also by the cost of running a high resolution model which covers an extensive area on a given computer. The modeler therefore inevitably has to end the numerical grid at artificial boundaries where the fluid motion is a priori unrestricted. Thus, an open boundary condition (OBC) has to be specified by the modeler at these boundaries. The question therefore immediately arises as to the requirements of such conditions and which criteria should be established in order to judge the performance of the chosen OBC.

The criteria for success or failure of a particular OBC must be based on some kind of objective requirements. These criteria naturally serve as guidelines for the modeler to choose a proper OBC as well as the location of the OB. Inherent in the the definition of an OB is first of all the requirement that disturbances should be allowed to leave the computational domain without disturbing or deteriorating the interior response. It should therefore be numerically stable and the accuracy should be adequate in comparison with the interior scheme. Further, it must be mathematically well posed (or at least well enough) and finally it should be suitable (i.e cost efficient) to the problem at hand. In short, an OBC should ideally satisfy the following requirements:

- Fluid motions such as propagating waves or advection should be unrestricted

- It should be numerically stable

- The accuracy should be adequate

- It should be mathematically well (enough) posed

- It should be suitable to the numerical model

The main limitations of an OBC is also inherent in the definition of the OB given above. Accordingly the time evolution of the model's dependent variables along or at the OB should evolve in harmony with the time evolution of its interior counterpart such that no reflections or disturbances are created due to the existence of the OB. Thus, an

OBC can not handle tidally forced boundaries or any specification of the model's dependent variables along or at the OB. In some models, such as tidally forced models, it is sometimes useful to allow waves hitting the OB from the interior to pass through. Such a boundary, however, constitutes an artificial boundary in which the open and specified boundary types have been blended.

The above discussion outlines briefly the need for, the requirements of and the limitations of OBCs in general. In Section 2 below follows first a review of some popular and recently suggested OBCs, together with a review of some key papers, and ends with a list of OBC candidates which a modeler is likely to consider. Section 3 provides some details on the numerical form and the implementation of a few OBCs with emphasis on the free and forced wave radiation conditions and the treatment of the oblique waves. In Section 4 the sensitivity of a simple barotropic model to different OBCs implemented at its open sea boundaries is considered, while Section 5 offers a short summary and some final remarks.

2. REVIEW OF SOME SUGGESTED OPEN BOUNDARY CONDITIONS

2.1 Types of open boundary conditions

An important feature of the OBC is that the model's dependent variables are allowed to evolve in time coherently with their interior counterparts so as not to cause any nonphysical or artificial reflections or disturbances. This may be accomplished by extrapolating the interior values of one or several of the model's dependent variables towards the OBC. Let Q denote any of a model's dependent variables, then most of the OBCs suggested in the literature may be condensed into the formula

$$Q^{n+1}{}_{b}=F(Q^{n}{}_{b},Q^{n-1}{}_{b},Q^{n+1}{}_{b-1},Q^{n}{}_{b-1},Q^{n}{}_{b-1},\ldots) \tag{1}$$

Here the subscript b entails evaluation of Q at the open boundary. Similarly the subscript b-1 entails evaluation of Q at the point next to the boundary. The superscripts n and n+1 entail evaluation of Q at time $n\Delta t$ and $(n+1)\Delta t$, respectively. The function F in (1) establishes a time-space extrapolation using the already computed values of the dependent variable either at the boundary itself (at previous time steps) or values at the grid points adjacent to the OB.

The OBCs reported to be successful in the literature may be divided in three main types:

- simple extrapolation formulas,
- radiation conditions based on the Sommerfeld condition, and
- sponge filters combined with a radiation condition.

Note, as will be alluded to below, that widely used conditions such as the clamped condition (e.g. zero surface elevation) and the gradient condition (e.g. zero normal gradient of the sea surface elevation) may be regarded as radiation conditions [see Chapman, 1985].

Several types of radiation conditions and one sponge filter will be considered in some detail below. The discussion will also include such popular conditions as the clamped condition, the gradient condition as well as the gravity wave radiation condition. Another type of OB which was alluded to above is the telescoped boundary. This latter type of boundary condition as well as ordinary simple extrapolation schemes [Sundström and Elvius, 1979; Miller and Thorpe, 1981] will not be considered below.

2.2 Review of certain key papers

For historical reasons the reader is urged to start by reading the paper by Charney et al. [1950]. Their computational boundary was artificial and partly specified and partly open in that they computed the vorticity at outflow boundaries by a linear extrapolation formula and specified the vorticity at inflow boundaries. Unfortunately, the extrapolation scheme they used caused the scheme to become unstable as shown by Platzman [1954] which serves to demonstrate the necessity of the stability requirement above.

Most of the OBCs suggested in the literature are based on or are variations of the Sommerfeld radiation condition [Sommerfeld, 1949]. This condition states that if Q is any of the models dependent variables then

$$Q_t+c_Q Q_n=0 \text{ at the OB.} \tag{2}$$

Here c_Q is the propagation speed of disturbances, t, time and n the coordinate normal to the OB. Subscripts t and n entail differentiation with respect to subscript. A widely used numerical implementation of this condition was presented by Orlanski [1976] and shown to be successful in the case of a collapsing bubble. Modifications and variations of the Sommerfeld radiation condition have been presented by various scientists. Camerlengo and O'Brien [1980] suggested a simplification of Orlanski's implementation and showed an improvement for outgoing Rossby waves. Miller and Thorpe [1981] essentially used Orlanski's implementation but tested it against several other implementations as well as more direct extrapolation formulas of higher order accuracy. They showed that the original scheme suggested by Orlanski [1976], which involved a leapfrog in time, could be made more accurate by use of the upstream method. Enquist and Majda [1977] noted that the Sommerfeld condition was valid only for waves with a normal incident and suggested a hierarchy of local boundary conditions which, in addition to guarantee stability, also minimized artificial (nonphysical) reflections due to the obliqueness of the incident disturbance or wave. This problem was also noted by Raymond and Kuo [1984] who suggested using the Sommerfeld equation in vector form in which the propagation velocity was projected in each coordinate direction. They also showed that this approach gave a significant improvement over the traditional Sommerfeld radiation condition. Røed and Smedstad [1984] noted that the Sommerfeld radiation condition was only valid for unforced waves and therefore suggested a modification of the Sommerfeld condition in which the forced and free wave part was treated separately. Israeli and Orszag [1981] showed that the combination of radiation conditions with sponge filters (i.e. viscous damping) offered a significant improvement over traditional radiation conditions.

Most of the papers reviewed above considered OBCs within a rotational framework. However several other studies, in which the problem of constructing useful OBCs in non-rotational cases have been considered, may also offer useful and viable approaches in the rotational cases. Some of these papers, for instance the paper by Wurtele et al. [1971] who suggested an open boundary condition for the storm surge equations (neglecting rotation), have had a large impact on the development of OBCs in general. Also notable, although not easily available, is the paper by Mungall and Reid [1978], who extended the condition of Wurtele et al. [1971] in order to minimize reflections due to obliquely incident waves hitting the OB.

Finally, the recent paper by Chapman [1985] should be noted. Here the responses of a barotropic coastal model to wind and pressure forcing are compared for eleven different OBCs (mostly radiation conditions with the exception of one sponge filter) at the models cross-shelf open boundaries. The reference list also provides references to several papers not reviewed above.

2.3 Possible candidates for the modeler to consider

As mentioned above, most of the successful OBCs used in rotational problems are derived from the Sommerfeld radiation condition (2). The more popular and recently suggested ones differ mostly in the way the propagation speed c_Q is specified. Table 1 offers a list of some common OBCs with their reference and comprise a set of possible candidates which most modelers are likely to consider. Most of these conditions was included in the intercomparison study by Chapman (1985). However, he left out important OBCs constructed to handle oblique waves and wind forcing at the boundary. The actual choice of OBC is somewhat determined by the problem at hand. Both the clamped and the gradient conditions give severe artificial reflection of incident waves but nevertheless may be very efficient in short term integrations due to their simplicity. The various radiation conditions with finite c_Q offer substantially less reflections, but differ in accuracy and their ability to handle forcing at the OB and are generally less cost efficient.

2.4 Brief discussion of stability, accuracy, well-posedness and suitability

As alluded to above, the condition imposed at the OB must satisfy certain mathematical constraints to guarantee a unique solution of the underlying differential equations used in the numerical model and in

Table I: Some possible candidates of OBCs

OBC type	c_Q	Reference
A. Free wave radiation condition		
Clamped	0	Various
Gradient	∞	Various
Gravity wave	$(gh)^{1/2}$	Pearson (1974)
Orlanski	computed from interior values	Orlanski (1976)
Modified Orlanski	- " -	Miller and Thorpe (1981)
Non-normal Orlanski	computed and projected	Raymond and Kuo (1984)
B. Forced wave radiation condition		
Damped gravity wave	$(gh)^{1/2}$	Blumberg and Kantha (1984)
Separated Orlanski	computed from interior values	Røed and Smedstad (1984)
C. Other conditions		
Sponge/radiation	- " -	Israeli and Orszag (1981)

order for the problem to be well-posed. Daubert and Graffe [1967], Kreiss [1970] and Verboom et al. [1983] discuss these issues for simple flow cases. A strict mathematical proof demonstrating that a given OBC together with the governing equations forms a properly posed problem is impossible for most practical problems. Consequently, modelers have generally relied on previous experience or comparisons to analytic solutions for simple test cases. The paper by Enquist and Majda [1977] provides an example of the problems involved.

The problem of numerical stability was already encountered by Charney et al. [1950] and is fairly common when implementing OBCs by using the straightforward direct, linear extrapolation formulas. A method of comparing the accuracy of the OBC is to carry out a Taylor expansion to evaluate the leading terms of the truncation error. This kind of analysis does not include the interaction of the errors inherent in the the coupling of the interior scheme with the boundary scheme. Sküllermo [1979] considered this more complicated problem for simple interior equations and concluded that the order of the truncation error of the boundary scheme should be of the same order or less than that of the interior scheme in order for the combined error to be small. Thus, small errors in the boundary condition scheme is a necessary but not sufficient requirement for the numerical solution to be an accurate representation of the true solution.

3. DETAILS OF SOME POPULAR AND PROMISING OBCS

3.1 Free wave radiation conditions

The free wave radiation conditions are all based on the Sommerfeld radiation condition as expressed by (2). An alternate form of the Sommerfeld radiation condition used for shallow water equations are

$$U=c\eta \quad ; \quad c=(gH)^{1/2} \tag{3}$$

where U is the volume flux component across the boundary, H the equilibrium depth (constant) and η the sea surface elevation. This expression is derived by writing the Sommerfeld condition for both U and η with $c_U=c_\eta=c$, then using the continuity equation to derive (3). It is important to realize that (2) and (3) only provide a perfect OBC for the problem of free waves in a one-dimensional, nonrotational channel [Foreman, 1985]. In problems involving rotation, external forcing (e.g. bottom stress or wind stress), topography changes or advection and diffusion the Sommerfeld condition does not apply in the physical sense and hence reduces to an intelligent extrapolation formula of the form (1). A practical and realistic implementation of the Sommerfeld condition has been suggested by Orlanski [1976]. He approximated (2) by a leapfrog scheme to find the new value of the dependent variable in question. The phase speed close to the boundary was evaluated by means of the already computed values of the dependent variable at previous time steps close to the boundary. Orlanski's procedure is easier to understand by implementing an upstream differencing of (2) [e.g. Miller and Thorpe, 1981]. In the time step Δt, a disturbance travelling with speed c_Q covers the distance $c_Q \Delta t$. Let $r = c_Q \Delta t / \Delta n$ denote the ratio of this distance and the grid space length Δn normal to the OB. Then an upstream differencing of (2) gives

$$Q^{n+1}{}_b = (1-r) Q^n{}_b + r Q^n{}_{b-1}, \qquad (4)$$

which demonstrates its basic character as an extrapolation formula. The essence of the numerical implementation of Orlanski [1976] is that r is numerically evaluated by rewriting (4) with r as the subject. For a non-trivial OBC this requires reference to interior grid point values at preceding times, e.g.

$$r = (Q^n{}_{b-1} - Q^{n-1}{}_{b-1}) / (Q^{n-1}{}_{b-2} - Q^{n-1}{}_{b-1}). \qquad (5)$$

Orlanski also required that $0<r<1$. If $r<0$ (4) is replaced by the simple formula $Q^{n+1}{}_b = Q^n{}_b$. It is important to realize that when (5) is used to substitute for r in (4), the latter becomes nonlinear.

Various modifications of the formula (4) have been suggested. For instance Camerlengo and O'Brien [1980] suggested using $r=1$ in (4) if $r>0$ in (5). In an equatorial test model their new implementation showed no reflection at the outflow of Kelvin waves through the OB, while some reflection at the outflow of Rossby waves was observed. Røed and Smedstad [1984] made a comparison of the Orlanski versus the Camerlengo and O'Brien approach for a flow in a rotational channel and concluded that the Camerlengo and O'Brien implementation handled dispersive waves better than the Orlanski implementation. As alluded to above, the Orlanski implementation is nonlinear while the Camerlengo and O'Brien implementation is linear. Thus, the truncation error of the former is $O(\Delta t^2)$ and $O(\Delta x^2)$ while the latter has truncation errors of $O(\Delta t)$ and $O(\Delta x)$. The latter is, however, more cost efficient.

In the study by Røed and Smedstad [1984] the upstream version (4) of the OBC (2) was utilized. However, both Orlanski [1976] and Camerlengo and O'Brien [1980] used a centered time-difference replacement for (2), in which the boundary point $Q^n{}_b$ was replaced by a time average to maintain stability in the presence of physical and computational modes. The main point is that the scheme used for (2) should be suited to the scheme used for the interior and the problem at hand. When comparing the Orlanski and the Camerlengo and O'Brien implementation, Røed and Smedstad [1984] used upstream differencing mainly because they used an explicit time-differencing scheme for the interior integration of their barotropic model.

3.2 Forced wave radiation condition

As noted by Chapman [1985] OBCs continue to be developed in order to handle different physical situations. Although the Sommerfeld radiation condition (2) appears to be successful in a number of cases, it is important to realize that it is basically just another extrapolation formula, which in certain cases satisfies the requirements of an OBC. As mentioned above (2) is in strict physical and mathematical terms only valid without forcing at the boundary. Foreman [1985] notes that when bottom friction is added to the one-dimensional shallow water equations, the reflection of long waves is severe, which significantly influences the accuracy and the stability of the radiation condition (2). This is to be expected since (2) requires that the Q_t and Q_x should be in phase, which is not true when bottom friction is present. This was also noted by Røed and Smedstad [1984], who argued that in the case of wind forcing at the boundary, the right hand side of (2) should be non-zero. In order to still use the radiation condition concept as outlined by Orlanski [1976] they suggested separating the solution in two parts close to the OB. One was termed the local part and was constructed from those terms in the governing equation which did not require an OBC. In a barotropic model this part is made up of the local Ekman transport and the geostrophic flow across the OB. The remaining part was termed the global part and was thought to be the carrier of information about interior events towards the OB. Since the latter part required an OBC to be imposed, the Orlanski implementation with the Camerlengo and O'Brien modification was applied on this part only. The former part did not require any OBC and was computed directly from their governing equations (i.e. a subset of the governing equation for the combined solution). They tested their OBC for a problem involving the integration of the two-dimensional shallow water equations. The fluid which was confined between two infinitely long straight and parallel solid walls was set to motion by applying a wind stress in the lower half portion of the channel for 8 hours. The governing equations are

$$\eta_t=-U_x-V_y, \tag{6}$$

$$U_t=fV-g\eta_x+\tau^x/\rho, \tag{7}$$

$$V_t=-fU-g\eta_y+\tau^y/\rho. \tag{8}$$

Here U,V are the horizontal components of the volume flux based on the depth integrated current, η the sea surface elevation and τ^x,τ^y the components of the wind stress acting on the sea surface, g the gravitational acceleration, and ρ the density of sea water. In the case in which the OB is along x=constant, the terms containing differentiation with respect to y mathematically speaking are those that require an OBC to be specified. In the governing equation for the local part these terms are, therefore, neglected to give

$$\eta^l{}_t=-U^l{}_x, \tag{9}$$

$$U^l{}_t=fV^l-g\eta^l{}_x+\tau^x/\rho, \tag{10}$$

$$V^l{}_t=-fU^l+\tau^y/\rho. \tag{11}$$

Here the superscript l denotes the local part of the dependent variable in question. The global part is found simply by subtracting out the local parts in (6), (7) and (8) by use of (9), (10) and (11). In the implementation of this OBC it is important to note that the governing equations for the global part is never integrated. In addition to integrate the governing equations (6), (7) and (8) of the combined solution for the interior the governing equations (9), (10) and (11) for the local part are integrated at those points close to the OB that is required in order to compute the global part (done by subtraction). The expressions (4) and (5) with $Q=U^g$ are then used to find the global part U^g of the volume flux at the OB. The local and global parts are then added to give the the new value of U at the OB. The results of Røed and Smedstad [1984] clearly demonstrates the success of this approach (look at Figure 2). Moreover, when the wind is shut off the spin-down of the flow, based on potential vorticity conservation arguments, seems realistic, demonstrating that the approach is valid also for the unforced problem. As noted by Røed and Smedstad [1984], this is to be expected, since when the wind ceases the implementation theoretically reduces to the Camerlengo and O'Brien [1980] implementation.

Blumberg and Kantha [1984] have suggested another modified form of (2) to account for the external forcing at the OB., viz.,

$$\eta_t + c\eta_x = -\frac{\eta - \eta_k}{T_f} \tag{12}$$

Here η_k is a known value which can contain tide, mean alongshore pressure gradients, etc., and T_f is a characteristic time scale.

3.3 The treatment of the oblique incident wave in radiation conditions

It is generally impossible to locate the OB so as to insure that all waves are normally incident. The consequence of this can be substantial. Enquist and Majda [1977] calculated the energy reflected by (2) to be 17% for a wave hitting the OB at an angle of 45°. This percentage rapidly increases as the wave becomes more parallel to the OB. To remedy the reflection generated by these waves Enquist and Majda [1977] developed a general form of a perfectly absorbing boundary condition

$$Q_x - (Q\sqrt{\omega^2 - l^2})_t = 0. \tag{13}$$

Here the OB is at x=0, l is the wave number in the y-direction, and ω is the wave frequency. They noted that to solve (12) is impractical because it requires a knowledge of the variables at all points on the OB at all times in order to advance the solution at one point on the OB. However, they presented three approximations of (12). The first order approximation is found by assuming normal incidence (l=0), which yields (2) (dimensionalized so that c=1). The second order approximation is found by using a Taylor or Padé expansion for the

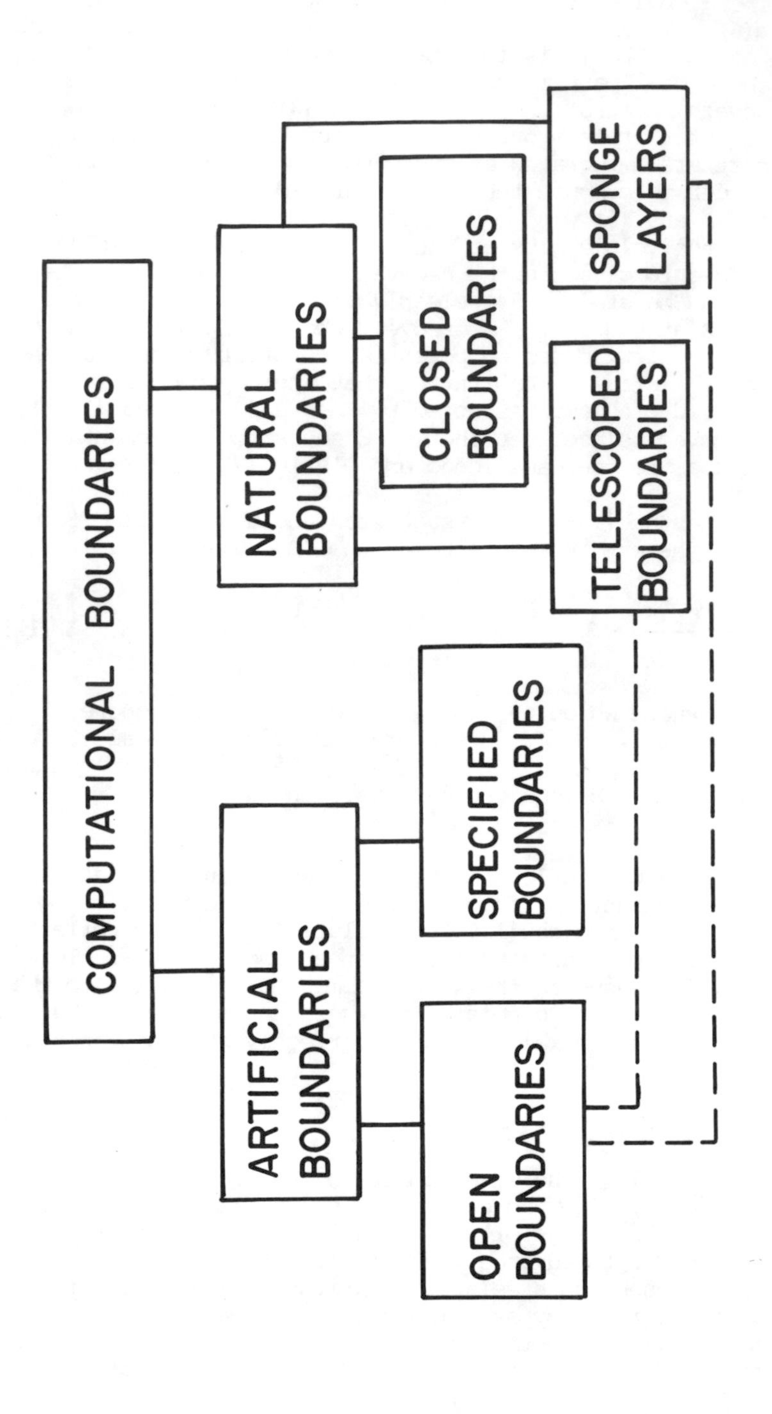

Figure 1: Hierarchy of computational boundaries. Note that sponge and telescoped boundaries may be both natural and artificial or a mixture of the two. For definition of boundary types see Section 1: Introduction.

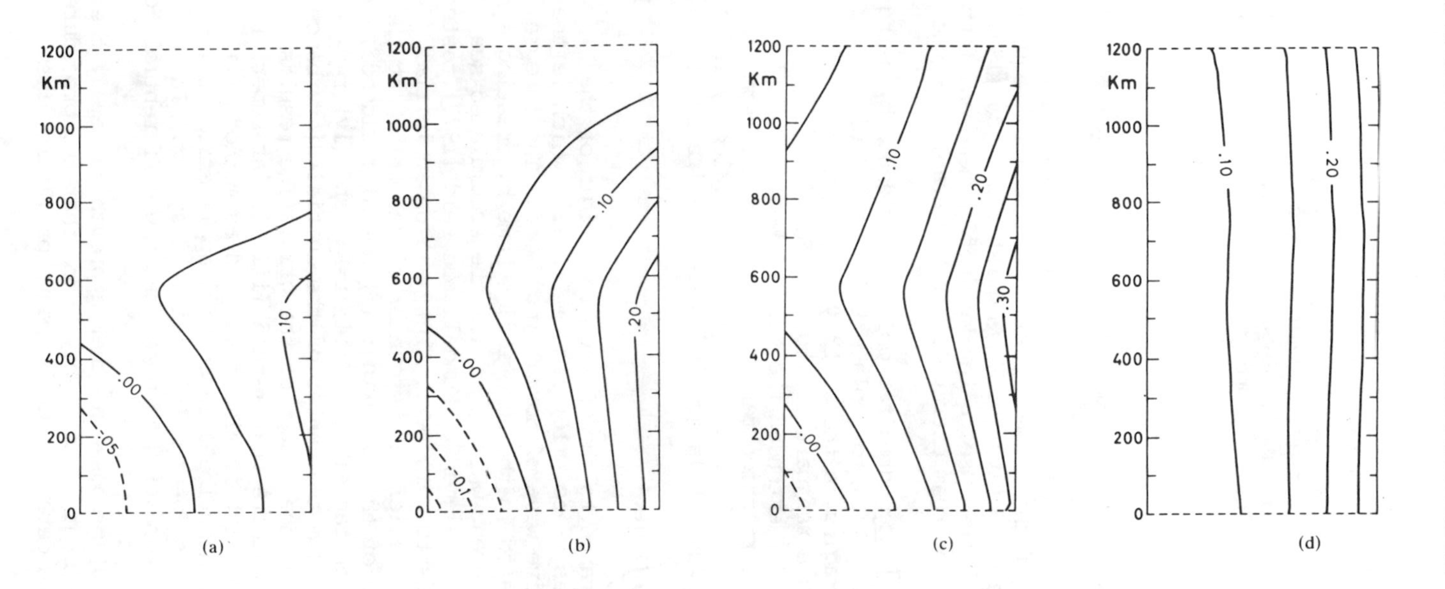

Figure 2: (From Røed and Smedstad, 1984) Solid (positive) and dashed (negative) curves give the deviations of the sea surface elevation away from its equilibrium position in meters. The open boundaries are shown by straight dashed lines, whereas the parallel channel walls are shown by solid straight lines. Solutions are shown after (a) t=3 hours, (b) t=5 hours, (c) t=8 hours and (d) t=16 hours. The upper open boundary is forced for 8 hours. The upper open boundary is unforced throughout the experiment. The forcing is confined to the lower half portion of the channel.

square root term to give

$$Q_{xy}-Q_{tt}+(1/2)Q_{yy}=0. \tag{14}$$

The third order approximation differs depending on whether a Taylor expansion or a Padé expansion is used. Enquist and Majda (1977) show that the Taylor expansion yields a strongly ill-posed problem while the Padé is properly posed and is written

$$Q_{xtt}-Q_{ttt}-(1/4)Q_{xyy}+(3/4)Q_{tyy}=0. \tag{15}$$

The improvements gained by using higher order theory is substantial. For the case considered by Enquist and Majda [1977] of a wave approaching at 45°, (14) reflects 3% of the incident wave while (15) reflects only 0.5%. Use of the higher order approximations requires the estimate of higher order derivatives. Enquist and Majda [1977] performed studies in which they used a Lax-Wendroff scheme and found that a relatively simple upstream extrapolation was stable for the cases they considered.

Wurtele et al. [1971] accounted for oblique incidence by using the local water speed, i.e. $\sqrt{U^2+V^2}$, to replace U in (3). Their limited tests were encouraging, although they did encounter instabilities when the wave ran parallel to the OB.

Mungall and Reid [1978] extended the work of Wurtele et al. [1971] by adding a higher order correction term to the (3). The term is developed from a relationship between U_r, the radial velocity and η for a radially spreading wave at large distances from its origin, viz.,

$$\eta=(U_r/c)-(c/2r)\int_0^t \eta\,dt. \tag{16}$$

Here r is the radial distance to the OB from the origin of the disturbance. Note that as r->oo, (16) reduces to (3), which is the case of a plane progressive wave of normal incidence. The integral in (16) can easily be evaluated in the numerical scheme by creating a new variable on a boundary element which contains the accumulated sea surface elevation, η at the element. Mungall and Reid [1978] tested (16) using several methods to compute c. They set up an initial disturbance with circular relief in the center of the grid. At the boundary they assumed one of the following: (i) normal incidence, (ii) an incidence angle based on $\tan^{-1}(V/U)$, (iii) $r=R_d$, and (iv) $r=R_c$ where R_d is the distance from the center of the initial disturbance and R_c is the distance from the center of the grid. The results of Mungall and Reid [1978] indicate that method (iii) is far superior to (i) and (ii), while method (iv) is almost as good as method (iii). Hebenstreit et al. [1980] describe a successful application of (16) to the study of tsunami response of a multiple-island system. In reference to (12) it is interesting to note that the differential form of (16) is (12) with $T_f=2r/c$ and $\eta_k=0$.

Raymond and Kuo [1984] proposed another approach to remedy the reflection of oblique incident waves. They noted that the procedure used by Orlanski [1976] to determine c_Q by means of (2) involves evaluating Q_t/Q_n just one point interior to a specified boundary point using values of Q previously calculated (cf. expressions (4) and (5)). These Q normally represent multi-dimensional flow, and even though (2) is valid for multi-dimensional flow, they noted that the determination of the apparent phase velocity in the direction normal to the OB by (2) depends on the spatial variation of Q in only one direction, i.e. normal to the OB. Consequently c_Q can range between plus and minus infinity as the spatial derivative varies over the cycle of a simple two-dimensional wave. Therefore in the two-dimensional case, they suggested to use two individual radiation conditions, viz.,

$$Q_t+C_xQ_x=0 \tag{17}$$

$$Q_t+C_yQ_y=0 \tag{18}$$

where the coefficients C_x and C_y represent the apparent phase velocities in the x,y coordinate directions measured relative to the mean currents in these directions. Note that C_x and C_y are not projections of the vector phase velocity (c_x,c_y) since the phase velocity does not satisfy the rules of vector decomposition. Thus, by adding (17) and (18) it follows that

$$Q_t=-\alpha_xC_xQ_x-\alpha_yC_yQ_y=-c_xQ_x-c_yQ_y \tag{19}$$

where α_x and α_y are weighting factors so that $\alpha_x+\alpha_y=1$ and hence given by

$$\alpha_x=(Q_x)^2/((Q_x)^2+(Q_y)^2), \tag{20}$$

$$\alpha_y=(Q_y)^2/((Q_x)^2+(Q_y)^2). \tag{21}$$

By use of (19) it then follows that

$$c_x=-Q_tQ_x/((Q_x)^2+(Q_y)^2) \tag{22}$$

$$c_y=-Q_tQ_y/((Q_x)^2+(Q_y)^2) \tag{23}$$

which leads to two expressions similar to (5) projected in each coordinate direction. Q may then be updated in a fashion similar to (4) by use of (19).

4. SENSITIVITY OF A NUMERICAL MODEL TO DIFFERENT OBCS

In order to show how sensitive the interior response of a numerical model even far away from the models OB's can be, some of the OBCs discussed above have been implemented at the OB's of a simple barotropic model [Martinsen et al., 1979]. The governing equations of the model are the linear shallow water equations including a linear bottom stress i.e. (6), (7) and (8) with the addition of the bottom stress terms. The equations are approximated by a finite difference scheme, where a staggered grid is used for the space discretization. The grid corresponds to lattice C of Mesinger and Arakawa [1976, p. 47]. The time differencing is accomplished by a simple forward-backward scheme usually credited to Sielecki [1968]. More details about the model may be found in Martinsen et al. [1979]. The model has been used in the past for storm surge prediction in Norwegian waters in which a clamped OBC was imposed at all the model's open boundaries. The sensitivity study described below is similar to the study by Chapman [1985], but adds the forced wave radiation condition by Røed and Smedstad [1984]. Moreover, it constitutes an independent study and may therefore offer support (or vice versa) to some of Chapman's conclusions.

4.1 The numerical experiments

The integration area is limited to a rectangle of dimensions 1000 km by 500 km (look at Figure 3). The computational boundaries are partly open and partly natural (closed). The closed boundary or shoreline runs parallel to one of the longer sides. The offshore boundary running parallel to the coast is clamped in all cases. At the OB's perpendicular to the coast five different OBCs have been implemented as follows (i) clamped, (ii) gradient, (iii) free wave radiation [Camerlengo and O'Brien, 1980], (iv) coupled sponge and free wave radiation [Israeli and Orszag, 1981] and (v) forced wave radiation [Røed and Smedstad, 1984]. The wind-forced response of the model is considered. This is done by specifying the stress at the sea surface. The stress is parallel to the coast (with the coast to the right looking downstream, northern hemisphere) and decays exponentially offshore with an e-folding distance of about 200 km. Thus its value at the outer offshore boundary is about 8 % of its value at the coast where the maximum stress (=0.1 N/m^2) occurs. This wind forcing is simple enough to permit analytic solution of the governing equations in the case of a flat bottom and an infinitely long straight coast as considered here. The staggered grid has a grid spacing of 20 km between adjacent elevation points and 10 km between an elevation point and a velocity point. The time step is 90 seconds, which is well below the CFL constraint. The experimental set-up will be identical for the five different OBC cases to allow for a direct intercomparison of the performance of each OBC. The wind is impulsively imposed at time equal zero and kept constant during the entire 96 hour simulation.

4.2 Numerical form of the OBCs

Let the coordinate system be defined such that y=0 coincides with the coastline and such that the cross-shore OB's are located at x=0 and x=1000 km, respectively. Then for all cases

$$V=0 \text{ at } y=0, \text{ and } \eta=0 \text{ at } y=500 \text{ km}. \tag{24}$$

At the cross-shore OB's the clamped and gradient conditions will be applied to the sea surface elevation while the free wave and forced wave radiation conditions will be applied to U. The sponge area

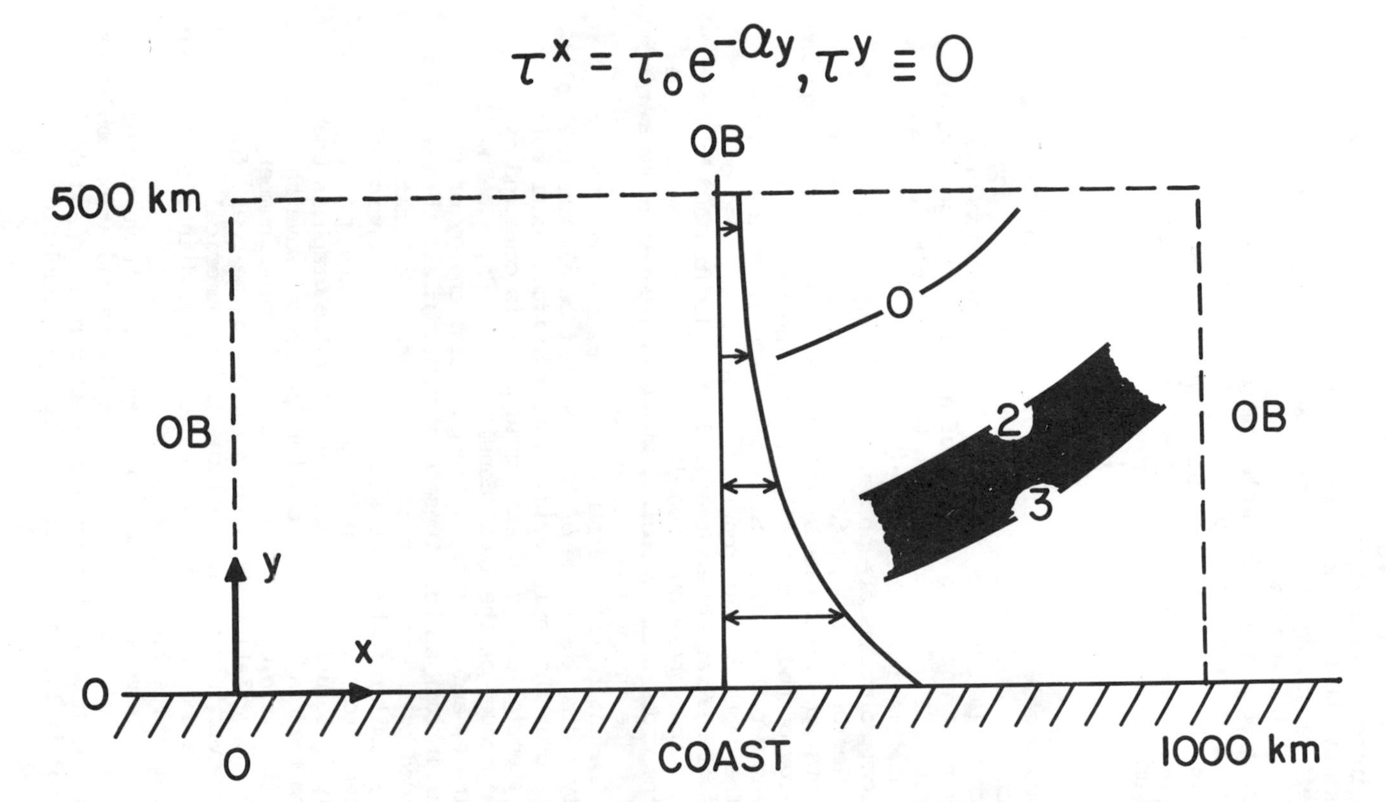

Figure 3: The integration area of the barotropic model for the sensitivity experiments of Section 4. Dashed lines indicate open boundaries. The coast is shown by a solid straight line. Also the forcing in terms of wind stress is shown. Note that the stress has its maximum at the coast and decays exponentially offshore to about 8% at the open boundary offshore. The stress direction is along the coast with no variation along the coast. The dimensions are 1000 km by 500 km. In the sponge case, this area is extended to include an area 200 km wide on each side along the coast in which the bottom friction coefficient increases exponentially to four times its interior value.

extends the integration area to include two frictional layers of depth 200 km on either side of the area of interest (1000 km by 500 km) so that the total integration area including the frictional sponge layers measures 1400 km by 500 km. The frictional parameter is the bottom stress drag coefficient, which increases exponentially to 4 times its interior value within the sponge.

The different OBCs have been implemented as follows. The clamped condition is given the numerical form $\eta^{n+1}_b=0$. Similarly the numerical form of the gradient condition is $\eta^{n+1}_b=\eta^{n+1}_{b-1}$. The numerical form of the free wave and forced wave radiation conditions are as in Røed and Smedstad [1984].

4.3 Discussion

By analytic means a solution is obtained that reveals no alongshore gradients, which is to be expected since the wind forcing does not promote any such variation. The response is a combination of a wind-forced Ekman current and a buildup of an alongshore geostrophic current which due to the bottom friction terms approaches a steady state solution as time approaches infinity.

The numerical solutions are presented in Figures 4 through 9 in form of time series of the sea surface elevation, the depth mean current components, the excess mass, the available potential energy and the kinetic energy of the depth mean current. The time series of the model's dependent variables (i.e. the sea surface elevation and the components of the depth mean current) are constructed based on the computed values of the variables at a grid point located 500 km away from each of the cross-shore open boundaries and 10 km away from the coast (except the cross-shore velocity, which is located 20 km away from the coast). The integral quantities are based on a numerical summation over the entire integration area. A measure of the excess mass is found by adding the value of the sea surface elevation at all elevation points over the entire grid. The individual contribution to the available potential energy at each grid point is computed by integrating the product of the local density with gzdz, z being the vertical coordinate, and subtracting the potential energy of the initial (or equilibrium) state. These contributions are then added and the result divided by the total integration area. The kinetic energy is computed in a similar fashion by adding the kinetic energy of the depth mean current from each grid point over the entire grid.

The analytic solution is also shown, but the graph is unable to separate it from the response depicted for the forced wave radiation condition case (i.e. the curves marked with the garbled number). As mentioned above, no variation should develop along the coast for a perfect OBC. However this may not be true in the numerical integrations, since the OBC chosen may cause such gradients to build up due to nonphysical reflections or advection.

In general the response is seen to be quite sensitive to the specified open boundary condition. The forced wave radiation condition (see the curves with a garbled number) suggested by Røed and Smedstad [1984] appears to simulate a perfect open boundary condition

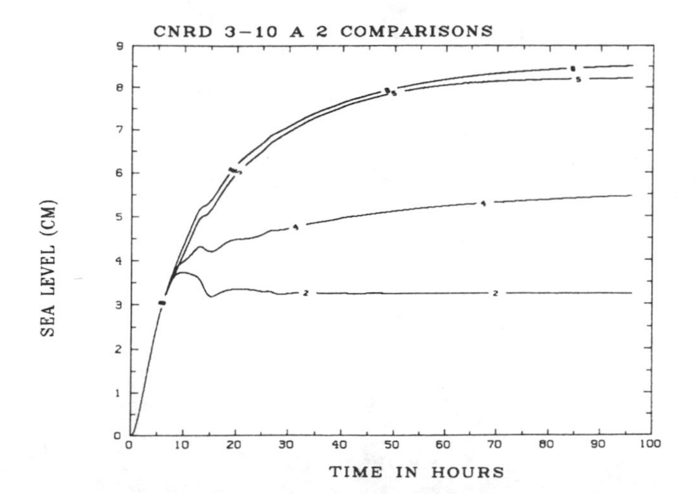

Figure 4: Solid curves depict time series of the sea surface elevation 10 km away from the coast and 500 km away from the open boundaries. Time in hours is shown along the horizontal axis and elevation is indicated in cm along the vertical axis. The numbers on the curves indicate the open boundary condition used so that 2=clamped, 3=gradient, 4=sponge combined with free wave radiation, 5=free wave radiation and 6=forced wave radiation. Also shown is the analytic solution which falls onto the solution for the forced wave radiation case and causes the number on this curve to be garbled.

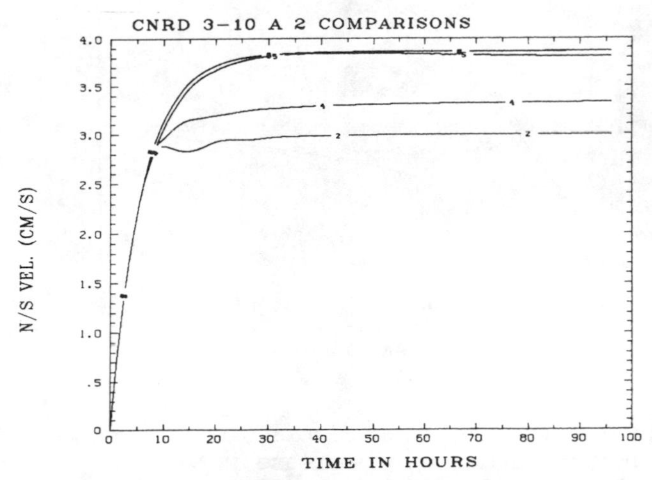

Figure 5: Same as Figure 4 but displays the alongshore (N/S) depth mean current component. Positive numbers indicate a current directed along positive x-axis (see Figure 3). Numbers along the vertical axis indicate speed in cm/s.

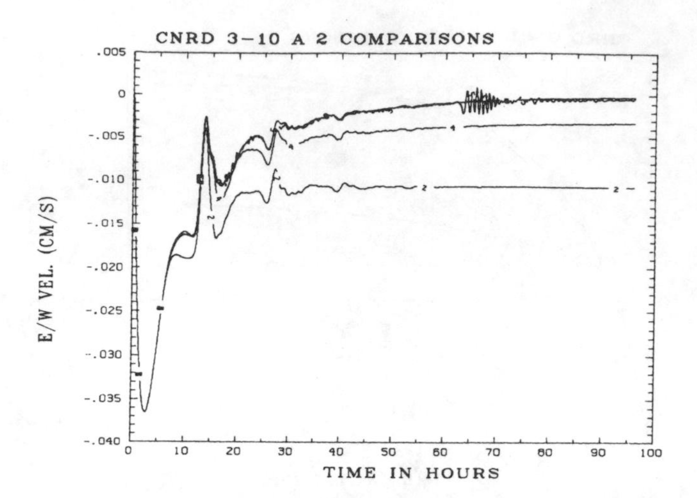

Figure 6: Same as Figure 4 but displays the cross-shore (E/W) depth mean current component. Positive numbers indicate a current directed away from the coast (i.e. along positive y-axis). Numbers along the vertical axis indicate speed in cm/s.

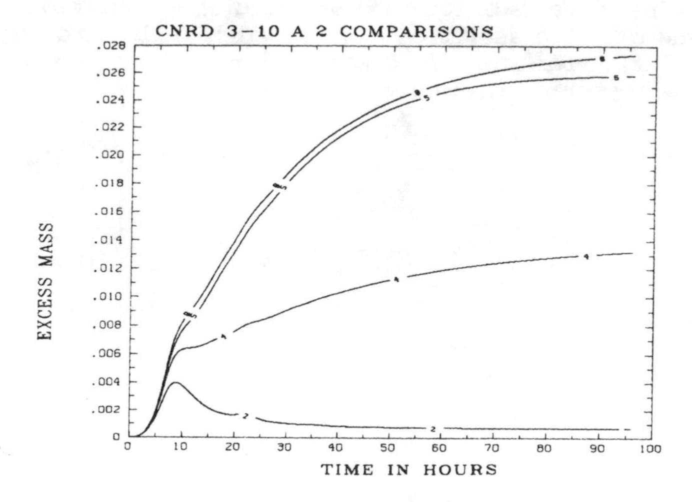

Figure 7: Same as Figure 4 but displays the excess mass. Positive numbers indicate an increase in volume. Numbers along the vertical axis indicate excess mass divided by the total integration area times density of sea water and is given in m. It is a measure of the average raise or depression of the sea surface over the entire integration area.

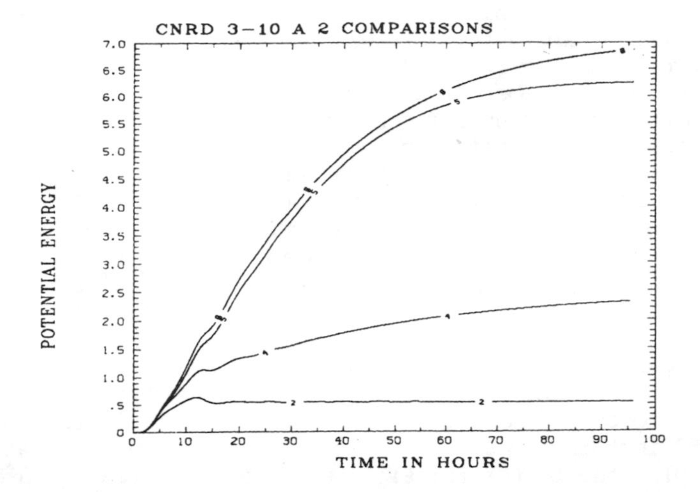

Figure 8: Same as Figure 4 but displays the available potential energy. Numbers along the vertical axis indicate available potential energy divided by the total integration area and is given in J/m^2.

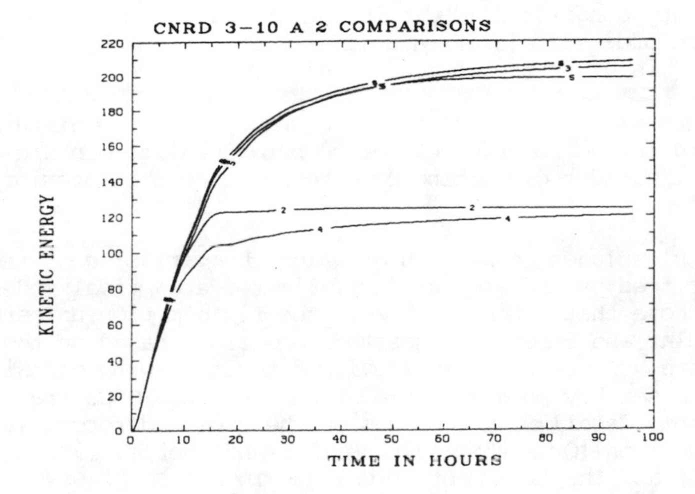

Figure 9: Same as Figure 4 but displays the kinetic energy of the depth mean current. Note that a perfect vertical recirculating current has zero kinetic energy. Numbers along the vertical axis indicate kinetic energy divided by the grid cell area (Δs^2) times equilibrium depth and is given in J/m^3.

in this particular case. This is hardly surprising since it is the only condition which explicitly acknowledges the fact that there is appreciable forcing at the OB's. Also the gradient condition (see the curves marked with the number 3) does fairly well in this case and can only be separated from the analytic solution when looking at the response in terms of the kinetic energy. The free wave radiation condition (see the curves marked with the number 5) does surprisingly well except for some notable reflections depicted in particular by the time series of the cross-shore depth mean current component (see Figure 6). The clamped condition (see the curves marked with the number 2) seems to be the worst choice for this particular physical set up. It develops a substantial alongshore variation in the dependent variables. This can be explained by the condition's inherent inability to produce any geostrophic current close to the OB which is an important part of the response. Also the mixed sponge-radiation condition (see the curves marked with the number 4) fails to simulate the correct OBC in this case. By inspecting the governing equations (equation (8) plus the bottom friction term) for the steady state it is seen that the depth mean current is inversely proportional to the bottom friction coefficient. The current is therefore gradually reduced in the sponge layers and the sponge fails to advect the correct amount of water towards or away from the OB in this case.

5. SUMMARY AND FINAL REMARKS

The open boundary conditions (OBCs) for numerical ocean models are considered. In order to clarify and limit the discussion, the term "open boundary" is given the following definition:

> An open boundary is a computational boundary at which disturbances originating in the interior of the computational domain are allowed to leave it without disturbing or deteriorating the interior solution.

This definition excludes boundaries at which the variables of the model have been specified by the modeler (i.e. limited-area tidal models), but includes those that attempt to leave the fluid motion unrestricted.

Some popular and recently suggested conditions based on the definition above are presented and reviewed together with certain subjectively chosen key papers. Some of the reviewed OBCs are presented in some detail with emphasis on the free and forced wave radiation conditions [Orlanski, 1976; Miller and Thorpe, 1981; Røed and Smedstad, 1984] and the treatment of oblique incidence of waves [Enquist and Majda, 1977; Raymond and Kuo, 1984]. Finally a sensitivity study with a barotropic model to the implementation of different OBCs is considered. A similar study has been presented by Chapman [1985] and the present study supports many of his conclusions. The present study also includes the forced wave radiation condition suggested by Røed and Smedstad [1984] which was neglected by Chapman [1985].

This sensitivity study reveals that:

- the response is highly sensitive to the implemented OBC even far away from the models open boundaries.

- the forced wave radiation condition for all practical purposes simulates a perfect OBC and recovers nicely the analytic solution.

- the gradient and free wave radiation condition also does fairly well in this particular case.

- the clamped condition is highly reflective and fails to develop important features of the response. It should probably be avoided in wind-forced cases.

- this is also true for the sponge type OBCs. This type of boundary condition which is shown to be successful in cases without forcing at the open boundary should be used with extreme care in cases with wind and/or bottom stress close to or at the open boundary.

In summary, this review reveals the importance of discriminating between the types of computational boundaries and the importance of constructing conditions to be used at a computational boundary with care so as to make them adequate to the problem at hand. Furthermore, it seems that so far the development of OBCs has been a trial and error type procedure and that most radiation type OBCs that has been suggested reduce to just another extrapolation formula. Finally, it should be noted that those OBCs that have been suggested to handle the oblique waves and the forced waves appear promising and to have a lot of potential for the future.

It may be fruitful for the future development of OBCs to consider the physics of the open boundary conditions in more detail. A viable approach may be offered by the method of characteristics in two dimensions and including rotation [Hartree, 1953; Freeman and Baer, 1957a,b; Lister, 1966; O'Brien and Reid, 1967; Wurtele et al., 1971; Røed and O'Brien, 1983].

Acknowledgment

One of us (LPR) would like to express his sincere thanks to Jim O'Brien for going through the trouble of staging and organizing the NATO ASI on Advanced Physical Oceanographic Modelling during the first two weeks of June 1985, which provided such a stimulating and scientifically rewarding environment for those of us who where lucky enough to attend. This research was supported by Conoco Norway Inc.. The support is gratefully acknowledged.

References

Arakawa, A., and Y. Mintz, The UCLA atmospheric general circulation model, UCLA Workshop notes, 25 March-4 April, 1974.

Blumberg, A.F., and L.H. Kantha, An open boundary condition for circulation models, J.Hydraulics, ASCE, (in press), 1984.

Charney, J.G., R. Fjørtoft, and J. von Neumann, Numerical integration of the barotropic vorticity equation, Tellus, 2, 237-254, 1950.

Camerlengo, A.L., and J.J O'Brien, Open boundary conditions in rotating fluids, J.Comp.Phys., 35, 12-35, 1980.

Chapman, D.C., Numerical treatment of cross-shelf open boundaries in a barotropic coastal ocean model, J.Phys.Oceanogr., 15, 1060-1075, 1985.

Chen, J.H., Numerical boundary conditions and computational modes, J.Comput.Phys., 13, 522-535, 1973.

Daubert, A., and O. Graffe, Quelques aspects des écoulements presque horizontaux a deux dimensions en plan et non-permanents application aux estuaires, La Houille Blanche, 8, 847-860, 1967.

Davies, H.C., Limitations of some common lateral boundary schemes used in regional NWP models, Mon.Wea.Rev., 111, 1002-1012, 1983.

Davies, A.L., and G.K. Furnes, Observed and computed M_2 tidal currents in the North Sea, J.Phys.Oceanogr., 10, 237-257, 1980.

Elvius, T., and A. Sundström, Computational efficient schemes and boundary conditions for a fine-mesh barotropic model based on the shallow-water equations, Tellus, 25, 132-256, 1973.

Enquist, B., and A. Majda, Absorbing boundary conditions for the numerical simulation of waves, Math.Comp., 31, 629-651, 1977.

Foreman, M.G.G., An accuracy analysis of boundary conditions for the forced shallow water equations, J.Comput.Phys., (in press), 1985.

Freeman, J.C. Jr., and L. Baer, Pseudo-characteristics, Trans.Amer. Geophys.Union, 38, 65-67, 1957a.

Freeman, J.C. Jr., and L. Baer, The method of wave derivatives, Trans. Amer.Geophys.Union, 38, 483-494, 1957b.

Hebenstreit, G.T., E.N. Bernard, and A.C. Vastano, Application of improved numerical techniques to the tsunami response of island systems, J.Phys.Oceanogr., 10, 1134-1140, 1980.

Hartree, D.R., Some practical methods of using characteristics in the calculation of non-steady compressible flow, Rep.AECU-2713, U.S. Atomic Energy Comm., Washington, D.C., 1953.

Israeli, M., and S.A. Orszag, Approximation of radiation boundary conditions, J.Comput.Phys., 41, 115-135, 1981.

Kreiss, H.O., Initial boundary value problems for hyperbolic equations, Comm.Pure Appl.Math., 23, 277-298, 1970.

Kreiss, H.O., Difference approximations for initial value boundary-value problems, Proc.R.Soc.Lond., A323, 255-261, 1971.

Larsen, J., and H. Dancy, Open boundaries in short wave simulations - a new approach, Coastal Engineering, 7, 285-297, 1983.

Lindman, E.L., "Free-space" boundary conditions for the time dependent wave equation, J.Comput.Phys., 18, 66-78, 1975.

Lister, M., The numerical solution of hyperbolic partial differential equations by the method of characteristics, in Mathematical Methods for Digital Computers, edited by A. Ralston and H.S. Wilf, John Wiley, New York, 1966.

Martinsen, E.A., B. Gjevik, and L.P. Røed, A numerical model for long barotropic waves and storm surges along the western coast of Norway, J.Phys.Oceanogr., 9, 1126-1138, 1979.

Mesinger, F., and A. Arakawa, Numerical methods used in atmospheric models, GARP Publ.Ser.no.17, 1, WMO-ICSU Joint Organizing Committee, 1976.

Miller, M.J., and A.J. Thorpe, Radiation conditions for the lateral boundaries of limited-area numerical models, Quart.J.R.Met.Soc., 107, 615-628, 1981.

Mungall, J.C.H., and R.O. Reid, A radiation boundary condition for radially-spreading non-dispersive gravity waves, TR 78-2-T, Dep. of Oceanogr., Texas A&M Univ., 1978.

O'Brien, J.J., and R.O. Reid, The non-linear response of a two-layer, baroclinic ocean to a stationary, axially symmetric hurricane, 1, Upwelling induced by momentum transfer, J.Atmos.Sci., 24, 197-207, 1967.

Orlanski, I., A simple boundary condition for unbounded hyperbolic flows, J.Comp.Phys., 21, 251-269, 1976.

Pearson, R.A., Consistent boundary conditions for numerical models of systems that admit dispersive waves, J.Atmos.Sci., 31, 1481-1489, 1974.

Platzman, G.W., The computational stability of boundary conditions in numerical integration of the vorticity equation, Arch.Met.Geophys. Biokl., Ser. A, 7, 29-40, 1954.

Raymond, W.H., and H.L. Kuo, A radiation boundary condition for multi-dimensional flows, Quart.J.Roy.Met.Soc., 110, 535-551, 1984.

Robinson, A.R., and W.G. Leslie, Estimation and prediction of oceanic fields, Progress in Oceanogr., 14, 485-510, 1985.

Robinson, A.R., and L.J. Walstad, Numerical modelling of ocean currents and circulation, Numerical Fluid Dynamics (to appear), 1985.

Røed, L.P., and J.J. O'Brien, A coupled ice-ocean model of upwelling in the marginal ice zone, J.Geoph.Res., 88(C5), 2863-2872, 1983.

Røed, L.P., and O.M. Smedstad, Open boundary conditions for forced waves in a rotating fluid, SIAM J.Sci.Stat.Comput., 5, 414-426, 1984.

Sköllermo, G., Error analysis of finite difference schemes applied to hyperbolic initial boundary value problems, Math.Comp., 33, 11-35, 1979.

Smith, W.D., A non-reflecting plane boundary for wave propagation problems, J.Comput.Phys., 15, 492-503, 1974.

Sielecki, A., An energy-conserving difference scheme for storm surge equation, Mon.Weath.Rev., 96, 150-156, 1968.

Sommerfeld, A., Partial differential equations: Lectures in Theoretical Physics, Vol. 6, Academic Press, 1949.

Sundström, A., and T. Elvius, Computational problems related to limited-area modelling, GARP Publ.Ser.no.17, 11, 1979.

Verboom, G.K., G.S. Stelling, and M.J. Officier, Boundary conditions for the shallow water equations. In Engineering Applications of Computational Hydraulics I, edited by Abbott and Cunge, Pitman, London, 1983.

Wurtele, M.G., J. Paegle, and A. Sielecki, The use of open boundary conditions with the storm surge equations. Mon.Weath.Rev., 99, 537-544, 1971.

DATA ASSIMILATION

David L. T. Anderson and Andrew M. Moore
Department of Atmospheric Physics
Clarendon Laboratory
Oxford, ENGLAND

ABSTRACT. Some of the problems which oceanographers may expect to encounter when trying to assimilate oceanographic data into models are illustrated. In the first part of the paper, simple pedagogical examples are given. In the second, results from a tropical ocean model are used.

1. INTRODUCTION

One of the objectives of the TOGA (Tropical Ocean Global Atmosphere) experiment is to determine the extent to which the climate is predictable over time scales of months to years.

The atmosphere is generally thought to have only limited memory and hence to be essentially unpredictable on the time scale of a few weeks or longer. The basis for longer term prediction is thought to reside in the boundary conditions, e.g., primarily the ocean, but possibly also the state of land conditions. The evidence over the past few years is that it is the tropical ocean which is responsible for large fluctuations in the tropical climate and that these fluctuations can also affect the climate at midlatitudes. On the other hand, the midlatitude atmosphere probably has a variability resulting from midlatitude SST anomalies also and could, even for fixed boundary conditions, display considerable variability on long time scales (weeks to months to years) if the system were almost intransitive [Lorenz, 1978].

Thus it is unclear the extent to which midlatitude climate will be predicatble, since it is influenced not just by the "coherent" tropical response but also by more stochastic processes. Regardless of the relative importance of local versus remote forcing, the main hope for predictability will reside in knowledge of the ocean state. This means we need to know the ocean state accurately. It is a prerequisite for a good climate forecast. To be sure, we need good coupled atmosphere-ocean general circulation models (CGCM's) also, but without a good initial state these models will be of no use for prediction purposes. The detail of the atmospheric state is probably

J. J. O'Brien (ed.), Advanced Physical Oceanographic Numerical Modelling, 437–464.

not very important, i.e., it does not matter whether we start a forecast integration today, last week, or next Friday, and in any event a routine analysis procedure already exists. What is needed is an ocean data assimilation/analysis procedure.

Meteorologists have considered initialisation problems for many decades. The oceanographic problem is similar in some respects but quite different in others. Similarities and differences will be indicated where appropriate.

In Section 2 we consider a number of simple problems to illustrate specific points within a framework which is readily understandable. In Section 3, we consider some simple geostrophic adjustment ideas and in Sections 4 and 5, consider results from numerical models.

2. PEDAGOGICAL EXAMPLES

Data assimilation consists of trying to achieve the best analysis of a system based on past incomplete and noisy data. The analysis is usually centered round a model which connects different variables in a physical way, and it is this which distinguishes it from straight contouring, for which no knowledge of the dynamics is required. In the event of a surfeit of data, a contouring routine would presumably do very well, but it is a most unlikely situation that oceanographers would ever find themselves with an excess of data (even though some aspects of the "analysis" may be over-determined).

2.1. A single linear oscillator

Consider the following dynamical system

$$\frac{dy}{dt} = wx; \qquad \frac{dx}{dt} = -wy \tag{1}$$

and suppose we can only ever measure y. Can we assimilate the y measurements in such a way as to estimate x?

Equation 1 represents simple harmonic motion (SHM) of period w, which we will take to be unity without loss of generality. Suppose this SHM represents a wave of period $2\pi/w$. Then practically we might anticipate sampling this wave every $\pi/2$ or $\pi/4$ and so on. The strategy is to measure y. We do not know x, so we keep the model value of x but include immediately the measured value of y. The results of doing this twice are shown in Figure 1a (upper) when we sample ever $\pi/6$ and in Figure 1b (lower) when we sample every $\pi/2$. It is clear that in the latter case, the error instantly drops to zero. On the other hand if the measurement is made at π, no useful information is gained at all. This illustrates, albeit very simply, two useful ideas. Firstly, for certain systems it is possible by using the dynamical constraints to gain information on variables we may never measure. Secondly, some measurements are more useful than others, and some may be totally useless.

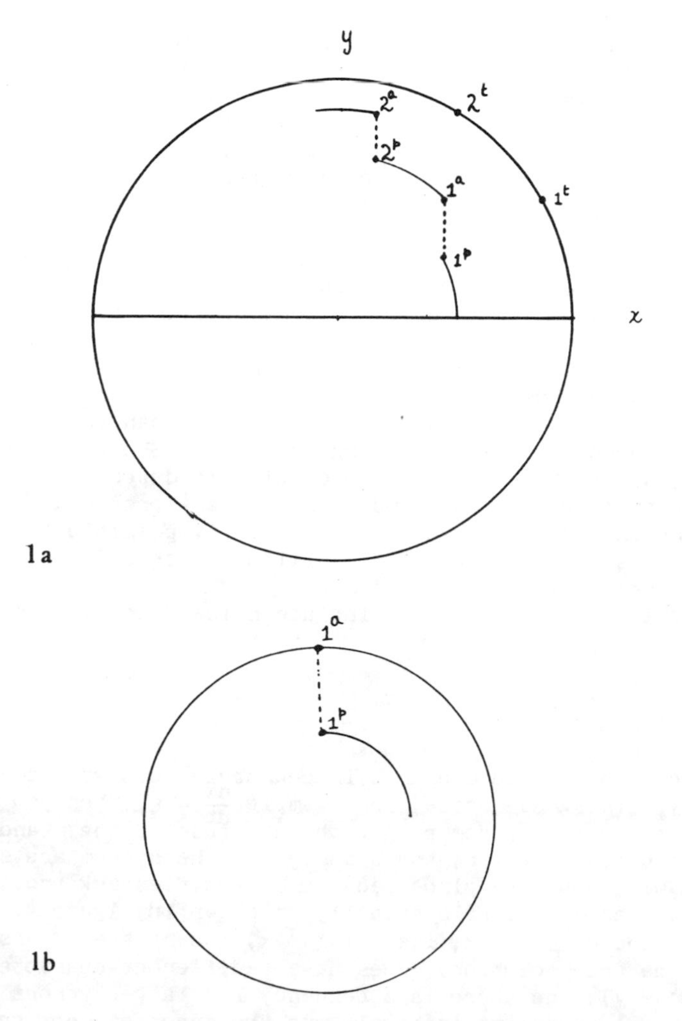

Figure 1a. Illustration of an assimilation procedure for the SHM equation (1). The true solution is given by the outer circle. Only y can be sampled. At t=0 (chosen to be the x axis) we know y, but not x. So we guess a value. The model is then integrated for $\pi/6$ units of time to point 1^p. y is again sampled and an analysis made 1^a (we keep the same value of x but adjust y to the new value). The model is then integrated for a further $\pi/6$ units of time to 2^p, an analysis made to 2^a, and the model integrated further, and so on. As can be seen clearly, the model solution is nudged successively closer to the true solution.

Figure 1b. As for 1a, but when the update is every $\pi/2$ units of time. The true solution is obtained immediately in this case.

An interesting side issue arises. It is clear that $\pi/2$ is the best sampling interval, but suppose we have measurements at much higher frequency, say $\pi/60$ or even higher. Should we use all this information or not? Figure 2b (lower) shows the error reduction when data are assimilated every $\pi/60$ units a time. Comparing this panel with the upper which corresponds to assimilating data every $\pi/6$ shows that the error decreases faster (in real time) with the less frequent update.

In the example quoted, we have a whole series of measurements, and the error decreases steadily as shown in Figure 2a as more information is accumulated. It is obvious however, (see Figure 3) that by integrating the model forwards to the next observing time, then backwards to the previous, then forwards again and so on, that the error again decreases with time and so more information is actually extracted from the data in this scheme than in the previous example. Of course what we are doing is solving SHM with 2 pieces of information, all that is required to completely determine the solution; in that respect this model is not very profound. However, the two approaches do mimic what is practically possible in less trivial examples, a direct forward assimilation or a forward-backward adjustment scheme.

The model can be extended to include noise. Consider the system

$$\frac{dy}{dt} = wx \qquad \frac{dx}{dt} = -wy + \varepsilon \tag{2}$$

where we consider y to be measured inaccurately (but with zero bias). If you integrate noise, for example $\frac{dx}{dt} = \varepsilon$, then you would expect x to execute a random path with amplitude of the meanders growing with time. However, the dynamics of the system are such that such growth will not take place because of the feedback provided by rotating, and the solution is actually as shown in Figure 4. The solution for the error in x, i.e., $(x^t - x)$ (where the superscript t relates to the true solution), does have a different character to that for $(y^t - y)$, and there is a tendency for larger errors to persist over a longer time interval, but the error can not grow like $t^{1/2}$.

The spectra of the error in x is plotted in Figure 5. One can see that whereas the errors in y are white, those on x are red. Thus high frequency measurement errors can lead to errors in the low frequency response of the model. It is interesting to note that the natural period w=1 does not show in the error spectrum. [It is also worth noting that we are not considering the solution to (2) without updating. In that case y is very different to the case where we do update y. If we take the finite Fourier transform of (2) by letting

$$y = \sum_n y_n e^{iw_n t}; \qquad x = \sum_n x_n e^{iw_e t}$$

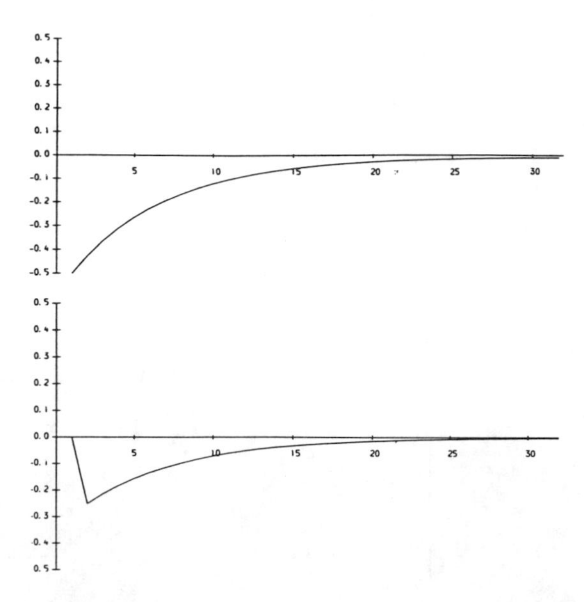

Figure 2a. The error in x (upper) and y (lower) as a function of the number of updates (made every $\pi/6$ units of time). The error in y is measured just before an update and is thus an envelope of the error, i.e., it does not show the sawtooth pattern which occurs following an update.

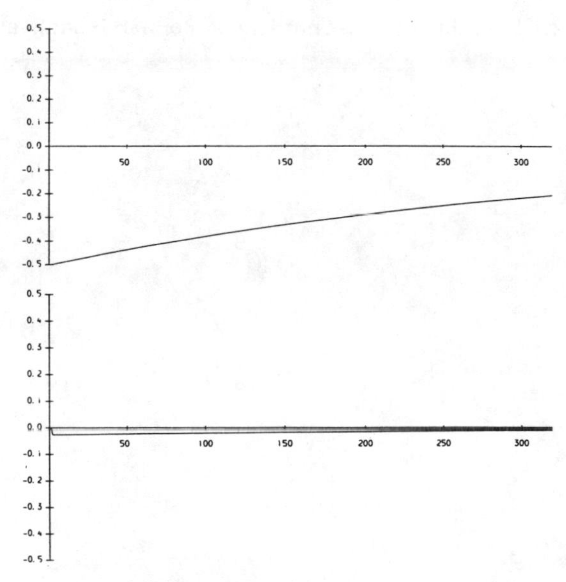

Figure 2b. As for a, but when the update is made every $\pi/60$ units of time, showing the much slower decrease in the error in x. (300 on this graph corresponds to the same real time as 30 on Figure 2a.)

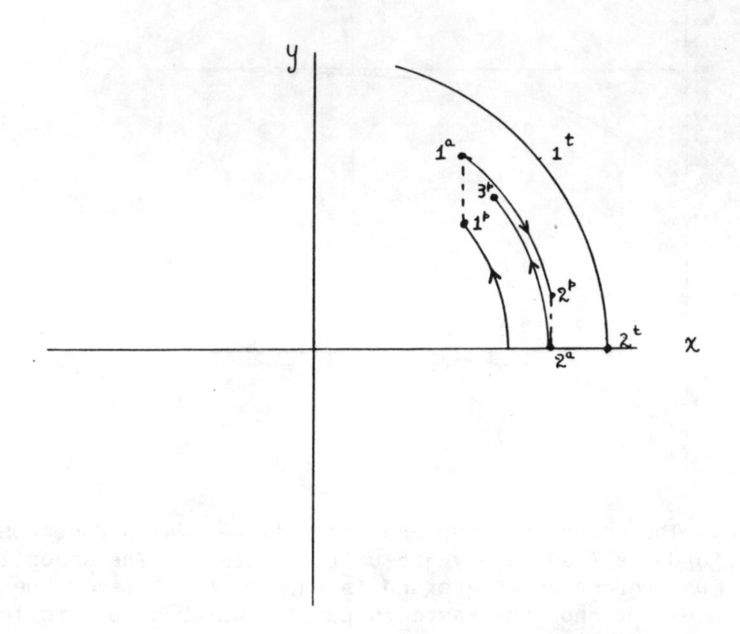

Figure 3. As for Figure 1, but illustrating a forward-backward assimilation scheme.

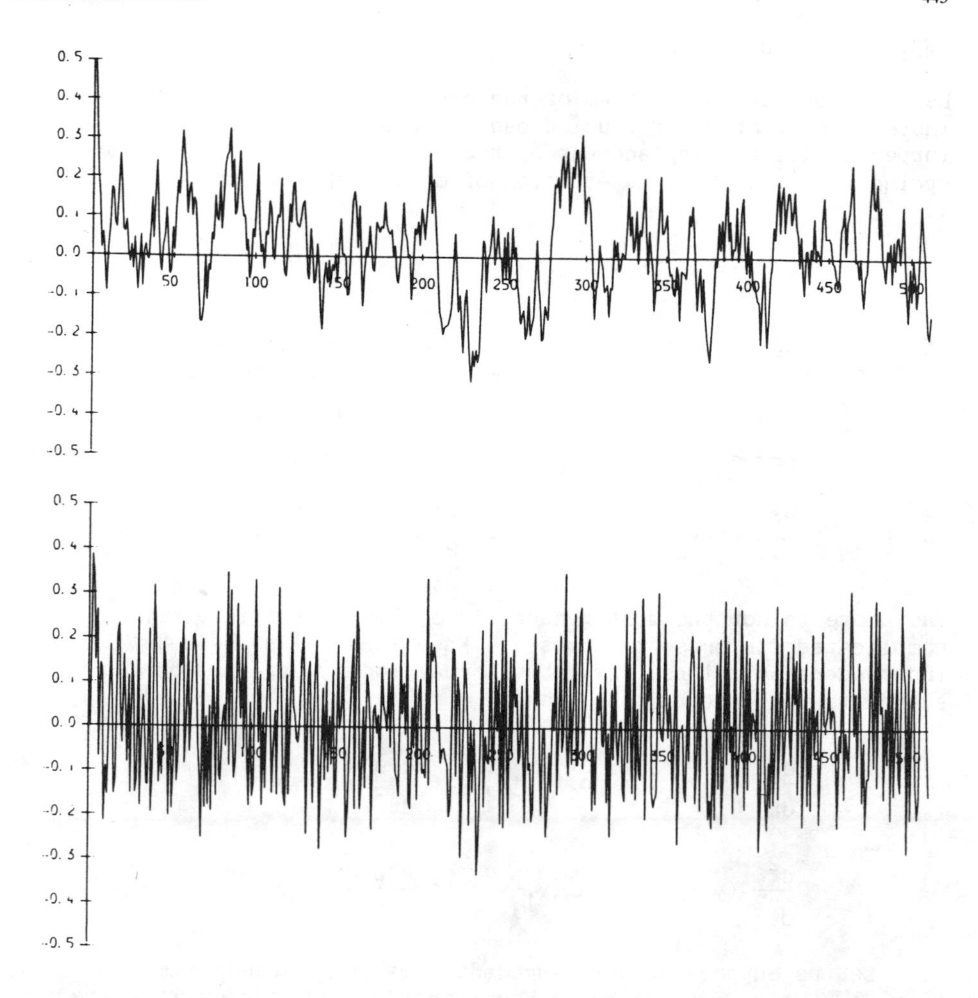

Figure 4. The error in x (upper) and y (lower) when white noise is added to the measurement of y. The errors in x are of a longer duration than those in y. The horizontal axis is the number of updates (made every $\pi/6$ units of time).

2.3. Two coupled oscillators

Let us rephrase the problem of the previous section and consider instead the problem of coupled oscillators as shown in Figure 9 (upper). If the displacement of the equilibrium position of the spring k is x_k, then the equation of motion of the spring k is

$$\ddot{x}_k = K(x_{k+1}-x_k)-K(x_k-x_{k-1})$$

If the springs have different properties, then the K will be different for different k. For two oscillators, the equations become

$$\frac{dy_1}{dt} = -x_1 \quad ; \quad \frac{dx_1}{dt} = \frac{-(\alpha_1+\alpha_2)y_1}{2} - \frac{(\alpha_1-\alpha_2)y_2}{2} \qquad (4)$$

$$\frac{dy_2}{dt} = -x_2 \quad ; \quad \frac{dx_2}{dt} = \frac{-(\alpha_1-\alpha_2)y_1}{2} - \frac{(\alpha_1+\alpha_2)y_2}{2}.$$

The above is not the most general form, but is sufficiently complicated for present purposes. Equations (4) possess two linearly independent solutions, viz., the normal modes, obtained by adding and subtracting the equations. Hence

$$\frac{d^2}{dt^2}(y_1+y_2) = -\alpha_1(y_1+y_2) \text{ ie } \frac{d^2\bar{u}}{dt^2} = -\alpha_1\bar{u}$$

$$\frac{d^2}{dt^2}(y_1-y_2) = -\alpha_2(y_1-y_2) \text{ ie } \frac{d^2\hat{u}}{dt^2} = -\alpha_2\hat{u}$$

Let us suppose at a measurement time, we can determine $y_1+y_2=c$. We do not know y_1 and y_2; they can lie anywhere on the line $y_1+y_2=c$, which we will call the data manifold. This is shown in Figure 6. By integrating equation 3, however, we will always have estimates of y_1^p and y_2^p (denoted by superscripts p for prediction). The predicted point is marked on Figure 6 by P. The question is: how do we get from the predicted point P to the data manifold. There are, of course, an infinite number of ways of doing this, but two suggest themselves. The first is to assume that y_1^p is held fixed and we adjust y_2 (then next time maybe hold y_2^p and adjust y_1) and so on. This gives the point R as the point on the data manifold. The second is to make the minimum correction to the predicted solutions, which corresponds to drawing the normal to the curve. This gives the point N.
Let us consider mode 1,(y_1+y_2), which oscillates at a higher frequency α_1, while mode 2,$(y +y_2)$ will oscillate at a lower frequency, α_2. For clarity, you may want to consider α_1 as

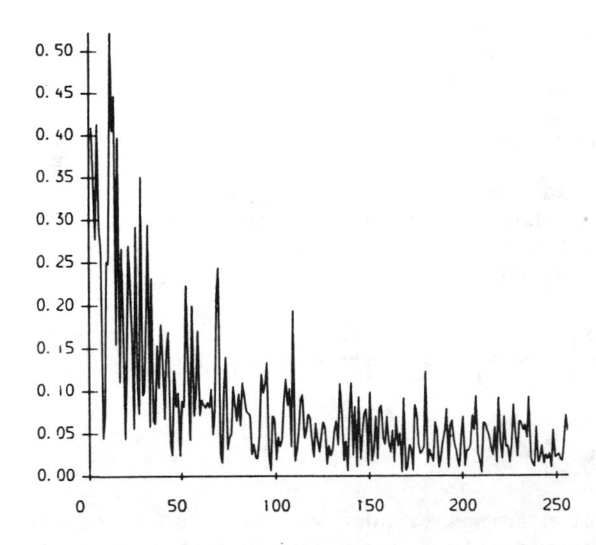

Figure 5. The Fourier spectrum of the error in x. It is red although the spectrum of the error in y (i.e. noise) is white. Thus high frequency errors in y can translate into low frequency errors in x.

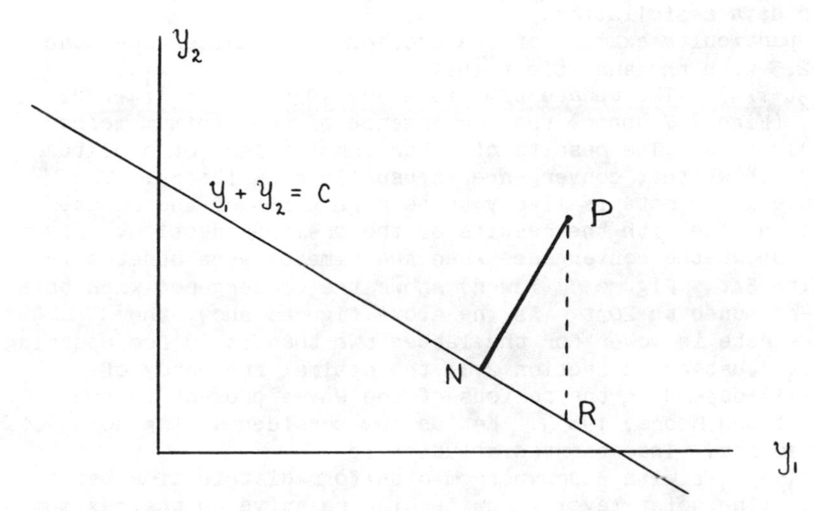

Figure 6. Plot of $y_1+y_2=c$. The point P is the model prediction. The analysis procedure is to correct P in such a way as to make the analysis of y_1 and y_2 consistent with the data, i.e. they must lie on the line $y_1+y_2=c$. Of the many ways of doing this, two are indicated. N is the minimum projection of P to the data manifold. (This was the scheme used in Figures 1 and 3.) An alternative (R) is to hold y_1 fixed and adjust y_2 (and the next time to hold y_2 fixed and adjust y_1).

and the noise $\varepsilon = \sum a_n e^{iw_n t}$, then we would obtain for x_n, the solution $(x_n)^2 = w_n^2 a_n^2/(w^2 - w_n^2)^{2n}$. This latter spectrum has a peak at $w = w_n$ and zero power at low frequency.]

2.2. Two linear oscillators: (the altimeter problem)

The previous example was particularly simple because there was no ambiguity over what the measurement corresponded to. But suppose we have two oscillators which, for the present, we consider to be linearly independent

$$\frac{dy_1}{dt} = w_1 x_1 \quad ; \quad \frac{dx_1}{dt} = -w_1 y_1 \qquad (3)$$

$$\frac{dy_2}{dt} = w_2 x_2 \quad ; \quad \frac{dx_2}{dt} = -w_2 y_2$$

The complication arises because we can measure only (y_1+y_2), say, not y_1 and y_2 independently. For a linear system like this we know we should need only 4 pieces of information to determine the solution. This could easily be measurements of (y_1+y_2) at 4 different times. While such a strategy could work very well for this problem, it is not practical for the more general problems to be considered later. So we will here consider only the approach cited in the previous section of data assimilation.

As a particular example of the procedures, consider the case $W_1=1$, $w_2=2.5$ with the analytic solution given by $y_{1a}=\text{Sin}w_1t$, $y_{2a}=\text{Sin}(w_2t+\pi/4)$. The update time is every $10\Delta t$, where $\Delta t=\pi/40$. In this case, (Figure 7 upper) the convergence of the minimum method is particularly fast. The results of other combinations of amplitudes and periods shows that convergence is usually oscillatory. The actual convergence rate is likely to be a function of the update frequency, in line with the results of the previous section. Figure 7 (middle) shows the convergence when the time between updates is shortened to $5\Delta t$. Figure 7 (lower) shows the convergence when this time is lengthened to $20\Delta t$. As the above figures show, the convergence rate is lower for the latter two than it is for updating every $10\Delta t$. Just as in Section 2.1, the optimal frequency of updating will depend on the periods of the waves present in the system [Webb and Moore, 1985]. Let us now consider adding noise to the measurements. The observed solution is now $y_{obs}=y_{1a}+y_{2a}+\varepsilon$ with ε drawn from a uniform distribution between -1 and +1. (The noise level is quite high relative to the maximum signal of 2.) Figure 8 shows the error spectrum for $x_1-x_1^P$ and $x_2-x_2^P$. In the simpler 1-dimensional oscillator problem, this was a red spectrum. In this case, it is rather different, as Figure 8 shows. Now an error spectrum results with maximum at $w=w_1$ and $w=w_2$ respectively. The spectrum for $y_1-y_1^P$ and $y_2-y_2^P$ are similar to those for x.

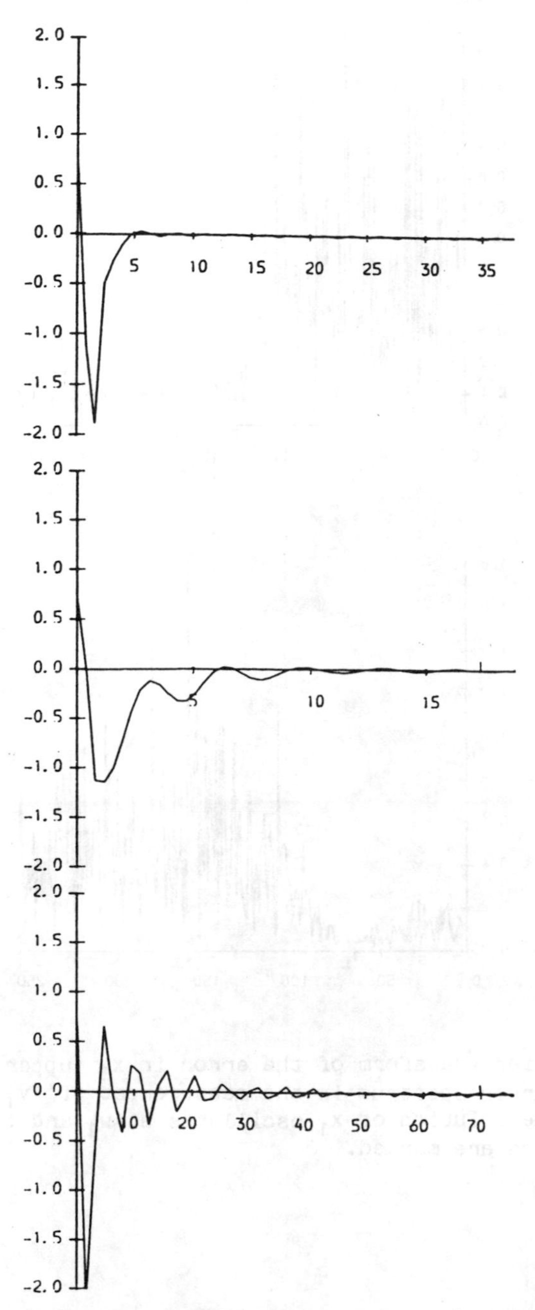

Figure 7. Plot of the envelope of the error in (y_1+y_2) as a function of time, when the update frequency is every 10 Δt (top) 5 Δt (middle) and every 20 Δt (bottom). The horizontal axis is time.

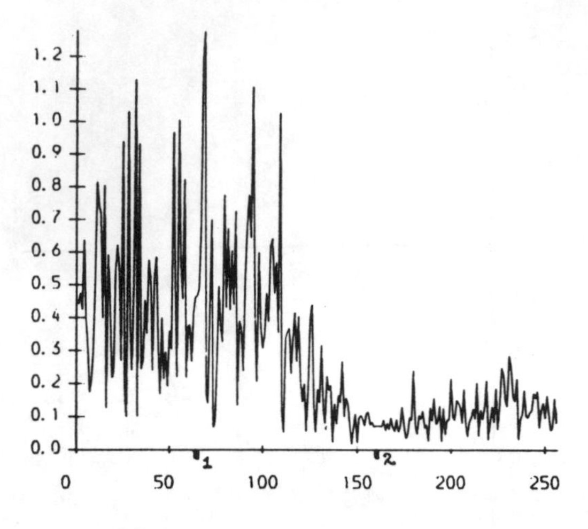

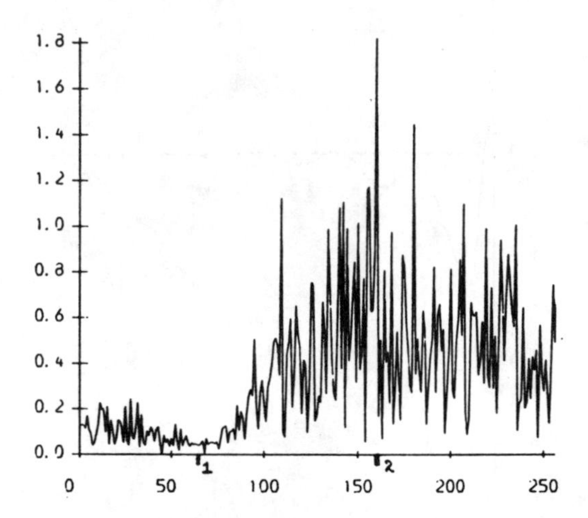

Figure 8. Fourier transform of the error in x_1 (upper) and x_2 (lower). The error spectrum in the measurement of (y_1+y_2) is white. The true solution of $\dot{x}_1$ oscillates at w_1 and for x_2 at w_2. These frequencies are marked.

equations

$$u_t - fv = -P_x \tag{5}$$

$$v_t + fu = -P_y \tag{6}$$

$$P_t + c^2(u_x + v_y) = 0 \tag{7}$$

If we define $u = -\psi_y + \chi_x$; $v = \psi_x + \chi_y$ where ψ and χ represent the rotational and divergent part of the velocity respectively, the following equations can be derived.

$$\nabla^2(\psi_t + f\chi) = 0 \tag{8}$$

$$\chi_{tt} + f^2 = c^2\nabla^2\chi \tag{9}$$

$$P_t + c^2\nabla^2\chi = 0 \tag{10}$$

$$\chi_t = f\psi - p \tag{11}$$

where the vorticity is given by $\nabla^2\psi$ and the divergence by $\nabla^2\chi$. Equation 10 shows the existence of gravity waves with a dispersion relationship $w^2 = f^2 + c^2(k^2 + \ell^2)$ when is decomposed into Fourier models,

$$\chi \propto e^{i(kx + \ell y - wt)}$$

Equations (5)-(7) or (8) and (10) conserve perturbation potential vorticity $(V_x - u_y - fp/c^2) = (\nabla^2\psi - fp/c^2)$.
The above system has no dissipation and so χ need not asymtote to zero. However, if some dissipation is added to (10), then potential vorticity is still conserved, and a steady state is possible in which χ_t, ψ_t and P_t are zero.
In this case, $\chi=0$ is a solution to (9) to (11) and $(\nabla^2\psi - fp/c^2)$ initial = $(\nabla^2\psi - fp/c^2)$ final.
Further, from (12), $\psi = \frac{p}{f}$.
This means that if we Fourier decompose ψ and p we get

$$-(k^2+\ell^2)\psi_s - \frac{f}{c}2P_s = [-(k^2+\ell^2) - f^2/c^2] \quad \psi_s$$

corresponding to gravity waves and α_2 to slower Rossby waves. Now suppose we can measure only y_2, which, let us say, has a value of -5. (Y_1 really has the value +5 at this time but cannot be measured. This means the true solution has u=0 at t=0 and therefore will remain zero, so the solution has only slow motion). But the observation $y_2=-5$ is consistent with $\bar{u}=y_1-5$ and $\hat{u}=y_1+5$ i.e. $\bar{u}=\hat{u}-10$.

The true solution is shown on Figure 9 (lower) together with the data manifold. If we really start at T, the true solution will be followed. However, if we start at S, a solution with high frequency waves will be present. It is clear that if the data manifold were steeper, in terms of getting the slow oscillation correct it would then be less critical where on the manifold we started. But we do need accurate measurements, i.e., we must have the correct data manifold, not one parallel to it. On the other hand, if the data manifold has a slack gradient, then it will be very important how the data is projected onto the slow and fast modes. Sometimes it is clear. For example, if we had a guiding principle that the fast mode did not appear or was present with only small amplitude, then we should seek to be near the point T on the diagram. [For a fuller discussion, see Daley, 1980].

A complication that might arise is that the system be nonlinear, governed, for example, by equations like

$$d\bar{u}/dt = -i\overline{wu}+i\varepsilon\hat{u}^2$$

$$\frac{d\hat{u}}{dt} = -i\hat{w}\hat{u}$$

If $\varepsilon=0$, then we have a system analogous to the previous system. However, if $\varepsilon\neq0$, then starting with $\bar{u}=0$ will not suppress the fast mode, because of the nonlinearity in the "gravity" wave equations. To suppress the fast mode, one wants to ensure not that $\bar{u}$ is zero, initially, but that $\frac{d\bar{u}}{dt} = 0$, which requires that $\bar{u} = \varepsilon u^2/\bar{w}$. Figure 10 shows solutions when (a) $\bar{u} = 0$ and (b) $\bar{u} = \frac{\varepsilon\hat{u}}{\bar{w}}$ initially. In the former case, high frequency waves are initially absent but quickly grow, while in the latter case they are initially present but their amplitude stays lower than that attained in the former.

3. SIMPLE GEOSTROPHIC ADJUSTMENT THEORY

An argument is frequently advanced in meteorology [Temperton, 1973] to indicate that at low latitudes, velocity field data are more useful for initialising models then mass field data. While this argument may have validity in an atmospheric context, it may be inappropriate for oceanographic applications because of the presence of boundaries and the greater importance of divergence in internal oceanographic waves. The argument is based on the free surface

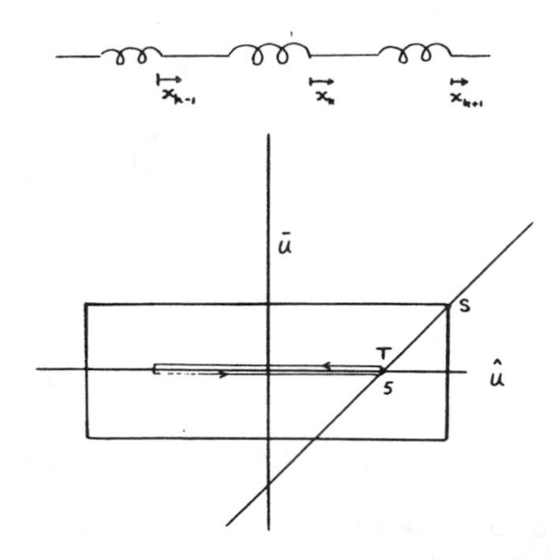

Figure 9. Upper, plot of coupled oscillators; lower, data manifold (straight line) in $\bar{u}$, $\hat{u}$ space. The point T is the initial condition, with the solution following the curve indicated. The box, with S at corner, is the envelope of the solution which has S as initial state. The actual solution would have both fast and slow oscillations with this box.

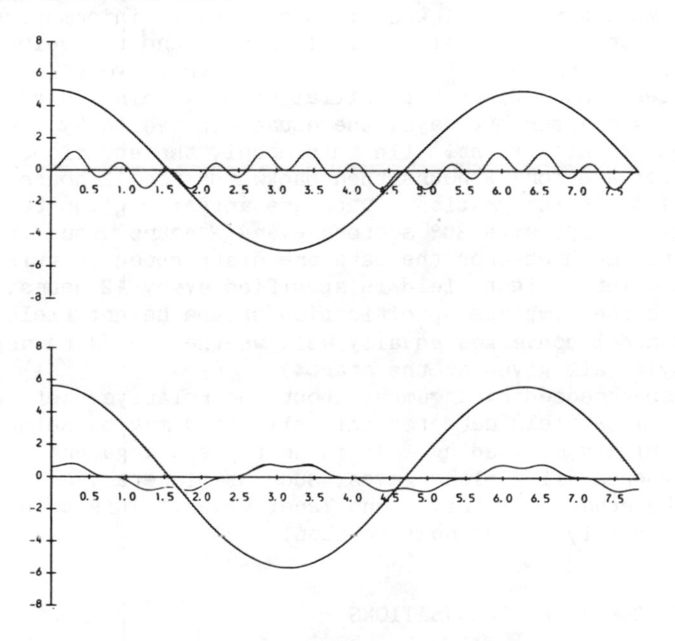

Figure 10. Plot of the real part of solutions to the nonlinear normal mode equations. In the upper diagram, the high frequency mode is set to zero initially but quickly grows. A smaller amplitude is obtained in (b) where sufficient of the high frequency mode is present in the initial state to make its tendency zero.

$$= -(k^2+\ell^2)\psi_i - \frac{f}{c}2P_i$$

$$\psi_s = \frac{(k^2+\ell^2)\psi i + \frac{f}{c}2P_i}{[k^2+\ell^2+f^2/c^2]} = \alpha\psi i + (1-\alpha)\frac{P}{f}i \quad (12)$$

$$\text{with } \alpha = \frac{k^2+\ell^2}{[k^2+\ell^2+\frac{f^2}{c^2}]} = \frac{1}{[1 + \frac{f^2}{c^2(k^2+\ell^2)}]}$$

This argument shows that for $\alpha\to 1$, the steady state for ψ is governed by ψ_i, whereas for $\alpha\to 0$ the final state ψ_s is dictated by P_i. $\alpha\to 1$ corresponds to $f\to 0$ or k,ℓ becoming large. This implies that the final solution will be governed by the initial pressure (or mass) field only if $\alpha\to 0$, i.e., we have large-scale features with both k and ℓ small. If we are at low latitudes or have small-scale features, then the final value of ψ is governed by the initial value of ψ, i.e., the velocity field.

Partial confirmation of these ideas appears to have come from the work of Morel and Talagrand [1974, Figure 3] reproduced here as Figure 11. When their model was given mass field information, the model appears to accept it at middle latitudes and the velocity field to adjust quickly there: at 45°N the rms error in velocity has dropped to less than 20% of the initial error within 3 days. On the other hand, even after 100 days, the equatorial velocity field has not adjusted to this extent. (In this model, the equations are integrated forward for 12 hours then backwards for 12 hours. This is denoted as 1 day's integration. Data are actually given to the model asynoptically (2 opposite 30° sectors every 2 hours), but it does not seem to matter much whether the data are distributed in this way or whether a complete height field is specified every 12 hours, equivalent to one complete specification of the height field/12 hours. The model converges equally well whether the data are fed in continuously or all given at the start.)

While the preceeding argument about the relative usefulness of velocity vs. mass field data for initialisation may be helpful, it can also be highly misleading. It is an f-plane argument and therefore ignores all β-plane waves, such as planetary or Rossby waves and the equatorial Kelvin and Yanai waves. This point is illustrated briefly in the next section.

4. RESULTS FROM MODEL CALCULATIONS

4.1. Description of model and analysis procedure

The dynamics of assimilation can be studied in fairly simple geometry, e.g. a box ocean or channel, or in more realistic configuration. For present purposes, we will consider first the more realistic calculation and later mention simpler models which explain the results.

The model used is a 1 or 2 layer model with the geometry of the Indian Ocean, the details of which are given in an appendix. Monthly mean wind stress from the low level cloud wind and ship reports for the FGGE year are used to drive the model. In case different regions of the ocean respond differently to data insertion, the domain was split into 5 regions shown in Figure 12.

A control version of the model is run for several years. An assimilating version is then started and supplied with either height field data or velocities. The root mean square difference between the assimilation and control runs is then calculated for the 5 areas plotted as a function of time. The Bay of Bengal region does not usually behave differently from other regions, so for convenience we will show only 4 panels.

4.2. The adjustment process

Here we consider whether mass field data or current data are dynamically the more useful.

Figure 13 shows the normalized rms error of H1 (layer depth) v time when the model is given velocity data every 10 days (solid curve) and the rms error in U1 (zonal velocity) when the model is given height field data (dashed curve). Figure 14a is a plot of the normalized rms error in H1 when the model is given H1 and Figure 14b is a plot of the rms error in U1 when the model is given velocity data. Let us consider Figure 13 first. This shows a very interesting result. First, contrary to Section 3, height field data do a much better job of reconstructing the velocity field than the velocity field does for height. In fact, for the equatorial region of Figure 13, it appears that the errors are not decreasing in the case when velocity data is supplied. (However longer integrations (not shown) do confirm convergence). Turning to Figure 14b, we see that following velocity insertion, the model quickly rejects the velocity information and in a time less than 10 days (actually ~ 1 day) returns to almost its original value; most of the velocity information is rejected by the model. In contrast, height field data are assimilated much better with relatively little rejection. The rejection is higher in the equatorial band than outside it (Figure 14a) but even so is relatively small. This is in marked contrast to the velocity case; the velocity rejection is highest in the equatorial region just where the theory of Section 3 suggests it should be least.

5. DISCUSSION OF DYNAMICS

5.1. The Dynamics of adjustment

In this section we consider further the results of Sections 3 and 4,

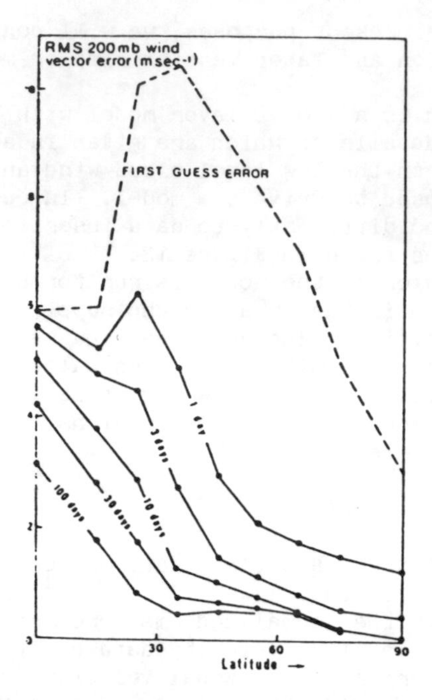

Figure 11. Plot of the rms errors in velocity in an atmospheric model as a function of latitude when the model is given height field data every 12 hours. Determination of the velocity field is much slower at low latitudes [Morel and Talagrand, 1974].

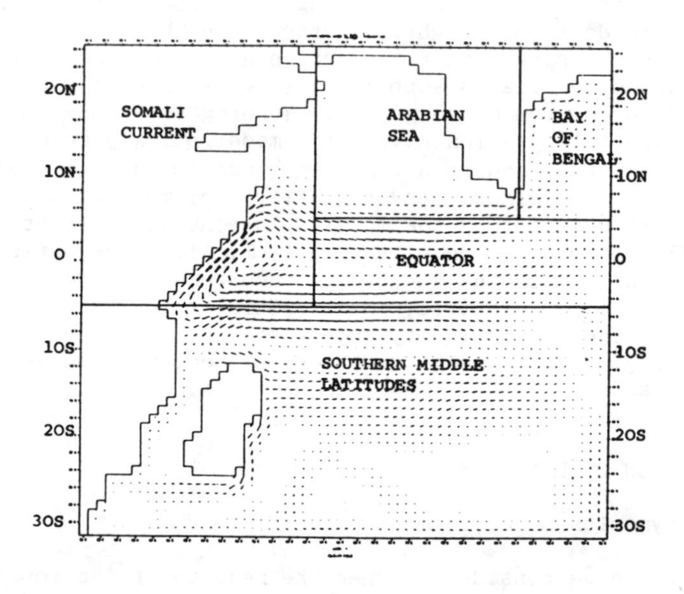

Figure 12. Indian Ocean reduced gravity model domain.

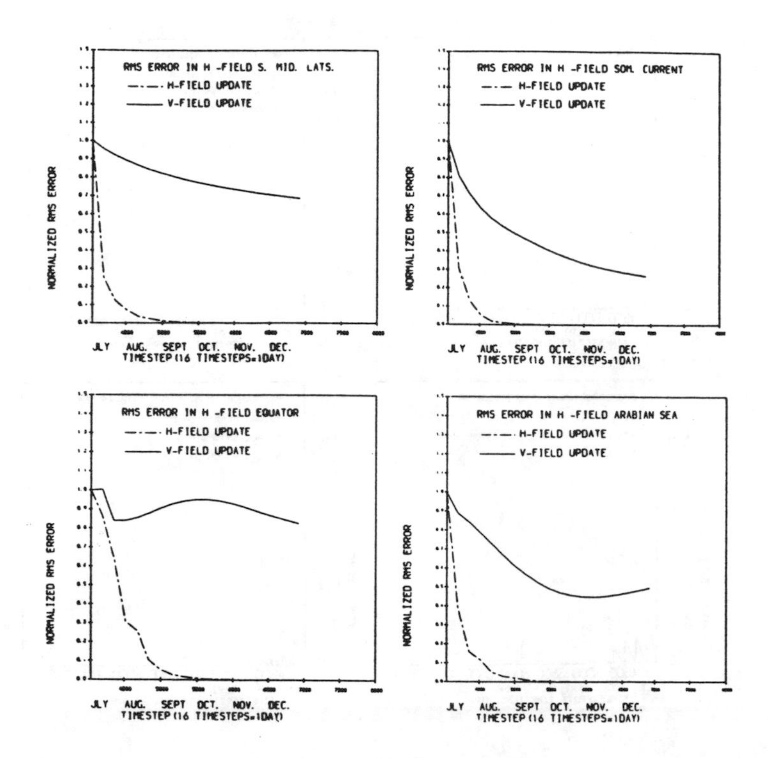

Figure 13. The root mean square errors in 4 regions in a model of the Indian Ocean. The dashed curve shows the 'error' in the zonal component of the velocity and represents the difference between a control run and one in which the model is given height field data only. Similarly, the solid curve shows the 'error' in the height field when the model is given velocity field data only.

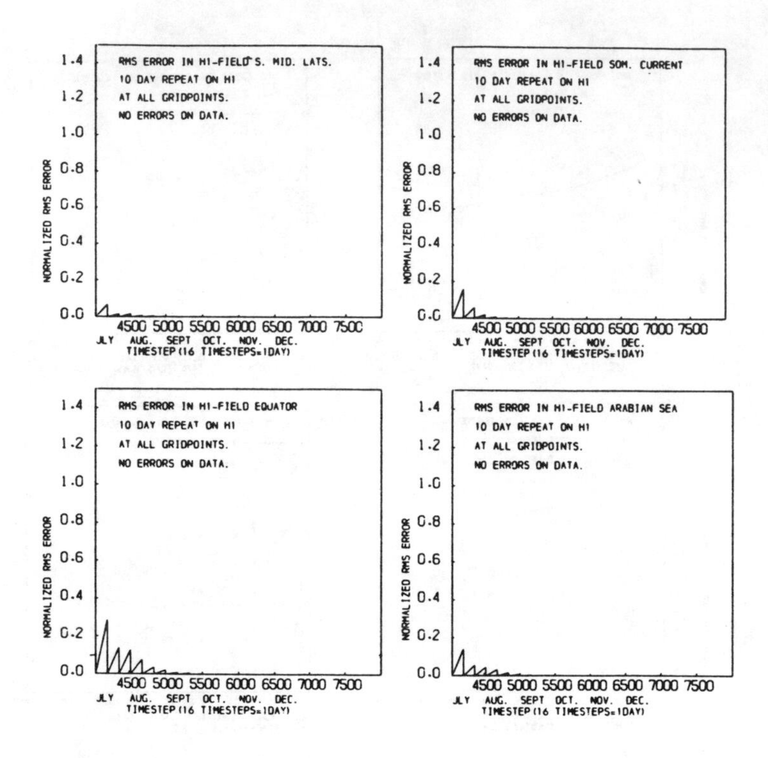

Figure 14a. The normalized root mean square error in height field in four regions of the Indian Ocean when the model is given height field data every 10 days at all grid points.

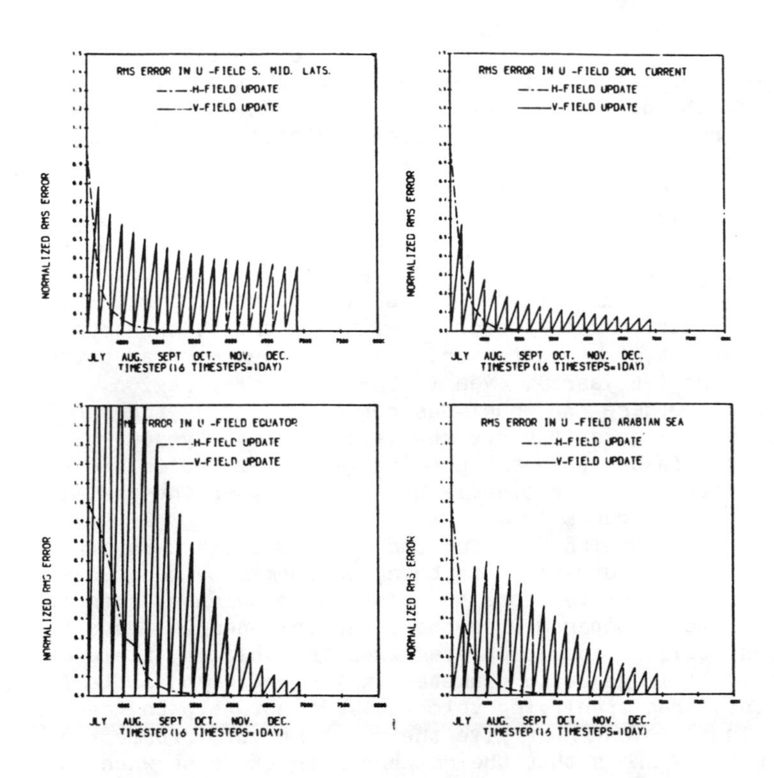

Figure 14b. The normalized root mean square error in the zonal velocity field component in four regions of the Indian Ocean when the model is given velocity field data. The velocity field error when the model is given height field data is shown for comparison.

especially the apparent lack of impact of velocity field data. In Figure 15 we plot the normalized rms difference of the height and zonal velocity fields between the control run and another started from rest and with the height field at its undisturbed value. By differencing the equations for these two calculations, we will derive an unforced damped set of equations in which the initial field represents the difference between the two models at the start of the perturbed run. Since we have a linear dissipative system, we know that ultimately the rms differences in all fields must tend to zero, but on what time scale? Clearly, the higher the friction, the shorter the time scale, but this time scale cannot be calculated without knowing the scale of the adjusting waves and that these will be a function of the forcing. In Figure 15 (solid curve) a value of 1×10^7 cm^2 s^{-1} was used for the dissipation, while for the dashed curve the friction is a factor 20 higher. Truly, this run converges faster, but not 20 times faster. The time scale of adjustment is several years (~6 years), even in the equatorial region, and in the southern hemisphere the model has not converged by the end of the integration. Figure 16 shows the velocity convergence. This is considerably faster than the h convergence, except in the southern middle latitudes. This clearly has a bearing on the assimilation experiments of Figures 13 and 14.

A possible explanation for the rapid velocity convergence is that much of the adjustment is being performed by long Rossby waves. If a system is close to geostrophic balance, then the kinetic energy is small compared with the potential energy. Thus the height field can still be adjusting even when the velocity field has adjusted. If large scale planetary waves are responsible for the adjustment, then specifying velocity will not help to identify these waves. To do so one must give the modal height information.

Figure 17 shows that the rms error in the case when current data are supplied at every grid point (solid line) does converge faster than the case when no current data are given at all, but the convergence rate is not very much faster. So, while the model is using some current data, it is also rejecting a lot. Closer examination reveals that after velocity field data are supplied the model velocity field quickly returns to close to its previous value and often overshoots. The restoration time is fast, of the order of 1 day. Figure 17 shows further that the region which shows least improvement to velocity data is the equatorial band - just that region where Section 3 suggests its impact should be greatest.

5.2. Wave Assimilations

One of the most important waves in the dynamics of low-latitude adjustment is the equatorial Kelvin wave. The energy of this wave is equipartitioned between potential and kinetic energy, and the previous arguments suggest that both height and velocity information would be equally important for initialising such a wave. This has been verified analytically.

To test these ideas, an equatorial channel version of the Indian

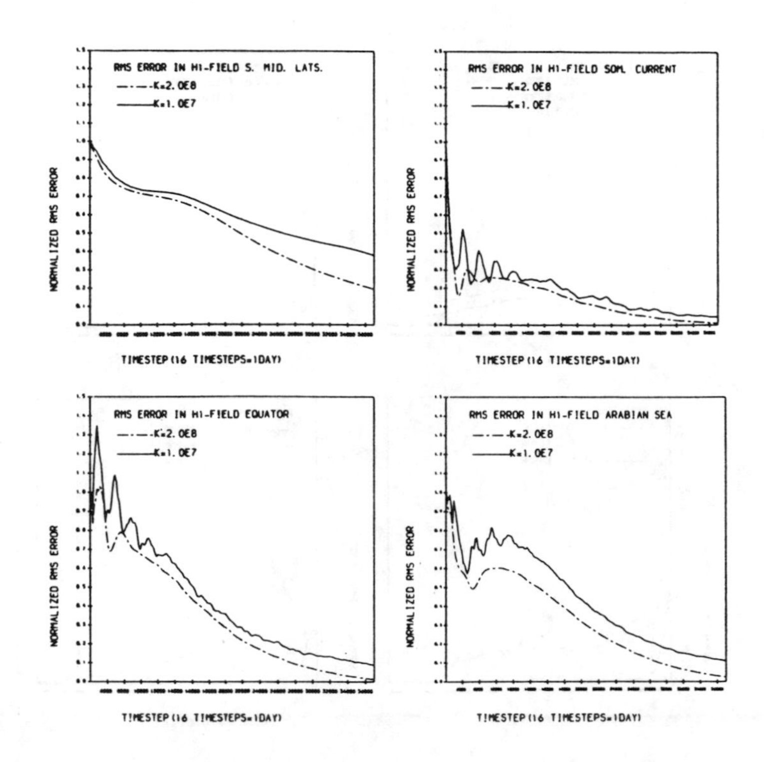

Figure 15. The normalized root mean square errors in the model height field when the model is given wind stress forcing data only. Two different cases are shown, namely when the coefficient of eddy viscosity is 2×10^8 (dashed) and 1×10^7 (solid).

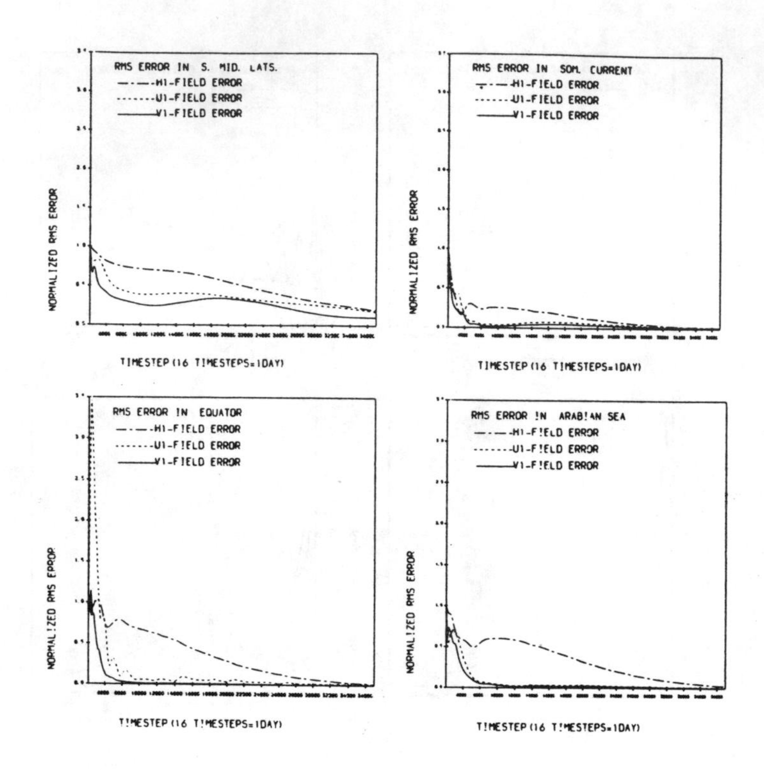

Figure 16. The normalized root mean square errors in height field (H1) and the velocity field components (U1,V1) when the model is given wind stress data only. Coefficient of eddy viscosity = 2 x 10^8.

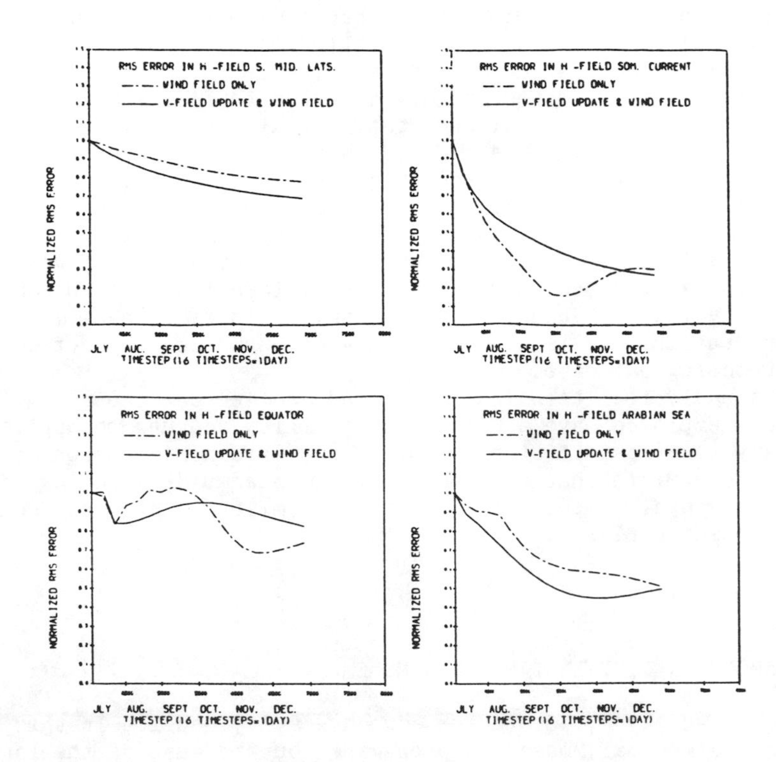

Figure 17. The normalized root mean square error in the height field when the model is given wind stress data only and velocity field every 10 days plus wind stress data.

Ocean model was constructed; i.e., the geometry was extended in the zonal direction, the ocean coastline being omitted. The resolution remained unchanged at 100km. As before, a control run was performed whereby an initial sinusoidal disturbance with the form of a Kelvin wave (i.e. Sin (x-ct) $e^{-\beta y^2/2c}$) was allowed to propagate along the equator. This wave is shown in Figure 18a 15 days after the initialisation of both the height and velocity fields simultaneously.

Following this control run, two initialisation experiments were performed, one in which the u-field is given (v is theoretically zero), and another in which the height field is given. The model height field is shown in Figures 18b and c for both these experiments 15 days after the initialisation. It is apparent that, apart from differences in phase of the planetary and inertia-gravity wave components excited, the model response is identical regardless of height or velocity field initialisation. This is in agreement with our expectations and is again contrary to the finds of Section 3 for the meteorological case.

The analytical technique mentioned earlier and the energy arguments employed have a wider applicability and can be applied to planetary wave assimilations also. These wave assimilation studies are very useful in that they promote a greater understanding of the underlying physical principles of the assimilation of data from more general oceanic fields.

APPENDIX

1. REDUCED GRAVITY INDIAN OCEAN MODEL

This is a two layer, adiabatic reduced gravity model. In this model two vertical normal modes can propagate, but because of the infinite total ocean depth, the barotropic mode propagates out of the system at infinite speed as in the box model leaving only two baroclinic modes.

2. MODEL EQUATIONS OF MOTION

$$u_t - fv = g'\eta_x + \tau^x + k\nabla^2 u \qquad \text{(A1)}$$

$$v_t + fu = g'\eta_y + \tau^y + K\nabla^2 v \qquad \text{(A2)}$$

$$\eta_t = H(u_x + v_y) \qquad \text{(A3)}$$

where H = undisturbed layer depth.

η = layer interforce displacement

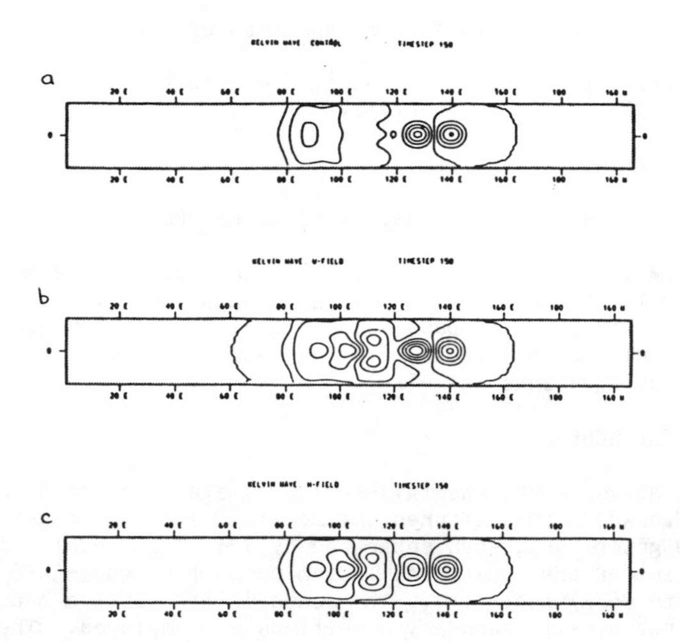

Figure 18. The height field in an equatorial channel model 15 days after the start of calculations in which the model is given height (mass) field data and velocity field data appropriate to an equatorial Kelvin wave (panel a). Panels b and c correspond to giving the model only height field or velocty field data respectively. In the latter two cases, the Kelvin wave response is the same, but only half of that of the control.

Panel a: Contour spacing = 200 cm.
Min = -800 cm, Max = 800 cm.

Panels b and c: Max, Min and contour spacing half that in panel a.

(u,v) = (zonal, meridional) components of velocity

$g' = g(\rho_2-\rho_1)/\rho_2$ where ρ_1 and ρ_2 are densities of upper and lower layers respectively.

K = coeffient of eddy·

(τ^x,τ^y) = (zonal, meridonal) wind stress forcing.

The sea surface wind stress (τ^x,τ^y) is applied as a body force over the whole of the upper layer. Horizontal friction is present also. The lower layer is passive, so there are no currents or pressure gradients in this layer. This assumption was made in the derivation of equations (A1)-(A3).

3. NUMERICAL SCHEME

This model solves equations (A1)-(A3) on a staggered grid and uses a leap-frog forward time differencing scheme. The entire grid domain is 63 x 59 gridpoints, each separated by 1°. This model incorporates the coastline of the equatorial Indian Ocean between (-34°S to 26°N) and (36°E to 96°E) inclusive. The southern and eastern boundaries are solid and no-slip boundary conditions are employed. The model is forced using observed monthly mean wind stress data (FGGE Data) linearly interpolated in time at each gridpoint to provide a continuous wind stress field in time. Unless otherwise specified, the following parameters are used: $K = 2 \times 10^8$, $H_1 = 10^4$cm, $g' = 3ms^{-2}$, where $g' = g(\rho_2-\rho_1)/\rho_2$. One tenth strength wind stress data was used initially to provide a linear model response, i.e., nonlinear terms are small since (u,v) are small.

REFERENCES

Daley, R., On the optimal specification of the initial state for deterministic forecasting, Mon. Wea. Rev., 108, 11, 1719-1735, 1980.

Lorenz, E., On the prevalence of aperiodicity in simple systems. In Global Analysis: Lecture Notes in Mathematics 735, Springer Verlag., Berlin, 1978.

Morel, P., and O. Talagrand, Dynamic approach to meteorological data assimilation, Tellus, 26, 3, 334-343, 1974.

Temperton, C., Some experiments in dynamic initialization for a simple primitive equation model, Quart. J. R. Met. Soc., 99, 303-319, 1973.

Webb, D. J. and A. M. Moore, On the assimilation of altimeter data into ocean models, submitted to J. Phys. Oceanogr., 1985.

DATA ASSIMILATION, MESOSCALE DYNAMICS AND DYNAMICAL FORECASTING

Allan R. Robinson
Harvard University
Center for Earth and Planetary Physics
Cambridge, Massachusetts

ABSTRACT

We first introduce the concepts and methods of optimal field estimation and data assimilation. The meteorologists example is mentioned and special considerations for oceanography discussed. The research of the Harvard group for open regions of the mid latitude ocean is overviewed including hindcasting the POLYMODE SDE data, a real time forecast in the California Current and the setting up of a ODPS in the Gulf Stream ring and meander region.

1. INTRODUCTION

Data assimilation is a technique for the description of ocean currents and related fields which is starting to be used by dynamical oceanographers, and which is expected to be increasing and of crucial importance to ocean science in the future. It is related to the 'four dimensional assimilation' methods of modern meteorology (Bengtsson, Ghil and Kallen, 1981) and the 'optimal estimation' methods used by e.g. electrical engineers and astronomers (Liebeldt, 1967). The use of the technique can be considered for application within any physical system which has been well enough studied previously so that a reasonably accurate general dynamical model exists. Simply stated, the method consists of melding model forecast output with new observations to obtain a combined estimate of the currents which is generally better than either the dynamical forecast fields alone or the observed fields alone. The two estimates are considered independent, and they are combined using coefficients weighted with respect to the relative accuracy of each estimate. The 'relative accuracies' are measured in terms of the error statistics of past forecasts and previous measurements. Thus a knowledge of the relevant statistics and a statistical model is also required. The weight coefficients are chosen to minimize the expected value of a preselected error norm, e.g. the root-mean square difference between the estimate and the true field.

J. J. O'Brien (ed.), Advanced Physical Oceanographic Numerical Modelling, 465–483.

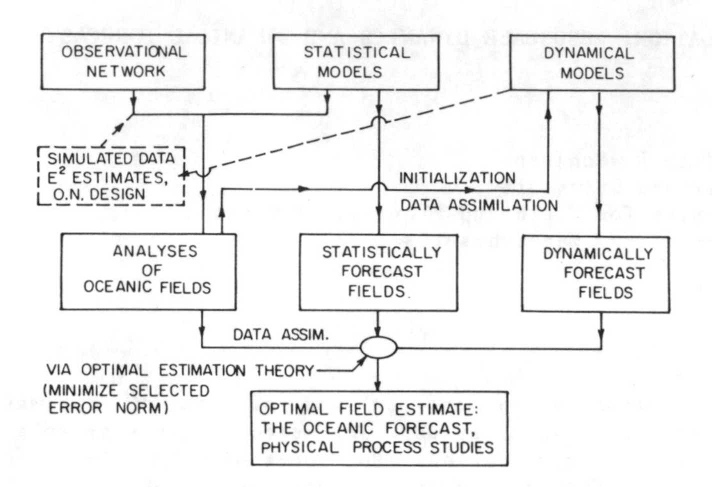

Figure 1 - A schematic of the components of an oceanic descriptive system.

We have discussed some of the advantages and problems associated with the introduction of the method into oceanography in terms of the concept of an Oceanic Descriptive Predictive System (ODPS) (Robinson and Leslie, 1985) which is schematized in Figure 1.

Consider the simplest case of estimating a field (ψ) with zero mean, in terms of two independent and unbiased estimates (ψ_1,ψ_2). With an overbar denoting an expectation value, the error variances are given by

$$E^2_{1,2} = \overline{(\psi_{1,2} - \psi)^2}$$

If the errors are uncorrelated, ie. $\overline{(\psi_1 - \psi)(\psi_2 - \psi)} = 0$, then the optimal estimate (ψ_0) which minimizes the expected mean square error, $E^2 = \overline{(\psi_0 - \psi)^2}$, is given by the linear combination

$$\psi_0 = \frac{E_2^2}{E_1^2 + E_2^2}\psi_1 + \frac{E_1^2}{E_2^2 + E_1^2}$$

with

$$E^2 = \frac{E_1^2 E_2^2}{E_1^2 + E_2^2}$$

In this simple case the relative accuracies are easily interpreted intuitively and the property that E^2 is less than either of $E^2_{1,2}$ is apparent. A recent and readable introduction to the general formalism may be found in Ghil et al. (1981).

Consider the case that ψ_1 is a field estimate obtained from a dynamical model forecast and that E_2 is a field estimate obtained from an observational data set. A number of technical and scientific issues are involved which are novel in oceanography. The ψ_2 field estimate will generally be made from measurements irregularly distributed in space and time, and taken by a variety of sensors with different accuracies. The physical field of interest must be estimated by combining the different kind of measurements. Moreover, a regularly gridded and synoptic estimate is required via some multivariate analysis scheme (Gustafsson, 1981; Carter and Robinson 1985). Relevant schemes involve optimal space time interpolations based on the statistics of historical data sets. For the ψ_1 estimate a verification must be made of a numerical dynamical model and a useful and stable procedure for initialization of the model with real data developed. These issues represent substantial and challenging research tasks. The verification procedure is iterative. As a model's behavior is revealed in the assimilation procedure, and as more data is collected, model improvements and developments naturally will occur. The initialization problem is known to be interestingly complex from both mathematical and physical viewpoints. Well- vs. Ill-posedness and the nonlinear mixing of environmental noise and signal are two examples. A special and knotty problem facing oceanographers is that of open boundary conditions, since data sets intensive enough for assimilation are and will be available essentially in limited areas only.

Finally we mention the need to know computationally, observational and physical error sources and the structure of error fields associated with all of the operations involved. Such direct information is generally inadequately available from real oceanographic data today, but two approaches are possible: i) the use of computer simulated operations on dynamical model generated simulated fields, and ii) the use of simple ad hoc error models with very few degrees of freedom.

The initialization and boundary condition problems and associated data requirements are somewhat different depending upon the relevant dynamical model i.e. linear or nonlinear, primitive equation, shallow-water, quasigeostrophic or equatorial approximation, etc. Oceanographers interested in data assimilation in estuarine, coastal, polar, mid-latitude or equatorial regions and high-frequency, mesoscale, low frequency or climatological phenomenon will be undertaking research problems with both considerable overlap and distinctive features (Robinson, 1985). Following the general discussion of this section, the remainder of the chapter will be devoted to an overview of current research by the Harvard group on data assimilation for the oceanic synoptic/mesoscale, most closely tied to the meteorological problem. An approach by Soviet oceanographers is presented by Timchenko (1984). Research on the equatorial problem is being carried out by Philander and Pacanowski (1984) and Cane and Patton (1984). The earliest work by oceanographers was probably in the area of higher frequency phenomena as illustrated by Schwiderski (1980) and Le Provost (1978). These references are intended to be exemplary rather than exhaustive.

Optimal field estimates are important in physical oceanography in order to provide the basis for fundamental studies of local dynamical processes. Such estimates are also important broadly throughout ocean science and applied marine science, since the physical fields transport and disperse dissolved chemicals, nutrients, pollutants, etc. The extension of the method to the optimal estimation of biological, chemical etc. fields is evidently desirable and possible. Data today is collected from a variety of sensors and platforms including ships, moored and free-floating buoys, and satellites. Such data sets are inherently gappy and asynoptic, and usually far from capable on their own of describing the complete regional or global fields of interest. Dynamical interpolation and extrapolation by nowcasting and forecasting with a dynamical model increases the usefulness of the data. Data assimilation procedures are involved in any method of deriving initialization fields for dynamical model runs. As oceanographers gain experience, they must be expected to emulate meteorologists by using sophisticated weights to merge new data continually with model integration fields between initialization and reinitialization times.

In overviewing the meteorological status, Dr. P. Morel (1981) lists five reasons why data assimilation is essential in the atmospheric context: i) geographical gaps in the twice-daily synoptic measurements of the conventional observations of pressure, temperature and wind, ii) the discrepancy between the conventional observations as point measurements and the true volume averages required by numerical models, iii) the inherently asynoptic character of remote measurements obtained from sensors borne by orbiting satellites, iv) the limitation of cloud motion winds to only one or two vertical levels, and v) the significant random and systematic errors involved in data processing and the reconstructing of atmospheric fields from remotely sensed data. He concludes:

> 'Any forecasting scheme must therefore be initiated or initialized at time t = 0 (say) by merging the new observations with the currently estimated meteorological fields, computed on the basis of earlier observations collected at times t < 0. Formally, the problem consists in optimizing the generalized N - dimensional 'trajectory' of the model, considered as a mechanical system with N degrees of freedom, while taking into account all available information at times t < 0, as well as the dynamical constraints between successive (model) states, specified by the governing dynamical equations. This process of merging new data with the ongoing integration of a numerical forecasting model is known as 'data assimilation' or equivalently '4-dimensional data assimilation' in consideration of the time-space distribution of the data base.'

The analogous reasoning for the ocean presents a stronger rationale for data assimilation, e.g. no conventional synoptic observational network exists at all, and the subsurface ocean is opaque to remote sensing from satellites. But there are conceivably some advantages to oceanographers. Notably the synoptic/mesoscale time scales (from several days to months) are much slower than atmospheric, allowing a much larger real time window for quasi-synoptic observations. Furthermore, if and when the ocean counterpart of the 'conventional observational network' is established the network design can take into account ab initio modern methods of remote sensing and computer modelling.

Dynamical oceanography today can be regarded as a specialty within geophysical fluid dynamics, which is itself part of modern nonlinear mechanics. The great successes of nineteenth and early twentieth century physics lay in the realm of linear systems. Nonlinear systems, and notably fluid dynamical systems provided formidable barriers to progress, especially to generalizations based on the occasionally (partially) solved problem. As we near the end of the twentieth century, however, the situation is changing. Experience has been gained in dealing with such complex partially random and partially deterministic systems in terms of some solved phenomenological examples and an evolving scientific methodology. Models as theories, data as nature, and the practice of simple hypothesis testing is no longer the whole story. Consider the subtle interplays between historical and ongoing observational data sets and evolving models within the framework of optimal estimation concepts. Powerful computers and the development of computational fluid dynamics are absolutely essential to numerical forecasting and modelling of atmospheric and oceanic

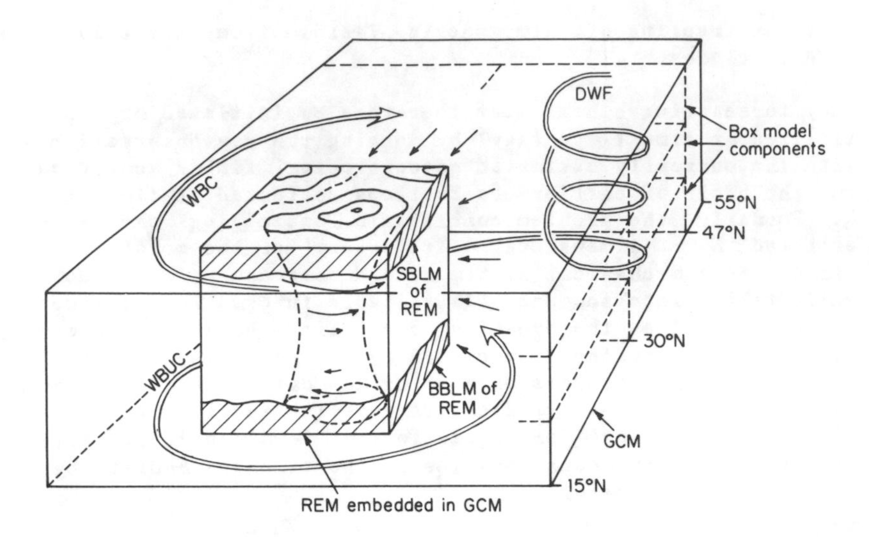

Figure 2 - A heirarchy of coupled models for the study of ocean circulation and dynamics. Included are a General Circulation Model (GCM), a Regional Eddy Resolving Model (REM), Surface (SBLM) and Bottom (BBLM) Boundary Layer Models. Forcing is represented by Direct Wind Forcing (DWF), a Western Boundary Current (WBC) and a Western Boundary Undercurrent (WBUC).

currents and circulations. Supercomputers are needed for four dimensional data assimilation, nowcasts and forecasts in the ocean. Only the advent of Class 7 supercomputers will allow the treatment of interestingly large oceanic regions with reasonably sophisticated techniques (e.g. Kalman-like filtering of the fields evolving about the slowly varying 'present' state of the ocean).

The two new tools that can be expected definitely to advance very substantially the state of ocean scientific knowledge are: satellites and supercomputers. The central role expected of data assimilation, in no small part, lies in its ability to link these two powerful tools and efficiently exploit their scientific utility.

2. MODELS, METHODS AND ISSUES

The Harvard group is carrying out research on the optimal estimation of mid-latitude oceanic synoptic/mesoscale fields which involves a central focus on real ocean data assimilation in dynamical ocean models. Optimal estimation involves the melding of field estimates from the three components of the ODPS system introduced in Figure 1. The high order statistics required for this process which are generally not yet known for the ocean are taken now from computer generated data obtained by simulation runs of the dynamical model. The 'dynamical' model component of the system is in fact a multiscale hierarchy of coupled models developed by the Harvard group in collaboration with Dr. M.G. Marietta's group at Sandia National Laboratories (Robinson and Marietta, 1984, 1985). The hierarchy is intended for studies of ocean current dynamics and of the general circulation, and for transport and dispersion studies over a wide range of scales. (The application of interest to the SNL group is associated with feasibility studies for subseabed disposal of high level nuclear wastes.)

Figure 2 shows the Harvard Open Ocean Model, a Regional Eddy Resolving Model (REM) with attached surface (SBLM) and bottom (BBLM) boundary layers embedded in a coarse resolution General Circulation Model (GCM). Central to our approach and the main topic of the remainder of this chapter is the assimilation of real and intensive data sets (hydrography, currents, sea-surface height, etc.) in the REM. The purposes are to study local dynamical processes and to verify regional model components, so as to construct a general circulation model from regional components which are consistent with all available data. The components of the hierarchy and the coupling mechanisms should be iteratively improved. The present GCM is a converged (2055 model years) prognostic Bryan-Semtner model in idealized geometry. The GCM component is being tuned to large scale hydrographic, geochemical and transient tracer fields, and is itself coarsened to a Box Model (BXM) for simplified chemical and biological applications.

The present REM is the Harvard Open Ocean Model, 'a portable' baroclinic, quasigeostrophic (QG) model which is shown schematically in Figure 3 (Robinson and Walstad, 1986). The numerical model integrates the potential vorticity equation which in nondimensional form is:

$$\frac{\partial}{\partial}t\,[\nabla^2\psi] + \frac{\partial}{\partial}t[\Gamma^2(\sigma\psi_z)_z] = -\alpha V.\nabla\nabla^2\psi - \alpha\,\Gamma^2\,\nabla\,.\,\nabla(\sigma\,\psi_z)_z - \beta\psi_x + f$$

and symbolically

$$\dot{Q} = \dot{R} + \dot{T} = \Delta F_R + \Delta F_t + \Delta F_p + f$$

where $V = -k\times\nabla\psi$, $\alpha = t_0 v_0/d$, $\beta = t_0\,\beta_0 d$, $\Gamma^2 = (f_0 d/N_0 H)^2$.

The second equation gives the notation we will use to refer to the corresponding terms in the first equation. R indicates relative vorticity, T, thermal vorticity (the stretching term), Q the total vorticity, Δ F indicates an advective flux divergence (relative, thermal and planetary and f is the filter. Γ^2 measures the relative importance of T and R processes and α measures the strength of nonlinear advections relative to local time changes.

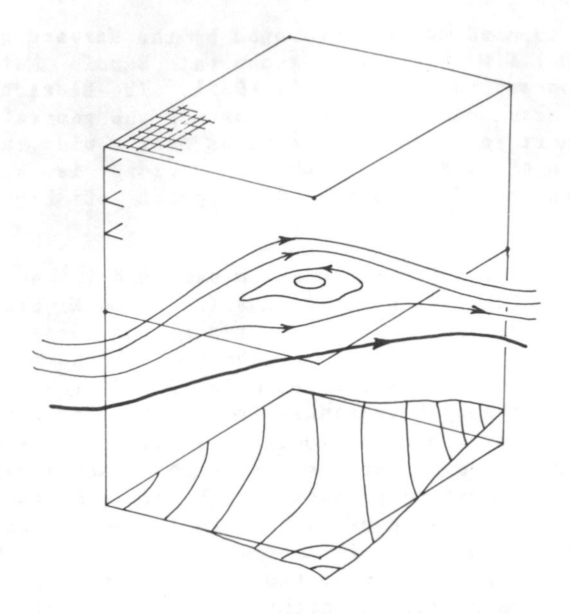

Figure 3 - The Harvard Open Ocean Model schematically showing the x,y grid, the vertical levels, inflow and outflow and bottom topography.

The arbitrary (water-water) boundaries are depicted as open, but recent work by Dr. J.A. Carton now allows for partially closed boundaries and interior domain islands. The mean stratification and local bottom topography describe the local environment. The fundamental initial-condition/boundary-condition problem (IC/BC) requires an initial QG stream function field which is equivalent to the geostrophic pressure with its associated vorticity field, and the specification of inflow-outflow on the open boundaries as well as vorticity on the inflow points at each level (Charny-Fjortoft-Von Neumann conditions). Numerically the model is finite-element in the horizontal (Haidvogel, Robinson and Schulman, 1980) and finite-difference or collocation in the vertical (Miller, Robinson, and Haidvogel, 1983).

The purposes of data assimilation in the model are: optimal field estimations, dynamical interpolation and forecasting, and model verification. Thus, it is essential to identify and attribute sources of error. Errors arise from i) computational operations, ii) the quantity and quality of observational data, iii) physical inadequacies of the dynamical and statistical models. The latter include explicitly resolved and sub-gridscale physics, the structure of the statistical model and the evaluation of statistical quantities. The interpretation of the difference-field between analyzed observations and dynamical-model forecasts/scientific runs is crucial. The model has been computationally calibrated against some exact advecting-Rossby wave solutions within round-off error and also against simulated data

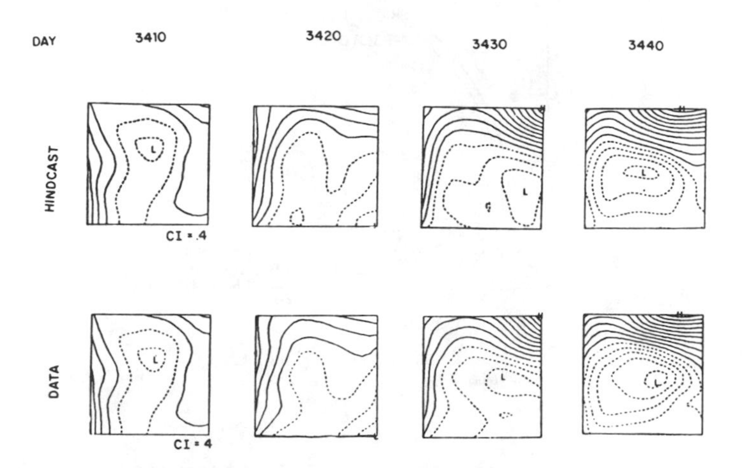

Figure 4a - A comparison of POLYMODE forecast results with observations. A benchmark hindcast uses the observations as initial conditions and provides new boundary conditions for each time step.

characteristic of real oceanic conditions of interest (Miller and Robinson, 1984). In conditions characteristic of the POLYMODE Synoptic-Dynamic Experiments (P-SDE), the model run in a forecast mode with 'true' simulated IC/BC data can maintain a NRMS interior error of only a few percent for longer than a year.

Initializing the model with real data requires the regular gridding in space and time of irregular, asynoptic and gappy data. This is accomplished in the horizontal by a multi-variate, anisotropic-mixed space-time objective-analysis scheme (Carter, 1983; Carter and Robinson, 1985) and in the vertical by projecting onto the QG dynamical modes or onto empirical orthogonal functions (Smith, Mooers, and Robinson , 1985). Jets, fronts, multiple types of eddies, etc. in complex oceanic regions require highly anisotropic, non-stationary and inhomogeneous statistical models which we are researching.

A real data forecast experiment from the P-SDE, shown in Figure 4a and Figure 4b summarizes the behavior of the model with respect to the quality of the boundary condition data made available. Fig. 4a is a 'benchmark' calculation in which the boundary conditions have been updated with new data at every time step. A 'Persistence' run is one in which the initial boundary conditions are maintained throughout the run. The statistically forecast B.C. run uses the same data as a persistence run, ie. one realization, but time extrapolates the boundary

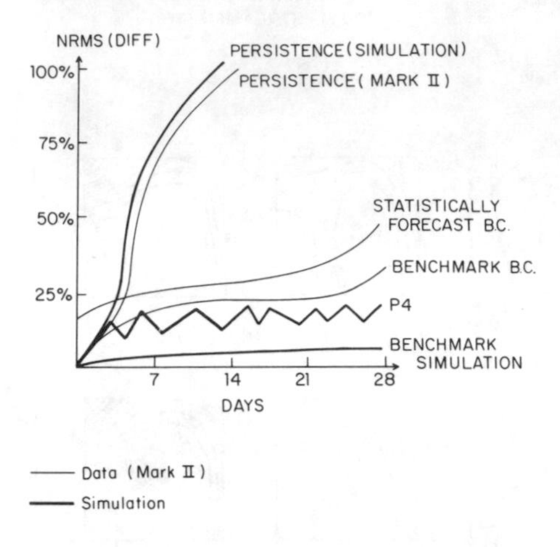

Figure 4b - Normalized root-mean squared streamfunction difference (NRMS (DIFF)) vs. time for the Harvard Open Ocean Model for various initial and boundary conditions. Benchmark continuously updates the boundary conditions, statistically forecast B.C. uses the initial conditions and predicts them forward in time, and persistence maintains the initial conditions. Mark II is POLYMODE data and simulation is a simulation of that data.

Figure 5 - A prototype mixed layer forecast using data from the California Current Region (OPTOMA XI).

condition via objective analysis. The result is close to benchmark for the month of integration. 'Simulation' is a simulated P-SDE data set and Mark-II is an analysis of actual P-SDE data. The notation P4 designates an experiment in which the boundary conditions are persisted for 4 days; note that the error growth is controlled well. An important model development recently completed is the attachment of an upper mixed layer model to the REM for dynamical studies and particularly for the assimilation of satellite IR-SST. Mr. L. Walstad accomplished (Figure 5) a first prototype forecast in real time of the California Current jet and eddy regime (Mooers and Robinson, 1984).

In order to use our system as a tool for learning local dynamical processes directly from oceanic data sets, an open-ocean energy and vorticity analysis scheme (EVA) has been developed (Pinardi and Robinson, 1985). The present version employs consistent quasigeostrophic energy equations

$$\frac{\partial}{\partial}t\ K_0 = -\alpha\nabla(u_0K_0) - \nabla_H\left[p_1u_o + p_0u_1\right] - (p_0w_1)_z + \rho_0w_1,$$

or symbolically

$$K = \Delta F_K + \Delta F_\pi + \gamma f_\pi - b,$$

and,

$$\frac{\partial}{\partial}tA = -\alpha(u_0A_0) - \rho_0w_1,$$

$$A = \Delta F_A + b.$$

K, A are respectively kinetic and available gravitational energies, b is buoyancy work, and the subscripts refer to the order of the contribution in terms of a Rossby number expansion. The horizontal pressure work flux divergences are evaluated entirely in terms of the zeroth-order geostrophic pressure field, since

$$\Delta F_\pi = -(p_0u_{0t}xk + \alpha p_0u_0\nabla(kxu_0) - \beta yu_0p_0)$$

$$\Delta F_\pi = \Delta F_\pi^t + \Delta F_\pi^\alpha + \Delta F_\pi^\beta$$

Running real data fields through the dynamical model as a quasigeostrophic filtering process allows the evaluation of high derivatives required for consistent energetics which is not possible in terms of simply analyzed data.

3. EXPERIMENTS AND SIMULATIONS

In the context of the OPTOMA program (Ocean Prediction Through

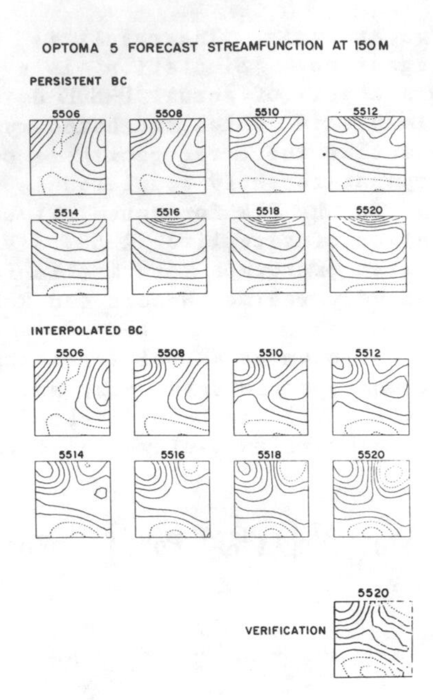

Figure 6 - Comparison of OPTOMA V Forecasts. Top: A real time 14 day forecast using persisted boundary conditions. Bottom: A 14 day forecast using boundary conditions linearly interpolated between days 5506 and 5520.

Observations Models and Analysis), conducted by the Harvard group together with Prof. C.N.K. Mooer's group at the Naval Postgraduate School, data is being collected and the ODPS developed and verified. Figure 6 shows a successful two-week real time forecast within a 150 kmj sq. region (Robinson, et al., 1984) in which two eddies merge to form a zonal jet via internal dynamical processes. An after the fact 'forecast experiment' with boundary condition updating (Figure 6) replicates the verification data very well and illustrates the power of dynamical model interpolation; the eddy merger could not be described by the two data sets alone. The boundary conditions were updated using a linear interpolation between days 5506-5520. EVA shows the merger to be a 'finite amplitude' barotropic instability event (Robinson, Carton, Pinardi and Mooers, 1986) and the vorticity balance is illustrated in Figure 7.

New strategies for realistic simulation of oceanic regions are being studied which involve running the dynamical model backward and forward between connected and disconnected field realizations. Based on such data sets, we have evolved via simulated experiments a <u>modular concept</u> of model initialization for nowcasting and forecasting.

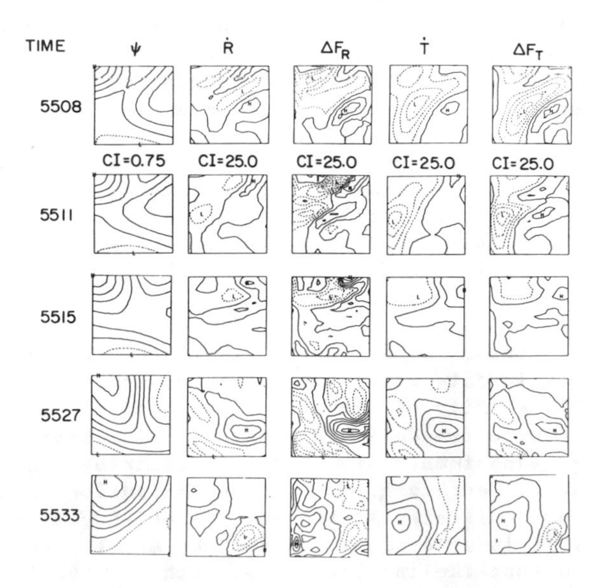

Figure 7 - OPTOMA V Vorticity Analysis - ϕ = streamfunction, $\dot{R}$ = time rate of change of thermal vorticity, ΔF_R = divergence of relative vorticity advective flux, ΔF_T = divergence of thermal vorticity advective flux.

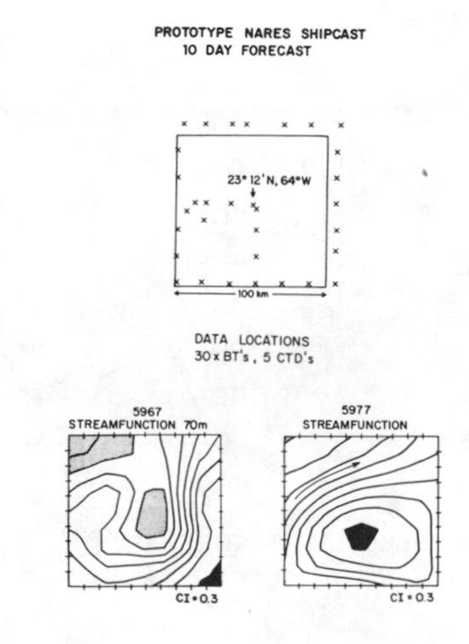

Figure 8 - A prototype at-sea forecast (shipcast) using data collected over the Nares Abyssal Plain showing the observations, the initial streamfunction field (5967), and the forecasted streamfunction field.

This involves i) dynamical interpolation of boundary data into a data empty module (as large as 75 km for the California Current region), ii) building up a large region out of connected modules, and iii) reinitialization of regions with past data and new boundary conditions before melding forecasts with new data. Taking advantage of hardware miniaturization and our growing real time experience, we are initiating 'shipcasting', i.e., dynamical forecasting at sea with shipboard computers. Including EVA will allow data acquisition in dynamically crucial regions via the real time evolution of experiments at sea. A prototype shipcast over the Nares Plain in the (Walstad, et al., 1985) Northwest Atlantic is shown in Figure 8.

Demey and Robinson (1985) have simulated the assimilation of satellite altimetric data using the P-SDE set and demonstrated (Figure 9) the remarkable ability of the dynamical model to generate the correct flow after a few days; the procedure involves projecting onto the EOF's. The experiments involve initializing the model by a B,P,E,A procedure: Benchmark, Projection, Extension, Altimeter Extension. The numbers 1,2 refer to the number of EOF's utilized. The deep level fills in after 2 or 3 weeks, by model generated nonlinear interactions, so that the initialization with 1 satellite sampled EOF rivals benchmark.

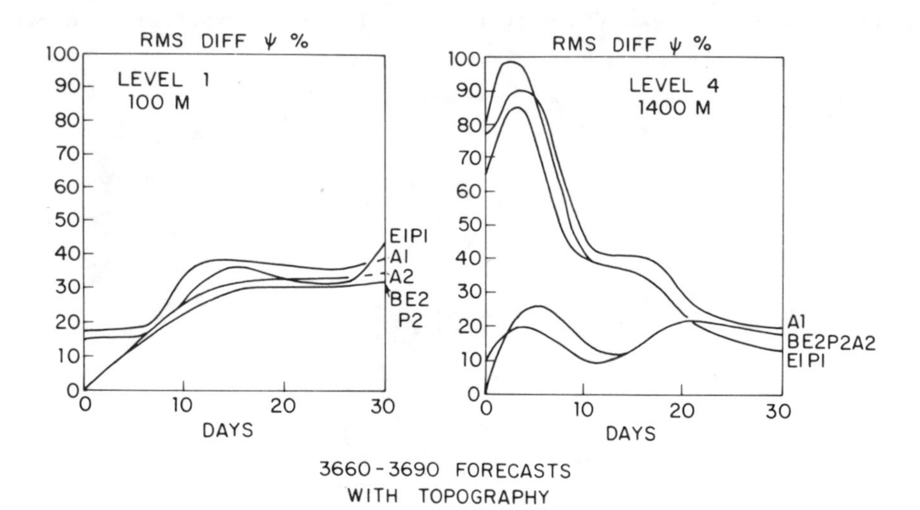

Figure 9 - Forecast results simulating the assimilation of satellite altimetric data. The letters refer to the procedure for introducing deep data; B,P,E,A: Benchmark, Projection, Extension, Altimeter Extension. The numbers refer to the number of EOF's utilized.

Finally, we have now set up the Harvard models and the ODPS in the Gulf-Stream region. Our simplest strategy involves using 'feature models' (Figure 10) for the stream and rings, so that IR identified stream meander and ring locations allow initialization. The assimilation procedure and forecast thus involves three steps: i) an initial dynamical adjustment of features, ii) dynamical interpolation between the features and then iii) the dynamical evolution of the flow. Using the NOAA analyzed SST alone, together with Dr. N. Pinardi and Mr. M. Spall, I initialized the dynamical model for November 23, 1984. As shown in Figure 11 the stream developed a deep sock meander and snapped off a cold ring, which was observed to occur in the NOAA analyzed SST alone; the dynamical model was initialized for November 23, 1984. As shown in Figure 11 the stream developed a deep sock meander and snapped off a cold ring, which was observed to occur in the NOAA analysis of December 1984. This scheme is being developed for research in the ONR sponsored SYNOPS program and will include the assimilation of GEOSAT data in collaboration with researchers at Johns Hopkins/APL (Dr. Jack Calman) and at Sandia National Laboratories (Dr. M. Marietta).

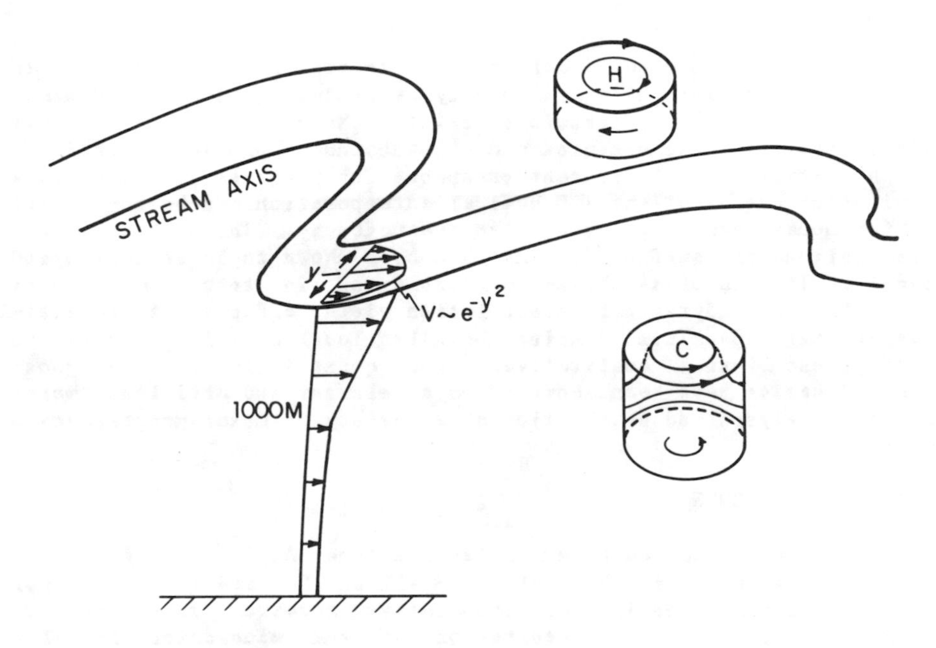

Figure 10 - Schematic of the feature models for Gulf Stream Region forecasts.

FORECAST STREAMFUNCTION AT 300M

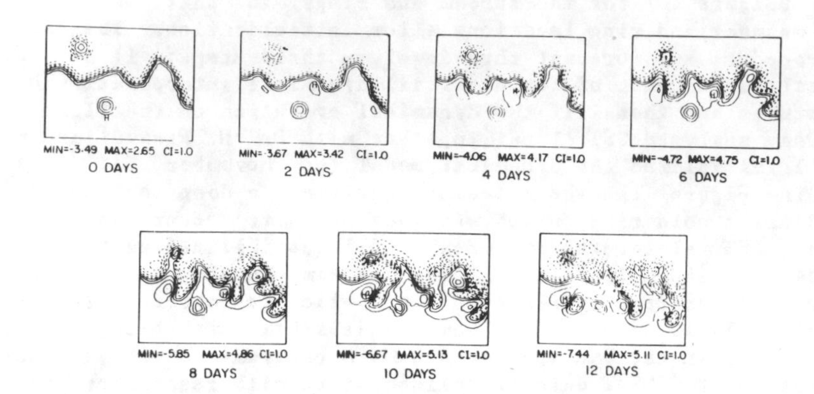

Figure 11 - Initialization of a Gulf Stream forecast from NOAA IR data showing formation of a sock meander (Days 8-10) and a cold ring (Day 12).

4. CONCLUSIONS

The Harvard dynamical model has been initialized with real data from several locations using various types of data. Real-time dynamical forecasting is now a practical reality. We learn about internal regional evolution via dynamics and about boundary propagation effects through statistics. The present prospects for progress in ocean data assimilation are excellent. Dynamical interpolation can be a powerful tool for modern synoptic/descriptive oceanography. The deep dynamical extrapolation of surface fields has been shown to be effective and accurate. The use of satellite measurements, in situ limited-area data sets, and dynamical interpolation yields efficient field estimates in large space/time domain. Detailed local dynamical processes affect these issues qualitatively and quantitatively. Quasigeostrophic dynamics have been shown to be a relevant and efficient means for the analysis and description of a variety of meso/synoptic scale phenomena.

5. ACKNOWLEDGEMENTS

I am deeply indebted to my colleagues James A. Carton, Wayne G. Leslie, Nadia Pinardi, Michael A. Spall and Leonard J. Walstad for stimulating interaction in our cooperative research ventures. Mr. Leslie's help in the preparation of the manuscript is also appreciated. It is a pleasure to thank the Office of Naval Research for their support under Contract Number N00014-84-C-0461.

6. REFERENCES

1. Bengtsson, L., M. Ghil and E. Kallen, (1981). 'Dynamic Meteorology: Data Assimilation Methods,' Springer-Verlag, New York, 330 pages.
2. Cane, M. and R. Patton, (1984). 'A Numerical Model for Low-Frequency Equatorial Dynamics,' Journal of Physical Oceanography, 14, (12), 1853-1863.
3. Carter, E.F. (1983). 'The Statistics and Dynamics of Ocean Eddies,' Reports in Meteorology and Oceanography, No. 18, Division of Applied Sciences, Harvard University.
4. Carter, E.F. and A.R. Robinson, (1985). 'An Analysis Model for the Estimation of Oceanic Fields,' to appear, Journal of Atmospheric and Oceanic Technology.
5. Demey, P. and A.R. Robinson. (1985). 'Simulations for the Assimilation of Satellite Altimetric Data at the Oceanic Mesoscale.'
6. Ghil, M., S. Cohn, J. Tavantzis, K. Bube and E. Isaacson, (1981). in Dynamic Meteorology: Data Assimilation Methods, (L. Bengtsson, M. Ghil and E. Kallen, eds.), Springer-Verlag, NY, NY.
7. Gustafsson, N, (1981). 'A Review of Methods for Objective Analysis,' in Dynamic Meteorology: Data Assimilation Methods, (L. Bengtsson, M. Ghil and E. Kallen, eds.), Springer-Verlag, NY, NY.
8. Haidvogel, D.B., A.R. Robinson, and E.E. Shulman. (1980). 'The Accuracy, Efficiency and Stability of Three Numerical Models with Application to Open Ocean Problems,' Journal of Computational Physics 34 (1), 38-70.
9. Le Provost, C., (1978). 'The Numerical Simulation of the Non-Linear Propagation of a Long Wave in a Channel of Constant Depth,' Comparison of Several Methods of Finite Differences, Oceanologia, 9, 95-113.
10. Liebeldt, P.B., (1967). 'An Introduction to Optimal Estimation,' Addison-Wesley, Reading, MA, 273 pages.
11. Miller, R.N., A.R. Robinson and D.B. Haidvogel. (1983). 'A Baroclinic Quasigeostrophic Open Ocean Model,' Journal of Computational Physics, 50 (1), 38-70.
12. Miller, R.N. and A.R. Robinson. (1984). 'Dynamical Forecast Experiments with a Baroclinic Quasigeostrophic Open Ocean Model,' in Proceedings of Conference on Predictability of Fluid Motions, (G. Holloway and B. West, eds.), American Institute of Physics, Proceeding No. 106, AIP, NY.

13. Mooers, C.N.K. and A.R. Robinson. (1984). 'Turbulent Jets and Eddies in the California Current and Inferred Cross-Shore Transports,' Science, 223, 51-53.
14. Morel, P., (1981). 'An Overview of Meteorological Data Assimilation,' in Dynamic Meteorology: Data Assimilation Methods, (L. Bengtsson, M. Ghil and E. Kallen, eds.), Springer-Verlag, NY, NY.
15. Philander, S.G.H. and R.C. Pacanowski, (1984). 'Simulation of the Seasonal Cycle in the Tropical Atlantic Ocean,' Geophysical Research Letter, 11, 802-804.
16. Pinardi, N. and A.R. Robinson. (1985). 'Local Quasigeostrophic Energy and Vorticity Analysis of Mesoscale,' Dynamics of Atmospheres and Oceans, (To Appear).
17. Robinson, A.R., (1985). 'Data Assimilation, Mesoscale Dynamics and Dynamical Forecasting,' in WOCE/TOGA Workshop on Inverse Modeling and Data Assimilation (R. Evans, Ed.), University of Miami, Miami, FL.
18. Robinson, A.R., J.A. Carton, C.N.K. Mooers, L.J. Walstad, E.F. Carter, M.M. Rienecker, J.A. Smith, and W.G. Leslie. (1984). 'A Real Time Dynamical Forecast of Ocean Synoptic/Mesoscale Eddies,' Nature, 309 (5971). 781-783.
19. Robinson, A.R. and M.G. Marietta, editors. (1984). Report of the Second Annual Interim Meeting of the Seabed Working Group, Physical Oceanography Task Group (POTG), Fontainebleau, France, 'Research, Progress, and the Mark A Box Model for Physical, Biological and Chemical Transport,' Sandia National Laboratories Report, SAND 84-0646.
20. Robinson, A.R. and W.G. Leslie. (1985). 'Estimation and Prediction of Oceanic Fields,' Progress in Oceanography, 14, 485-510.
21. Robinson, A.R. and M.G. Marietta, editors. (1985). Report of the Third Annual Scientific Workshop meeting of the Seabed Working Group, Physical Oceanography Task Group (POTG), Neuchatel, Switzerland, 'Research, Progress and the Description, Modelling Simulation and Dispersal Characteristics of Potential Disposal Sites in the North Atlantic,' Sandia National Laboratories Report, SAND 85-1729.
22. Robinson, A.R., J.A. Carton, N. Pinardi, C.N.K. Mooers. (1985). 'Dynamical Forecasting and Dynamical Interpolation: an Experiment in the California Current,' submitted.
23. Robinson, A.R. and L.J. Walstad. (1985). 'Numerical Modelling of Ocean Currents and Circulation,' Numerical Fluid Dynamics, (to appear).
24. Smith, J.A., C.N.K. Mooers, and A.R. Robinson. (1985). 'Estimation of Baroclinic, Quasigeostrophic Model Amplitudes from XBT/CTD Survey Data,' Journal of Atmospheric and Oceanic Technology, (to appear).

25. **Swiderski, E.W., (1980). 'On Charting Global Ocean Tides,' Review Geophysical Space Physics, 18, 243-268.**
26. **Timchenko, I.E., (1984). 'Stochastic Modeling of Ocean Dynamics,' Harwood, Switzerland, 311 pages.**

SENSITIVITY STUDIES AND OBSERVATIONAL STRATEGIES FROM A NON-LINEAR FINITE-DIFFERENCE OCEAN CIRCULATION MODEL

Jens Schröter
Max-Planck-Institut für Meteorologie
Bundesstrasse 55
2000 Hamburg 13
FRG

ABSTRACT. Any dynamical model driven by "observed" forcing fields (e.g. the wind) has a true solution uncertainty owing to observational errors in the driving. This uncertainty is usually hidden from view because conventional numerical methods do not easily calculate it. We have explored with a finite difference, non-linear model the uncertainties in interesting flow properties (western boundary current transport, potential and kinetic energy . . .) owing to the uncertainty in the driving surface boundary conditions. The procedure is based upon non-linear optimization methos. The same calculations permit us to study the importance of the addition of new information, in particular from altimetry, tomography, current meters, as a function of region of measurement, and accuracy, leading to inferences about optimal observing strategies.

SENSITIVITY STUDIES

We have studied the uncertainties in the steady state solution of a dynamical ocean circulation model that are introduced by uncertainties (e.g. observational errors) in the driving forcing field. A whole set of solutions exists the members of which obey the dynamics of the flow and fit the forcing field within prescribed bounds. Many inverse techniques find a unique solution within this set by minimizing some kind of norm of both the solution and the differences between observed and calculated "data" (e.g. least squares fitting). The aim of these techniques is to find a most probable solution that is "smooth" in some sense.

Contrary to this "best estimate" approach we explore extreme solutions. In many cases one is not interested in the solution of the model, that is the oceanic circulation in all its detail, but in some properties of the solution, e.g. meridional heat transport, sea surface temperature gradients, western boundary current transport, etc. We model these properties directly and look for solutions which maximize or minimize their values by varying the forcing field only within the limits set by the prescribed bounds of the uncertainties. The resultant solution is neither necessarily smooth nor is there a preference for the calculated "data" to lie close to the observed data (here the measured forcing field). If the maximum and minimum values of the property we are interested in lie close together we might be satisfied with the result and disregard the indeterminacy that remains in the details of the solution.

The mathematical procedure we apply to the problem is a non-linear

J. J. O'Brien (ed.), Advanced Physical Oceanographic Numerical Modelling, 485–494.

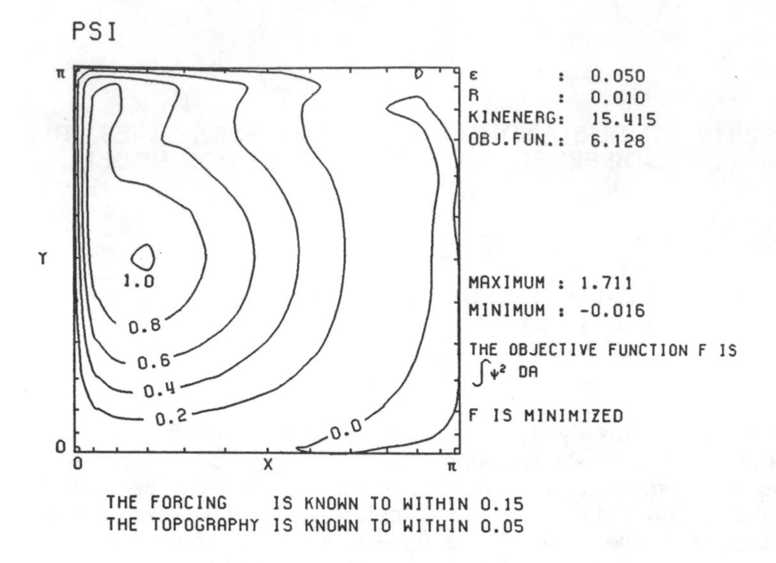

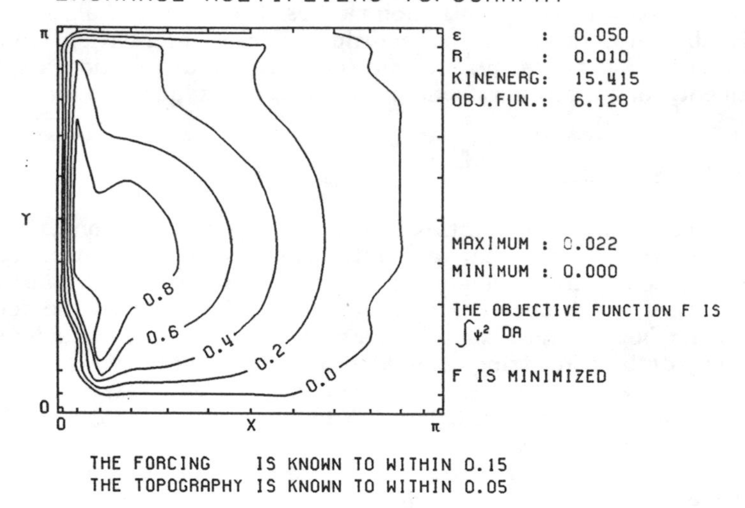

Fig. 1: One layer model:

a. Streamfunction that minimizes the potential energy **subject to** I) the uncertainty in the forcing is not larger than 0.15 and II) uncertainty of the sea surface topography is not larger than 0.05.

b. Sensitivity of the extreme value of the objective function F to small variations in the uncertainty of the sea surface topography.

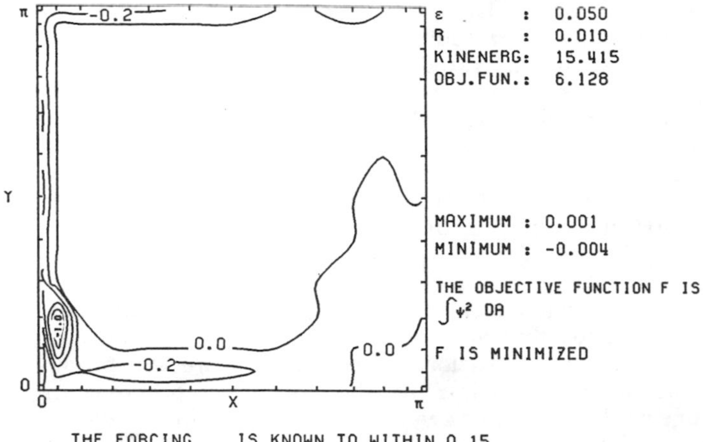

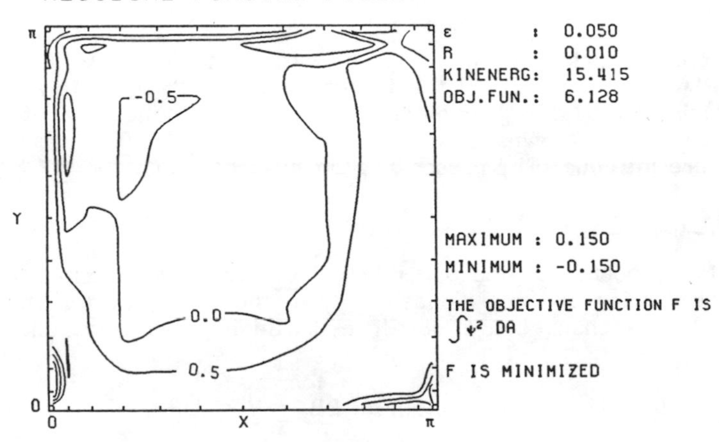

c. Sensitivity of the extreme value of F to small variations in the bounds of the uncertainty of the wind-forcing.

d. Deviation of the forcing that corresponds to the optimal solution from the measured forcing. Values in the interior are close to zero.

optimization method ("non-linear programming"). This method enables us to maximize or minimize any linear or non-linear diagnostic function that can be calculated from the ocean circulation model. The modelled dynamics may be non-linear, too. Furthermore, additional data such as sea surface topography, density, etc. can easily be tested for compatibility with the extreme solutions. The importance of this new information as well as the previous information, as a function of region of measurement and accuracy is calculated quantitatively. Thus, we have at hand a powerful method for exploring optimal observational strategies.

We will demonstrate the non-linear programming technique with two simple finite difference models of the wind driven circulation. These models were chosen deliberatly for their simplicity. They are not meant to give a realistic picture of the oceanic circulation. Because of their straightforward dynamics which are well understood we will be able to distinguish between features of the solution that are due to the modelled dynamics and those that are introduced by the optimization procedure. Thus we will gradually build up an intuition about the behaviour of the non-linear programming algorithm. This knowledge will be of great value when, at a later stage, we apply the method to a complete and realistic general circulation model.

First we consider the barotropic flow in a square basin of constant depth D and side lenghts ℓ . The nondimensional vorticity equation for the steady state reads

$$R \quad J(\psi . \nabla^2 \psi) + \frac{\partial \psi}{\partial x} + \quad \varepsilon \nabla^2 \psi \quad - \text{forc} = 0 \tag{1}$$

The nondimensional friction parameter ε was set to 0.05, the non-linearity parameter R was set to 0.01 and the forcing term "forc" equal to -sin x sin y. Equation (1) is solved at n gridpoints. We have thus a collection of n non-linear equations in n unknowns ψ which we denote $\underline{M}(\underline{\psi}) = \underline{0}$. Uncertainties in the forcing are introduced by prescribing bound-constraints for $\underline{M}(\underline{\psi})$

$$\underline{LB} \leq \underline{M}(\underline{\psi}) \leq \underline{UB} \tag{2}$$

Note that lower bound $\underline{LB}$ and upper bound $\underline{UB}$ directly describe the uncertainty in the forcing and not deviations from the steady state. The 2n inequalities of (2) are changed to n equalities by adding bounded slack variables to $\underline{M}(\underline{\psi})$:

$$\underline{M}(\underline{\psi}) + \underline{Y} = \underline{0} \qquad \text{subject to } \underline{LB} \leq \underline{Y} \leq \underline{UB} \tag{3}$$

Equation (3) is conventionally written as

$$\underline{C}(\underline{X}) = \underline{0} \qquad \text{subject to } LB_i \leq X_{n+i} \leq UB_i,\ i = 1 \ldots n \tag{4}$$

where the first n components of $\underline{X}$ are the previous unknowns $\underline{\psi}$ and the second n components are the slack variables $\underline{Y}$.

We have studied the following characteristic properties of our modelled circulation:

a) western boundary current transport (ψ_{max})

b) potential energy ($\int \psi^2 da$)

c) kinetic energy ($\int (\nabla \psi)^2 da$

These properties will be called "objective function" and denoted $F(\underline{X})$. The mathematical procedure that finds extreme values of $F(\underline{X})$ subject to the constraints given in (4) will be described briefly:

We minimize the augmented Lagrangian function L defined by:

$$L(\underline{X}, \underline{\lambda}, \rho) = F(\underline{X}) - \underline{\lambda}\, \underline{C}(\underline{X}) + \rho\, \underline{C}(\underline{X})^T\, \underline{C}(\underline{X}) \tag{5}$$

here ρ is a positive penalty parameter that has to exceed some threshold value. $\underline{\lambda}$ is the vector of the Lagrange multipliers that describe the sensitivity of the objective function F to variations in $\underline{C}(\underline{X})$, which in turn are variations of the prescribed bounds on the slack variables (i.e. the bounds on the uncertainties).

$$\lambda_i = \frac{\partial F(X)}{\partial C_i(\underline{X})} = \frac{\partial F(X)}{\partial LB_i} \qquad \left(\text{or } \frac{\partial F(X)}{\partial UB_i}\right) \tag{6}$$

$\lambda_i = 0$ indicates that the slack variable X_{n+i} is not equal to one of its bounds. The $\underline{X}$ and $\underline{\lambda}$ that minimize L are found with an unconstrained minimization routine which uses an 'active set strategy' to overcome problems that are introduced by the simple bounds on $\underline{X}$. For a maximization of F simply replace $F(\underline{X})$ by $-F(\underline{X})$ in (5).

Additional constraints such as knowledge about the sea surface topography are introduced by adding further components to $\underline{C}$ with the simultaneous addition of the apropriate slack variables to $\underline{X}$.

As an example we will solve the model (1) for minimal potential energy while putting bounds for the uncertainty in the forcing equal to $\pm$ 0.15 and the uncertainty in the sea surface topography equal to 0.05. Results are shown in Figs. 1a to 1d. Fig. Ib shows that the topographic constraints are "active" (i.e. reach their bounds) almost everywhere in the basin. Only in regions where ψ is small to topographic constraints contain no additional information, slack variables are not at their bounds indicating zero sensitivity of F to small variations of the bounds. The region of maximum importance of the knowledge about the sea surface topography lies in the southern part of the western boundary current (in our scenario). It is here where we would want to measure the topography with greatest care if we are interested in the potential energy of the circulation. If we maximize F under the same conditions the region of highest sensitivity shifts somewhat to the north due to an increased flow and an increased non-linearity. Fig. 1c shows that the Lagrange multipliers that correspond to the forcing constraints are mainly zero. The bounds prescribed to the forcing have been set too high to control the flow effectively. If we cannot measure the forcing better than $\pm$ 0.15, we will find the measurement of the sea surface topography much more important for the problem considered here.

In another example we will explore the relative importance of temperature (or density) measurements versus wind stress information. To do so we introduce a second layer to our model

$$RJ(\psi_1, \nabla^2 \psi_1) + H_1 J(\psi_1, \psi_2 - \psi_1) + \frac{\partial\psi}{\partial x} + \varepsilon_1 H_1 \nabla^2 (\psi_1 ' \psi_2) - \text{forc} = 0 \tag{7}$$

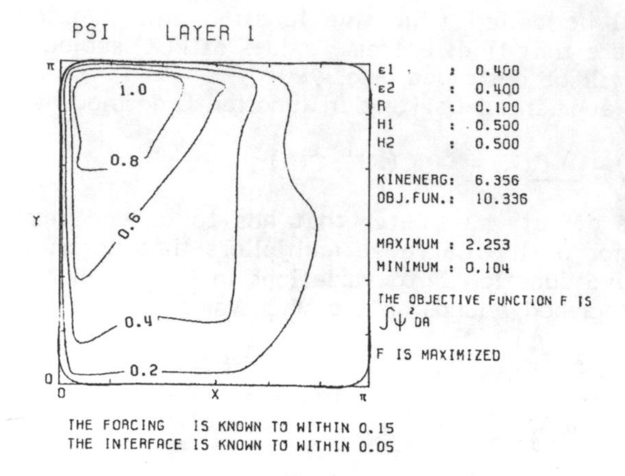

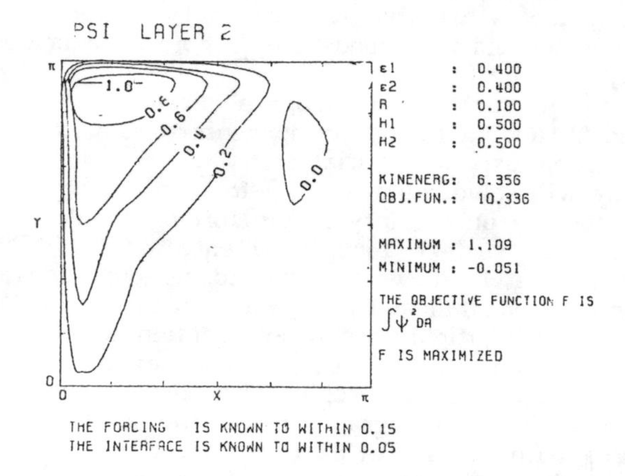

Fig. 2: Two layer model:
a.+b. Streamfunction that maximizes the potential energy subject to I) the uncertainty in the forcing is not larger than 0.15. II) the uncertainty in the position of the interface is not larger than 0.05.

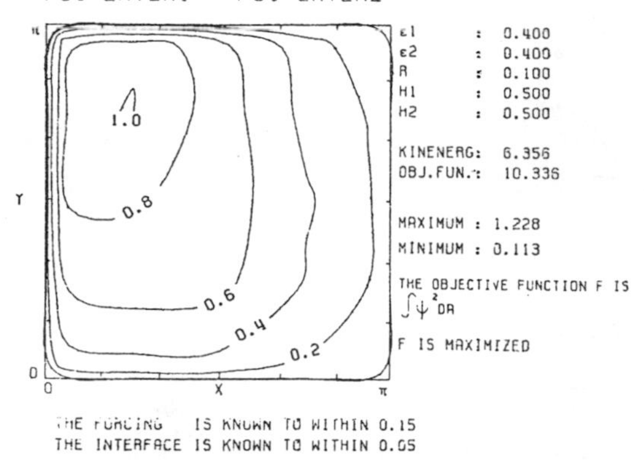

LAGRANGE MULTIPLIERS LAYER 2

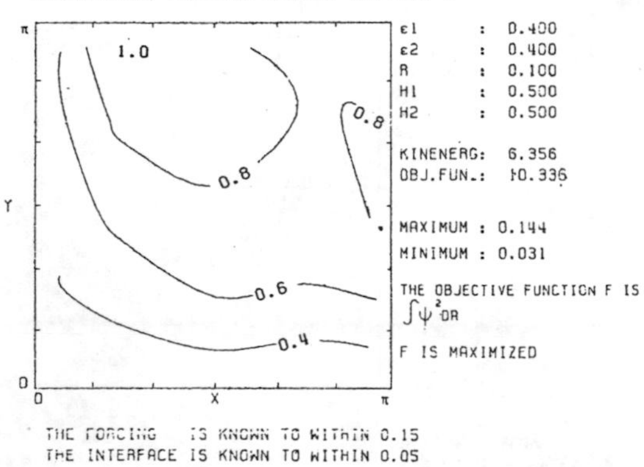

c. Position of the interface between the two layers, interpreted here as temperature.

d. Sensitivity of the objective function F to violations of the dynamics of the lower layer (e.g. wrong parameterization of the bottom friction).

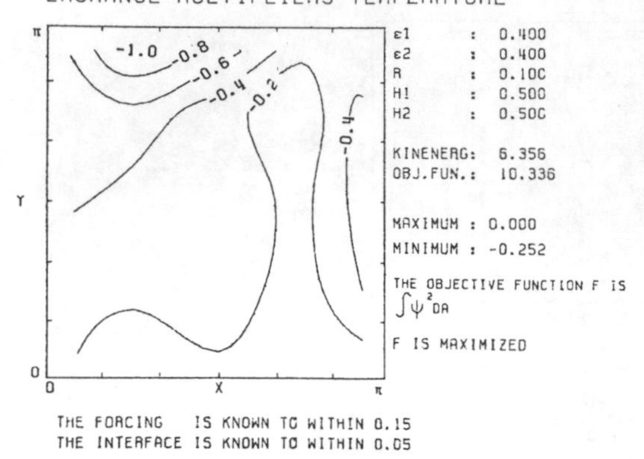

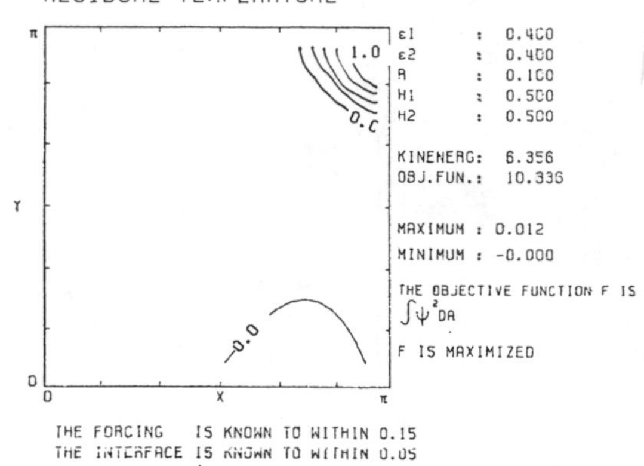

e. Deviation of the forcing that corresponds to the optimal solution from the measured forcing.

f. Sensitivity of the objective function F to small variations in the bounds of the uncertainty of the forcing. The sensitivity is nonzero only in regions where the deviation from the measured forcing (see 2e) reaches one of the bounds.

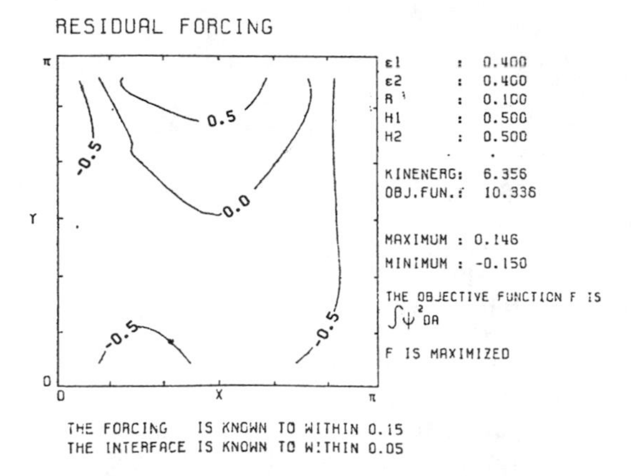

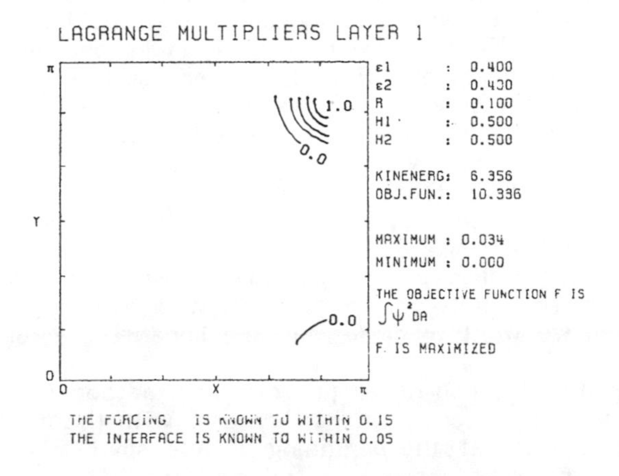

g. Sensitivity of the extreme value of F to small variations in the bounds of the uncertainty in the measured "temperature".

h. Deviation of the "temperature" that corresponds to the optimal solution from the measured "temperature". Values are at their lower bound (-0.05) everywhere except in the two regions where the sensitivity (see 2g) is zero.

$$RJ(\psi_2, \nabla^2 \psi_2) + H_2 J(\psi_2, \psi_1 - \psi_2) + \frac{\partial \psi_2}{\partial x} + \varepsilon_1 H_2 \nabla^2(\psi_2 - \psi_1) + \varepsilon_2 H_2 \nabla^2 \psi_2 = 0 \tag{8}$$

Subscripts 1 and 2 denote upper and lower layer respectively. Values for the nondimensional parameters are:

nonlinearity: $R = 0.1$, friction coefficient: $\varepsilon_1 = \varepsilon_2 = 0.4$,
depths of layers: $H_1 = H_2 = 0.5$

Forcing "forc" and its uncertainty are identical to the first example. The objective function F is again the potential energy. F is maximized. "Temperature" constraints are introduced by prescribing that $\psi_1 - \psi_2$ (i.e. the position of the interface) is known to within 0.05.

Results are shown in Figs. 2a to 2h. The streamfunction that maximizes $F(\underline{X})$ is depicted in Figs. 2a and 2b, the "temperature" field shown in Fig. 2c. As we do not allow any uncertainty in the dynamics of the lower layer (eq. (8)), the corresponding Lagrange multipliers (Fig. 2d) are only meaningful if we consider violations of the prescribed dynamics, e.g. the bottom friction and in what regions we would want to improve its formulation. The difference between measured and optimal forcing (Fig. 2e) is small in the interior of the basin and reaches the bounds only in the two eastern corners. On the other hand the "temperature" constraints reach their bounds practically everywhere (Fig. 2h) and their highest sensitivity is encountered in the northern boundary current (Fig. 2g). We conclude that for this example with its prescribed measurement accuracies the measurement of the temperature is more important than that of the wind forcing.

In the examples considered here we did not make use of the capability of our method to deal with regionally varying uncertainties. This capability is of great importance when we want to model e.g. the impact of errors in geoid estimation.

It is planned to use this non-linear programming method in conjunction with a GCM. As the number of computations increases with the number of unknowns cubed, we will start at the beginning with a small number of variables that describe for instance the amplitudes of the EOF's of the windstress. From these the GCM will calculate several thousand values for currents, temperature, salinity etc. which in turn are used to calculate an objective function such as sea surface temperature gradient or meridional heat transport.

THE TREATMENT OF MIXING PROCESSES IN ADVECTIVE MODELS

D. Adamec
Department of Meteorology, Code 63Ac
Naval Postgraduate School
Monterey, California 93943
USA

ABSTRACT. The technique for including mixing processes in advective models is presented. Included in the discussion are convective and dynamic adjustment schemes, parameterizations of diffusion, and the inclusion of bulk mixed-layer dynamics in ocean circulation models. A two-dimensional version of an ocean model with two different parameterizations of mixing is used to simulate the response of an ocean density front to atmospheric forcing. The model results indicate a strong interdependence between vertical turbulent mixing and advection of heat.

1. INTRODUCTION

Given precise knowledge of the surface boundary conditions, a variety of mixed-layer models have been able to simulate the one-dimensional temperature structure at Ocean Weather Ships for time-scales from a month [Denman and Miyake, 1973] to a year [Garwood and Adamec, 1982]. However, there are instances when non-linear effects force the ocean out of local thermal balance with the atmosphere as in the case of the oceanic response to strong storm forcing or low frequency adjustments in the ocean. Numerical models of oceanic features are generally tailored to the time and space scales of the phenomenon being studied. Numerical models that are able to simulate both the high frequency response of the mixed layer to local forcing and the lower frequency response of advective effects have only recently been introduced.

The purpose of this paper is to review some of the techniques used to *couple* mixing and advective effects in numerical models. The first part of this paper reviews the rationale and techniques that are used in some coupled models. Two different methods of including mixing effects in general circulation models are compared in the second part of this paper which examines the simulated response of an ocean density front to atmospheric forcing.

J. J. O'Brien (ed.), Advanced Physical Oceanographic Numerical Modelling, 495–510.

2. INCLUSION OF MIXING

The changes in temperature in many numerical models of the ocean are governed by a relationship such as

$$\frac{\partial T}{\partial t} = -\vec{V}\bullet\nabla T - w\frac{\partial T}{\partial z} - K_M\nabla^2 T - \frac{\partial}{\partial z}(K_T\frac{\partial T}{\partial z}) + \frac{Q}{\rho_o C_p} + T_{mix} , \quad (1)$$

where K_M and K_T are diffusion coefficients, Q is the diabatic heating rate, and the rest of the terms have their usual meaning. The local rate of change in temperature is due to advective, diffusive, diabatic, and turbulent mixing effects. The inclusion of mixing in advective models is usually accomplished through different parameterizations of vertical diffusion and turbulent mixing. Some of those parameterizations are discussed below.

2.1. Convective Adjustment

A simple form of mixing can be accomplished through convective adjustment which has been used in a number of advective models [Bryan and Cox, 1967; and Haney, 1971]. Convective adjustment (or free convection) occurs when the density profile becomes gravitationally unstable, i.e. heavy water overlying lighter water. Implementing a convective adjustment scheme requires checking the vertical profile of density and identifying those areas where $\partial\rho/\partial z > 0$ (z positive upwards) or, if density is a function of temperature alone, areas where $\partial T/\partial z < 0$. If the density (temperature) profile is found to be unstable, then heat is assumed to be mixed vigorously between levels so that total heat is conserved and the local vertical gradient is at least neutrally buoyant. The new temperature at each level is calculated using

$$T'_k = T'_{k+1} = T_{new} = \frac{T_k\,\Delta z_k + T_{k+1}\,\Delta z_{k+1}}{\Delta z_k+\Delta z_{k+1}} , \quad (2)$$

where the subscripts refer to the specific model levels being considered, and ′ refers to the value after adjustment. After checking the entire profile, and after convective adjustment has taken place, the profile is then rechecked in the event that the adjustment has produced new areas where the profile is gravitationally unstable. Faster convergence to a stable profile can be obtained by assuming that the free convection produces a profile which is slightly stable.

2.2. Dynamic adjustment

A generalization of the familiar convective adjustment schemes is presented in Adamec et al. [1981] based on the value of the local gradient Richardson number. Dynamic stability is imposed by ensuring that the value of the gradient Richardson number is always greater than or equal to 0.25 and is implemented as follows. If a mixing ratio is denoted by

$$m = \frac{\xi_k-\xi'_k}{\xi_k-\bar{\xi}_k} = \frac{\xi_{k+1}-\xi'_{k+1}}{\xi_{k+1}-\bar{\xi}_{k+1}} \quad (3)$$

where the subscript refers to a model level, $\bar{\xi}$ is the weighted average of the quantity ξ between levels k and k+1, and ξ' is the value of ξ after adjustment. The quantity ξ can be temperature or horizontal velocity. The adjustment assumes that the mixing ratio for all quantities ξ are equal between layers, and requires conservation of heat and momentum. With some algebra, it can be shown that the new adjusted values which satisfy these conditions are

$$\xi'_k = \bar{\xi} + \beta\gamma\Delta\xi \quad , \tag{4}$$

$$\xi'_{k+1} = \bar{\xi} - \beta\gamma\Delta\xi \quad , \tag{5}$$

where

$$\beta = \frac{\Delta z_k}{\Delta z_{k+1}} \quad , \tag{6}$$

$$\gamma = \frac{Ri}{Ri_{cr}(1+\varepsilon)} \qquad 0 < Ri < Ri_{cr} \quad , \tag{7}$$

$$\gamma = 0 \qquad Ri < 0 \quad , \tag{8}$$

ε is a small number (~ 0.05) to assure fast convergence of the adjustment scheme and $\Delta\xi$ is ξ_k - ξ_{k+1}. Thus, there is complete mixing between levels when the profile is gravitationally unstable. As in the traditional convective adjustment, the profile is rechecked if adjustment has taken place to ensure that the new adjusted profile satisfies the stability criterion at all levels.

2.3. K-theory

A different parameterization of mixing effects assumes that mixing is accomplished entirely through diffusion. Rapid vertical exchange of heat and momentum is accomplished through growth of the diffusion coefficients. This type of model is sometimes referred to as a K-theory model.

Some mixing formulations assume that the turbulent exchange of heat is directly related to the vertical gradient in the turbulent Reynolds heat flux

$$\frac{\partial T}{\partial t} = \frac{\partial}{\partial z}(-\overline{w'T'}) \quad . \tag{9}$$

A traditional K-theory formulation is implemented by setting

$$-\overline{w'T'} = L\,q\,S_T \frac{\partial T}{\partial z} \quad , \tag{10}$$

where L is a characteristic vertical length scale, q^2 is twice the turbulent kinetic energy, and S_T is stability factor dependent on the vertical gradient of temperature. This parameterization is identical to setting

$$K_T = L\,q\,S_T \quad , \tag{11}$$

in (2). A similar parameterization can also done for momentum.

One of the first descriptions of a K-theory implementation is given in Munk and Anderson [1948]. In their formulation, the diffusion coefficient is computed as a reference value multiplied by a stability factor. The temperature and momentum diffusion coefficients are given by

$$K_T = K_o (1 + \beta_T R_f)^{-3/2} \quad , \tag{12}$$

$$K_M = K_o (1 + \beta_M R_f)^{-1/2} \quad , \tag{13}$$

where β_T is a constant, 3.33, and β_M is 10.0. The stability factors from Munk and Anderson's [1948] formulation are plotted in Figure 1 as functions of flux Richardson number (R_f). Both diffusion coefficients become very large when the Richardson number becomes less than zero (gravitational instability). Notice that momentum is diffused more rapidly than heat in this formulation when the profile is unstable.

Mellor and Durbin [1975] implement their K-theory formulation by assuming that S_T is functionally dependent on the flux Richardson number. In their formulation, the flux Richardson number is estimated from the gradient Richardson number (Ri)

$$Ri = \frac{\alpha g \Delta T \Delta z}{\Delta u^2 + \Delta v^2} \quad \text{and} \quad , \tag{14}$$

$$R_f = 0.725\,(Ri + 0.186 - (Ri^2 - 0.316\,Ri + 0.0346)^{1/2}) \tag{15}$$

The stability factor, S_T, is then evaluated using

$$S_T = 3A\,(\gamma_1 - \gamma_2 \Gamma) \quad , \tag{16}$$

where

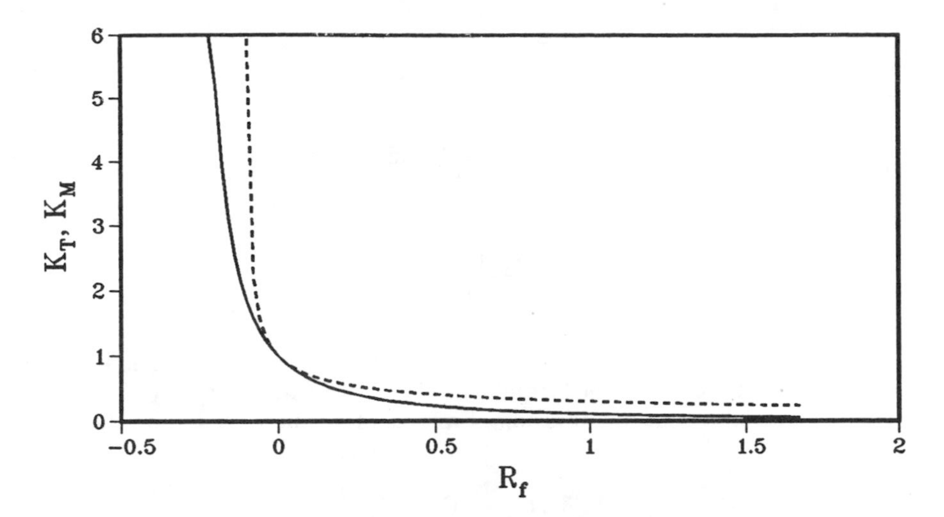

Figure 1. Dependence of the vertical heat (solid line) and momentum (dashed line) diffusion coefficients on the flux Richardson number as given by Munk and Anderson [1948].

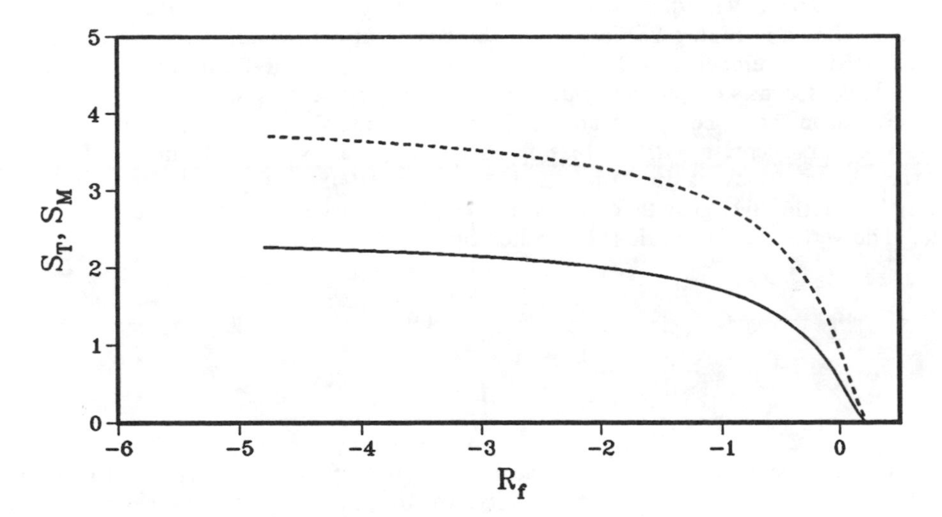

Figure 2. Dependence of the vertical heat (solid line) and momentum (dashed line) stability coefficients on the flux Richardson number as given by Mellor and Durbin [1975].

$$\Gamma = \frac{R_f}{1 - R_f} , \tag{17}$$

$$\gamma_1 = \frac{1}{3} - \frac{2A}{B} \quad \text{and} \tag{18}$$

$$\gamma_2 = \frac{C + 6A}{B} . \tag{19}$$

The constants A, B, and C have values 0.78, 15.0, and 8.0, respectively. The functional forms and constants were determined from neutral turbulent flow data.

A similar formulation for the diffusion of momentum is used by setting the stabilty factor

$$S_M = \frac{S_T \, 3A \left(\gamma_1 - D - \frac{A\Gamma}{B}\right)}{\gamma_1 - \Gamma\gamma_2 + \frac{3A\Gamma}{B}} , \tag{20}$$

where D = 0.056. The dependence of these stability factors on the flux Richardson number is shown in Figure 2. Notice that there is a cut-off for diffusive turbulence at a flux Richardson number of 0.21 (corresponding to a gradient Richardson number of 0.23). This is consistent with a number of laboratory studies which show that flow becomes turbulent once the gradient Richardson number drops below 0.25. The stability factors increase with decreasing Richardson number, but do not tend toward the very large values as in Munk and Anderson [1948]. Large diffusion rates are accomplished through growth of the turbulent kinetic energy and the vertical length scale. The vertical length scale is computed by

$$L = \alpha \frac{\int_{-\infty}^{0} z\, q\, dz}{\int_{-\infty}^{0} q\, dz} , \tag{21}$$

which is the ratio of the first moment of the turbulent kinetic energy to the zeroth moment; α is a constant of proportionality. In an earlier version of an atmospheric model, Mellor and Yamada [1974] used $L \sim \kappa z$ near the surface layer, which produces a logarithmic profile of the velocities in the near surface layer.

3. COUPLED MODELS

Three methods to include turbulent processes in advective models have been briefly described. These methods do not make assumptions about the density and momentum profiles but rather would predict the formation of a well-mixed layer and a thermocline. An alternative approach to this type parameterization is that of bulk mixed-layer models which assume that density (and sometimes momentum) are uniform down to some depth h (mixed-layer depth) and that there exists a level at which the turbulent kinetic energy vanishes and a discontinuity develops in the density profile. This is a markedly different approach from the methods described above and is more complicated to include in advective models because the mixed layer is an extra parameter which must be predicted, and the assumption of uniformity throughout this layer and the predicted discontinuity must somehow be incorporated into the advective model.

Most bulk mixed-layer models are extensions of the Kraus and Turner [1967] model. Bulk models assume that the buoyant damping is balanced by a convergence of turbulent kinetic energy within an entrainment zone, with the turbulent kinetic energy usually modelled as a function of the shear and buoyancy production of turbulent kinetic energy and the distance over which it must be transported. The depth at which the turbulent kinetic energy disappears is the mixed-layer depth over which density is assumed to be uniform. Hence, turbulence is a dominant feature within the mixed layer and is confined solely to the mixed layer.

One of the more important parameters in predicting the evolution of the mixed layer is the stability at its base. If density is assumed to be a function of temperature alone, then accurate predictions of temperature at the base of the mixed layer are critical. A model that uses a multi-level formulation of the primitive equations predicts the average of the prognostic variables between two specified depths. However, the base of the mixed layer rarely lies exactly on a specific level. Requiring a sufficiently fine resolution to prevent errors in the thermal and potential energy balances due to the truncation of the mixed-layer depth to a model level is unreasonable both in terms of computer storage and computational time. The problem becomes to provide for a meaningful communication between the fixed-level advective portion of the model, and the mixed-layer part of the model.

Two rationales for *embedding* bulk mixed-layer physics into advective models are described below. The first is a formulation by Adamec et al. [1981] that endeavors to provide a formulation for general applications and the second by Schopf and Cane [1983] presents a highly accurate solutions for more specific applications.

3.1. The Adamec et al. [1981] formulation

The coupling of the advective model can be considered in two stages in the Adamec et al. [1981] model. First, the changes in the upper ocean due to advective and diffusive processes are calculated and put into the form for the mixed-layer model. Second, the changes due to surface fluxes and entrainment mixing are calculated in the mixing part of the model and transmitted to the dynamical part of the model. In both these stages, a special treatment is required for the model level which contains the base of the mixed layer.

The first stage of the embedding is accomplished by first calculating the tendencies due to advective and diffusive processes for all the layers for the prognostic variables (u,v,T, and h). The prognostic variables are then stepped forward in time and checked for dynamic stability as described in Section 2.2. The special treatment of the layer containing the new mixed-layer depth is now described with the aid of Figure 3. The dashed line is the vertical structure used by the advective part of the model, and the solid line is the vertical structure used by the mixing part of the model. Both the advective and mixing part of the model use the same vertical structure at all levels which lie entirely within the mixed-layer. The variable ξ can represent u, v, or T. If the base of the mixed layer lies within the kth model layer, the value of ξ in the mixed layer portion is denoted by ξ_1, and the value of just below the mixed layer is denoted by ξ_2. The integrated average over the model layer is denoted by $\overline{\xi}$, and the mixed-layer depth is denoted by h. The change $\delta\overline{\xi}$ due to the advective and diffusive processes is assigned the value calculated for the general circulation model immediately above (the layer bounded by $z_{k-1/2}$ and $z_{k-3/2}$). The base of the mixed layer is always constrained to lie at or below the base of the first model level. The associated change $\delta\xi_2$ is then calculated from the requirement that the weighted average of $\delta\xi_1$ and $\delta\xi_2$ be equal to the change $\delta\xi$ predicted for the layer by the dynamical part of the model; or,

$$\Delta z_k \delta\overline{\xi} = \Delta z_1 \delta\xi_1 + \Delta z_2 \delta\xi_2 \tag{22}$$

where Δz_1 is the portion of the layer above the new mixed layer and below $z_{k-1/2}$. The tendency $\delta\xi_1$ is assumed to be equal to $\delta\xi$ for the k-1 level.

In the dynamical model, the advective and diffusive changes are based on the average properties in each layer. However, there can be a large velocity and temperature jump across the interface h. The use of a layer mean advecting velocity in the dynamical part of the model necessarily introduces error in the calculation of $\delta\xi$ which, by this formulation, is introduced in $\delta\xi_2$. This error in the layer average of the horizontal advection of ξ can be shown to be

$$\overline{u\xi} - \overline{u}\ \overline{\xi} = \frac{\Delta z_1 \Delta z_2}{\Delta z_k} \Delta u \Delta\xi \tag{23}$$

where $\Delta u = u_1 - u_2$ is the jump in the advecting velocity and $\Delta\xi = \xi_1 - \xi_2$. Because Δz_1 is correctly treated as a portion of the well-mixed layer, the error is entirely absorbed in the Δz_2 layer. The error is small when either Δz_1 or Δz_2 is small and is largest when h is near the center of a model level. An additional complexity would be introduced by calculating a differential advection within the Δz_2 layer, especially if the mixed-layer depth at adjacent grid points is not within the same level of the dynamical model. The scheme described of using a layer mean advecting velocity does satisfy integral constraints for the predicted properties. An analogous but slightly more complex formulation is used whenever the dynamically predicted δh causes the base of the mixed layer to move into an adjacent layer.

Before proceeding to the mixing portion of the model, the values of ξ throughout the mixed layer (including ξ_1) are replaced by the vertical average of ξ over the new

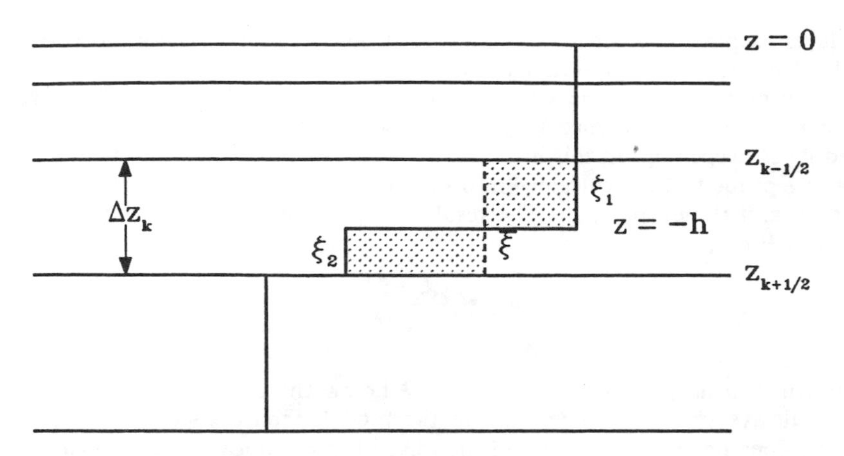

Figure 3. Schematic drawing of the Adamec et al. [1981] embedding scheme. The solid line is the profile used in the mixing portion of the model, and the dashed line is the profile used in the advective portion of the model.

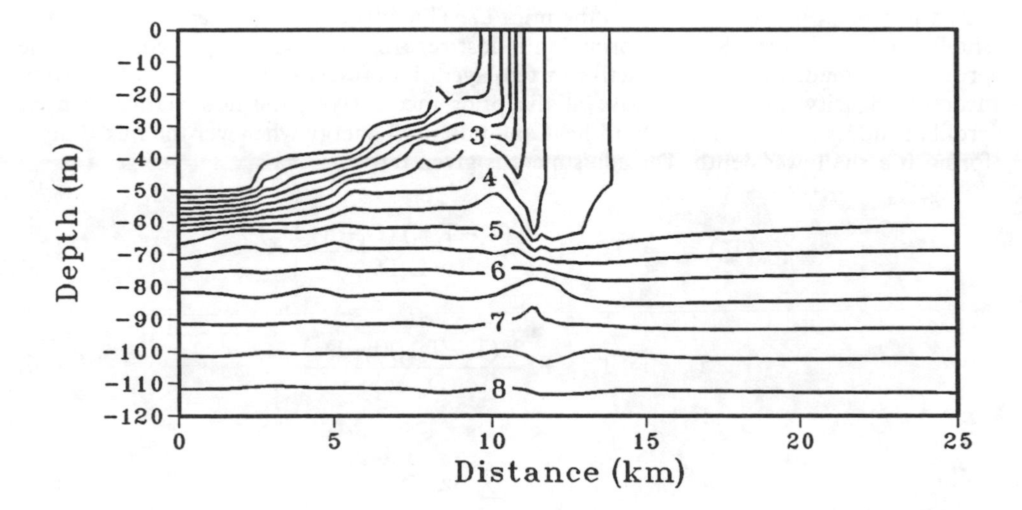

Figure 4. Initial density profile used by Adamec and Garwood [1985] in frontal simulations. West is to the right and east is to the left. The contour interval is 0.5 σ_t.

mixed-layer depth. The small amount of turbulent kinetic energy that would be expended (or released) in performing this vertical mixing has been neglected.

The second stage of the coupling is the mixing portion. In the case of an entraining mixed layer, the mixed layer model will predict an entrainment heat flux and a mixed-layer deepening. In this instance, a volume of water (per unit area) equal to δh and having property ξ_2 is entrained into the mixed layer having volume (per unit area) equal to h and the property ξ_1. The result is a new mixed layer of depth $h+\delta h$ and property $\xi_1+\delta\xi_1$, where

$$\delta\xi_1 = \frac{(\xi_1-\xi_2)\delta h}{h+\delta h} \tag{24}$$

Entrainment mixing produces no change in ξ below the new mixed layer. If $\xi_1 > \xi_2$, which is always true for temperature in the model, then $\delta\xi_1 < 0$. If the entrainment deepening does not cause the base of the mixed layer to move into an adjacent model layer, then the magnitude of the property jump across the mixed layer, $\Delta\xi$, will be decreased. In the model, a separate array of ξ_2 values must be stored to permit the jump values to be always specified for the mixing process. This value, and the mixed-layer depth, h, are the only additional variables which must be retained in coupling the mixing model to the dynamical model.

In the case of mixed-layer shallowing, the heat and momentum flux across the surface are distributed over the new, shallower depth to produce a new value of ξ_1. A new value of ξ_2 is determined from the prior ξ_1. The advective step operates on the new shallower mixed layer. Some information is lost regarding the deeper structure because provision is made for only one active mixed-layer. To preserve some of the information near the density jump at the base of the prior mixed layer, the new vertical density profile is adjusted to conserve both heat and potential energy whenever the mixed layer forms at a shallower depth. The adjustment is given by

$$(T'_i)_{i=1,k} = T_{k+1} + \frac{h\,(T_k-T_{k+1})\,(D_k-h+D_M)}{D_k D_M} \tag{25}$$

$$T'_{k+1} = T_{k+1} + \frac{h\,(T_k-T_{k+1})(h-D_M)}{D_k\,\Delta z_k} \tag{26}$$

$$D_k = \sum_{i=1}^{k} \Delta z_i \tag{27}$$

$$D_M = \max\,(h', D_{k-1}) \tag{28}$$

where k is the model level which contained the original mixed-layer depth before shallowing and the primes denote quantities after the shallowing.

There are objections to the Adamec at al. [1981] embedding scheme, some of which have already been alluded to, but will be repeated here for completeness. First, there is an inherent inefficiency in the way the model predicts advection in the mixed layer. The advective tendencies within the mixed layer only need to be computed once for a given point since the tendencies would be equal down to the level of the mixed-layer depth. Also, an error in the dynamical tendencies enters in the layer immediately below the base of the mixed layer, which is largest when the base of the mixed layer is between model levels. However, the scheme does preserve the integral quantities of the advective tendencies. Finally, the scheme has come under some criticism because of the coarse vertical resolution used in the mixing portion of the model. Sensitivity tests by Adamec et al. [1978] have shown that the mixing model gives satisfactory results provided the vertical resolution is not more than 10 m near the surface, and more than 20 m between depths of 60 and 150 m.

3.2. The Schopf and Cane [1983] formulation

Schopf and Cane [1983] overcome the problem of the base of the mixed layer rarely coinciding with a model level by redefining the vertical coordinate such that h is a function of x,y,t, and the new vertical coordinate s. They define $h(x,y,s,t) = \partial z/\partial s$, with $z = 0$ at $s = 0$. The method is similar to the σ-coordinates which are sometimes used in meteorology and tidal modelling. By redefining the vertical coordinate in this way, the vertical velocity is replaced by

$$w_e = w - \frac{\partial z}{\partial t} + \vec{V} \bullet \nabla z \tag{29}$$

where w_e is an entrainment velocity which is related to the net acquisition of mass within the mixed layer. In the Adamec et al. [1981] model, the balance of the entrainment velocity is given by

$$w_e = w + \frac{dh}{dt} \tag{30}$$

In the Schopf and Cane [1983] model, the entrainment velocity is directly related to the vertical velocity, the change in time of the old vertical coordinate, z, as the mixed-layer evolves, and the angle of the horizontal flow (which is now along a constant-s surface) to a z surface.

The vertical structure used in the Schopf and Cane [1983] model is an idealized representation of the ocean. The model is divided into three layers: the upper-most layer is homogeneous in buoyancy and momentum; the second layer, in which the temperature is linearly stratified and the velocity is homogeneous; and the bottom layer in which the pressure gradients are assumed to adjust so that there is no motion within that layer. A discontinuity in the temperature occurs between the first and second, and the second and third layers. By choosing the vertical coordinate such that the upper layer thickness coincides with the thickness of the mixed layer, the problem with the calculation of the advective tendencies immediately below the mixed layer does not appear as in the Adamec et al. [1981] formulation.

The governing equations for the model assume an integrated average for the velocities in the upper two layers, the horizontal flow is now along a constant-s surface and derivatives are replaced using the chain rule

$$\frac{\partial Q}{\partial x}\Big)_z = \frac{\partial Q}{\partial x}\Big)_s + \frac{\partial Q}{\partial s}\frac{\partial s}{\partial x} \tag{31}$$

$$\frac{\partial Q}{\partial z} = \frac{\partial s}{\partial z}\frac{\partial Q}{\partial s} \tag{32}$$

The equations are derived in Schopf and Cane [1983] and are not repeated here.

As in Adamec et al. [1981], there is a potential energy gain and loss of kinetic energy at the base of the mixed layer when entrainment occurs. In the case of boundary layer shallowing, Schopf and Cane [1983] specify an artificial mixing of momentum and buoyancy to insure that the stratification in the second layer remains linear. Recall a similar artificial mixing is done in Adamec et al. [1981] when the base of the mixed layer shallows to a different model level.

The Schopf and Cane [1983] formulation represents a highly accurate method of solution to idealized case studies in which the ocean is assumed to have a specific vertical structure. The Adamec et al. [1981] formulation is not as accurate a solution immediately below the mixed-layer depth ,however, it is not constrained by a specific vertical structure.

4. SIMULATIONS WITH EMBEDDED MIXING

Adamec et al. [1981] test their model by looking at the response of an ocean that is initially motionless to a stationary hurricane. The response to the hurricane was chosen because of the large horizontal gradient response in the advective and mixing processes that was expected. The results indicated a strong interdependence between vertical turbulent mixing and advection of heat. An interesting result of that study is that the results from the model with the embedded bulk mixed-layer parameterization gave very similar results to a simulation which included the generalized convective adjustment scheme discussed in Section 2.3 only. This result is not surprising since the dynamic stability condition does lead to an estimate of the thickness of the entrainment zone. For cases with a constantly deepening mixed layer, the embedded model and convective adjustment scheme would be expected to give similar predictions for u,v, and T. A different situation is more likely to occur when the mixed layer shallows (i.e. during periods of heating). The width of the entrainment zone is not necessarily related to the mixed-layer depth as assumed in a mixed-layer scheme based on the value of the Richardson number.

Adamec and Garwood [1985] performed numerical simulations of the response of an upper ocean density front to different atmospheric forcings, including heating. The initial conditions for the model are obtained from an idealization of the Maltese front observed by O.M. Johannessen [unpublished report, 1975]. Although the density difference across the Maltese front is mainly due to salinity effects, density was chosen as the prognostic variable to treat the more general effects of surface heating and cooling. The initial condition for the density is shown in Figure 4. The surface front spans 2 km in the horizontal, with a density increase of 4 σ_t at the surface. The initial mixed-layer depth is 15 m near the surface manifestation of the front and deepens to 50 m over a 10 km distance on the western side of the front. On the eastern side of the front, the mixed-layer depth increases abruptly to 50 m. The initial currents are prescribed to be in geostrophic balance consistent with the estimated Rossby number of 0.1. Thus, the cross-front flow is initially zero, and the along-front flow at the surface is negative (out of page).

The wind stress is spun up from zero over the first 6 hours of each integration, and is applied in the positive-y direction only. The maximum magnitude of the wind stress in each case is 1 dPa. The buoyancy flux includes a diurnal signal: negative and sinusoidal for the first 12 hours and then a positive constant for the next 12 hours so that there is no net input over 24 hours. The value for the maximum of the buoyancy flux (175 W m^{-2}) was chosen such that the shear production and the buoyancy flux terms in the turbulent kinetic energy budget would be comparable.

The time variation of the surface isopycnals for the simulation with an along-front wind stress into the page and diurnal surface buoyancy flux is shown in Figure 5. The mixed layers reform at a shallow depth and the u components (advecting velocities) increase very quickly. The initial shallow mixed layers and buildup of an eastward Ekman flow allow an overriding plume to develop. A similar simulation without a surface buoyancy flux did not produce a buoyant plume as the action of the wind stirring mixed the less dense water as it was advected eastward. Here, the mixing does not mix all the buoyant water and the plume is advected eastward. The advection of the plume is slowed at hour 12 when the buoyancy flux becomes positive (cooling). However, at hour 24, when the buoyancy flux is again negative, there is a slight increase in the advection of the plume as evident from a change in slope of the isopycnals at hour 24. At hour 36, the advection of the plume is again retarded.

A cross-section of the density field at hour 48 is shown in Figure 6. The overriding plume has spread far eastward. The subsurface layers have been protected by the buoyancy flux (when it is negative) and the shield of the stable overriding plume. Below the spreading plume at hour 48 is a well-mixed region that is a fossil of the cold-side mixed layer that existed two days earlier.

The simulation was re-run with a model which included dynamic adjustment only. A cross-section of the density field at hour 24 is shown in Figure 7. The response is very different from the simulation with mixed-layer dynamics included. The overriding plume has already been advected farther to the east after 24 hours in this simulation than it was in 48 hours in the previous simulation. Notice also that there is net a net buoyancy gain in the surface layer even though the net surface buoyancy flux over 24 hours is zero. The reason for the large differences between density cross-sections is due to the strength of the Ekman advection. In the embedded model, momentum is input

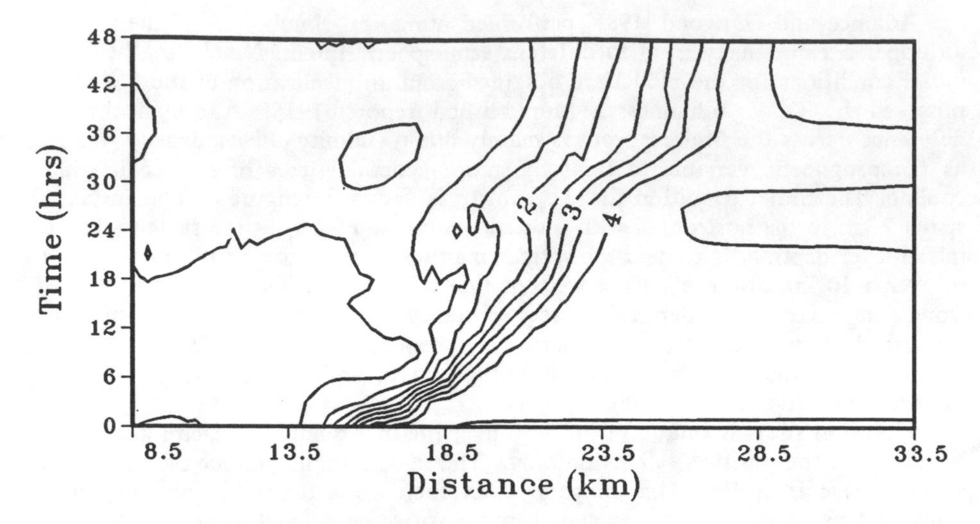

Figure 5. Time dependence of the surface isopycnals for the simulation with a northward wind stress and diurnal heat flux. [from Adamec and Garwood, 1985]. The contour interval is 0.5 σ_t.

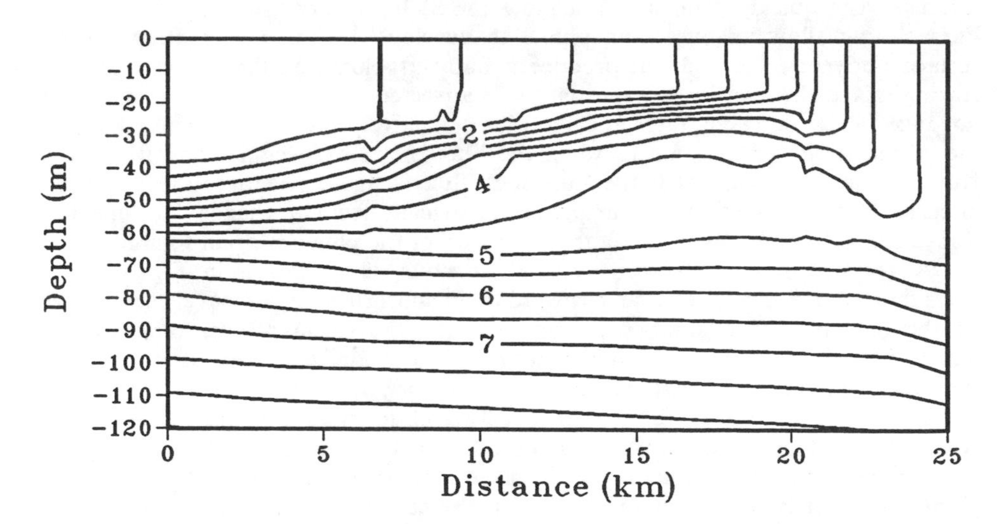

Figure 6. Cross-section of density at hour 48 for the simulation with a northward wind stress, diurnal heat flux, and using the Adamec et al. [1981] embedding scheme. The contour interval is 0.5 σ_t.

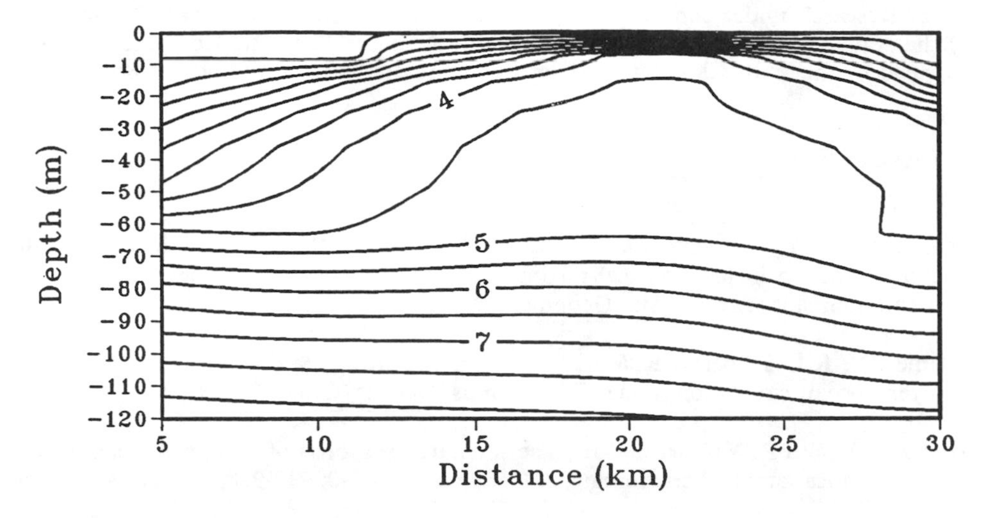

Figure 7. As in Figure 6 except the dynamic stability condition is used in lieu of the embedding scheme.

over the mixed-layer depth, which during times of heating is near 15 m. In the model without bulk mixed-layer dynamics, the momentum is being input over the thickness of the top layer of the model (6 m). Since the Ekman advection is inversely related to the depth to which the surface stress is penetrating, the model with bulk mixed-layer dynamics develops higher advecting velocities, and low density water is being more efficiently advected from the west. The surface cooling is not able to overcome the advective tendency, a very shallow, very stable layer is always present, and the rate of advection of the plume remains constant.

The response of the simulation without bulk mixed-layer dynamics is very dependent on the grid size. In this case, the lack of turbulent processes forces the momentum to remain confined in a very shallow layer. A much more realistic simulation would have occurred if the uppermost layer were thicker. Prescribing a thick upper layer necessarily removes some of the accuracy of the event that is being simulated. The use of a dynamic stability condition in a simulation which includes heating requires care in the choice of the vertical resolution near the surface.

Much progress has been made toward the treatment of mixed-layer dynamics in general circulation models over the past years. Detailed verifying data sets for these models should bring about an even greater understanding of the how mixed-layer processes affect the general circulation of the ocean.

5. ACKNOWLEDGEMENTS

Russ Elsberry, Bob Haney, and Bill Garwood played an active role throughout much of the research presented in this paper. The research was sponsored by the Office of

Naval Research under contract number NR083-275. Some funding was also provided by the Naval Postgraduate School Foundation Research Program. Computer time was provided by the W.R. Church Computer Center at the Naval Postgraduate School.

6. REFERENCES

Adamec,D., R.L. Elsberry, R.W. Garwood Jr. and R.L. Haney, Developmental experiments to include vertical mixing processes in numerical model simulations of ocean anomalies, Trans. Am. Geophys. U., 59, 1115, 1978.

Adamec,D., R.L. Elsberry, R.W. Garwood Jr. and R.L. Haney, An embedded mixed layer - ocean circulation model, Dyn. Atmos. Oceans, 6, 69-96, 1981.

Adamec, D., and R.W. Garwood Jr., The simulated response of an upper-ocean density front to local atmospheric forcing, J. Geophys. Res., 90, 917-928, 1985.

Bryan, K., and M.D. Cox, A numerical investigation of the oceanic general circulation, Tellus,19, 54-80, 1967.

Denman, K.I., and M. Miyake, Upper layer modification at Ocean Station Papa: observations and simulations, J. Phys. Oceanogr., 3, 185-196, 1973.

Garwood, R.W. Jr., and D. Adamec, Model simulation of seventeen years of mixed-layer evolution at Ocean Station Papa, Tech. Rep. Naval Postgraduate School, NTIS U-205,132, 1982.

Haney, R.L., Surface thermal boundary condition for ocean circulation models, J. Phys. Oceanogr., 1, 241-248, 1971.

Kraus, E.B., and J.S. Turner, A one-dimensional model of the seasonal thermocline. II: The general theory and its consequences, Tellus, 19, 98-109, 1967.

Mellor, G.L., and P.A. Durbin, The structure and dynamics of the ocean surface mixed layer, J. Phys. Oceanogr., 5, 718-728, 1975.

Mellor, G.L., and T. Yamada, A hierarchy of turbulent closure model for planetary boundary layers, J. Atmos. Sci., 31, 1791-1806, 1974.

Munk, W., and E.R. Anderson, Notes on a theory of the thermocline, J. Mar. Res., 7, 276-295, 1948.

Schopf, P.S., and M.A. Cane, On equatorial dynamics, mixed layer physics and sea surface temperature, J. Phys. Oceanogr., 13, 917-935, 1983.

ABOUT SOME NUMERICAL METHODS USED IN AN OCEAN GENERAL CIRCULATION MODEL WITH ISOPYCNIC COORDINATES

J.M. Oberhuber
Max-Planck-Institute for Meteorology
Bundesstrasse 55
2000 Hamburg 13
Federal Republic of Germany

ABSTRACT. Some numerical techniques are developed which allow an efficient formulation of an ocean circulation model based on primitive equations. Density surfaces determine the vertical coordinates in the interior ocean. The proposed scheme allows an accurate treatment of density surfaces even if an arbitrary topography is chosen. The underlying idea is that each layer has its own horizontal boundary which may change its position with time. Vectorization can be easily done. The upper ocean is represented by a mixed layer. To illustrate how the coupling between the mixed layer and the interior ocean is treated the effects of a seasonal cycle are investigated. Finally a time step scheme is presented which allows an unconditionally stable and rather simple formulation of the primitive equations by splitting up the problem into a prediction and several correction steps. Each of these steps consists of a semi-implicit time step scheme.

1. INTRODUCTION

The numerical solution of differential equations describing problems in fluid dynamics requires methods which take into account both the specific behavior of the physical system and the limitations of computer resources. In the interior ocean, density of a particle is nearly conserved because of the absence of strong mixing processes. In an Eulerian coordinate system, numerical diffusion caused by discretization errors or by explicit diffusion necessary to ensure numerical stability acts like eddy mixing. The explicitly included diffusion operator can be rotated so that it does not contribute to this error; however, discretization errors cannot be calculated and therefore cannot be eliminated exactly. An obvious idea then is to take isopycnals, levels of equal density, as coordinates. This leads to a multi-layer formulation. Particles in this coordinate system go "a priori" along the isopycnals even if the equations are discretisized. A further advantage of the isopycnic coordinates is that a front in the vertical below the mixed layer, for example, can be resolved with only a few layers. Because this front can vary its position, many levels in

J. J. O'Brien (ed.), Advanced Physical Oceanographic Numerical Modelling, 511–522.

z-coordinates are needed.

However, isopycnals can intersect boundaries such as the sea surface in regions of near-surface baroclinity or the topography in the interior ocean. Because of technical reasons, it is suitable on a vector computer to let the layers run with "zero layer" thickness along the surface or the topography towards the model boundary when they have physically disappeared somewhere. However, this technique does not allow an arbitrary topography. In an ocean model this would be a severe deficiency because only topography allows conversion from the barotropic to the baroclinic modes due to the bottom torque. Bleck and Boudra [1981] have circumvented that problem concerning "zero layers" and topography by taking a variable coordinate system and by allowing only isopycnic layers if it is possible with their technique. A further technical disadvantage of the isopycnic coordinates is that they do not allow an accurate calculation of the SST. The resolution of the SST would be dependent on the number of layers, which is typically much less than the number of grid points in the horizontal direction. Because of the fact that turbulent processes due to the mixed layer dynamics dominate near the surface, it is more suitable to couple a mixed layer model for the upper boundary layer with an isopycnic coordinate model for the interior ocean. But this leads to the difficulty how to couple two different coordinate systems. In the interior ocean the density is constant in each layer; in the mixed layer it is variable.

As a final consideration, a model should be formulated as generally as possible so that it can be used for a wide range of problems. This makes it necessary to use primitive equations in order to simulate the physics in the equatorial region, for instance. The primitive equations limit the time step, and therefore the efficiency, considerably in respect to quasi-geostrophic models. Time steps of up to 1 week for a general circulation model would be desirable, but only a few hours are mostly used in primitive equation models.

Here three techniques will be described: how to treat the isopycnals near the topography; how to couple a mixed layer to an isopycnic coordinate model; and how problems concerning restrictions on the time step can be circumvented.

2. MODEL FORMULATION

2.1 Equations

First primitive equations in a multi-layer formulation are taken assuming that in the horizontal and the vertical direction density is constant in each layer except the uppermost two layers, the mixed layer and the layer below, where the density varies in the horizontal direction. Therefore, a prediction equation for density is required. In all other layers only the horizontal flux and the layer thickness are predicted. In the momentum equation, Coriolis and pressure gradient forces are included. No simplifications in respect to the barotropic mode like the rigid-lid approximation are made. Further, in all equations advection and diffusion terms are added. The mixed layer physics is represented by a model from the type Kraus-Turner [1967].The equations are formulated on the sphere. Wind stress is included by a simple bulk parameterization, the heat flux with a Newtonian cooling formulation. Neither vertical shear stress nor any dissipation between the layers are added because it must be the aim first to get a model which does contain as little dissipation as possible. This concept guarantees that the behavior of the ocean model is like the one of an ideal fluid and, due to explicit and numerical dissipation, only mixing along isopycnals occur, but not cross-isopycnal mixing takes place.

2.2. Isopycnic coordinates

Difficulties arising when isopycnic coordinates are used are first of technical nature. Layers are not physically present in the whole domain, but disappear at the topography or below one of the uppermost two layers. In this model surfacing of isopycnals is prevented by the fact that a mixed layer is included. (See further comments later.) Additionally, the horizonal extension of a physically present layer, this means that the layer thickness is greater than zero, can vary with time. This happens, for instance, when an interface of two layers changes its height above ground near the topography due to a divergent component of the flux. Then, due to the slope of the topography, the point where the interface intersects the topography, varies not only its vertical but also its horizontal position.

If a vector computer is to be used efficiently, all layers must be defined in the whole domain. Therefore, it was an obvious idea for a long time to let a layer run along the topography towards the model boundary with a "zero layer" thickness, when this layer has disappeared physically somewhere. Ignoring for the moment the hatched regions, Figure 1 shows the mathematical layout of such a model in principle. If we forget the mixed layer the aspired model should handle a situation as shown in Figure 2. An obvious idea then would be to search for techniques which treat the situation in Figure 2 only as an extreme case of Figure 1. However, difficulties will arise due to discretization errors at those points where an isopycnal touches the topography. There the isopycnic coordinates are strongly curved. Therefore it is first necessary to find out the reasons in detail why it is not enough for an isopycnal model to allow "zero layers" and to understand the proposed method of "moving walls".

2.2.1 "Zero layer" technique and limitations. The first problem in respect to the "zero layer" technique is that a "zero layer" must be conserved. Let us take a B-grid. If a velocity point is considered which is surrounded by massless points (h=0) and if we take the flux form of the equations of motion, the flux does not change because all terms are multiplied with h. If a mass point is surrounded by such velocity points, then there is no divergence and, therefore, no change in the layer thickness. This again leads to no change of the fluxes in the surrounding velocity points and, consequently, is an endless loop.

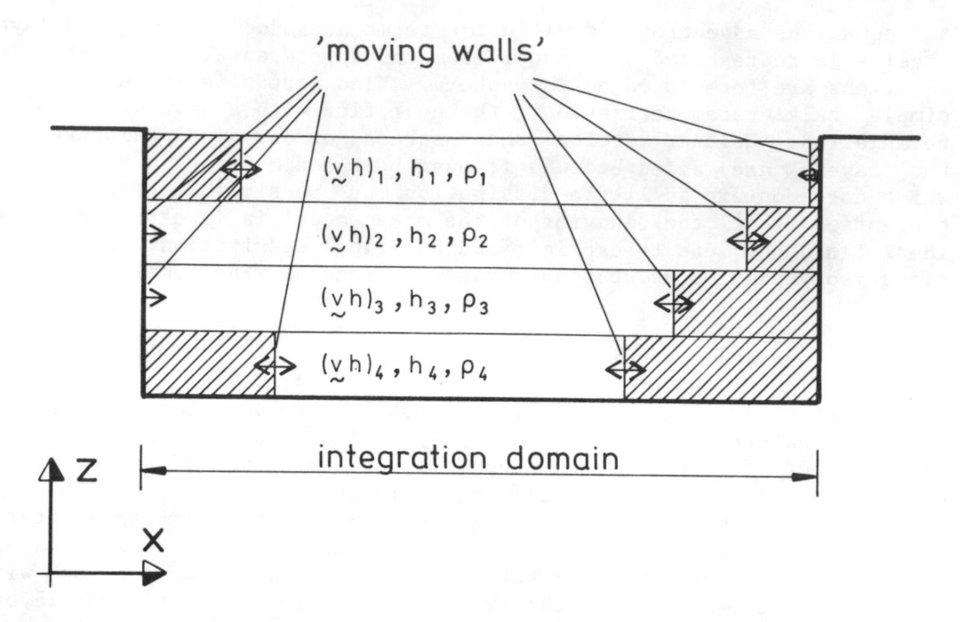

Figure 1. Cross section of the mathematical layout of the ocean model. The multi-layer model is formulated in a rectangular basin. "Moving walls" determine the interior ocean (blank), the remaining regions are those of "zero layer" thickness (hatched).

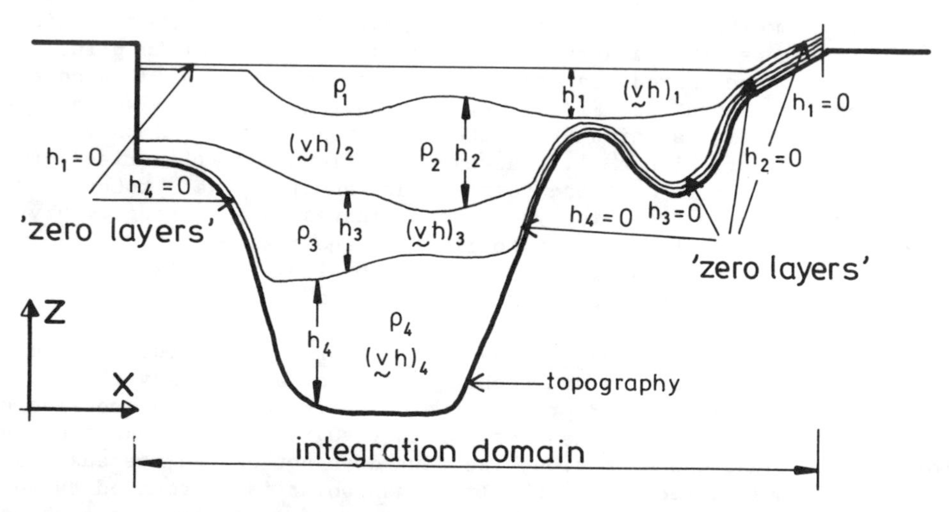

Figure 2. Physical situation in the ocean model (cross section). The distribution of "zero layers" corresponds roughly to Figure 1. Layers with h = 0 are drawn as a thin layer.

However, if an arbitrary topography is included in a model taking the "zero layer" technique, then flows occur even if there is no forcing. The origins of this unphysical behavior are those velocity points which are positioned between a "zero layer" and a "nonzero layer" thickness point. As illustrated in Figure 3, layer thickness gradients cause pressure gradients exactly at that velocity point, where the layer disappeares at the topography. Because the mean layer thickness at the considered velocity point is larger than zero, acceleration leads to a nonzero flux. In case 1, the pressure gradient is mostly dominated by the slope of the topography and, therefore, probably without any physical meaning. In contrast case 2 shows a situation where a reasonable physical balance must exist; however, it is not clear how strongly the pressure gradient is influenced by the topography. If we start the model from rest (without any density gradients in the first two layers and constant height of the surface and interfaces in the interior ocean) and add no forcing, only case 1 occurs. The topography gradient leads to a geostrophically balanced flow at the considered velocity point and, obviously, if the slope of the topography increases, the flow also increases. If one is willing to accept this discretization error in principle but wants to hold them neglegibly small, then only a very smooth topography may be chosen. This, however, contradicts the aim of allowing an arbitrary topography. The failure of the "zero layer" technique is due to the fact that, for instance, in case 1 at the first "zero layer" point, divergence due to acceleration in the momentum equation at the neighbouring velocity point is predicted and, consequently, a negative layer thickness is generated. A flux correction scheme can prevent the model from getting negative layer thicknesses but note that in a balanced state there is a strong geostrophic flow which allows a motion free of divergence at the considered mass point.

2.2.2 <u>Technique of "moving walls".</u> The basic problem in the "zero layer" technique is that negative layer thicknesses can occur and that pressure gradients are influenced by the topography. This leads to a misinterpretation of the physical situation due to discretization errors. Consequently, we now search for a scheme, which does not only avoid the negative layer thicknesses, but also does not lead to any unphysical flow. One could forbid flows at those points which contribute to a divergence but allow flow if they contribute to a convergence at "zero layer" points. In this method it would be the problem how to determine the distribution of "land" and "sea" points. If some criteria were found, an arbitrary topography could be included in such a model. Because B-grid is used, only some expressions in the code must be multiplied with "0" for "land" and "1" for "sea". The corresponding array in the code is defined on the velocity points, which allows a very simple formulation of boundary conditions.

The criterion whether a velocity point is a "sea" or "land" point now is illustrated in a one-dimensional example in Figure 4. (For better understanding, layer thickness is drawn in steps in contrast to Figure 1.) If all surrounding layer thickness points contain mass, then the velocity point between is a "sea" point. If all surrounding layer

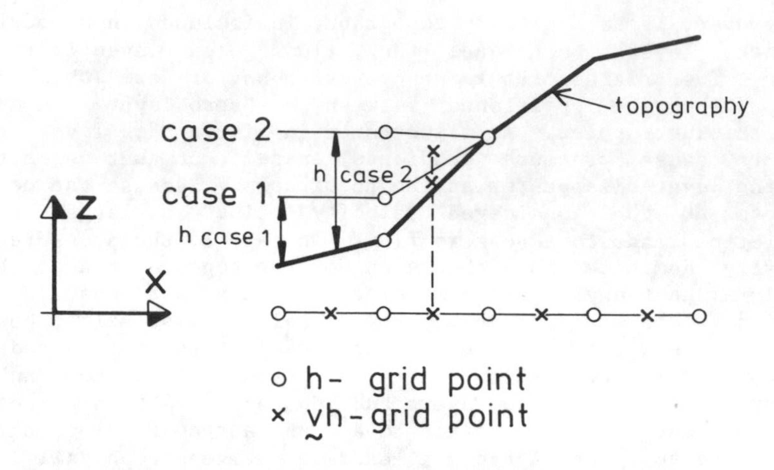

Figure 3. Cross section nearby a point where a layer disappeares at the topography. Layer thickness is linearly interpolated. For explanation of case 1 and 2, see text.

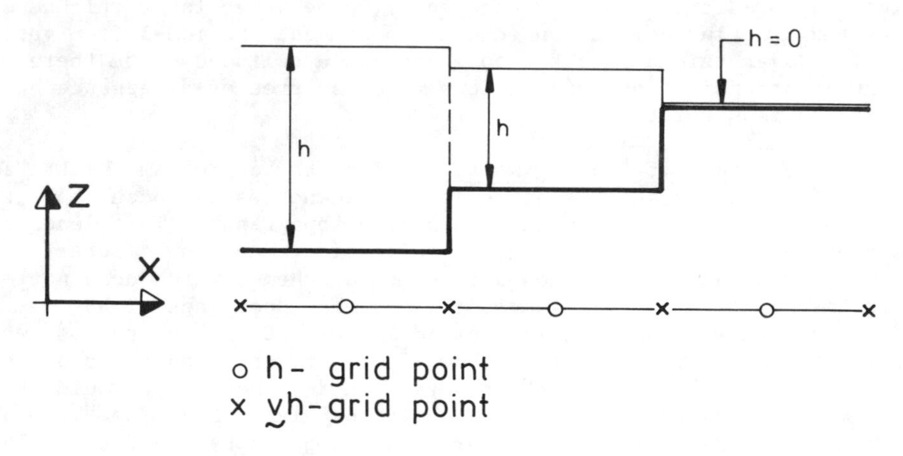

Figure 4. Cross section nearby a point where a layer disappeares at the topography. In opposite to the interpretation in Figure 3 the layer thickness now is drawn in steps. For further explanation see text.

thickness points do not contain any mass, then the velocity point between is a "land" point. These two decisions can be made without any doubt. Finally, there are velocity points where some of the surrounding layer thickness points contain mass and the remaining do not. In Figure 4 there is one velocity point which is surrounded by a "nonzero layer" and a "zero layer" point. At this point the flux at the time level n+1 is calculated explicitly forward from the complete momentum equation. This gives the sign of the divergence of the flux considering the continuity equation at the neighbouring "zero layer" point. If one of the four surrounding velocity points contributes to a divergence at the "zero layer" point then all those surrounding velocity points are defined as "land" points. If all of the four surrounding velocity points contribute to a convergence at the "zero layer" point then all surrounding velocity points may be defined as "sea" points if this does not contradict with the decision at a neighbouring "zero layer" point. In other words the distribution of "sea" and "land" points is chosen so that at no "zero layer" point divergence and, consequently, no negative layer thickness can occur at the next time level.

The specific property of this method is that only mass points can be subtracted from or added to the interior ocean when they have exactly no mass. Thus, the scheme is mass conserving. In an implicit formulated model values at the new time level are not known when "sea" and "land" points are updated. Thus, only an estimate for the time dependency of the variables and, therefore, only an estimate for the new "sea" and "land" point distribution can be made. It must be pointed out that three kinds of error sources leading to a negative layer thickness are possible. The first was explained in the last section and concerns errors in the pressure gradient near a steep topography. This error is excluded by the method of "moving walls". The second one is where a positive layer thickness becomes negative through a simple time extrapolation error. The third type of error occurs when convergence is estimated at a "zero layer" point but divergence has actually occured, which may happen in a few cases because of the implicit technique. However, all those reasons for the occurence of negative layer thicknesses concerning the last two error sources are small and can be avoided by some kind of flux correction scheme. Only one insufficiency of this method remains. If a velocity point becomes a "land" point, the momentum at this point is lost. During the reverse process no momentum is gained. Therefore, the scheme is stable. However, the error above does not occur at every time step and at every grid point but, rather, is a rare event. In the mean over the whole model integration it is therefore believed that the loss of momentum when a velocity point is switched off from the interior ocean is neglegibly small compared with other errors.

We may conclude that the method of "moving walls" allows a very efficient formulation of a model with isopycnic coordinates. As illustrated in Figure 1, a simple multi-layer model is formulated in a rectangular basin. Such a model can be vectorized easily. Areas with "zero layer" thickness are treated in the same way as "land" points. Thus, the inclusion of any coastline geometry is simply extended to all layers. Because the physically present part of the layers can change their horizontal extension due to the slope of the topography those "coastlines" are updated every time step. The increase in computer time is neglegible when this technique is employed in a multi-layer model in a rectangular basin.

2.3. Coupling between mixed layer and isopycnal model

To get realistic sea surface temperatures it is obviously necessary to include a mixed layer at the top of the ocean. In the present model the parameterization is taken from Kraus and Turner [1967] including shear stress at the bottom of the mixed layer. The essential problem now is not the calculation of the mixed layer density itself but the coupling between the mixed layer and the interior ocean. In the mixed layer density must be allowed to vary in the horizontal direction; however, in the isopycnic layers density is constant "a priori". Consequently, there are two different coordinate systems which must be coupled. The actual problem is what happens if the surface layer becomes heavier than the layer below due to cooling, for instance. To explain the coupling of the two different coordinate systems, the mixed layer and the isopycnic part of the model, a whole seasonal cycle will be discussed now.

Let us start with a vertical density distribution as in Figure 5a. If cooling occurs, then the mixed layer parameterization predicts entrainment at the bottom of the uppermost layer. Cooling at the surface leads to convective overturning producing turbulent kinetic energy, which penetrates toward the bottom of the mixed layer and at the end is converted into potential energy. This means that heavier water is transferred into and mixed throughout the upper layer. Its layer thickness thereby increases (see Figure 5b). After some time the second layer is dried up and physically no more present. From this moment the first layer gets its mass from the third layer (see Figure 5c). Again, when the third layer has lost its whole mass to the mixed layer, then water will be entrained from the fourth layer and so on. During entrainment situations it is unimportant for the technique what the density of the entrained water is. The density in the surface layer is calculated simply by mixing the water of the surface layer with the entrained water.

During the detrainment phase, however, we must be careful. Detraining water from the mixed layer , which has an arbitrary density and in all cases a smaller density than in the layer below, into an isopycnic layer, where density is prescribed "a priori", makes it necessary to balance the resulting density change by transfering a certain amount of mass from the third into the second layer. On the one hand a third layer is not always present because of the topography; on the other hand such a mechanism for maintaining the density in an isopycnic layer cannot be justified by physical arguments. It does not conserve potential energy. To circumvent that problem we allow the density in the second layer to vary. If we now start from a situation like that in Figure 5d, the second layer will then be flooded during

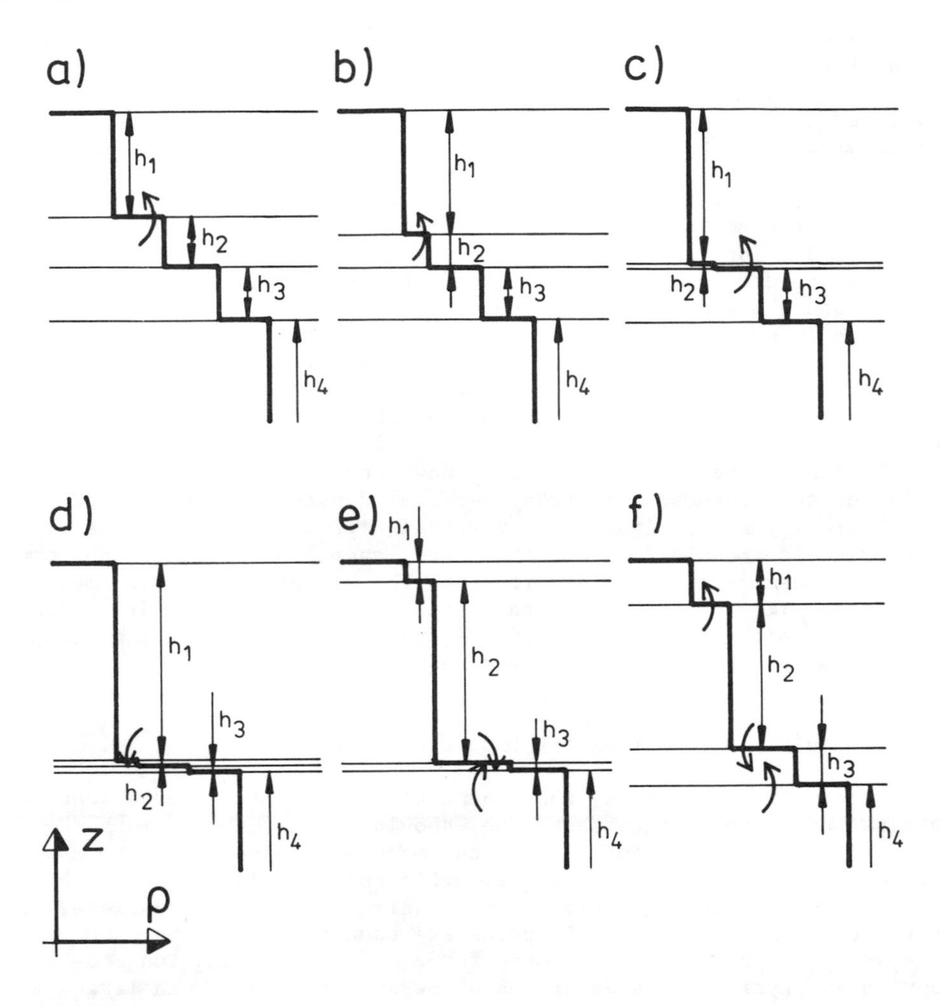

Figure 5a-f. Situations of layer thicknesses (thin horizontal lines) and vertical density distribution (thick lines) during a seasonal cycle. "Zero layers" are drawn as very thin layers. Cases a-d show situations in the model during the entrainment phase with a disappearing second and third layer. Case e shows the treatment of detrainment in the model and case f the decomposition of a "zero layer" due to vertical dissipation with a formation of a new mixed layer.

detrainment. Because detrainment is mostly connected with warming at the surface, a very shallow mixed layer with smaller density is generated. This leads to a situation like Figure 5e. It must be mentioned that the detrainment rate may not be calulated by the parameterization itself. In reality the old mixed layer is not retained, but a new one occurs at the surface. To simulate this correctly the new mixed layer thickness could be calculated diagnistically from the Monin-Obukov length. However, the mixed layer physics is coupled, for example, to wave physics. In considering this, the detrainment rate is computed from the difference of the Monin-Obukov length and the old mixed layer thickness divided by the time step. If the mixed layer physics is not disturbed by other processes the Monin-Obukov length is reached exactly in one time step. This has the advantage that we do not have to switch between prognostic to diagnostic treatment of the equations which would be hard for the whole set of equations especially if they are solved implicitly.

To finish the seasonal cycle we now need some kind of vertical diffusion to decompose the strong vertical density gradient represented by a "zero layer" in Figure 5e. Vertical diffusion in this case means that mass is transferred into the third layer from the layer above and below considering that the density of the isopycnic layer must be conserved. If a reasonable parameterization for vertical diffusion is included, a situation like that in Figure 5f, a state similar to the starting point (Figure 5a) of the seasonal cycle, is reached.

2.4. Multiple-semi-implicit time step scheme

Using primitive equations, the amount of computer time for a long term integration is so large that one should think about methods for increasing the efficiency of such models. Because of the aspired generality of the underlying model, efficiency should not be achieved by simplifying the equations to quasi-geostrophic equations, for instance. The main problem in primitive equation models is the small time step of typically a few hours for a general circulation model. In many cases physical considerations allow one to take a much larger time step of, say, up to 1 week to resolve physical processes like a seasonal cycle. Usually only the barotropic mode is treated implicitly and therefore only the external waves are unconditionally stable. However, fast equatorial Kelvin and Rossby waves or other mechanisms like advection, diffusion or entrainment in some parts of the ocean can lead to numerical instability when large time steps are taken. The time-splitting method developed by Marchuk [1965] allows the implicit treatment of all terms in the equations. However, it has some severe disadvantages. The most important one is that the stationary state is dependent on the time step. Further, time-splitting instability can occur. The first point does not allow a variable time step like that very often used in mathematical computations because not only the fluctuations but also the mean state would change. Now the basic idea of the time-splitting scheme is to split the whole set of equations up into many small sets where an implicit solution can be found. The difficulties described above occur because in each step of the time-splitting method those terms are ignored which are not treated implicitly. To circumvent this problem all missing terms can be added to all time-splitting steps in explicit formulation thus getting semi-implicit steps (For details see Robert [1972]). However, this still makes no sense, because all intermediate values of the time-splitting scheme are now guesses for the next time level. The explicit part in each semi-implicit step can cause instability if the time step is large enough. But, if we retain the values of all those terms which were treated implicitly in one of the steps before rather than recalculate them explicitly, the system is stable. This should be explained in the following example. If we take the equation

$$a_t = f(a) + g(a) \;,$$

where f and g are arbitrary operators, then we formulate the first semi-implicit step; implicit with respect to f, explicit with respect to g:

$$a^{*n+1} - \frac{\Delta t}{2} f(a^{*n+1}) = a^n + \frac{\Delta t}{2} f(a^n) + \Delta t\, g(a^n)$$

a^{*n+1} now is a first guess for a^{n+1}. If the solution for a^{*n+1} is found then the next semi-implicit step can be formulated:

$$a^{**n+1} - \frac{\Delta t}{2} g(a^{**n+1}) = a^n + \frac{\Delta t}{2} g(a^n) + \frac{\Delta t}{2} (f(a^{*n+1}) + f(a^n))$$

Notice that f is not treated explicitly forward in time but, rather, that the implicitly calculated value $f(a^{*n+1})+f(a^n)$ from the first step is taken. Therefore, the whole right side of the second step can be calculated explicitly. If the implicit solution for a^{**n+1} is found and there are no further terms which must be treated implicitly, we take the second guess as final value a^{n+1}. To circumvent the problem that all intermediate guesses for a^{n+1} have to be stored and all terms have to be calculated in each semi-implicit step, the first step is subtracted from the second one leading to the following expression:

$$a^{**n+1} - \frac{\Delta t}{2} g(a^{**n+1}) = a^{*n+1} - \frac{\Delta t}{2} g(a^n)$$

It now clearly can be seen that in the stationary state all guesses are identical. Thus from the first step it can be seen that $f(a^n)+g(a^n)=0$. Consequently, the stationary state is independent on the time step and only dependent on the discretization schemes in space. Now the time step may be variable.

Stability is always guaranteed by the mathematics, so there are no rules how to split up the set of equations. However, if some quantity should be conserved, all those terms which are responsible for that conservation must be treated together. The construction of such a scheme is possible if done very carefully. The underlying model conserves mass and momentum exactly. To reduce the time discretization errors in a transient state it is necessary to treat the most important terms in the first step implicitly. This is a prediction step. In the following correction steps the guesses are changed a little bit but enough to ensure stability. In the underlying model, the pressure gradient and Coriolis terms in the momentum equation and the divergence term in the continuity equation are treated implicitly leading to a rather sophisticated method to get a stable solution when time steps larger than 1 day are taken. From the convergence condition of the resulting wave equation an optimal time step is computed. In the first correction step the mixed layer physics is treated implicitly, followed by correction steps in respect to advection and finally diffusion.

Summarizing, it can be said that the Multiple-Semi-Implicit (MSI) scheme is an extension of the time-splitting method combined with the semi-implicit technique. It allows an unconditionally stable formulation and a variation of the time step taking always the largest possible one allowed by the physics rather than the numerics. However, the resulting code is rather complex and still there is the problem of finding the solution in each semi-implicit step.

3. CONCLUSION

The described methods above make it possible to develop a rather efficient ocean circulation model on a vector computer in spite of taking isopycnic coordinates for the interior ocean. The coupling to a mixed layer model as upper boundary condition for the isopycnic part of the ocean model allows coupled experiment with atmospheric models because the SST is predicted directly. Finally, using the proposed time step scheme, primitive equations can be formulated with a relatively high efficiency. Runs with 1 month time step for a global version of this OGCM were successfully done. Consequently, such a type of model can be used for a wide range of problems. By using the isopycnic coordinate concept there are some advantages to models with Eulerian coordiantes; however, limitations occur if phenomenons like double diffusion in the deep ocean should be included.

4. REFERENCES

Bleck, R., and D.B. Boudra, Initial testing of a numerical ocean circulation model using a hybrid (quasi-isopycnic) vertical coordinate, J. Phys. Oceanogr., 11, 755-770, 1981.

Kraus, E.B., and J.S. Turner, A one-dimensional model of the seasonal thermocline, Tellus, 1, 88-97, 1967.

Marchuk, G.I., Numerical methods for solving weather-forecasting and climate-theory problems, Redaktsionno - Izdatel'skii Otdel Sibirskogo Otdeleniia Akademii Nuuk, SSSR, 1965.

Robert, A., J. Henderson, and C. Turnbell, An implicit time step scheme for baroclinic models of the atmosphere, Mon.Wea.Rev.,100, 329-335, 1972.

TOWARDS A NONLINEAR 2-MODE MODEL WITH FINITE AMPLITUDE TOPOGRAPHY AND SURFACE MIXED LAYER

D. van Foreest and G.B. Brundrit
Department of Oceanography
University of Cape Town
7700 Rondesbosch
South Africa

ABSTRACT. The forced, linear equations of motion with a free surface and finite amplitude topography can be transformed into a set of coupled modal equations which are independent of z. This is achieved through the use of a Galerkin procedure which used the vertical normal modes to represent the vertical structure. The vertical normal modes cannot adequately describe the mixed surface layer and consequently do not allow density changes at the surface.

New vertical structure functions are derived. These allow for density changes at the surface and discontinuity in u, v, and ρ at the base of the mixed layer, but they are identical to the vertical normal modes in the stratified water below the mixed layer. It is then possible to extend the Galerkin procedure to realistically include advection of momentum and density in the equations of motion.

1. INTRODUCTION

The equations of motion for a stratified geophysical fluid such as the ocean are very difficult to solve because they are nonlinear and four dimensional (3 space dimension and time). It is common practice to alleviate this by reducing the dimensionality of the equations. This can be done by approximating the vertical density structure with two or more layers by constant density. There can be little doubt that such two layer models have contributed immensely towards explaining many ocean circulation phenomena.

Another approach, which retains the continuously stratified nature of the ocean, is to separate the vertical structure from the time and horizontal structure in the equations of motion. This latter approach is generally restricted to the inviscid equations of motion linearised about a static background density for a flat bottom ocean. Examples of analytical models that use this approach are numerous [Moore and Philander, 1977; Gill and Clarke, 1974; Pollard, 1970; Lighthill, 1969]. These linearised equations are generally referred to as the modal equations because the separation of the vertical structure from horizontal

J. J. O'Brien (ed.), Advanced Physical Oceanographic Numerical Modelling, 523–555.

and time dependent structure results in an infinite number of uncoupled sets of equations for the horizontal and time dependent structure and a corresponding set of orthogonal modes for the vertical structure (1 barotropic and infinite number of baroclinic modes). As shown by McCreary [1981] inclusion of vertical mixing of heat and momentum and separation into normal modes is also possible as long as the vertical diffusion coefficients are restricted to specific forms. Analytical studies based on this separation procedure are restricted to a flat bottom ocean as the separation procedure does not work when topography is included.

It is possible however, to remove the z dependence in more complete equations of motion, while retaining the continuously stratified nature, i.e. including topography nonlinear and diffusion terms, through the use of a Galerkin (Strang and Fix) procedure. Galerkin procedures in ocean modelling have been used by, for example, Davies [1985] and Flierl [1978]. For the quasigeostrophic equations, Flierl advocates the use of the standard vertical normal modes both as basis functions, in which the dependent variables are expanded, and as test functions in the Galerkin procedure. In this way Flierl managed to construct a set of horizontal and time dependant quasigeostrophic coupled equations, for barotropic and baroclinic modes, which include order Rossby number topographic variations and nonlinear terms, but which are independant of z. Following essentially the ideas of Flierl [1978], van Foreest and Brundrit [1982] have shown that it is possible to construct horizontal and time dependent sets of linearised equations for the barotropic and baroclinic modes which include finite amplitude topography. By using the vertical normal modes as test functions in the Galerkin procedure it is possible to separate the original differential equations into a set of coupled differential equations for the barotropic and baroclinic modes. In our experience the use of the vertical normal modes as test functions in the Galerkin procedure has two advantages. Firstly, because the test functions are physically meaningful and because it is sufficient for many applications to consider barotropic and first baroclinic modes, one need only take the first two terms in the expansion of the dependent variables into account. If one had used arbitrary orthogonal test functions the expansion would presumably have to take many more terms into account in order to resolve barotropic and first baroclinic modes. The second practical advantage is that the resulting differential equations have been separated in barotropic and baroclinic modes so that one can take advantage of the large time-steps that can be used in numerical modes for the baroclinic modes.

In this paper we analyse the Galerkin procedure in some detail and extend the work of van Foreest and Brundrit [1982] to include advection of momentum and density. The approach is based on the use of discontinuous vertical structure functions which are slightly different from the standard vertical normal modes. The use of the new structure function also allows density changes at the surface.

2. THE FUNDAMENTAL PROCEDURE

In order to illustrate the fundamental procedure, a dynamically simple case is considered. Its governing equations are the equations of motion linearised about a static background density, $\hat{\rho}(z)$, and an equilibrium state of no motion. External forcing is provided by a surface wind stress, τ_s. The equations take the form

$$u_t - fv = -\frac{1}{\bar{\rho}} p_x + \frac{1}{\bar{\rho}} \tau^x_z \qquad (1a)$$

$$v_t + fu = -\frac{1}{\bar{\rho}} p_y + \frac{1}{\bar{\rho}} \tau^y_z \qquad (1b)$$

$$u_x + v_y + w_z = 0 \qquad (1c)$$

$$\rho_t + ws = 0 \qquad (1d)$$

$$p_z + g\rho = 0 \ . \qquad (1e)$$

Subscripts denote partial derivatives, superscripts denote vector components and the symbols have the following meaning

Independent variables	x	positive east
	y	positive north
	z	positive up
	t	time
Dependent variables	u	x velocity component
	v	y velocity component
	w	z velocity component
	p	dynamic pressure
	ρ	density anomaly
Physical parameters	g	gravity
	f	Coriolis parameter
	$\bar{\rho}$	mean density
	$\hat{\rho}$	background density
	$s = \hat{\rho}_z$	vertical density gradient
Forcing	τ^x, τ^y	horizontal (x,y) components of stress

Equations (1), for the flat bottom case, are often used in analytical ocean modelling [Gill and Clarke, 1974; Moore and Philander 1977]. Their success in the flat bottom case is due to the fact that the vertical structure can be separated from the horizontal and time dependent structure. The separation results in uncoupled sets of equations similar to (1) but for the horizontal and time dependent structure, one set for the barotropic and one set for each baroclinic mode, and a set of orthogonal functions for the vertical structure usually referred to as normal vertical modes (for more detail see Appendix). The application range of equations (1) is limited and can be determined by careful scale analysis [Moore and Philander, 1977]. It is sufficient here to note that their applicability requires the Rossby number to be small.

To illustrate the power of the Galerkin procedure we show that finite amplitude topography can be included in equations (1) without affecting the useful decomposition in horizontal time dependent sets of equations for the barotropic and baroclinic modes. The surface and bottom boundary conditions are

$$p_t = \bar{\rho} g w \qquad \text{at the free surface} \quad z = H$$

$$w = uB_x + vB_y \qquad \text{at the bottom} \quad z = B \,. \tag{2}$$

The horizontal momentum equations (1a,b) show that the horizontal velocities and the pressure must have the same vertical dependence. The detail of this dependence can be established by eliminating the vertical velocity and density anomaly from the last three equations of (1). Together with the vertical boundary conditions (2) this takes the form

$$u_x + v_y + \left(\frac{p_{tz}}{sg}\right)_z = 0 \tag{3}$$

with $$p_t = \bar{\rho} g \left(\frac{p_{tz}}{sg}\right) \qquad \text{at the free surface } z = H$$

$$p_{tz} = uB_x + vB_y \qquad \text{at the bottom} \quad z = B \,.$$

Rather than deal directly with differential equations (1a, b) and (3), a Galerkin approach is taken which requires the construction of an equivalent integral form. A test function, F, is introduced into a weak form of the governing equations which exchanges z-derivatives between the dependent variables and the test function. The appropriate function spaces need careful definition, and Strang and Fix [1973] are closely followed.

A function F has finite energy if the inner product

$$\langle F,F \rangle = \int_{z=B}^{z=H} F^2 dz < \infty \,.$$

The space of such functions F is denoted by H^o. Finite energy functions may be discontinuous and lack convergence properties in the Fourier sense. More restricted spaces are those containing functions whose derivatives are required to have finite energy. The space H^n contains functions whose nth derivatives have finite energy. The (n-1)th derivative of such a function from H^n will possess uniform convergence properties, even though the nth derivative will merely have finite energy. A subscript on the space, e.g. H_E, will refer to a particular set of boundary conditions E which are to be satisfied by all members of the space. For a Galerkin approach, it is the essential rather than the natural boundary conditions which will be referred to in this manner.

We deal first with equation (3). A test function, F, is used to construct a suitable equivalent integral form of equation (3).

$$\langle u_x+v_y, F\rangle + \langle p_t, \left(\frac{F_z}{sg}\right)_z\rangle = \Big[(uB_x+vB_y)F\Big]_{z=B} . \tag{4}$$

It is now required to find

$$u_x,\ v_y \in H^o$$

$$p_t \in H^2$$

which satisfy the equivalent integral form (4) for all test functions $F \in H^2_{E1}$ with E1:

$$F = \bar{\rho}g \frac{F_z}{sg} \quad \text{at the surface } z = H$$

$$F_z = 0 \quad \text{at the bottom } z = B . \tag{5}$$

Equivalence of the integral form (4) with the original differential equation (3) can be demonstrated by twice integrating (4) by parts with respect to z to give

$$\langle u_x+v_y+\left(\frac{p_{tz}}{sg}\right)_z, F\rangle + \left[\frac{p_tF_z}{sg} - \frac{p_{tz}F}{sg}\right]_{z=B}^{z=H} = (uB_x+vB_y)F\Big|_{z=B} \tag{6}$$

in which the use of the essential boundary conditions E1 results in

$$\langle u_x+v_y+\left(\frac{p_{tz}}{sg}\right)_z, F\rangle + \left(p_t-\bar{\rho}g\frac{p_{tz}}{sg}\right)\frac{F_z}{sg}\Big|_{z=H} + \left(\frac{p_{tz}}{sg} - (uB_x+vB_y)\right)F\Big|_{z=B} = 0. \tag{7}$$

Each term must be separately zero, giving the original differential equation and boundary conditions (3).

In order to solve the equivalent integral form (4) we need to define suitable test functions and, according to the Galerkin procedure, expand the dependent variable, u, v, p in terms of the test functions. We insist that the test functions be identical to the normal flat bottom vertical modal functions, F, which satisfy

$$\left(\frac{F_z}{sg}\right)_z + \lambda F = 0 \quad \text{and boundary condition E1} \quad . \tag{8}$$

Detailed derivations of (8) can be found in, for example, Gill [1982], Moore and Philander [1977] or van Foreest and Brundrit [1982]. Equation and boundary conditions (8) form a Sturm Liouville problem and for each eigenvalue

$$\lambda^i \quad i = 0,1,2 \ldots$$

the solutions

$$F^i \quad i = 0,1,2 \ldots$$

form a complete orthonormal set with

$$\langle F^i, F^j \rangle = \delta_{ij} \tag{9}$$

In the flat bottom case the solutions F^i are independent of the horizontal coordinates and depend only on z. In the case here where finite amplitude topography is present, the F^i are functions of x,y, and z as they are derived using the depth H - B(x,y) but assuming the bottom is locally flat. Consequently the orthogonality relation (9) will only hold locally.

By expanding the dependent variables in terms of F^i, i.e.

$$u = \sum_i U^i(x,y,t)F^i(x,y,z)$$

$$v = \sum_i V^i(x,y,t)F^i(x,y,z)$$

$$p = \sum_i P^i(x,y,t)F^i(x,y,z) \tag{10}$$

and using the orthogonality property (9) the equivalent integral form (4) becomes

$$U_x^j + \sum_{i=o}^{n} U^i\langle F_x^i, F^j\rangle + V_y^j + \sum_{i=o}^{n} V^i\langle F_y^i, F^j\rangle - \frac{\lambda^j}{\bar{\rho}} P_t^j$$

$$= \sum_{i=0}^{n} (U^i B_x + V^i B_y) F^i F^j \Big|_{z = B} \quad . \tag{11}$$

The barotropic mode is found from (11) by taking $j = 0$ and the baroclinic modes by taking $j = 1,2,3 \ldots$. In the case where only the barotropic and first baroclinic modes are considered ($n = 1, j = 0,1$) the remaining innerproducts in (11) can be simplified [van Foreest and Brundrit, 1982].

We now deal with the horizontal momentum equations (1a,b). We show the procedure only for the x momentum equation as the y-momentum equation can be handled in the same way. The equivalent integral form of (1a) is

$$\langle u_t - fv + \frac{1}{\bar{\rho}} p_x, F\rangle + \frac{1}{\bar{\rho}} \langle \tau^x F_z\rangle = \frac{1}{\bar{\rho}}(\tau_s^x F\Big|_{z=H} - \tau_B^x F\Big|_{z=B}) \quad . \tag{12}$$

On the right hand side of (12) are the vertical boundary conditions for the stress

$$\tau^x = \tau_s^x \text{ at } z = H \text{ where } \tau_s^x \text{ is the x component of the surface wind stress} \tag{13}$$

$$\tau^x = \tau_B^x \text{ at } z = B \text{ where } \tau_B^x \text{ is the x component of the bottom stress}$$

Equivalence of (12) with the original differential equation (1a) can be demonstrated by integrating (12) by parts, with respect to z, once to give

$$\langle u_t - fv + \frac{1}{\bar{\rho}} p_x - \frac{1}{\bar{\rho}} \tau_z, F\rangle + \frac{1}{\bar{\rho}} (\tau^x F - \tau_s^x F)\Big|_{z=H} - \frac{1}{\bar{\rho}} (\tau^x F + \tau_B^x F)\Big|_{z=B} = 0 \quad . \tag{14}$$

Each term must be separately zero giving the original differential equation (1a) and the boundary conditions (13).

Again, we can solve the equivalent integral form (12) by expanding u, v, and p in term of F as before. From (1a) we can see that the first derivative with respect to z of the stress, τ, must equal the vertical dependence of u, v, p. So we should expand τ not in terms of F^i but in terms of functions G^i.

$$\tau^x = \sum_{i=o}^{n} T^{x^i} G^i$$

where $G^i_z = F^i$. (15)

Making use of the orthogonality property of the F^i, (9), we can write (12) as

$$U^j_t - fV^j + \frac{1}{\bar{\rho}} P^j_x + \frac{1}{\bar{\rho}} P^i \langle F^i_x, F^j \rangle + \frac{1}{\bar{\rho}} \langle T^{x^i} G^i F^j_z \rangle$$

$$= \frac{1}{\bar{\rho}} \left(\tau^x_s F^j \Big|_{z=H} - \tau^x_B F^j \Big|_{z=B} \right) \quad . \qquad (16)$$

The innerproduct for the stress term can be rewritten as

$$T^{x^i} \langle G^i F^j_z \rangle = T^{x^i} \Big[G^i F^j \Big]^{z=H}_{z=B} - T^{x^i} \langle G^i_z, F^j \rangle \quad ; \qquad (17)$$

using (15) and (9) we finally write (17) as

$$T^{x^i} \langle G^i F^j_z \rangle = \left(\tau^x_s F^j \Big|_{z=H} - \tau^x_B F^j \Big|_{z=B} \right) - T^{x^j} \quad , \qquad (18)$$

Substitution of (18) back into (16) results in

$$U^j_t - fV^j + \frac{1}{\bar{\rho}} P^j_x + \frac{1}{\bar{\rho}} P^i \langle F^i_x F^j \rangle - T^{x^j} = 0 \quad . \qquad (19)$$

The above expression for the x-momentum equation can be obtained in a more straightforward manner [Gill and Clarke, 1974], but the above procedure is consistent with the Galerkin procedure. In the above no particular assumption on the vertical structure of the stress, τ, has been made. Once a realistic form for τ has been taken it is easy to calculate T^i

$$<\tau_z^x, F^j> = T^{x^j} \tag{20}$$

If only the barotropic and first baroclinic modes are taken into account (n = 1, j = 0,1), the remaining innerproduct in (19) can be simplified [van Foreest and Brundrit, 1982]. The resulting momentum equations for this case are listed in Table I.

TABLE I

Linear 2 mode equations

Barotropic mode

x - momentum

$$U_t^o - fV^o = -\frac{1}{\bar{\rho}}\left(P_y^o + \frac{1}{2}P^o(F^o)^2 B_x\Big|_{z=B}\right) + \frac{\tau^x}{\bar{\rho}}F^o - \frac{1}{\bar{\rho}}P^1F^oF^1B_x\Big|_{z=B}$$

y-momentum

$$V_t^o + fU^o = -\frac{1}{\bar{\rho}}\left(P_y^o + \frac{1}{2}P^o(F^o)^2 B_y\Big|_{z=B}\right) + \frac{\tau^x}{\bar{\rho}}F^o - \frac{1}{\bar{\rho}}P^1F^oF^1B_y\Big|_{z=B}$$

hydrostatic

$$\bar{\rho}R_t^o + P_t^o = 0$$

conservation of density

$$R_t^o + W^o = 0$$

continuity

$$U_x^o + V_y^o - \lambda^o W^o = \frac{1}{2}(U^oB_x + V^oB_y)F^{o^2}$$

First baroclinic mode

x-momentum

$$U_t^1 - fV^1 = -\frac{1}{\bar{\rho}}\left(P_x^1 + \frac{1}{2}P^1(F^1)^2B_x\Big|_{z=B}\right) + T^{x^1}$$

y-momentum

$$V_t^1 + fU^1 = -\frac{1}{\bar{\rho}}\left(P_y^1 + \frac{1}{2}P^1(F^1)^2B_y\Big|_{z=B}\right) + T^{y^1}$$

hydrostatic

$$\bar{\rho}R_t^1 + P_t^1 = 0$$

conservation of density

$$R_t^1 + W^1 = 0$$

continuity

$$U_x^1 + V_y^1 - \lambda^1 W^1 = \frac{1}{2}(U^1 B_x + V^1 B_y)F^{1^2}\Big|_{z=B} + \frac{1}{2}(U^o B_x + V^o B_y)F^o F^1\Big|_{z=B}$$

In the procedure above we have concentrated on the solution for u, v, and p by eliminating w and ρ from (1c),(1d),(1e). This has led to a continuity equation which contains a second derivative of p with respect to z while no z derivatives are present for u and v. We therefore required

$$u,\ v \in H^o$$
$$p \in H^2$$

and consequently the test functions F^j (21)

$$F^j \in H^2_{E1} \quad .$$

This implies that the vertical structure functions F^j must have a continuous second derivative which in turn implies that approximated solutions for u, v, and p must have continuous second derivatives with respect to z. These continuity requirements can be relaxed if equations (1c),(1d),(1e), are treated separately (i.e. their coupled low order form).

3. COUPLED LOW ORDER FORM

The fundamental procedure was conveniently illustrated, in the previous section, using a combined continuity, hydrostatic, and conservation of density equation by eliminating ρ and w. It may not be clear, however, at this stage, how one should obtain the equivalent integral form, nor is it obvious from the previous section how one should solve for ρ and w. Also the elimination of ρ and w from the continuity equation has led to a restricted test function space in the sense that the second derivative with respect to z has to be continuous. By treating equations (1c, d, e) in their coupled low order form the continuity properties for the test functions can be relaxed.

In practice we start by choosing a set of functions to use as test functions in the equivalent integral form and in which to expand the dependent variables. These functions should have suitable properties, i.e. orthogonal. In addition, we feel there is much to be gained by using physical meaningful functions as these should improve the convergence of the expansion. An obvious choice to make is the vertical normal modes [Lighthill, 1969; Gill and Clarke, 1974] which possess the required orthogonality properties. The vertical normal modes are found by separating the dependent variables in the unforced flat bottom form of equations (1) into a function of z and a function of x, y, t. If u, v, p, w and ρ are expanded as follows

$$\begin{aligned} u &= \Sigma\, U(x,y,t)\, F(z) \\ v &= \Sigma\, V(x,y,t)\, F(z) \\ p &= \Sigma\, P(x,y,t)\, F(z) \\ w &= \Sigma\, W(x,y,t)\, G(z) \\ \rho &= \Sigma\, R(x,y,t)\, sG(z) \end{aligned}$$

the vertical normal modes are found as eigenfunctions for eigenvalues λ of the Sturm Liouville problem (see Appendix).

$$\frac{F_z}{sg} = \frac{G}{\bar{\rho}} \qquad F,G \in H^1_{E2} \qquad G_z = -\lambda F \tag{22}$$

with E2: $F = gG$ at the surface $z = H$
$G = 0$ at the bottom $z = B$.

We can now use the eigenfunctions F and G as test functions for the equivalent integral form of equations (1) bearing in mind that we will be expanding the dependent variables in terms of the test functions F,G later (the Galerkin approach is to restrict the solution space to the same space as the test function space). The momentum equations (1a,b) can be handled as before and we concentrate on the remaining three equations (1). We first write equations (1c,d,e) in their weak form by taking the innerproduct of the original differential equation with one of the test functions.

$$<u_x+v_y+w_z, F^j> = 0 \quad (23a)$$

$$<\rho_t+sw, G^j> = 0 \quad (23b)$$

$$<p_{tz}+g\rho_t, G^j> = 0 \quad (23c)$$

It is difficult to give a general rule for how to decide what test function to use in the innerproduct. In equation (23a) we take the innerproduct with F because u, v will be expanded in terms of F and, because w is expanded in terms of G, w_z is in some sense expanded in functions proportional to F.

In equation (23b) we take the innerproduct with G because both ρ and w are expanded in terms of G. In equation (23c) we take the innerproduct with G because ρ is expanded in terms of G and because p_{tz} is expanded in functions proportional to F_z. The overriding consideration of whether to choose F or G for a particular equation is the simplification that results after substitution of the expansions.

Because we will be substituting expansions in terms of vertical structure functions, it is necessary to transfer all z-derivatives onto the test functions (Gibbs effect). This can be done through integration by parts. Equations (23a) takes the form

$$<u_x+v_y, F^j> - <w, F_z^j> + \left[wF^j\right]_{z=B}^{z=H} = 0 \, . \quad (24)$$

Two boundary terms, one at the free surface and one at the bottom, have appeared. The free surface condition is part of the essential boundary conditions for the test function. The bottom boundary condition $w = uB_x + vB_y$ is not taken into account in the derivation of the test functions and can therefore be enforced at this stage. The resulting form for (24) is

$$<u_x+v_y, F^j> - <wF_z^j> + wF^j\Big|_{z=H} - (uB_x+vB_y)F^j\Big|_{z=B} = 0 \quad . \quad (25)$$

After integrating the first term in equation (23c) by parts once we obtain

$$<g\rho_t, G^j> - <p_t, G_z^j> + \left[p_t G^j\right]_{z=B}^{z=H} = 0 \quad . \quad (26)$$

Again two boundary terms appear. The natural free surface boundary condition has been taken into account in the test function and need not be enforced again although one can do so if desired. It is not possible to enforce the natural bottom boundary condition because of the essential bottom boundary E2: $G^j = 0$. The resulting equivalent integral forms of (23a, b, c) are therefore

$$
\begin{aligned}
&\langle u_x+v_y, F^j\rangle - \langle wF^j_z\rangle + wF^j\Big|_{z=H} - (uB_x+vB_y)F^j\Big|_{z=B} = 0 \\
&\langle \rho_t+sw, G^j\rangle = 0 \qquad (27)\\
&\langle g\rho_t, G^j\rangle - \langle p_t, G^j_z\rangle + p_tG^j\Big|_{z=H} = 0 \ .
\end{aligned}
$$

It is often easier to write down the equivalent integral form directly and to show equivalence with the original differential equations through integration by parts.

It is now required to find

$$
\begin{aligned}
&u,v,\rho \in H^o \\
&p,w \in H^1 \qquad (28)
\end{aligned}
$$

which satisfy the above integral (27) form for all

$$F,G, \in H^1_{E2} \ .$$

The solution is obtained by substituting the expansions for u, v, p, and w in terms of F, G (27) and by making use of the orthogonality etc., properties of the F, G (see Appendix). The solution can be written as

$$
\begin{aligned}
&\langle (U^iF^i)_x, F^j\rangle + \langle (V^iF^i)_y, F^j\rangle - \lambda^jW^j = (U^iF^iB_x+V^iF^iB_y)F^j\Big|_{z=B} \\
&R^j_t + W^j = 0 \qquad (29)\\
&\bar{\rho}R^j_t + P^i_t = 0 \ .
\end{aligned}
$$

Again the remaining innerproducts can be simplified when only barotropic and first baroclinic modes are taken into account. The equations for the 2-mode case are listed in Table I.

In the above we have managed to reduce the continuity properties of the solution by taking the test functions F, $G \in H^1_{E2}$, while before $F \in H^2_{E1}$. This implies that F, G must be functions whose first derivative with respect to z has to be continuous. As the solution for u, v, and ρ is obtained as a series expansion in the functions F or G they will also have a continuous first derivative with respect to z (although in theory it is possible to approximate a discontinuous function in terms of a series of continuous functions if enough terms in the expansion are taken into account). As the original differential equations only require u, v, and $\rho \in H^0$ (possibly discontinuous functions) we should try to relax the test function space such that the test functions used for the expansion of u, v, ρ are also $\in H^0$.

4. INCORPORATION OF MIXED LAYER DYNAMICS

The contrast in continuity properties permitted for u, v and ρ as against w and p (see (28)) is consistent with more realistic dynamics. That this is so is highlighted when a surface mixed layer, driven by wind induced turbulence, is introduced into the dynamical picture [Pollard et al. 1973]. Within such a layer, the background vertical gradient of density s is zero. There can also be an abrupt discontinuity in density at the base of the mixed layer. Such a discontinuous structure is shared by the horizontal velocity components and the density anomaly. On the other hand, the vertical velocity and the dynamic pressure must be continuous throughout the water column. For a mixed layer of thickness h, with a background density discontinuity $\Delta\hat{\rho}$ at the base (z = H-h indicated by *), the governing linear unforced equations are taken as

$$u_t - f_v = \frac{-1}{\bar{\rho}h} \int_{z=H-h}^{z=H} p_x \, dz$$

$$v_t + fu = -\frac{1}{\bar{\rho}h} \int_{z=H-h}^{z=H} p_y dz$$

$$u_x + v_y + w_z = 0 \tag{30}$$

$$p_t + w^* \frac{\Delta\hat{\rho}}{h} = 0$$

$$p_z + \rho g = 0$$

with $p_t = \bar{\rho} g w$ at the free surface z = H

w and p continuous at the interface z = H-h .

Under upwelling conditions (i.e. w > 0) the density changes in the mixed layer ari e through the entrainment of density from the stratified water below into the surface mixed layer above. When w < 0, $\rho_t = 0$.

The wind induced turbulent mixing ensures that the horizontal velocity components and the density anomaly remain independent of depth in the mixed layer. In the water below the mixed layer, the standard modal equations (i.e. unforced flat bottom version of equations (1)) still hold.

In order to accommodate this combined vertical structure in a Galerkin procedure on a more complete set of equations (nonlinear, finite amplitude topography) the manner in which the test functions are constructed is crucial. We proceed by separating the horizontal and time variation from the vertical variation in the surface mixed layer equations by setting

$$
\begin{aligned}
u &= U(x,y,t)L \\
v &= V(x,y,t)L \\
w &= W(x,y,t)G_m(z) \\
p &= P(x,y,t)F_m(z) \\
\rho &= R(x,y,t)K
\end{aligned}
\qquad (31)
$$

The subscripts m are not partial differentiations but merely refer to F and G in the mixed layer.

In (31) K and L are constants. Substitution into equations (30) and separation of the resulting equations such that the horizontal equations are identical to horizontal equations that result from the standard separation of variables in the stratified water, lead to the following relations for the vertical structure in the mixed layer (see Appendix)

$$
\begin{aligned}
(F_m)_z &= \frac{Kg}{\bar{\rho}} \\
(G_m)_z &= -\lambda L \\
K &= G_* \frac{\Delta\hat{\rho}}{h} \\
L &= \frac{1}{h} \int_{z=H-h}^{z=H} F_m \, dz
\end{aligned}
\qquad (32)
$$

with $F_m = gG_m$ at the free surface $z = H$

$F_m = F_s = F_*$ at the base of the mixed layer $z = H-h$

$G_m = G_s = G_*$ at the base of the mixed layer $z = H-h$.

In the above F_s and G_s are the vertical structure functions in the stratified water below the mixed layer (see (22)). In the stratified water the same relations as before hold between F_s and G_s, but we now have to give the boundary condition at the mixed layer base rather than at the free surface. The relations that hold in the stratified water are (see (22))

$$\frac{F_z}{sg} = \frac{G}{\rho}$$

$$G_z = -\lambda F \qquad (33)$$

with E3: $G = 0$ at the bottom $z = B$

$G_s = G_m = G_*$ at the base of the mixed layer $z = H-h$

$F_s = F_m = F_*$ at the base of the mixed layer $z = H-h$

Equations (32) and (33) form an eigenvalue problem (not quite a Sturm Liouville problem), and for the eigenvalues λ^k, $k = 0,1,2,...$ it is possible to find the modal vertical structure functions

$$F_m^k,\ G_m^k,\ K^k,\ L^k,\ F_s^k,\ G_s^k \qquad k = 0,1,2,...$$

More detail can be found in the Appendix. It is now possible to construct composite vertical structure functions for the surface mixed layer and the stratified water beneath

$$A^k = [L^k, F_s^k] \in H^0$$

$$B^k = [F_m^k, F_s^k] \in H^1 \qquad (34)$$

$$C^k = [K^k, sG_s^k] \in H^0$$

$$D^k = [G_m^k, G_s^k] \in H^1 \quad .$$

As the combined vertical structure function relations and boundary conditions do not form a Sturm Liouville problem, the baroclinic modes are not orthogonal. Fortunately the barotropic mode is normal to the baroclinic modes. If we restrict the vertical structure function space (34) to barotropic and first baroclinic modes only, we can make use of the orthonormality relationships between them. Details are given in the Appendix.

We can now use the vertical structure functions (34) as test functions in a Galerkin procedure on equations (1) in the presence of finite amplitude topography. Because of the introduction of a mixed surface layer, it has become necessary to provide boundary conditions for the stress at the base of the mixed layer. The complete boundary conditions for equations (1) with discontinuities in u, v, ρ at the bottom of mixed surface layer and in the presence of finite amplitude topography are

$$\tau^y = \tau^y_s, \ \tau^x = \tau^x_s \qquad \text{at the surface } z = H$$

$$\tau^x = \bar{\rho}w\Delta u, \ \tau^y = \bar{\rho}w\Delta v \qquad \text{at the base of the mixed layer } z = * \tag{35}$$

$$w = uB_x + vB_y \qquad \text{at the bottom } z = B$$

$$\tau^x = \tau^y = 0 \qquad \text{in the stratified water column}$$

We follow Pollard et al [1974] and assume that the stress at the bottom of the mixed layer is used entirely to bring the momentum of the entrained water up to that of the mixed layer. It is convenient to assume that the stress is zero everywhere below the mixed layer, although this is not necessary. In the Galerkin procedure the dependent variables are expanded in terms of the vertical structure functions as

$$u = \sum_{i=o}^{1} U^i A^i \qquad w = \sum_{i=o}^{1} W^i D^i$$

$$v = \sum_{i=o}^{1} V^i A^i \qquad \rho = \sum_{i=o}^{1} R^i C^i \tag{36}$$

$$p = \sum_{i=o}^{1} P^i B^i$$

so that the degree of continuity in the vertical structure functions (test functions) matches that required in the solution of the equivalent integral form. The equivalent integral representation is now to find

$$u,v,\rho \in H^o$$
$$w,p \in H^1$$

which satisfy

$$<u_t - fv + \frac{1}{\rho} p_x, B^j> + \frac{1}{\rho} <\tau^x B^j_z> = \frac{1}{\rho}(\tau^x_s B^j\Big|_{z=H} - \tau^x_B B^j\Big|_{z=B}) - w\Delta u B^j\Big|_{z=*} \quad (37a)$$

$$<v_t + fu + \frac{1}{\rho} p_y, B^j> + \frac{1}{\rho} <\tau^y B^j_z> = \frac{1}{\rho}(\tau^y_s B^j\Big|_{z=H} - \tau^y_B B^j\Big|_{z=B}) - w\Delta v B^j\Big|_{z=*} \quad (37b)$$

$$<u_x + v_y, B^j> - <wB^j_z> + wB^j\Big|_{z=H} - (uB_x + vB_y)B^j\Big|_{z=B} = 0 \quad (37c)$$

$$<\rho_t + ws, D^j> = 0 \quad (37d)$$

$$<p_t, D^j_z> - <g\rho_t D^j> = p_t D^j\Big|_{z=H} = 0 \quad (37e)$$

for all $A^j, C^j \in H^o$

$B^j, D^j \in H^1$.

Equivalence of (37) with the original differential equations (1) can be shown by integrating (37a, b, c, e) by parts once. As an example we will prove the equivalence of (37a). Integrating (37a) by parts once gives

$$<u_t - fv + \frac{1}{\rho} p_x - \frac{1}{\rho} \tau^x_z, B^j> + \frac{1}{\rho}\Big[\tau^x B^j\Big]_{z=B}^{z=*} + \frac{1}{\rho}\Big[\tau^x B^j\Big]_{z=*}^{z=H}$$

$$= \frac{1}{\rho}(\tau^x_s B^j\Big|_{z=H} - \tau^x_B B^j\Big|_{z=B}) - w\Delta u B^j\Big|_{z=*} \quad ; \quad (38)$$

this can be rewritten as

$$<u_t - fv + \frac{1}{\rho} p_x - \frac{1}{\rho} \tau^x_z, B^j> +$$

$$+ \frac{1}{\rho} (\tau^x - \tau^x_s) B^j\Big|_{z=H} \quad (39)$$

$$- \frac{1}{\rho} (\tau^x - w\Delta u) B^j\Big|_{z=*} = 0 \quad .$$

Each term will have to be separately zero giving the original differential equation (1a) and the boundary conditions (35). To obtain (39) we made use of the assumption that $\tau^x = 0$ everywhere in the stratified water.

After substituting the expansions (36) in the equivalent integral form, the z independent equations for barotropic and first baroclinic mode can be obtained and are listed in Table II. We note that the barotropic mode is exactly as before, except for the continuity equation which now contains a coupling term due to the slope of the base of the mixed layer. The dynamic pressure terms in the momentum equations for the first baroclinic modes contain terms representing the dynamic pressure change in the mixed layer due to changes of density (entrainment) in the mixed layer. Again the first baroclinic continuity equation contains a coupling term due to the slope of the mixed layer base.

TABLE II

Linear 2 mode equations with surface mixed layer

Barotropic mode

x-momentum

$$U^o_t - fV^o = -\frac{1}{\bar\rho}\left(P^o_x + \frac{1}{2}P^o(F^o)^2 B_x\Big|_{z=B}\right) + \frac{\tau^x}{\bar\rho}F^o - \frac{1}{\bar\rho}P^1F^oF^1B_x\Big|_{z=B}$$

y-momentum

$$V^o_t + fU^o = -\frac{1}{\bar\rho}\left(P^o_y + \frac{1}{2}P^o(F^o)^2 B_y\Big|_{z=B}\right) + \frac{\tau^y}{\bar\rho}F^o - \frac{1}{\bar\rho}P^1F^oF^1B_y\Big|_{z=B}$$

hydrostatic

$$P^o_t + \rho R^o_t = 0$$

conservation of density

$$R^o_t + W^o = 0$$

continuity

$$U^o_x + V^o_y - \lambda^o W^o = \frac{1}{2}(U^oB_x + V^oB_y)F^{o^2} - F^o(F^1_* - L^1)(U^1h_x + V^1h_y)$$

First baroclinic mode

x-momentum

$$U^1_t - fV^1 = -\frac{1}{\bar{\rho}}\left\{P^1_x\left[1 + \frac{h^3}{12}\left(\frac{K^1 g}{\bar{\rho}}\right)^2\right] + P^1\left[\frac{1}{2}\left(\frac{h^3}{12}\left(\frac{K^1 g}{\bar{\rho}}\right)^2\right)_x + \frac{1}{2}F^{1^2}B_x\right]\right\} + T^{x^1}$$

y-momentum

$$V_t + fU^1 = -\frac{1}{\bar{\rho}}\left\{P^1_y\left[1 + \frac{h^3}{12}\left(\frac{K^1 g}{\bar{\rho}}\right)^2\right] + P^1\left[\frac{1}{2}\left(\frac{h^3}{12}\left(\frac{K^1 g}{\bar{\rho}}\right)^2\right)_y + \frac{1}{2}F^{1^2}B_y\right]\right\} + T^{y^1}$$

hydrostatic

$$P^1_t + \bar{\rho}R_t = 0$$

conservation of density

$$R^1_t + W^1 = 0$$

continuity

$$U^1_x + V^1_y - \lambda^1 W^1 = (U^0 B_x + V^0 B_y)F^0 F^1\Big|_{z=B} + \frac{1}{2}(U^1 B_x + V^1 B_y)F^{1^2}\Big|_{z=B} - \left(\frac{1}{2}L^{1^2} + \frac{1}{2}F^{1^2}_* - L^1 F^1_*\right)(U^1 h_x + V^1 h_y)$$

5. INCLUSION OF NONLINEAR TERMS

With the new vertical structure functions (see section 4), it is possible to extend the Galerkin procedure for a 2 mode model to include advection of momentum and density. The equations of interest are then

$$u_t + (uu)_x + (vu)_y + (wu)_z - fv = -\frac{1}{\bar{\rho}}p_x + \frac{1}{\bar{\rho}}\tau^x_z \qquad (40a)$$

$$v_t + (uv)_x + (vv)_y + (wv)_z + fu = -\frac{1}{\bar{\rho}}p_y + \frac{1}{\bar{\rho}}\tau^y_z \qquad (40b)$$

$$u_x + v_y + w_z = 0 \qquad (40c)$$

$$\rho_t + u\hat{\rho}_x + v\hat{\rho}_y + ws = 0 \qquad (40d)$$

$$p_z + g\rho = 0 \qquad (40e)$$

with

$$\left.\begin{aligned} p_t &= \bar{\rho} g w \\ \tau^x &= \tau^x_s \\ \tau^y &= \tau^y_s \end{aligned}\right\} \text{ at the free surface } z = H$$

$$\left.\begin{aligned} \tau^x &= \bar{\rho} w \Delta u \\ \tau^y &= \bar{\rho} w \Delta v \end{aligned}\right\} \text{ at the base of the mixed layer } z = *$$

$$\left.\begin{aligned} w &= uB_x + vB_y \\ \tau^x &= 0 \\ \tau^y &= 0\,. \end{aligned}\right\} \text{ at the bottom } z = B$$

The nonlinear terms in the momentum equations are written in conservative form. The conservation of density equation is special in that it is linearised about the background density field $\hat{\rho}(x,y,z)$. In practice the background density field will have to be redefined when the density anomaly becomes large. This point is taken up further in the discussion. Using the vertical structure functions A, B, C, D from section 4 as test functions we can write down the equivalent form of (40) as

$$\langle u_t + (uu_x) + (vu)_y + \frac{1}{\bar{\rho}} p_x - fv, B^j\rangle - \langle wuB^j_z\rangle + \frac{1}{\bar{\rho}} \langle \tau^x B^j_z\rangle$$

$$= -\frac{up_t}{\bar{\rho}g} F^j_m\Big|_{z=H} + u(uB_x + vB_y)F^j_s\Big|_{z=B} + \frac{1}{\bar{\rho}} \tau^x_s F^j_m\Big|_{z=H} \tag{41a}$$

$$\langle v_t + (vu)_x + (vv)_y + \frac{1}{\bar{\rho}} p_y + fu, B^j\rangle - \langle wvB^j_z\rangle + \frac{1}{\bar{\rho}}\langle \tau^y B^j_z\rangle$$

$$= -\frac{vp_t}{\bar{\rho}g} F^j_m\Big|_{z=H} + v(uB_x + vB_y)F^j_s\Big|_{z=B} + \frac{1}{\bar{\rho}} \tau^y_s F^j_m\Big|_{z=H} \tag{41b}$$

$$\langle u_x + v_y, B^j\rangle - \langle wB^j_z\rangle + wF^j_m\Big|_{z=H} - (uB_x + vB_y)F^j_s\Big|_{z=B} = 0 \tag{41c}$$

$$\langle \rho_t + u\hat{\rho}_x + v\hat{\rho}_y + ws, D^j\rangle = 0 \tag{41d}$$

$$\langle p_t, D^j_z\rangle - \langle g\rho_t, D^j\rangle - p_t D^j\Big|_{z=H} = 0 \tag{41e}$$

Equivalence of (41) with the original differential equations (40) can again be shown through integration by parts. The procedure is straightforward for equations (41c, d, e). It is instructive to illustrate the equivalence of (41a) with (40a).

Integrating (41a) by parts with respect to z once gives

$$<u_t+(uu)_x+(vu)_y+(wu)_z-fv+\frac{1}{\bar\rho}p_x-\frac{\tau^x_z}{\bar\rho}, B^j>$$

$$-\Big[wuF_m\Big]_{z=*}^{z=H}-\Big[wuF_s\Big]_{z=B}^{z=*}-\frac{1}{\bar\rho}\Big[\tau F_m\Big]_{z=*}^{z=H}-\frac{1}{\bar\rho}\Big[\tau F_s\Big]_{z=B}^{z=*}$$

$$-\frac{up_t}{\bar\rho g}F_m^j\Big|_{z=H}+u(uB_x+vB_y)F_s^j{}_{z=B}+\frac{1}{\bar\rho}\tau^xF_m^j{}_{z=H}=0$$

recalling that $F_s = F_m$ at $z = *$ we can write the above as

$$\begin{aligned}&<u_t+(uu)_x+(vu)_y+(wu)_z-fv+\frac{1}{\bar\rho}p_x-\frac{1}{\bar\rho}\tau^x,B^j>-\\&-u\left(w-\frac{p_t}{\bar\rho g}\right)F_m^j\Big|_{z=H}\\&+u\,(w-(uB_x+vB_y))F_s^j\Big|_{z=B}\\&-w(\Delta u-\frac{1}{\bar\rho}\tau^x)F_*^j\Big|_{z=*}=0\end{aligned}\tag{42}$$

Each term in (42) must be separately zero giving the original differential equation and boundary conditions. In constrast to the linear equations (see section 3), the stress condition at the base of the mixed layer appears naturally in the nonlinear momentum equation.

The equivalent integral form can be solved for a 2 mode case barotropic and first baroclinic) after substitution of expansions (36). Due to the presence of the nonlinear terms, the resulting equations for barotropic and first baroclinic modes contain many coupling terms. Each coupling term will contain the product of three vertical structure functions, and very little simplification is possible. However the triple products and the integrals of the triple products are only calculated at the start. In areas where the density anomaly is large the structure functions and the various innerproducts will have to be recalculated at regular intervals (see discussion).

A computer program to solve the two-dimensional two-mode form (barotropic + first baroclinic) of equations (41) has been developed. The program updates the vertical structure functions every 10 timesteps (1 timestep = 60 seconds) at those grid points where the density anomaly is large [van Foreest and Brundrit, 1982]. A Richardson number criterion (Pollard et al., 1974) is used to govern the depth of the mixed layer which was initially set at 10 m. Some preliminary results for a flat bottom ocean are shown in Figures 1 and 2. Figure 1 displays the density structure and first baroclinic longshore velocity structure after a longshore wind of 0.2 N/m**2 has blown for 10 days. The figure clearly shows a surface front with associated longshore velocity structure which has advected nearly 30 km offshore. The surface mixed layer depth has increased inshore of the front. For example at a point 2.5km offshore, the mixed layer depth has reached 30 m at day 10. Figure 2 shows the offshore profiles of longshore first baroclinic surface velocity for days 1 - 10. The maximum first baroclinic longshore velocity (36 cm/s) is reached after 2 days, the maximum in the first baroclinic longshore velocity profile then moves offshore and reaches a constant value of 19 cm/s after approximately 8 days. As the nonlinear two-dimensional two-mode model does not include bottom friction or an imposed longshore pressure gradient, the barotropic mode develops unrealistically large barotropic longshore velocities (180 cm/s at day 10). In the first baroclinic mode the front and associated velocity structure is continuously advected offshore and therefore accelerating water initially at rest. In that way the baroclinic velocity maximum is not allowed to grow continuously.

6. DISCUSSION

We have shown that the forced, linear equations of motion with a free surface and finite amplitude topography can be transformed into a set of coupled modal equations which are independent of z. This was achieved through the use of a Galerkin procedure which used the standard, vertical, normal modes as test functions and in which the dependent variables are expanded in terms of the vertical normal modes. The vertical normal mode test functions cannot adequately describe the mixed surface layer and consequently do not allow density changes at the surface.

New vertical structure functions were derived which resolve some of the detail of the mixed layer. These vertical structure functions are identical to the vertical normal modes in the stratified water below the mixed layer. The vertical structure functions allow for a discontinuity in u, v and ρ at the base of the mixed layer. The use of these new vertical structure functions in the Galerkin procedure on the linear forced equations of motion in the presence of finite amplitude topography allows for density changes in the mixed layer and at the surface. It was then possible to extend the procedure to realistically include advection of momentum and density in the equations of motion. The new vertical structure functions are only useful when one restricts the model to 2 modes, the barotropic and first baroclinic mode. This is a result of the fact that the vertical structure functions for the baroclinic modes are not orthogonal. As the vertical structure functions are always derived from the background density, their use in the Galerkin procedure is, strictly speaking, limited to situations where the density anomaly is sufficiently small. In practice this restriction is easily overcome by redefining the vertical structure functions at regular time intervals using the most recent density structure (previous background density + density anomaly) [van Foreest and Brundrit, 1982].

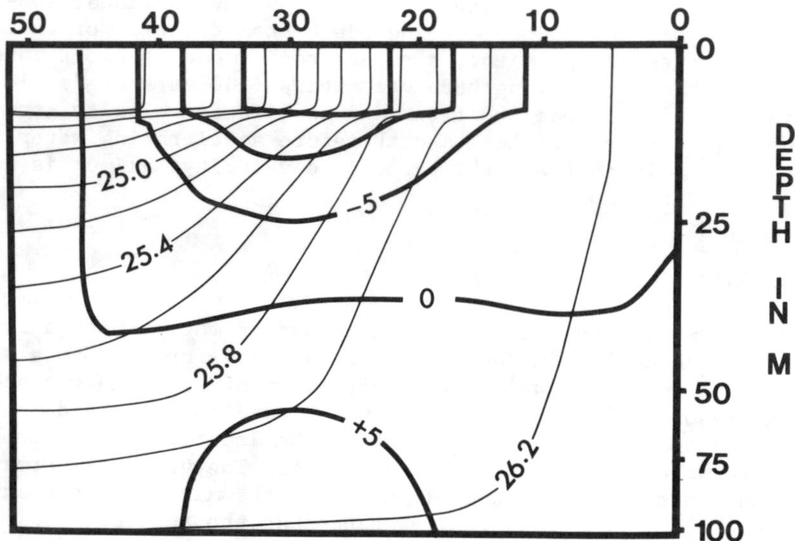

Figure 1. A vertical section of density (thin lines) and longshore velocity (heavy lines) obtained at day 10 with the nonlinear two-mode two-dimensional model. Contour intervals are 5 cm/s for velocity and 0.2 sigma-t for density.

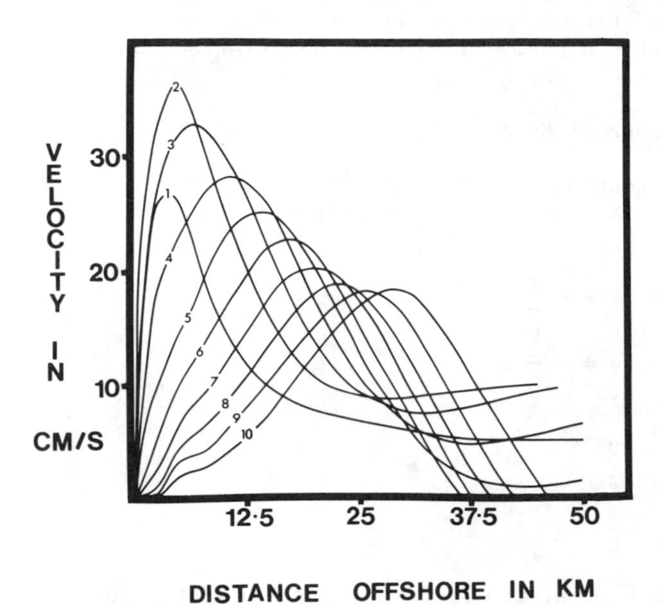

Figure 2. Development of the longshore first baroclinic surface velocity profile as a function of time and distance offshore. The profiles are shown for days 1-10.

7. APPENDIX

Here we give a brief account of the derivation of the vertical structure functions that are used as test functions in the Galerkin procedure described in this paper. Various useful relations between vertical structure functions are given.

7.1 Vertical Normal Modes

The unforced equation of motion linearised about a static background density for a flat bottom ocean with free surface takes the form

$$
\begin{aligned}
&u_t - fv = -\frac{1}{\bar{\rho}} p_x \\
&v_t + fu = -\frac{1}{\bar{\rho}} p_y \\
&u_x + v_y + w_z = 0 \\
&\rho_t + ws = 0 \\
&p_z = \rho g = 0
\end{aligned}
\qquad \text{(A1)}
$$

with $p_t = \bar{\rho} g w$ at the surface and $w = 0$ at the bottom.

If we assume

$$
\begin{aligned}
&u(x,y,z,t) = U(x,y,t)\ F(z) \\
&v(x,y,z,t) = V(x,y,t)\ F(z) \\
&p(x,y,z,t) = P(x,y,t)\ F(z) \\
&w(x,y,z,t) = W(x,y,t)\ G(z) \\
&\rho(x,y,z,t) = R(x,y,t) s(z)\ G(z)
\end{aligned}
\qquad \text{(A2)}
$$

and substitute in equations (A1) we find that equations (A1) can be separated into a horizontal and time dependent part:

$$
\begin{aligned}
&U_t - fV = -\frac{1}{\bar{\rho}} P_x \\
&V_t + fU = -\frac{1}{\bar{\rho}} P_y \\
&U_x + V_y - \lambda W = 0 \\
&R_t + W = 0
\end{aligned}
\qquad \text{(A3)}
$$

$$p + \bar{\rho}R = 0$$

and a z dependant part:

$$F_z = G\frac{sg}{\bar{\rho}}$$

$$G_z = -\lambda F \qquad \text{(A4)}$$

with $G = 0$ at the bottom $z = 0$

$F = gG$ at the free surface $z = H$.

Equations (4A) with their boundary conditions form an eigenvalue problem. For each eigenvalue λ^i the F^i $(i = 0,1,2,...)$ form a complete orthogonal set with

$$\langle F^i F^j \rangle = \delta_{ij} \qquad \text{(A5)}$$

Because the density difference across the sea surface is so much greater than the density differences within the ocean, one mode has a different character from the others. This is the barotropic mode $(i = 0)$ for which, to a good approximation,

$$F^o = \frac{1}{\sqrt{H}}$$

$$G^o = -\frac{\lambda^o}{\sqrt{H}}z \qquad \text{(A6)}$$

$$\lambda^o = -\frac{1}{gH}$$

with $F^o = gG^o$ at the free surface

$G^o = 0$ at the bottom .

We note that the barotropic mode is independent of the stratification parameter s. The other modes, baroclinic modes, depend critically on the presence of a background stratification s. To a good approximation the free surface boundary condition simplifies to $G = 0$ while the other solutions (A4) remain identical. It follows therefore that

$$\int_{z-0}^{z=H} F^i dz = 0 \qquad \text{(A7)}$$

Some relations that are useful when solving the equivalent integral formulations are given below.

Eliminating G from A4 gives

$$\left(F^i_z \frac{\bar{\rho}}{sg}\right)_z + \lambda^i F^i = 0 \quad . \tag{A8}$$

Eliminating F from A4 gives

$$G^i_{zz} + \lambda^i G^i \frac{sg}{\bar{\rho}} = 0 \quad . \tag{A9}$$

Using the above relations and boundary conditions A4 it can be shown that

$$\langle sG^i G^j \rangle = \frac{\bar{\rho}}{g} F^i G^j \Big|_{z=H} + \lambda^j \frac{\bar{\rho}}{g} \langle F^i F^j \rangle \quad .$$

As $G^j = 0$ at the surface for all baroclinic modes, the only contribution arises from the barotropic mode j = 0. From (A6) it follows that $G^o F^o = -\lambda^o$ so that the above simplifies to

$$\langle sG^i G^j \rangle = \lambda^j \frac{\bar{\rho}}{g} \qquad j = 1,2,3,\ldots$$

$$\langle sG^i G^o \rangle = 0 \qquad j = 0 \quad . \tag{A10}$$

7.2 Vertical Structure Functions with Mixed Layer

The background structure here is assumed to consist of a turbulent mixed surface layer of depth h, separated by a discontinuity $\Delta\hat{\rho}$ in background density $\hat{\rho}(z)$, from a continuously stratified fluid of depth H-h. In the mixed layer equations (30) hold

$$u_t - fv = -\frac{1}{\bar{\rho}h} \int_{z=H-h}^{z=H} p_x \, dz$$

$$v_t + fu = -\frac{1}{\bar{\rho}h} \int_{z=H-h}^{z=H} p_y \, dz$$

$$u_x + v_y + w_z = 0 \tag{A11}$$

$$\rho_t + w_* \frac{\Delta\hat{\rho}}{h} = 0$$

$$p_z + \rho g = 0 \quad .$$

As before equations (A1) hold in the stratified water. The boundary conditions on the two sets of equations are

$p_t = \bar{\rho} g w$ at the free surface $z = H$

$w = 0$ at the bottom $z = B$

w, p continuous at the base of the mixed layer $z = H-h$ (indicated by *) .

Separation of variables in the stratified water proceeds exactly as before. We now denote the vertical structure function F, G in the stratified fluid as F_s, G_s

If we assume

$$
\begin{aligned}
u(x,y,z,t) &= U(x,y,t)\ L \\
v(x,y,z,t) &= V(x,y,t)\ L \\
p(x,y,z,t) &= P(x,y,t)\ F_m(z) \\
w(x,y,z,t) &= W(x,y,t)\ G_m(z) \\
\rho(x,y,z,t) &= R(x,y,t)\ K
\end{aligned}
\tag{A12}
$$

In the above, K and L are constants. Substitution of (A12) into (A11) and separation of the resulting equations such that the horizontal and time dependent part are identical to (A3) leads to the following relations for the vertical structure

$$
\begin{aligned}
L &= \frac{1}{h} \int_{z=H-h}^{z=H} F_m \, dz \\
(F_m)_z &= \frac{Kg}{\bar{\rho}} \\
K &= G_* \frac{\Delta\bar{\rho}}{h} \\
(G_m)_z &= -\lambda L
\end{aligned}
\tag{A13}
$$

For the boundary conditions we obtain

$F_m - gG_m = 0$ at the free surface $z = H$

$F_m = F_s = F_*$ at the base of the mixed layer $z = *$

$G_m = G_s = G_*$ at the base of the mixed layer $z = *$.

Again the solution for the barotropic mode takes a special form which to a good approximation is given by

$$
\begin{aligned}
K^o &= G^o_* \frac{\Delta\hat{\rho}}{h} \\
L^o &= F^o_m = -\frac{a}{\lambda^o} \\
G^o_m &= az \\
\lambda^o &= -\frac{1}{gH}
\end{aligned}
\qquad \text{(A14)}
$$

with $F^o = gG^o$ at the free surface $z = H$.

The baroclinic modes, for which to a good approximation $G = 0$ at the free surface, are easily obtained from (A13) to give

$$
\begin{aligned}
K^i &= G^i_* \frac{\Delta\hat{\rho}}{h} \\
L^i &= \frac{1}{2} K^i \frac{gh}{\bar{\rho}} + F^i_* \\
G^i_m &= -\lambda^i L^i (z-H+h) + G^i_* \\
F^i_m &= \frac{K^i g}{\bar{\rho}} (z-H+h) + F^i_*
\end{aligned}
\qquad \text{(A15)}
$$

with $G^i_m = 0$ at the surface $Z = H$ $i = 1,2,3,\ldots$

Using (A15) it is possible to express λ^i in terms of G^i_* and F^i_* as

$$\lambda^i = \frac{G^i_*}{\left(\frac{1}{2} G^i_* \frac{\Delta\hat{\rho}}{\bar{\rho}} g + F^i_*\right) h}$$

We can now combine the vertical structure functions for the mixed layer and the stratified water below as in (34). The combined vertical structure functions for various mixed layer depth are shown in Figure 3. When the mixed layer depth is zero the vertical structure functionsare identical to the vertical normal modes. The dependent variables can then be expanded in terms of these combined vertical structure functions as given by (36).

Unfortunately no general orthogonal relation exist for the combined vertical structure functions. It can be shown however that when one restricts the expansion to barotropic and first baroclinic modes only that

$$\langle A^iA^j\rangle = \delta_{i,j}$$

$$\langle B^iB^j\rangle = \begin{cases} 0 & i \neq j \\ 1 & i = j = 0 \\ 1 + \frac{1}{12}\left(\frac{kg}{\bar{\rho}}\right)^2 h^3 & i = j = 1 \end{cases} \qquad \text{(A16)}$$

$$\langle A^iB^j\rangle = \delta_{i,j}$$

where $i = 0,1$ $\quad j = 0,1$

We show below that the baroclinic modes are not orthogonal. We attempt to prove that

$$\langle A^i,A^j\rangle = 0 \qquad i \neq j \quad (i,j > 1) \quad .$$

As L^i is a constant we know

$$\int_{z=*}^{z=H} L^iL^j \, dz = L^iL^jh \quad .$$

From van Foreest and Brundrit [1982], Appendix A, it follows that

$$\int_{z=B}^{z=*} F_s^iF_s^j \, dz = -\frac{\bar{\rho}}{sg}\left(\frac{1}{\lambda^i-\lambda^j}\right)\left[(F_s^i)_zF^j-(F_s^j)_zF^i\right]_{z=B}^{z=*} \quad .$$

Using the previous relations

$$(F_s^i)_z = G_s^i\frac{sg}{\bar{\rho}} \text{ and } \lambda^i = \frac{G_*^i}{\left(\frac{1}{2}G_*^i\frac{\Delta\hat{\rho}}{\bar{\rho}}g+F_*^i\right)h} = \frac{G_*^i}{L^ih}$$

we can write this as

$$\int_{z=B}^{z=*} F_s^iF_s^j \, dz = -L^iL^jh\,\frac{G_*^iF_*^j-G_*^jF_*^i}{G_*^iL^j-G_*^jL^i} \quad .$$

As $F_* = L$ we have to conclude that

$$\langle A^iA^j\rangle \neq 0 \qquad i \neq j \quad (i,j > 1) \quad .$$

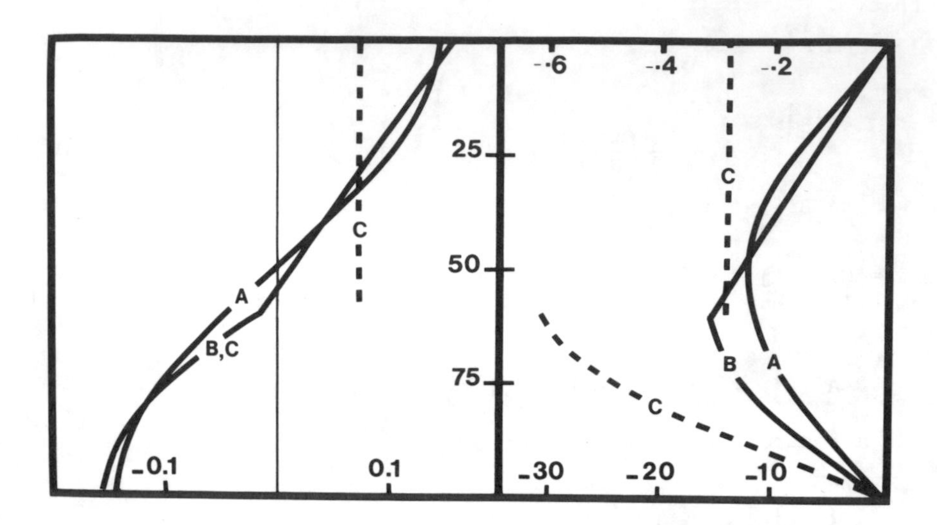

Figure 3. First baroclinic structure functions obtained for a constant stratification of 0.04 kg/m**2. Curves A-Mixed layer depth is zero. The left hand diagram shows the vertical structure for horizontal velocity and pressure. The right hand side diagram shows the vertical structure for the vertical velocity. Curves B-Mixed layer depth is 60 m. The left hand side diagram shows the vertical structure for the pressure. The right hand side diagram shows the vertical structure for the vertical velocity. Curves C-Mixed layer depth is 60 m. The left hand side diagram shows the discontinuous vertical structure for the horizontal velocity. Below 60 m curves B and C are the same. The right hand side diagram shows the discontinuous vertical structure for the vertical velocity.

REFERENCES

Davies, A.M., A three dimensional modal model of wind induced flow in a sea region, Progress in Oceanography 15, 2, 72-128, 1985.
Flierl, G.R., Models of vertical structure and the calibration of two layer models, Dyn. Atmospheres and Oceans, 2, (4), 341-383, 1978.
Gill, A.E., and A. J. Clarke, Wind induced upwelling, coastal currents and sea-level changes, Deep Sea Research, 21, 325-345, 1974.
Lighthill, M.J., Dynamic response of the Indian Ocean to the onset of the southwest monsoon, Phil. Trans. Roy. Soc. Lond., A, 265 45-93, 1969.
McCreary, J. P., A linear stratified ocean model of the coastal undercurrent, Phil. Trans. R. Soc. Lond., 302, 385-413, 1981.
Moore, D.W., and S.G.H. Philander, Modelling of the tropical oceanic circulation, The Sea, 6, edited by E.D. Goldberg et al., John Wiley and Sons, New York, 1977.
Pollard, R.T., On the generation by winds of inertial waves in the ocean, Deep Sea Research, 17, 795-812, 1970.
Pollard, R.T., P.B. Rhines, and R.O.R.Y. Thompson, The deepening of the wind mixed layer, Geophys. Fluid Dyn., 3, 381-404, 1973.
Strang, G., and G.J. Fix, An analysis of the finite element method Prentice Hall, Englewood Cliffs, New Jersey, 1973.
van Foreest, D., and G.B. Brundrit, A two mode numerical model with applications to coastal upwelling, Prog. Oceanog., 11, 328-392, 1982.

ON THE USE OF FINITE ELEMENT METHODS FOR OCEAN MODELLING

C.Le Provost
Institut de Mecanique de Grenoble
BP 68 - 38402 St Martin d' Heres - Cedex
France

ABSTRACT. The aim of this contribution is to present the way finite element methods can be used for ocean modelling. A short history of the early papers published in the field is followed by a presentation of the solution of a Poisson problem through a variational formulation and a finite element approximation. The oceanic example chosen is the classical mid-latitude wind-driven barotropic circulation in a rectangular basin on a beta-plane, under several ranges of parameters (linear and non-linear), with different mechanisms of dissipation (lateral dissipation or bottom friction), and different boundary conditions (slip, no-slip or neutral). The efficiency of the method is shown by comparing the numerical solution with the analytical solution available for the linear cases [Stommel and Munk], and with the classical finite difference solutions presented in the literature for non-linear cases [Veronis, Bryan, Blandford, Holland]. As the corresponding solutions include a western boundary current with intense velocity gradients, particular interest is devoted to the use of the finite element method flexibility to increase the resolution along the western wall of the basin. The conservative properties of the method are also underlined .

1. INTRODUCTION

Since the first attempts at modelling by Hansen [1949] for tides, and Sarkisyan [1962] and Bryan [1963] for ocean circulations, most of the numerical models developed in coastal dynamic simulations and in general ocean circulation studies have been based on finite difference methods (FDMs). However, within the last decade, a significant number of papers have proposed to apply finite element methods (FEMs) to ocean modelling. These methods are attractive because they allow the use of grids of variable size, shape, and direction, which is particularly useful for a good fit of the coastline and bottom topography, and for refinement of the meshes in areas where greater resolution is required. These methods are also interesting because of their properties of conservation and precision.

J. J. O'Brien (ed.), Advanced Physical Oceanographic Numerical Modelling, 557–580.

1.1. Early Beginnings in Tidal Modelling

The earliest finite element tidal models were developed by Grotkop [1973], Connor and Wang [1974],and Taylor and Davis [1975], who present ways of solving the primitive shallow water equations by that technique. Some practical attempts were then performed: for the North Sea by Brebbia and Partridge [1976], for the Massachussets Bay by Wang [1978], for the Hamaishi Bay and Tokyo Bay by Hasegawa [1978]. After a phase of careful analysis and criticism, in the beginning of the 1980s ,the FEMs are now becoming more and more popular in the engineering field for solving shallow water equations. However, most of these models are time consuming and , consequently, lose a part of their advantage over the more classical ones based on FDMs.

For the specific problem of tidal modelling, another approach is possible which considerably reduces the computer costs. The basic idea is that, when looking for periodic motions, important computer costs resulting from the succession of time-step integration can be avoided by replacing the time-dependant equations of motion by an equivalent set of modal equations corresponding to the significant tidal wave components present in the tidal spectrum of the real solution. That approach was investigated simultaneously in the late 1970s by Kawahara and al. [1977], Pearson and Winter [1977], Le Provost and Poncet [1977], Jamart and Winter [1978], Lynch [1978]. One particular point of the modal approach is that, when transposing the problem from the time domain into the spectral domain, the hyperbolic time-dependant problem of tidal propagation is transformed into a set of elliptic problems of Helmholtz type. A variational formulation can be derived for each of these modal problems. Thus, FEMs appear to be the natural way for numerical integrations in realistic complex areas. Such a method has been successfully applied, for example, by LeProvost ,Rougier and Poncet [1981] for the determination of the main components of the tides in the English Channel, and by Platzman, Curtis, Hansen and Sleter [1981] for the real ocean free modes.

1.2. Early Beginnings in Ocean Modelling

The properties of finite element techniques for modelling ocean dynamic problems have been investigated by Fix [1975] in the quasi-geostrophic approximation. He has shown that they are of interest on several points: precision, conservation of energy and enstrophy, natural treatment of boundary conditions, and flexibility of triangulation. He has also proved the stability and convergence of the semi-discrete finite element formulation of the problem, and established the conservative properties of its numerical approximation for energy, vorticity and enstrophy, independent of the irregularity of the grid used for spatial integration. On the basis of these conclusions, some attempts have been made to introduce FEMs in the field of ocean modelling. As with the development of FDMs, the barotropic problem has been first investigated. Haidvogel, Robinson and Schulman [1980] compared the precision of a finite difference model, a finite element model and a spectral model for applications to open ocean problems, and showed the value of FEM. Dumas, Le Provost and Poncet [1982] studied the performances of a FEM for solving the problem of oceanic wind-driven general circulation in a closed basin, and compared the precision of their results favorably to the ones obtained by more classical FDMs. Marchuk and Kuzin [1983] combined finite element and splitting-up methods to solve the same problem and gave an example of preliminary results for the world ocean circulations. More recently, Miller, Robinson and Haidvogel [1983] have developed two methods for treating the depth dependance of the flow in such models and demonstrated their feasability and efficiency for modelling realistic mid-oceanic mesoscale eddy flow regimes. A two layer model has also been formulated and tested by Le Provost [1984] for the general oceanic circulation in closed basins. A preliminary practical application to the barotropic and baroclinic circulations induced in the Mediterranean Sea by the flows in the Gibraltar and Sardinia straits have been presented by Loth and Crepon [1983].

Although less commonly used than FDMs, it appears from the preceding brief review that FEMs represent an interesting alternative for modelling complex areas such as estuaries, coastal seas, or oceanic basins. The following pages focus on the presentation of one example of such an application: the quasi-geostrophic modelling of the wind-driven circulation in a mid-latitude oceanic basin .

2. RESOLUTION OF A POISSON PROBLEM BY A FINITE ELEMENT METHOD

2.1. Variational Formulation

Let us compute the function u over a domain Ω limited by a boundary Γ such that:

$$\nabla^2 u = f \text{ on } \Omega \quad \text{with } u=0 \text{ on } \Gamma \qquad (1)$$

Our purpose is to search a solution of (1) in a weak sense, i.e. in a Sobolev space of order 1, $H^1(\Omega)$), where every function v and its first derivatives are square integrable on Ω:

$$[\ v \in L^2(\Omega)\ ,\ \partial v/\partial x_i\ L^2(\Omega)\ ]$$

To do this, we multiply (1) by a testing function v and integrate over the domain Ω:

$$\iint_\Omega \nabla^2 u . v \, d\Omega = \iint_\Omega f.v \, d\Omega \qquad (2)$$

With the notation:

$$\{ d, d' \} = \iint_\Omega d.d' \, d\Omega$$

and using the Green formula:

$$\{ \nabla^2 u , v \} = -\{ \nabla u , \nabla v \} + \int_\Gamma \partial u/\partial n \, v \, d\Gamma$$

we obtain:

$$- \{ \nabla u , \nabla v \} + \int_\Gamma \partial u/\partial n \, v \, d\Gamma = \{f,v\}. \qquad (3)$$

However, as u must be zero along Γ, the solution is to be searched in $H^1_0(\Omega)$, i.e. so that v=0 along Γ. Consequently (3) reduces to:

$$\{\nabla u,\nabla v\} = -\{f,v\}. \qquad (4)$$

By calling:

$$a(u,v) = \{\nabla u,\nabla v\}$$

$$L(v) = -\{f,v\}$$

we can thus formulate the following variational problem :

"Find u in $H^1_0(\Omega)$ such that:

$$a(u,v) = L(v) \quad \forall v \in H^1_0(\Omega) \text{ ."} \qquad (5)$$

The solution of such a problem corresponds in fact to a minimization of the energy over Ω, when taking U as the stream function and f as the vorticity of a two dimensional flow:

$$(1) \Rightarrow \nabla^2 \psi = \zeta. \qquad (5\text{ bis})$$

Indeed, the solution ψ^* of problem (5bis) is also the unique solution of the following problem of minimization:

"Find $\psi^* \in H^1_0(\Omega)$ so that $\mathcal{L}(\psi^*) = \text{Min } \mathcal{L}(\psi)$

with $\mathcal{L}(\psi) = -\frac{1}{2} a(\psi,\psi) + L(\psi)$."

As $L(\psi^*) = a(\psi^*,\psi^*)$, it follows that:

$$\mathcal{L}(\psi^*) = \frac{1}{2}\{\nabla\psi^*,\nabla\psi^*\}$$

$\mathcal{L}(\psi^*)$ is the total kinetic energy of the flow integrated over the studied area.

2.2. Finite Element Approximation

Given a regular triangulation of Ω, we consider a finite element space S_Δ on that grid , in which any function Φ can be written:

$$\Phi^\Delta(x,y) = \sum_{j=1}^{N} \Phi_j \varphi_j (x,y) \qquad (6)$$

where $\varphi_j(x,y)$ are basic functions and j is the index of the nodes (of number N). Let us present as an example Lagrange polynomial finite elements of degree k, with (k+1).(k+2)/2 nodes by triangle. For Lagrange elements of degree 1 (P1), one set S_1 of 3 nodes is defined on each triangle (see Figure 1.1), with the following basic functions:

$$\varphi_1(x,y) = 1-x-y$$
$$\varphi_2(x,y) = y$$
$$\varphi_3(x,y) = x$$

For Lagrange elements of degree 2 (P2) (see Figure 1.2), the basic polynomials are in that approximation:

$$\varphi_1(x,y) = (1-x-y).(1-2.(x+y))$$
$$\varphi_2(x,y) = x.(2x-1)$$
$$\varphi_3(x,y) = y.(2y-1)$$
$$\varphi_4(x,y) = 4x.(1-x-y)$$
$$\varphi_5(x,y) = 4xy$$
$$\varphi_6(x,y) = 4y.(1-x-y)$$

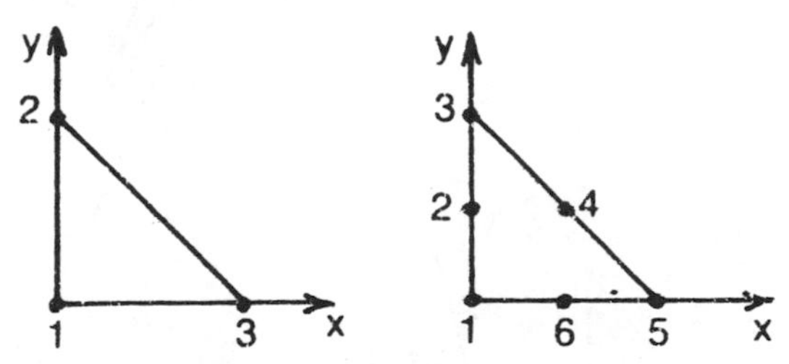

Figure 1. Reference triangles for P1 (1.1) and for P2 (1.2)

Notice that in all the cases:

$$\varphi_i(M_j) = 0 \text{ for } M_j \neq M_i$$
$$= 1 \text{ for } M_j = M_i$$

Figure 2 is an illustration of the shape of the polynomials φ_1 and φ_6. To determine the function Φ solution of problem (1), the weights Φ_j must be computed. This is done by taking in the variational formulation the basic functions as testing functions:

$$v = \varphi_q \text{ for } q = 1,\ldots N$$

This leads to a linear system of N equations of N unknowns:

$$a\left(\sum_{p=1}^{N} \Phi_p \varphi_p(x,y) , \varphi_q\right) = L(\varphi_q)$$

$$\text{i.e. } \sum_{p=1}^{N} \Phi_p \{\nabla\varphi_p , \nabla\varphi_q\} = -\sum_{p=1}^{N} f_p \{\varphi_p ,\varphi_q\}$$

i.e. in matrix notation: A.X = B , with:

$$X = (\Phi_i)_{i=1,N}$$
$$B = (b_i)_{i=1,N} \text{ with } b_i = \sum_{j=1}^{N} f_j \{\varphi_j, \varphi_i\} \qquad (7)$$
$$A = (a_{ij})_{i,j=1,N} \text{ with } a_{ij} = \{\nabla\varphi_i, \nabla\varphi_j\}$$

A is symetric and positive, as the bilinear form {.,.} is a scalar product. Moreover, it can be a band matrix, if the nodes are judiciously classified, because $a(\varphi_i, \varphi_j) = 0$ if i and j are not from the same triangle.

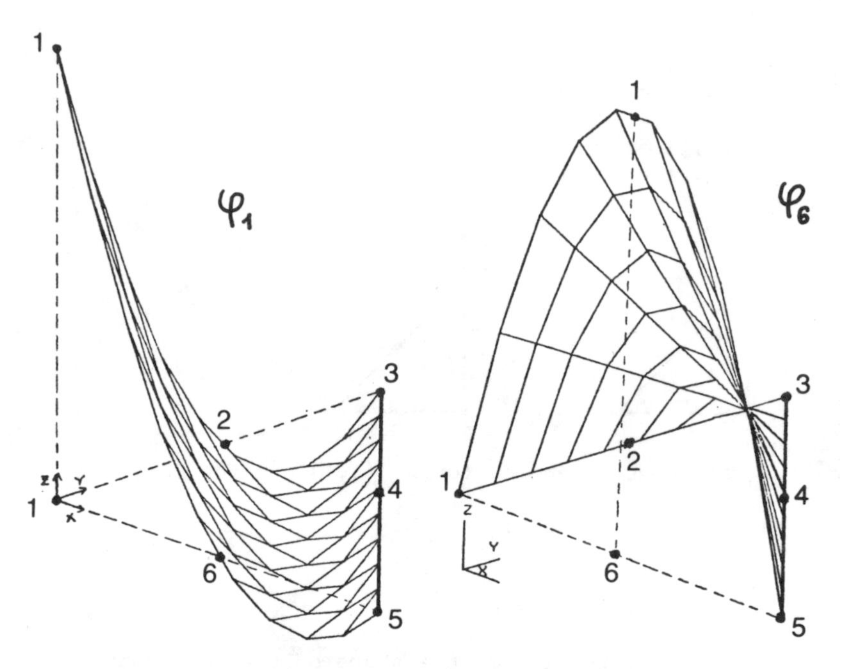

Figure 2. Perspective view of polynomials φ_1 and φ_6 in P2 Lagrange approximation.

3. FEASIBILITY OF FEMS FOR WIND-DRIVEN OCEAN CIRCULATIONS MODELLING

3.1. Formulation of the Test Model

We consider the barotropic vorticity equation of a quasi-geostrophic flow driven by a wind curl in a closed basin on a β-plane:

$$\zeta_t + R_o\, J(\psi,\zeta) + \psi_x = 1/\pi \text{ curl } \tau + R_o/R_e\, \nabla^2\zeta - \epsilon\, \zeta \qquad (8)$$
$$\zeta = \nabla^2\psi$$

In these adimensional equations on ζ vorticity and ψ stream function, the characteristic numbers are:

the Rossby number $R_0 = U/\beta L^2$
the Reynolds number $R_e = UL/A$
the frictional number $\epsilon = \kappa/\beta L$

with the typical scales L (horizontal length), $T = (\beta L)^{-1}$ (time), and $U = \pi \tau_0 (\rho h \beta L)^{-1}$ (typical velocity scale) and the notations:

τ_0: wind stress at the surface of the ocean
L : size of the basin
h : depth of the layer
ρ : sea water density
A : lateral friction coefficient
κ : bottom drag coefficient

Let us remember that corresponding to these typical numbers are typical widths of western recirculation flow:

$\delta_I = R_0^{1/2} L$ for the inertial boundary layer
$\delta_A = 2 (R_0/Re)^{1/3} L$ for the viscous boundary layer
$\delta_F = \epsilon L$ for the frictional boundary layer

Associated with this classical problem, the following boundary conditions can be examined:

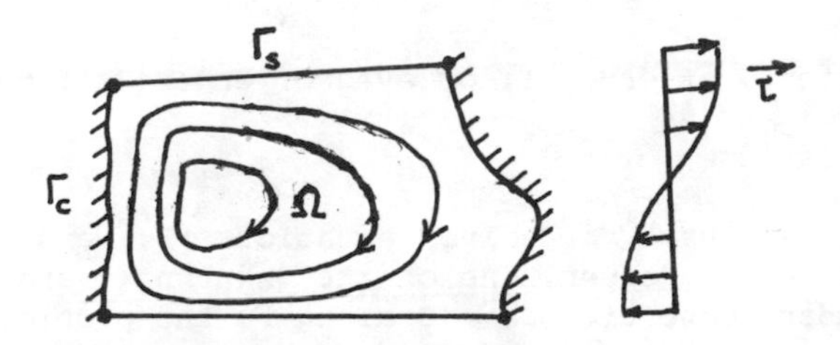

Figure 3. Geography of the Test Mode

Γ is a stream line: $\psi_\Gamma = 0$. Along the northern and southern limits, a slip condition is applied: $\zeta_{\Gamma_s} = 0$; along the coasts, several kinds of boundary conditions can be taken:

slip : $\zeta_{\Gamma_c} = 0$
no-slip : $\partial\psi/\partial n = 0$
neutral : $\partial\zeta/\partial n = 0$.

3.2. Time Discretization

A traditional discrete formulation of the vorticity equation is obtained by using second order centered finite differences [Crank-Nicholson]. Advection and Coriolis terms are taken at step n and bottom friction as the average of steps n-1 and n+1; the horizontal dissipation term is taken implicitly ($\gamma \in [0,1[$):

$$(\zeta^{n+1} - \zeta^{n-1})/2\,dt + R_o\,J(\psi^n,\zeta^n) + \psi_x^n =$$
$$(1/\pi)\,\mathrm{curl}\,\tau - (\epsilon/2)\,(\zeta^{n+1} + \zeta^{n-1}) +$$
$$+ (R_o/R_e)\,\gamma\,\nabla^2\zeta^{n+1} + (R_o/R_e)\,(1-\gamma)\,\nabla^2\zeta^{n-1}$$

i.e.:

$$\alpha\,\zeta^{n+1} - \beta\,\nabla^2\zeta^{n+1} = g^{n+1} \qquad (9)$$

with: $g^{n+1} = [1 - \epsilon\,dt + 2\,(R_o/R_e)\,(1-\gamma)\,dt\,\nabla^2\,]\,\zeta^{n-1} -$

$$- 2\,R_o\,dt\,J(\psi^n,\zeta^n) - 2\,dt\,[\psi_x^n - (1/\pi)\,\mathrm{curl}\,\tau\,]$$

$$\alpha = 1 + \epsilon\,dt$$

$$\beta = 2\,\gamma\,(R_o/R_e)\,dt$$

The stream function is computed from the vorticity at time n+1 by equation:

$$\nabla^2\psi^{n+1} = \zeta^{n+1} \qquad (10)$$

This scheme is consistent, of order 1, or even of order 2 if $\gamma=1/2$.

3.3 Variational Formulation

For each time step, we have to solve equations (9) and (10), two elliptic equations of second order, one of the unknown ζ^{n+1} and the other of ψ^{n+1}. With the boundary condition $\psi^{n+1} = 0$ along Γ, the problem to solve is:

* Stream function:

"Find $\psi^{n+1} \in H_o^1(\Omega)$, so that:

$$\{\nabla\psi^{n+1}, \nabla\psi'\} = -\{\zeta^{n+1}, \psi'\} \qquad (11)$$

$\psi' \in H_o^1(\Omega)$ "

* Vorticity equation:

We multiply equation (9) by a test function ζ' and integrate over Ω. The problem is then:

"Find $\zeta^{n+1} \in H^1(\Omega)$, so that:

$$\{[a - \beta \nabla^2]\, \zeta^{n+1}, \zeta'\} = \{g^{n+1}, \zeta'\} \qquad (12a)$$

$\zeta' \in H^1(\Omega)$"

i.e. with the use of the Green formula:

$$a\{\zeta^{n+1}, \zeta'\} - \beta[-\{\nabla\zeta^{n+1}, \nabla\zeta'\} + \int_\Gamma \partial\zeta^{n+1}/\partial n \; \zeta' d\Gamma] = \{g^{n+1}, \zeta'\}$$

This relation can be simplified by cancelling the curvilinear integral along the boundary if:

- either $\zeta' = 0$: we search for solutions of (9) in $H^1_0(\Omega)$, i.e. with $\zeta = 0$ along Γ. This is the classical slip condition used in most of the QG models.
- or $\partial\zeta/\partial n = 0$: this appears as a natural boundary condition, from the variational formulation and corresponds to no enstrophy dissipation along the boundary.

Equation (12a) then reduces to:

$$a\, \{\zeta^{n+1}, \zeta'\} + \beta\, \{\nabla\zeta^{n+1}, \zeta'\} = \{g^{n+1}, \zeta'\} \qquad (12b)$$

Note that when $\gamma = 0$ (explicit scheme), equation (9) is reduced to zero order and thus can be solved without any boundary condition on ζ.
When the boundary condition is a no-slip one, a difficult problem arises, because the condition is on the stream function ψ and not on the vorticity. We shall see later how to relax this difficulty.

3.4. Finite Element Approximation for the Case with Slip or Neutral Boundary Conditions.

The domain Ω is subdivided into triangles including N nodes and we consider an approximation of the vorticity and the stream function for problem (11, 12-b) of the form:

$$\begin{aligned} \hat{\zeta}(x,y,t) &= \sum_{j=1}^{N} \hat{\zeta}_j(t)\, \varphi_j(x,y) \\ \hat{\psi}(x,y,t) &= \sum_{j=1}^{N} \hat{\psi}_j(t)\, \varphi_j(x,y) \end{aligned} \qquad (13)$$

in the way presented in paragraph 2.2. The problem is thus to determine the weights $\hat{\zeta}_j(t)$ and $\hat{\psi}_j(t)$ for each time step, by using the variational formulations (11) et (12b). Let us take the matrix notations:

$$\hat{\zeta}^{n+1} = |\, \hat{\zeta}_j((n+1)dt) \,| \qquad \hat{\psi}^{n+1} = |\, \hat{\psi}_j((n+1)dt) \,|$$

$$M = |\, \{\varphi_i, \varphi_j\} \,| \qquad K = |\, \{\nabla\varphi_i, \nabla\varphi_j\} \,|$$

$$B^{n+1} = \Big|\sum_{j=1}^{N} \{g_j^{n+1}, \varphi_i\}\Big| \qquad C^{n+1} = \Big|-\sum_{j=1}^{N} \hat{\zeta}_j((n+1)dt)\, \{\varphi_j, \varphi_i\}\Big|$$

(M and K are usually called the mass and stiffness matrices). Consequently, the equations (11) and (12b) can be written :

$$(\alpha M + \beta K) \zeta^{\Delta, n+1} = B^{n+1}$$
$$K \psi^{\Delta, n+1} = C^{n+1} \qquad (14)$$

It is important to notice that the matrices M and K are built at the beginning of the iterations and do not vary while iterations occur. Thus the two matrices on the left hand side of (14) are inverted only once, at the beginning of the computation, and the problem reduces at each time step to the computation of the second members of equations (14) and the resolution of these linear systems by direct subsitutions. This is the reason why the present approach requires very reasonable computational work per time step.

Following Fix [1975], it is possible to prove the stability and the convergence of the semi discrete finite element system (14), at least when dissipative terms of friction, lateral diffusion and wind stress forcing terms are set to zero. It is also possible to show that within this formulation, potential vorticity, potential enstrophy, and energy are conserved, independently of the irregularity of the grid dividing Ω.

3.5. Resolution for the Case with No-slip Boundary Conditions

As noted before, the case with no-slip boundary condition is not easy because one condition is missing on the vorticity along Γ to solve equation (12a). Equations (9) and (10) are now coupled by the fact that the two boundary conditions are specified on the stream function along Γ_c ($\psi_c = 0$ and $\partial\psi/\partial n|_{\Gamma_c} = 0$). Indeed, it is possible to solve the corresponding 4-th order problem on ψ directly, but very sophisticated algorithms are then necessary, which are difficult to use. Another way has been developed by Telias [1983] following a method proposed by Glowinski and Pironneau [1977] for the biharmonic problem. The problem is to solve at each time step the system:

$$\alpha \zeta^{n+1} - \beta \nabla^2 \zeta^{n+1} = g^{n+1}$$
$$\nabla^2 \psi^{n+1} = \zeta^{n+1} \qquad \text{on } \Omega \qquad (15)$$
$$\text{with } \zeta^{n+1}_{\Gamma_s} = \psi^{n+1}_{\Gamma_s + \Gamma_c} = \partial\psi^{n+1}/\partial n\Big|_{\Gamma_c} = 0$$

To decouple the system, it is necessary to prescribe a condition on ζ . Such a condition is built as follows.

Let us define $H^{1/2}(\Gamma)$ as the space of functions which are the image on Γ by a trace operator, of a function $H^1(\Omega)$, $H^{-1/2}$ as its dual, and consider the operator A such that:

$$(A\lambda, \mu)_\Gamma = (\partial\psi_\lambda/\partial n, \mu)_\Gamma \qquad \forall \mu \in H^{1/2}(\Gamma) \qquad (16)$$

where ψ_λ is the solution of the problem:

$$\begin{aligned} a\,\zeta_\lambda - \beta\,\nabla^2\,\zeta_\lambda &= 0 \text{ on } \Omega \text{ with } \zeta_\lambda\big|_\Gamma=\lambda \\ \nabla^2\,\psi_\lambda &= \zeta_\lambda \text{ on } \Omega \text{ with } \psi_\lambda\big|_\Gamma=0. \end{aligned} \tag{17}$$

Note: Glowinski and Pironneau (1979) have shown that such an operator A is positive and symmetric. These properties are numerically important because they are conserved by the space discretization, and thus allow the use of a Cholesky method to invert the operator A.

Thus, if ζ_0 and ψ_0 are solution of:

$$\begin{aligned} a\zeta_0 - \beta\nabla^2\zeta_0 &= g^{n+1} \text{ with } \zeta_0\big|_\Gamma = 0 \\ \nabla^2\psi_0 &=\zeta_0 \text{ with } \psi_0\big|_\Gamma= 0 \end{aligned} \tag{18}$$

and $\lambda_0^* \in H^{1/2}(\Gamma)$ solution of:

$$\begin{aligned} &(A\lambda_0^*,\mu)_\Gamma = (-\partial\psi_0/\partial n,\mu)_\Gamma, \ \forall \mu \in H^{1/2}(\Gamma) \\ &\text{and } \lambda^* =\lambda_0^* \text{ along } \Gamma_c \\ &\lambda^*= 0 \quad \text{along } \Gamma_s \end{aligned} \tag{19}$$

then:

$$\zeta^{n+1} = \zeta_0 + \zeta_{\lambda^*}$$
$$\psi^{n+1} = \psi_0 + \psi_{\lambda^*}$$

are solutions of (15), because:

$$a.(\zeta_0 + \zeta_{\lambda^*}) - \beta.(\nabla^2\,\zeta_0 + \nabla^2\,\zeta_{\lambda^*}) = g^{n+1}$$
$$\nabla^2\,(\psi_0+\psi_{\lambda^*}) = \zeta_0 + \zeta_{\lambda^*}$$

and

$$\partial\psi^{n+1}/\partial n = \partial\psi_0/\partial n + \partial\psi_{\lambda^*}/\partial n = \partial\psi_0/\partial n + A\lambda^* = 0$$

(in a weak sense).

Let us consider the discrete formulation of problems (17),(18) and (19). Problems (17) and (18) are identical to problem (11-12b) and are solved in the way presented in section 3.4 . Problem (19) is discretized as follows:

$$(A\lambda_0^*,\mu^h)_\Gamma = (-\partial\psi_0^h/\partial n,\mu^h)_\Gamma \quad ,\forall \mu^h \in X_h \subset H^{1/2}(\Gamma) \cap H^{-1/2}(\Gamma) \tag{20}$$

where X_h is the subspace trace on Γ of the finite element space S_A.

We consider $\lambda_0^* = \sum_{i=1}^{q} \lambda_i \mu_i$ with

μ_i : basic functions of X_h

q : number of nodes on Γ_c

Thus (20) becomes the following system:

$$\sum_{i=1}^{q} \lambda_i (A\mu_i, \mu_j)_\Gamma = (-\partial\psi_0^h/\partial n, \mu_j) \text{ for } j=1,\ldots q \tag{21}$$

Given the definition (16), and using the Green formula, we can write:

$$(A\mu_i, \mu_j)_\Gamma = (\partial\psi\mu_i/\partial n, \mu_j)_\Gamma = \{\nabla\psi\mu_i, \nabla\bar{\mu}_j\} + \{\nabla^2\psi\mu_i, \bar{\mu}_j\}$$
$$= \{\nabla\psi\mu_i, \nabla\bar{\mu}_j\} + \{\zeta\mu_i, \bar{\mu}_j\}$$

where $\bar{\mu}_i$ is the function of $H^1(\Omega)$ of trace μ_i on Γ.
Similarly, for the second member of (21) :

$$(-\partial\psi_0^h/\partial n, \mu_j)_\Gamma = -\{\nabla\psi_0^h, \nabla\bar{\mu}_j\} + \{\zeta_0^h, \bar{\mu}_j\}$$

Thus we have to solve the linear system:

$$\sum_{i=1}^{q} \lambda_i [\{\nabla\psi\mu_i, \nabla\bar{\mu}_j\} + \{\zeta\mu_i, \bar{\mu}_j\}] = -\{\nabla\psi_0^h, \nabla\bar{\mu}_j\} - \{\zeta_0^h, \bar{\mu}_j\} \tag{22}$$

where $\psi\mu_i$ and $\zeta\mu_i$ have been computed once at the beginning from (17) by solving:

$$\alpha\, \zeta\mu_i - \beta\, \nabla^2 \zeta\mu_i = 0 \text{ with } \zeta\mu_i|_\Gamma = t_i$$

$$\nabla^2 \psi\mu_i = \zeta\mu_i \text{ with } \psi\mu_i|_\Gamma = 0$$

for i=1,...q and $t_k = 0$ except for k=i.

As in section 3.4, it is important to summarize the computing procedure. At the initial step, matrices M and K are built, the matrices $\alpha M + \beta K$ and K are inverted , A is built from (22), by solving q times problem (23), i.e. by q direct substitutions, as $\alpha M+\beta K$ and K are already inverted, and finally A must be inverted. At each time step, the system (22) has to be solved (2q substitutions), and then the two systems related to (17) and (18).

We see that globally the computer cost for solving the case with a no-slip boundary condition is more than twice that in the case with a slip or a neutral boundary condition.

3.6. Numerical Tests

The problem of mid-latitude wind-driven barotropic ocean circulations in closed basins is very classical in oceanography. In the case with no lateral viscosity and slip boundary conditions, Stommel [1948] has given an analytical solution for the linear case, and Veronis [1966] has intensively studied the non-linear cases, through analytical developments and numerical simulations (based on FDMs). In the case with no bottom friction and no-slip boundary conditions, Munk [1950] has given an analytical solution for the linear approximation and Bryan [1963] has numerically investigated the unsteady non-linear cases. This set of analytical and numerical solutions offers thus a good frame for testing the feasability of the finite element models presented in the preceding paragraphs.

3.6.1. Wind driven barotropic ocean circulations with no lateral eddy viscosity and slip boundary conditions.

Let us first consider tests of precision with the linear case, using Stommel's steady solution. When the domain Ω of integration is a square of size L, and the wind field taken as follows:

$$2\,\tau_x^* = -\,\tau_0 \sin(\pi x^*/L)\cos(\pi y^*/L)$$

$$2\,\tau_y^* = \tau_0 \cos(\pi x^*/L)\sin(\pi y^*/L)$$

$$\text{so that : } \quad \text{curl}\ \tau^* = -(\pi/L)\,\tau_0 \sin(\pi x^*/L)\sin(\pi y^*/L)$$

problem (8) reduces to the equation:

$$\psi_x + \epsilon\,\nabla^2\psi = -\sin(\pi x)\sin(\pi y) \text{ with } \psi = 0 \text{ on } \Gamma.$$

and the analytical solution given by Stommel is:

$$\psi_s(x,y) = (1/\pi)\,[\sin(\pi y)/(1 + 4\pi^2\epsilon^2)]$$
$$\{\,2\pi\epsilon\sin(\pi x) + \cos(\pi x) +$$
$$+\,[1/(e^{R_1} - e^{R_2})]\,[(1 + e^{R_2})e^{R_1 x} - (1 + e^{R_1})e^{R_2 x}]\}$$

$$\text{with:} \quad R_{1,(2)} = [-1\,(\pm)\,(1 + 4\pi^2\epsilon^2)^{1/2}]/2\epsilon$$

A first series of numerical tests can thus be realized with that linear non-viscous case. Starting from an initial situation where the fluid is at rest in the basin, the wind curl (24) is applied and we look at the time-evolving solution of the stream functions until a steady state is obtained. The corresponding solution is then compared to the analytical solution (26). This solution is an anticyclonic gyre with a narrow boundary layer along the western coast of the basin, of typical width $\delta_F = \epsilon L$, i.e. related to the intensity of the friction parameter ϵ. Consequently, it is obvious that the precision of the numerical solution is particularly sensitive to the mesh size of the discretization along the western wall. Dumas et al. [1982] have carried a series of computations over a large range of values for ϵ, and tested the influence of the refinement of the mesh and of the Lagrange polynomial finite elements used for the computations. Let us review their results and conclusions. The two triangulations used are presented on Figure 4.

The improvement of the solution, when using finer grids and higher degree basic elements, is clearly illustrated in Figure 5.

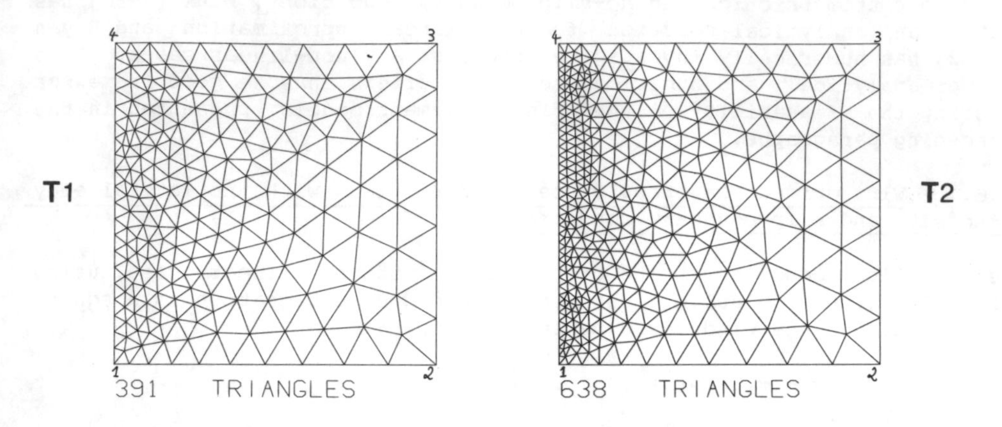

Figure 4. Triangulations used for the Stommel's solution tests. T1 (391 triangles - 220 nodes with P1 - 830 nodes with P2) and T2 (638 triangles - 353 nodes with P1 - 1340 nodes with P2). Typical size of the mesh: on the left 0.15, on the right 0.05 for T1 and 0.02 for T2.

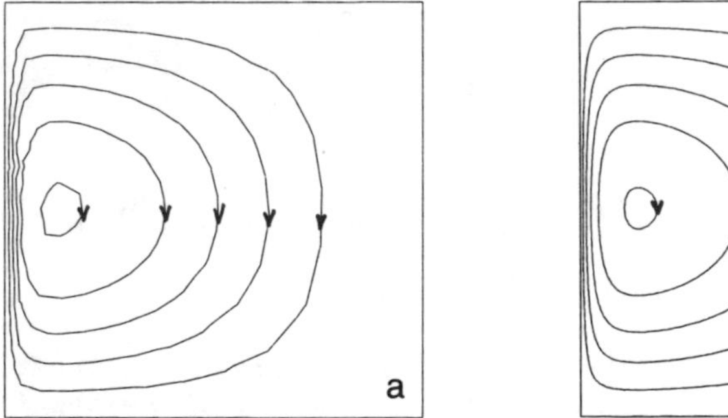

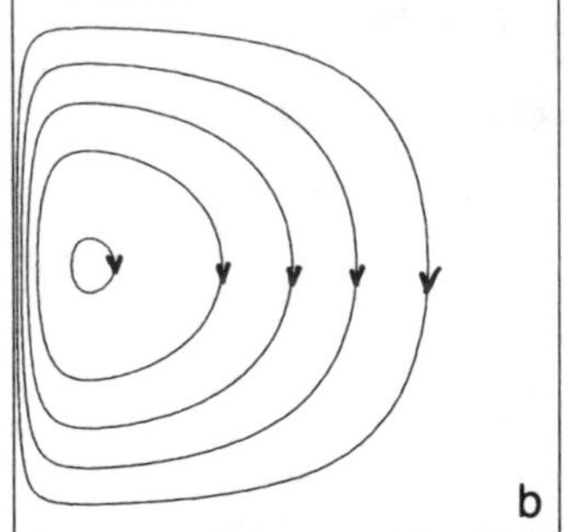

Figure 5. Examples of steady state solution for the Stommel's linear case with dissipation by bottom friction (ϵ = 0.03) and slip boundary conditions.

a - with triangulation T1 and P1 approximation.

b - with triangulation T2 and P2 approximation.

The solution is clearly improved along the western boundary for application (b).

It is possible to quantify this improvement by evaluating an RMS error between the numerical steady state solution and the analytical one, defined as follows:

$$RMS(\psi) = [\iint_{\Omega} (\psi-\psi_S)^2 \, d\Omega \; / \; \iint_{\Omega} \psi_S^2 \, d\Omega]$$

Figure 6 summarizes the corresponding results: the RMS of the stream function computed for different values of ϵ, with the two triangulations and P1 and P2 Lagrange polynomials, are plotted on the same graph. It can be seen that, with a given triangulation,

(1) the precision becomes dramatically bad when the size of the triangles along the western wall is bigger than the order of magnitude of the western boundary layer, and

(2) the precision is considerably improved when using quadratic elements in the FEM: for a mesh size h of the order of δ, the RMS(ψ) is 1% with P1 and 1 ‰ with P2, i.e. one order of magnitude of difference. When it is known that the CPU cost per time step is approximately the same for the computations (T1-P2) and (T2-P1), this point is very important to underline.

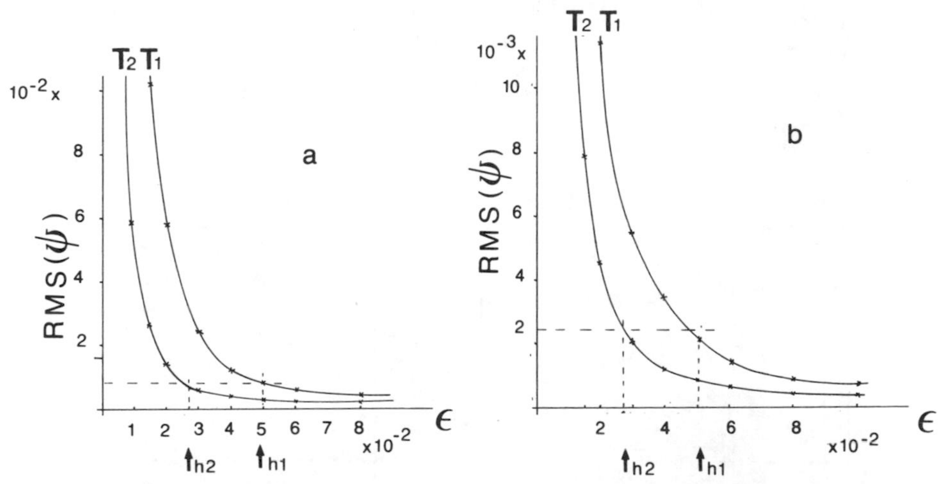

Figure 6. RMS of the stream function as a function of ϵ, with triangulations T1 and T2, for P1 (a) and P2 (b) approximations of the solutions. The typical widths of the western boundary layer corresponding to the size of the meshes along the western walls are indicated by an arrow.

We are also interested in tests of feasibility for non-linear cases. For the non-linear problems, analytical solutions are not available for comparisons, but we can refer to the previous results obtained by Veronis [1966] with FDMs. It is known that the non-linearities introduce a shift toward the north of the area of maximum negative vorticity, which is situated at (x=0, y=L/2) in the linear case, and to the production of a cell of positive vorticity near the northwest corner of the basin: correlatively, a northern boundary current develops in that corner, with a meandering southward flow before the particles again find the Sverdup recirculation. Results obtained with the present finite element model are presented on Figure 7 for the non-linear cases, when the inertial boundary layer thickness, $\delta_I = R_o^{1/2} L$, is six tenths or equal to the frictional boundary width (ϵ=0.02). These results are in agreement with those of Veronis and confirm the good behaviour of the presented FEM, when applied to the non-linear problems.

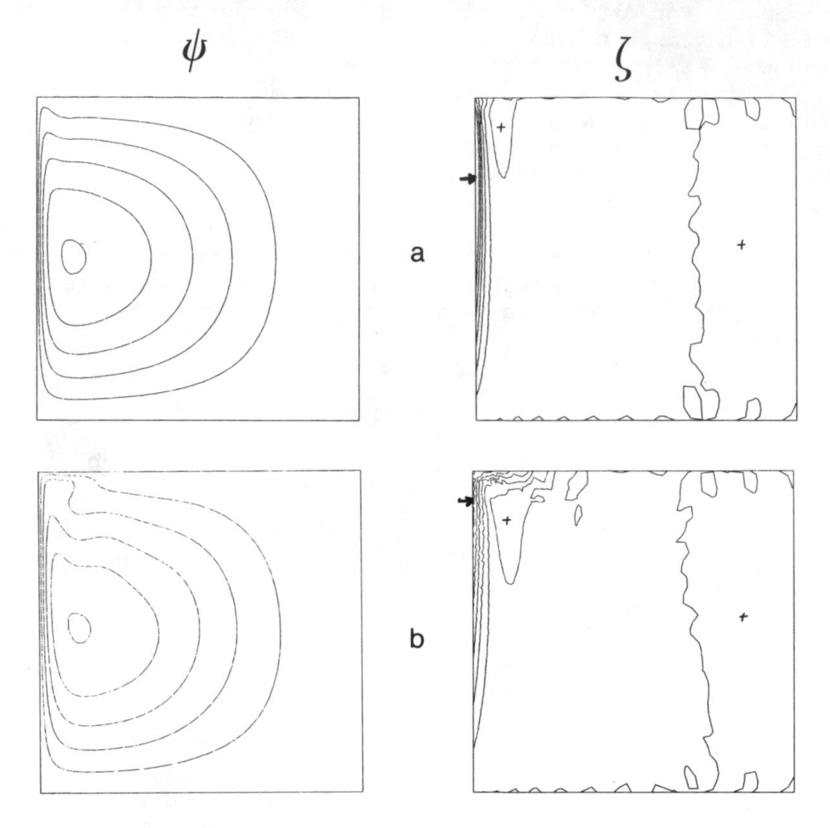

Figure 7. Contour lines of ψ and ζ for the non-linear cases with a frictional boundary layer width δ_F= 0.02 L and inertial boundary layer width (a) $\delta_I = 0.6\delta_F$ or (b) $\delta_I = \delta_F$. The inertial effects are correctly simulated in the northwest corner of the basin.

It is interesting to compare computer costs of FEMs and FDMs. By reference to the Stommel solution, we have seen that it is possible to give an estimate of the quality of the numerical solution. This has been used to compare the performance of the present FEM to more classical FDMs: the Hockney cyclic reduction method, and the relaxation method. Under the following conditions, the three methods applied to the case ϵ = 0.05 have given the following benchmarks:

F.E. T1-P2 - 830 nodes - RMS(ψ) = 0.8‰ - CPU/step: 27 sec

F.D. Relax.- 80*80 = 6400 nodes - RMS(ψ) = 1.1‰ - CPU/step: 69sec

F.D. Cycl.Red. - 80 *80 nodes - RMS(ψ) = 1.3‰ - CPU/step: 7sec

(the computations have been carried out on an HB68-DPS 7 computer).

Thus it appears that for the same order of precision, the FE-T1-P2 method is 4 times as long as the FD cyclic reduction, but 2.5 times faster than FD relaxation. Although these estimates are very qualitative, and perhaps even subject to controversy, they allow the conclusion that the present FEM is not out of scope by reference to the classical FDMs. This is somewhat surprising. It is probably due to the fact noted earlier that the mass and stiffness matrices do not vary during the iteration cycle, so that the matrices K and α M + β K in (14) can be inverted only once at the beginning of the computation.

3.6.2. Wind driven barotropic circulations with no bottom friction, and no-slip boundary conditions. Let us consider first tests of precision in the linear case, using Munk's steady solution. When taking a rectangular domain Ω_2 =]0,L[×]0,1[, and a wind field defined by:

$$\tau_x = -\sin(\pi y)$$
$$\tau_y = 0$$

and no-slip boundary conditions along the western and eastern sides of Ω, an approximate analytical solution can be given:

$$\psi(x,y) = -L\cos(\pi y)\ \{ -K\, e^{-\frac{1}{2}kx} \cos[\sqrt{3}/2\ (kx + 1/kL) - \pi/6] + 1 - (1/kL)\ (kx - e^{-k(L-x)} - 1)\}$$

with: $K = 2/\sqrt{3} - \sqrt{3}/(kL)$ and $k = (R_e/R_0)^{1/3}$

This solution has been used by Telias [1983] to test the numerical solutions obtained by applying the FEM presented in section 3.5. The triangulation built for these applications is displayed in Figure 8. Numerical results are presented in Figures 9 and 10 for the case R_0/Re = 6.10^{-5}, i.e. a viscous boundary layer δ_A = 0.08. Except in the Stommel's case, it is impossible here to reach a steady state solution, when starting from rest and time-stepping the integration. The horizontal eddy dissipation is mainly active only along the western boundary, because of the velocity gradient and the no slip condition: consequently, the planetary Rossby waves generated during the spin-up phase are travelling through the domain and are very slowly damped by lateral friction. When compared to Munk's steady solution, no instantaneous numerical solution fits exactly, as shown in Figure 9: the western boundary layer is very well obtained, but oscillations of the numerical solution around the analytical solution can be observed in the Sverdrup recirculation because of the presence of a Rossby wave. These results, however, show the efficiency of the numerical method.

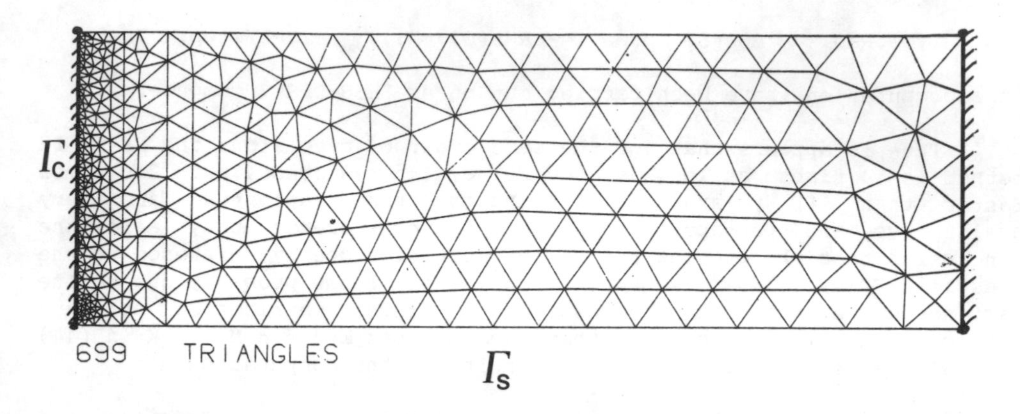

Figure 8. Triangulation used for the Munk's solution test over Ω_2.

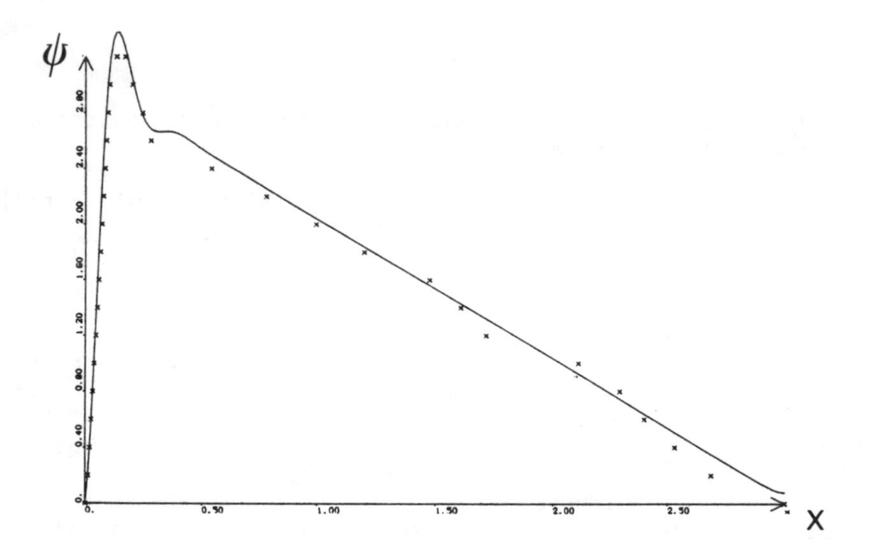

Figure 9. Comparison of the numerical solution (X) with the Munk's analytical solution (full line), along the mid-latitude of the basin. The presence of planetary oscillations can be clearly observed in the numerical solution.

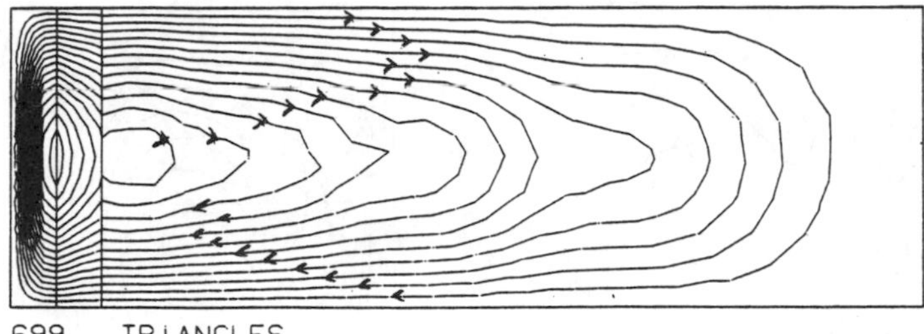

Figure 10. Instantaneous stream-function for the quasi-steady state solution of the Munk's linear problem, with $\delta_A = 0.08$ L.

Let us consider tests of feasability for the non-linear cases. The presence of these Rossby waves can also be seen on the results of the non linear simulation illustrated in Figures 11 and 12. When integrating the problem in a square basin, for a weakly non-linear case ($\delta_I = 0.5\ \delta_A$, with $\delta_A = 0.06$L), well developed Rossby waves fill the whole domain, superimposed to the general circulation pattern. The expected development of a northern boundary layer and meanders in the north west corner of the basin are also well reproduced, showing that the F.E method presented in section 3.5 to solve that problem is working satisfactorily.

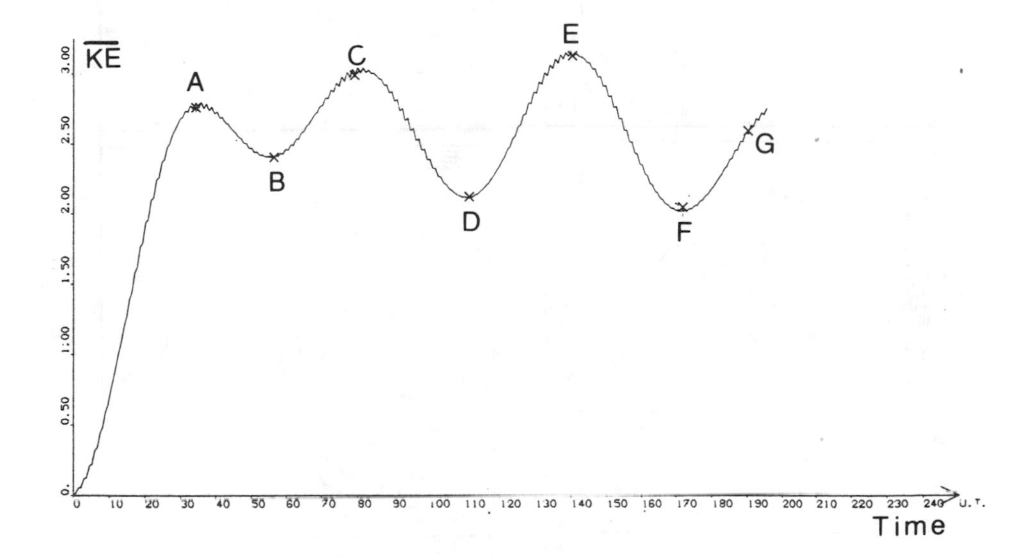

Figure 11. Time evolution of the total kinetic energy integrated over a square domain Ω_3, for the non-linear case and lateral eddy dissipation with no-slip boundary conditions. The oscillation corresponds to a basin mode Rossby wave of 16 days, if L is taken as 2000 km.

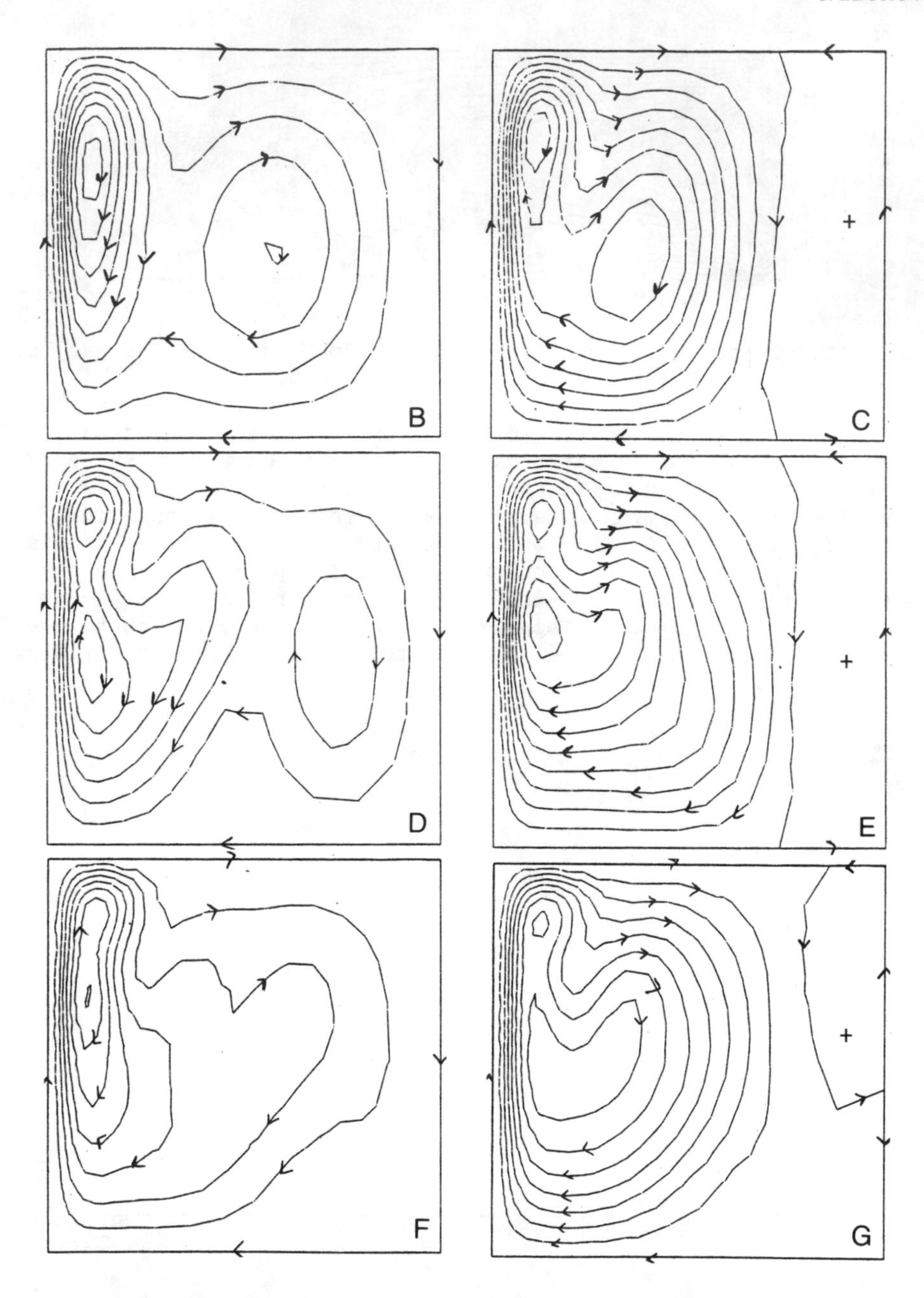

Figure 12. Stream function patterns obtained at the different times plotted on Figure 11. The problem is weakly non-linear: $\delta_I = 0.5\ \delta_A$, with $\delta_A = 0.06$ L. The inertial effects can be observed in the northwest corner of the basin. The presence of a basin mode 1 Rossby wave is clearly shown.

4. CONCLUSIONS

Finite element methods appear to be an interesting alternative to the classical finite difference methods classically used in ocean modelling. They offer interesting properties of flexibility, accuracy and efficiency which can help to resolve real ocean dynamic problems. A quick look at the recent bibliography on the subject shows that these FEMs are now more and more popular in the engineering field for coastal and shelf sea studies. Their applications for ocean circulation investigations are less developed, but several authors have recently clearly demonstrated their feasibility for mesoscale and large scale problems.

An example of a finite element model has been presented here for the integration of the quasi-geostrophic vorticity equation under a variety of boundary conditions (slip, no-slip, or neutral) and dissipative process parametrisation (lateral eddy viscosity, and bottom friction). Some illustrations of its feasability have been given for the computation of the wind-driven circulations in closed basins, with comparison to the analytical solutions of Stommel and Munk for the linear case,and to the classical results of Veronis and Bryan for the non-linear applications. These tests lead to important conclusions about the relation between space resolution and precision for the resolution of the western boundary layers, about the superiority of second order element approximations for the computed solutions, and about computer costs compared to more classical FD methods. Beside the reputation of high cost for FEMs,it appears that the present model,when applied to the preceding tests, can prove even cheaper than some FDMs for the same accuracy. The high level of flexibility offered by FEMs for the use of grids of variable size, shape and direction thus becomes very attractive for investigations over real oceanic domains. Furthermore, given the specific properties of conservation of the method, and the peculiarity of the present formulation leading to constant mass and stiffness matrices allowing vectorisation, we are convinced that this technique is a good candidate for ocean modelling, in spite of its more complex implementation. This presentation has been limited to the barotropic problem, but several baroclinic quasi-geostrophic ocean models have also been implemented, calibrated and demonstrated to be feasible and efficient for applications to realistic problems.

REFERENCES

Blandford, R. R.: Boundary conditions in homogeneous ocean models.*Deep Sea Research,***18**,739-751,1971.

Brebbia, C. A., and P. W Partridge: Finite-element simulation of water circulation in the North Sea.*J. Appl. Math. Model,* ,101-107,1976.

Bryan, K.: A numerical investigation of a non-linear model of wind-driven ocean. *J. of Atmosph. Sciences,***20**,594-606,1963.

Connor, J. J., and J. D.Wang : Finite element modeling of hydrodynamic circulations,*Numerical Methods Fluid Dynamics,* Pentech Press,1974,355-367.

Dumas, E., Le Provost, C. and A. Poncet: Feasibility of finite element methods for oceanic general circulation modelling. *4th Int. Conf. on Finite Element in Water Resources,* Springer-Verlag,New York, 5-43-55,1982.

Fix, G.J. : Finite element models for ocean circulation problems.,*SIAM J. Appl. Math.* ,**29-3**,371-387,1975.

Glowinski, R. and O. Pironneau: Sur une methode quasi directe pour loperateur biharmonique et ses applications a la resolution de Navier-Stokes,*Ann. Sc. Math. Quebec.*,**1**,No.2, 231-245,1977.

Glowinski, R. and O. Pironneau: Numerical method for the biharmonic equation and for the two dimensional "Stokes problem" ,*SIAM,* **21**, 167-212,1979.

Grotkop, G.: Finite element analysis of long-period water waves, *Comp. Meth. Appl. Mech. Eng.* ,**2**,147-157,1973.

Haidvogel, D.B. ,Robinson, A.R. and Schulman, E.E: The accuracy, efficiency and stability of three numerical models with application to open ocean problems,*J. of Comp. Phys.* ,**34**,1-53,1980.

Hansen, W.: Die halbtaggen Geseiten in Nordatlantischen Ozean,*Dt. Hyd. Zeit.* ,**2**,44,1949.

Holland, W. R.: On the wind-driven circulation in an ocean with bottom topography.*Tellus,***19**,582-600,1967.

Jamart, B.M. and D.F. Winter: A new approach to the computation of tidal motions in estuaries., *Hydrodynamics of Estuaries and Fjords,*Elsevier Oceanography Series,1978.

Kawahara, M. and K. Hasegawa: Periodic Galerkin finite element method of tidal flow.,*Int. J. Num. Methods Eng.* ,**12**,115-127,1978.

Kawahara, M. and Y. Kawanaga: Periodic tidal flow analysis by finite element perturbation method ,*Comput. fluids,***5**,175-189,1977.

Le Provost, C. and A.Poncet: Sur une methode numerique pour calculer les marees oceaniques et littorales,*C.R.Acad.Sci.*,**B285**,349-352,1977.

Le Provost, C. and A.Poncet: Finite element method for spectral modelling of tides,*Int. J. Num. Methods Eng.* ,**12**,853-871,1978.

Le Provost, C.,G. Rougier and A.Poncet: Numerical Modeling of the harmonic constituents of the tides , with application to the English Channel,*J. Phys. Oceanogr.* ,**11**,No.8,1123-1138,1981.

Le Provost, C.,1984: An application of finite element methods for modelling wind driven circulations in a stratified ocean.,*5th International Conference on Finite Elements in Water Resources*,Springer-Verlag,New York,567-576,1984.

Loth L. and M. Crepon: A quasi-geostrophic model of the circulation of the Mediterranean Sea.,*Hydrodynamics of the ocean,*,Elsevier Scientific Publishing Co.,Amsterdam,1984 .

Lynch, D. R.: Finite element solution of shallow-water equations ,*Ph.D thesis,*Princeton University,1978.

Marchuk G.I. and V.I.Kuzin: On the combination of finite element and splitting up methods in the solution of parabolic equations,*J. Comp. Phys,***52**No.2,237-272,1983.

Miller R.N., A.R.Robinson and D.B. Haidvogel: A baroclinic quasi-geostrophic open ocean model,*J. Comp. Phys.,***50**,No 1,38-70,1983.

Munk, W.H.: On the wind driven ocean circulation,*J. of Meteorol.,***7**,79-93,1950.

Pearson, C.E. and D.F. Winter: On the calculation of tidal currents in homogeneous estuaries,*J. Phys. Oceanogr,***7**,520-531,1977.

Platzman G.W., G.A. Curtis, K.S. Hansen and R.D. Slater: Normal Modes of the World Ocean.Description of modes in period range 8 to 80 hours.*J. Phys. Oceanogr.,***11**,No.5,579-603,1981.

Sarkisyan, A.S.: On the dynamics of the origin of wind driven currents in the baroclinic ocean.,*Okeanologie,***11**,393-409,1962.

Stommel, H. : The westward intensification of wind-driven ocean currents.,*Trans. of the American Geoph. Union,***29**,202-206,1948.

Veronis, G.: Wind-driven ocean circulation.*Deep Sea Research,***13**,17-29,31-55,1966.

Taylor, C. and J.M.Davies : Tidal propagation and dispersion in estuaries,*in Finite Elements in Fluids,*edited by Gallagher R.H., Oden J.T., Taylor C., Zinkiewicz O.C.,**1**,p 95.,John Wiley,New York,1975.

Telias, M. : Une condition de non glissement pour un modele quasi-geostrophique de circulation oceanique par elements finis,*These,*Universite de Grenoble,Grenoble,France,Juin,1983.

Wang, J.D. : Real time flow in unstratified shallow water, *J. Waterway Port, Coastal and Ocean Div., ASCE,* **104**, 53-68, 1978.

BOTTOM STRESS AND FREE OSCILLATIONS

Bruno M. Jamart, José Ozer, and Yvette Spitz
Management Unit of the Mathematical Models
of the North Sea and the Scheldt Estuary
Institut de Mathématique
Avenue des Tilleuls, 15
4000 Liège
Belgium

ABSTRACT. Three hydrodynamical models (2-D, 3-D, and 2.5-D) applicable to storm surge simulation are briefly described. A test problem (enclosed rectangular basin, uniform wind) is used to compare the results of the 2-D and 3-D models. A satisfactory overall agreement is observed between the patterns of surface elevation and depth-mean currents. However, it appears that long gravity waves propagate at different speeds in the two models. Part of the discrepancy can be ascribed to the formulation of the bottom boundary condition, i.e., to the modeling of the bottom stress. In models that resolve the vertical structure of the flow, the divergence of the bottom stress has two contributions. The first one reduces the "effective depth" of the water column, thereby increasing the periods of the free modes. The second contribution attenuates the amplitude of the oscillations.

1. INTRODUCTION

As part of a study of the turbulent bottom boundary layer (the Bottom Stress Experiment, alias the "BSEX" project), we have undertaken a series of numerical experiments aimed at comparing the results of various hydrodynamical models. The experiments are limited to a simple geometric configuration -- an enclosed rectangular basin -- and use schematic wind forcings. The models used sofar in this study are not boundary layer models but models that describe larger scale flow dynamics. Such models will be used to provide the upper boundary condition(s) of -- or will be coupled with -- specific bottom boundary layer models later on.

The purposes of the numerical experiments are :

1. to "validate" or verify the implementation of the numerical algorithms by comparing the results of independently developed computer programs used under the same conditions;
2. to gain insight into the intrinsic properties of the numerical

J. J. O'Brien (ed.), Advanced Physical Oceanographic Numerical Modelling, 581–598.

schemes by analyzing the details of the calculated solutions to simple problems.

Our approach to these goals is semi-empirical.

In section two, the models currently used within our group are described and the main aspects of the corresponding numerical procedures are summarized. Preliminary results of a comparison experiment are discussed in section three. It is shown that despite a satisfactory overall agreement between the solutions, long gravity waves do not propagate at the same speed in the 2-D (vertically-integrated) and the 3-D models. This discrepancy is then investigated in section four by means of numerical experiments on the seiche problem. The formulation of the bottom boundary condition and the actual numerical values taken on by the bottom stress are assessed. Concluding remarks are offered in the final section.

2. DESCRIPTION OF THE NUMERICAL MODELS

For this study, we consider three hydrodynamical models which we refer to as the "2-D", "3-D" and "2.5-D" model, respectively.

2.1. Vertically-integrated, 2-D model

The model equations are the conventional time-dependent, nonlinear shallow-water wave equations expressing conservation of mass and horizontal momentum for a homogeneous fluid of density ρ. With η denoting the instantaneous sea surface elevation with respect to mean sea level, and $\vec{u}$ the depth-averaged horizontal velocity vector, the governing equations are :

$$\frac{\partial \vec{u}}{\partial t} + (\vec{u} \cdot \vec{\nabla})\, \vec{u} + f\, \vec{e}_z \wedge \vec{u} = - \vec{\nabla} \left(\frac{p_a}{\rho} + g\eta\right) + \frac{\vec{\tau}_s}{\rho H} - \frac{\vec{\tau}_b}{\rho H} \tag{1}$$

$$\frac{\partial \eta}{\partial t} + \vec{\nabla} \cdot (H\, \vec{u}) = 0 \tag{2}$$

where $\vec{\nabla}$ is the gradient operator, $H = D(x,y) + \eta\,(x,y,t)$ is the total water depth, f the Coriolis parameter, $\vec{e}_z$ the unit vertical vector (positive upwards), g the acceleration due to gravity, and p_a the atmospheric pressure. The latter is assumed uniform henceforth.

The surface stress, $\vec{\tau}_s$, is assumed to be a known function of space and time. The bottom stress, $\vec{\tau}_b$, is parameterized as :

$$\vec{\tau}_b = \rho k_1 \mid \vec{u} \mid \dot{\vec{u}} + \rho k_2 \vec{u} - m \vec{\tau}_s \tag{3}$$

where k_1 and k_2 are frictional coefficients, and m is a corrective factor sometimes used in storm surge modeling [Ronday, 1976].

In practice, only one of the two frictional coefficients of equation (3) is taken as non-zero. We remark that the vertically-integrated approach requires the introduction of the concept of a bottom stress and its formal parameterization. We also note that such a formulation implies that the fluid must be allowed to slip along the bottom boundary.

Along the lateral "solid" boundaries of the basin, the normal component of the transport is set equal to zero. Since the convective terms are included in the equations, an additional boundary condition is necessary at those times when the water flows towards the interior of the domain : we assume that the normal velocity gradient vanishes at those times.

The numerical procedure is based on finite differences approximations, using the so-called "Arakawa-C" type of grid. In this grid, the two velocity components and the elevation of the free surface are defined respectively at the mid-points and at the vertices of two interwoven and staggered rectangular meshes.

The first step of the computation consists of calculating the new position of the free surface by solving explicitly the nonlinear continuity equation (2). The total water depth, needed to evaluate the divergence term, is calculated at the velocity points by linearly interpolating the mean depth in one direction (e.g. the y-direction for a "u-point") and the free surface elevation in the other direction. In the second step, the two components of the velocity at the new time level, t^{n+1}, are calculated from equation (1) where all the terms are evaluated at time level t^n except for those involving the elevation of the free surface. The latter terms, i.e., the pressure gradient term and the stresses, are calculated using the values of η at t^{n+1}. Total depths, needed for the stress terms, are again linearly interpolated. The calculation of the Coriolis term in, for example, the x-momentum equation requires the knowledge of the y-component of the velocity; the latter is taken as the mean of the four closest v-points. The same value is also used for the advective terms, which are evaluated by upwind differencing.

These necessary interpolations, due to and combined with the staggered nature of the grid, lead to inaccuracies in the solution which are of different origin than the well-known truncation errors. We have attempted to quantify these effects by investigating the integrated energy or power balance for several test cases and found that the calculated contributions to such a balance are slightly different from the theoretical ones. These findings will be reported separately and not discussed here. Suffice to say that the so-called "numerical energy dissipation" does not play a significant role in the results reported in this paper.

2.2. Primitive equations, 3-D model

We have implemented a version of the numerical model for three-dimensional, variable-density hydrodynamic flows developed by Paul and co-workers [Paul and Lick, 1981] . This model has recently been used to calculate the wind-driven circulation in and around the Santa Barbara Channel [Jamart et al., 1982] . Baroclinic effects have not yet been considered in the present study, so that the variable density capability of the 3-D model will not be discussed here.

The governing equations are the three-dimensional, time-dependent, nonlinear Navier-Stokes equations. In order to simplify and close the system of equations, it is necessary to introduce several assumptions and approximations. Assuming that the pressure is hydrostatic, the density and the Coriolis parameter constant, and accepting that the turbulent fluxes of momentum be parameterized or approximated by means of eddy coefficients, the equations read :

$$\frac{\partial u}{\partial t} + \frac{\partial (u^2)}{\partial x} + \frac{\partial (u\ v)}{\partial y} + \frac{\partial (u\ w)}{\partial z} + f\ v = -\frac{1}{\rho}\frac{\partial p}{\partial x} + \frac{\partial}{\partial x}(A_h \frac{\partial u}{\partial x}) + \frac{\partial}{\partial y}(A_h \frac{\partial u}{\partial y}) + \frac{\partial}{\partial z}(A_v \frac{\partial u}{\partial z}) \quad (4)$$

$$\frac{\partial v}{\partial t} + \frac{\partial (u\ v)}{\partial x} + \frac{\partial (v^2)}{\partial y} + \frac{\partial (v\ w)}{\partial z} - f\ u = -\frac{1}{\rho}\frac{\partial p}{\partial y} + \frac{\partial}{\partial x}(A_h \frac{\partial v}{\partial x}) + \frac{\partial}{\partial y}(A_h \frac{\partial v}{\partial y}) + \frac{\partial}{\partial z}(A_v \frac{\partial v}{\partial z}) \quad (5)$$

$$\frac{1}{\rho}\frac{\partial p}{\partial z} = g \quad (6)$$

$$\frac{\partial u}{\partial x} + \frac{\partial v}{\partial y} + \frac{\partial w}{\partial z} = 0 \quad (7)$$

with the x-, y-, and z-axes pointing northward, eastward, and downward respectively. The horizontal diffusion coefficient, A_h, has been assumed constant and isotropic, whereas the vertical coefficient, A_v, can be made a function of time and depth.

The boundary conditions along the bottom and solid lateral boundaries are that all three components of the velocity vanish (no-slip). At the interface between air and water, a tangential stress has to be specified and either a "rigid-lid" assumption is made (i.e., the vertical velocity is set equal to zero), or the linearized kinematic boundary condition is implemented. The latter case is referred to as the "free surface option". The rigid-lid assumption is commonly used in circulation modeling to circumvent the timestep restriction imposed by stability constraints on the calculation of surface gravity waves. A less restrictive approach, developed by Paul (personal communication) is to calculate the position of the free surface in an implicit (or semi-implicit) fashion.

The original equations are first transformed from a Cartesian coordinate system into a "sigma-coordinate" system. This transformation makes the variable bottom a coordinate surface, which provides for better vertical resolution in the shallower areas. The numerical procedure for the solution of the transformed equations is based on the so-called simplified marker and cell method of Amsden and Harlow [1970]. Details can be found in Paul and Lick [1981].

Briefly stated, the time and space derivatives are approximated by finite difference analogs; all terms, except the diffusive flux in the vertical direction and the Coriolis term, are calculated from the results of the previous timestep. Hence, the discretized equations are linear. The spatial distribution of the variables is such that u and v are defined at the vertices of a rectangular grid, whereas w and p (or η) are defined at the center of each cell.

In order to solve for all variables (i.e., the velocities and the pressure or the elevation) at the new time level simultaneously, the horizontal velocities (and therefore the momentum equations) are formally decomposed into two components. The first component of each velocity satisfies the corresponding momentum equation without the pressure term, whereas the second component is proportional to the unknown pressure gradient. By taking the divergence of the vertically integrated momentum equations and using the vertically integrated continuity equation, one can derive an equation of the Poisson type for the surface pressure or the elevation. The latter equation can be solved at the new time level and the results used to calculate the new horizontal velocities. The vertical velocities are then readily obtained by integrating the continuity equation.

2.3. Vertical plane, 2.5-D model

This model is a simplified version of the 3-D, primitive equations model. Therefore, the assumptions listed above still hold. In this model the physical domain is a single vertical plane, and the additional, simplifying assumption is made that all variables, except the pressure, are uniform in the direction normal to the vertical plane of interest. The denomination "2.5-D model" is proposed because the equations include the Coriolis term and all three components of the velocity vector are calculated.

The determination of the barotropic and baroclinic pressure gradients normal to the vertical section is site and/or problem specific. The normal pressure gradient has to be either specified or calculated on the basis of some appropriate assumption or observations. It can also be specified on the basis of the results of another model.

The main advantage of the 2.5-D model, as compared to the 3-D version, is that it requires less computer resources. It can therefore be quite useful in the development phase where various formulations and parameterizations can be tested without excessive expenses. Also, if most of the variability in the direction normal to the vertical plane of interest is related to the barotropic response of the system, it is conceivable that the 2.5-D model, coupled to a vertically-integrated model, could yield almost as much information at specific locations as a full-blown 3-D model.

3. NUMERICAL INTERCOMPARISON EXPERIMENT : PRELIMINARY RESULTS

In the simplest test problem for which comparisons are currently in progress, we consider a rectangular basin closed on all sides, of dimensions 600 X 1200 km and of uniform depth taken as 100 m. The Coriolis factor corresponds to a latitude of 55 degrees North. The motion is forced only through a uniform surface stress parallel to the long axis of the basin. The strength of this stress is taken as 0.1 N/m^2. The stress is applied impulsively at time zero, the water being initially at rest and the free surface horizontal, and remains constant thereafter.

The results of the 2-D model, calculated with a large bottom friction term linear in the depth-mean current velocity ($k_1 = m = 0.$, and $k_2 = 2.4 \; 10^{-3}$ m/s in eqn (3)), show that the sea surface response consists of a superposition of damped longitudinal and transverse seiche modes. Kelvin- and Poincaré-type waves, which are solutions of the unforced linearized set of equations, are also present. A sample result from this 2-D calculation, performed with a timestep $\Delta t = 6$ min, is shown in Figure 1 where the elevation and the depth-averaged currents after 12 hours are displayed. The horizontal spacing for this calculation is 40 km. Note that the currents in Figure 1 are not plotted at the η-points, as is usually done, but at the vertices of the grid on which u an v are defined. This convention facilitates the comparison of the results with those of the 3-D model.

As time passes, the free waves become increasingly dampened and the solution tends towards a steady state wherein all velocities vanish and the pressure gradient due to a uniform slope of the free surface exactly balances the surface stress.

To our knowledge, no complete analytical solution to this apparently simple problem has been found, despite several attempts [Renouard, 1980] . The results, however, are very similar to those obtained by other numerical modellers [Heaps, 1971] , and we believe they are correct.

Numerical experiments have shown that, as expected, the role of the nonlinear terms is insignificant in this problem. This result is not surprising, since the time-mean depth is much larger than the perturbation of the surface elevation and all possible free waves are *de facto* excited by the impulsive forcing.

A similar calculation was performed with the 3-D model, using a constant and vertically uniform eddy viscosity $A_v = 100$ cm^2/sec, a horizontal diffusivity A_h of $10^7 cm^2/sec$, a timestep of 30 min, a uniform horizontal spacing of 40 km, and 10 levels in the vertical. The results show patterns of sea surface elevations and depth-mean currents similar to those of the 2-D model. This is exemplified by Figure 2 which shows those two fields at the same time as in Figure 1. This result is of course encouraging. The extrema of surface elevation calculated in all corners are higher in the 3-D results than in the 2-D model, **which is consistent with the findings of Heaps [1971].**

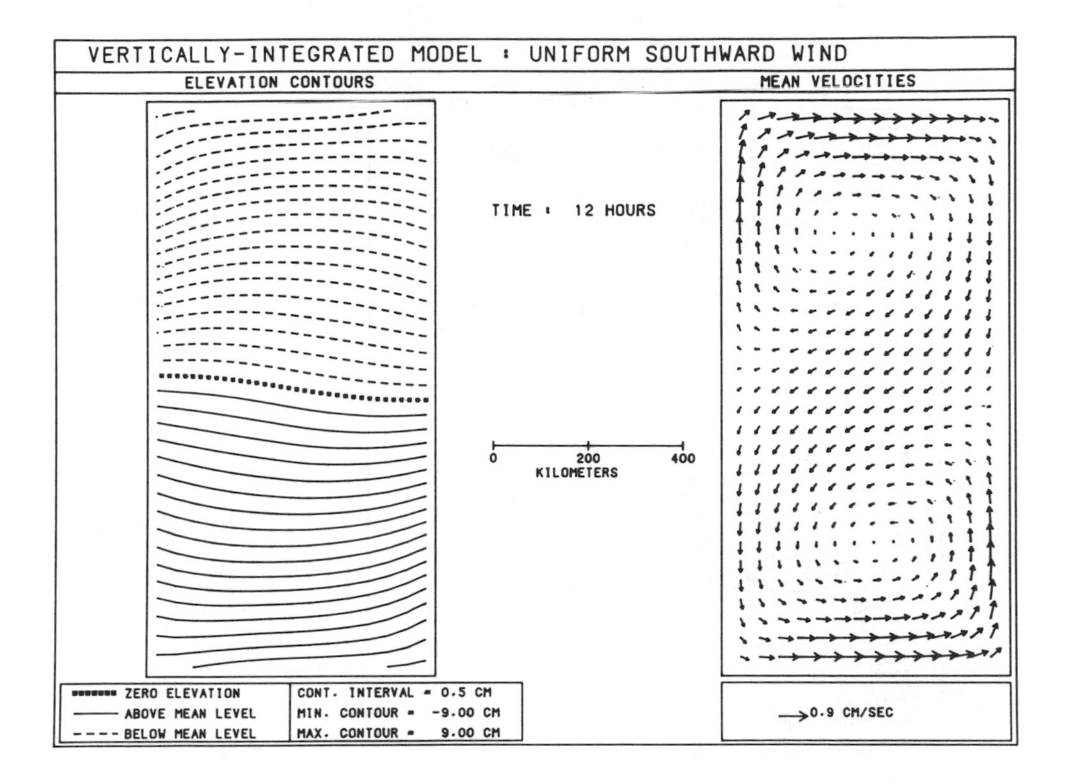

Figure 1. Results of the vertically-integrated, 2-D model after 12 hours of integration. Left : isolines of sea surface elevation; right : instantaneous depth-averaged velocity field.

The depth-averaged current field calculated with either the 2-D or the 3-D model is completely different from the flow field calculated with the 3-D model at any depth. An example of such differences is given in Figure 3, where the horizontal velocities at the surface and at 90 m are shown at t = 12 hrs. This observation is not surprising since the components of the current that are directly induced by the applied surface stress dominate the wave-driven part of the motion at most times and depths. In particular, we emphasize that if the flow field close to the bottom is of interest (e.g., to drive a boundary layer model), it would be difficult to deduce the currents calculated in the 3-D simulation from the results of the vertically-integrated model.

The comparison of auxiliary 2-D and 3-D power balance calculations shows that the physical understanding of the system response that can be derived from the study of simple test problems depends largely on the viewpoint.

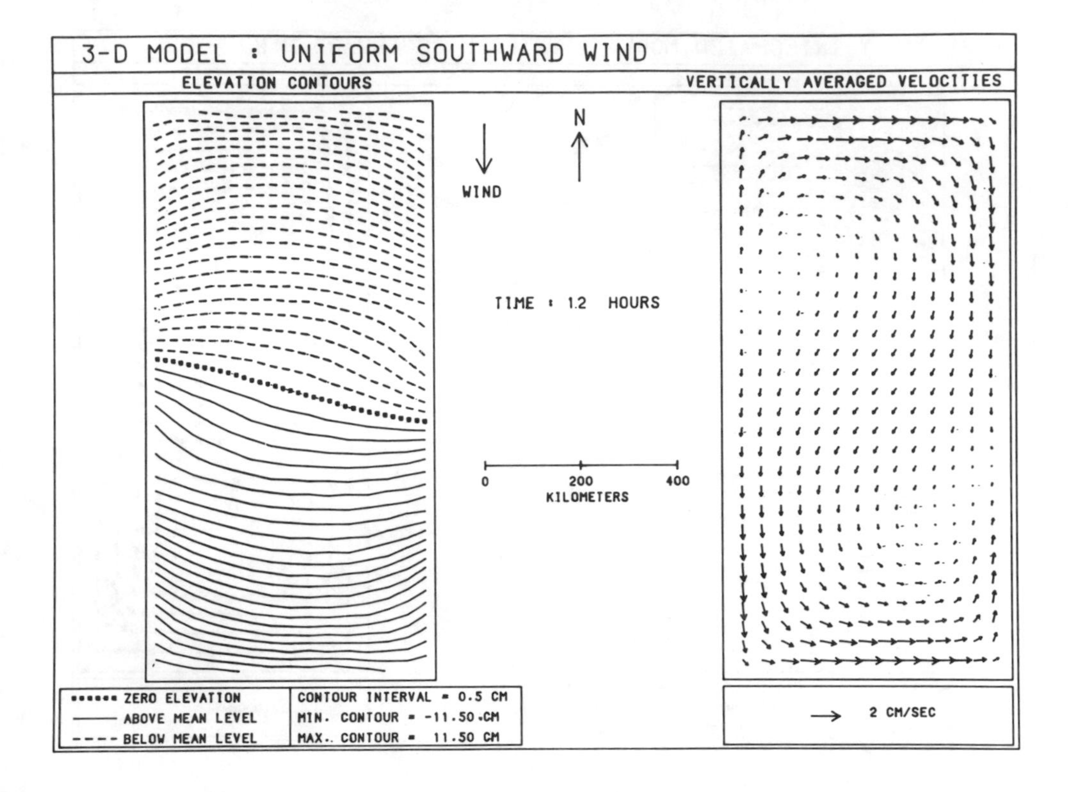

Figure 2. Results of the 3-D model for the same test problem as in Figure 1, after 12 hours of simulation. Left : isolines of sea surface elevation; right : vertically averaged horizontal velocity field.

Since the evolution of the sea surface elevation is quite similar in both models, the time rate of change of the total available potential energy is also basically the same. The kinetic energies differ by an order of magnitude, but their time rate of change becomes rapidly insignificant in the power balance. Note, however, that the total kinetic energy approaches zero in the 2-D model and tends toward a finite value in the 3-D calculation.

After some initial transients, the dominant terms of the power balance in the 2-D simulation are the time derivative of the potential energy and the power input by wind. The latter term, which is the product of the surface stress by the vertically-averaged velocity, changes sign with time at the frequency of the gravest longitudinal mode. In the 3-D model, on the other hand, the two largest terms are the dissipation associated with vertical friction and the power drawn from the wind stress, the difference between these terms being balanced by the change in potential energy. Since the Ekman surface layer is approximately resolved in this case, the power input from the wind stress is always positive. This might explain in part why the extrema of surface elevation are larger in the 3-D model than in the 2-D. Another part of the explanation is that no attempt has been made to use frictional coefficients that would lead to equivalent damping rates in the two models. This point will be discussed further in the next section.

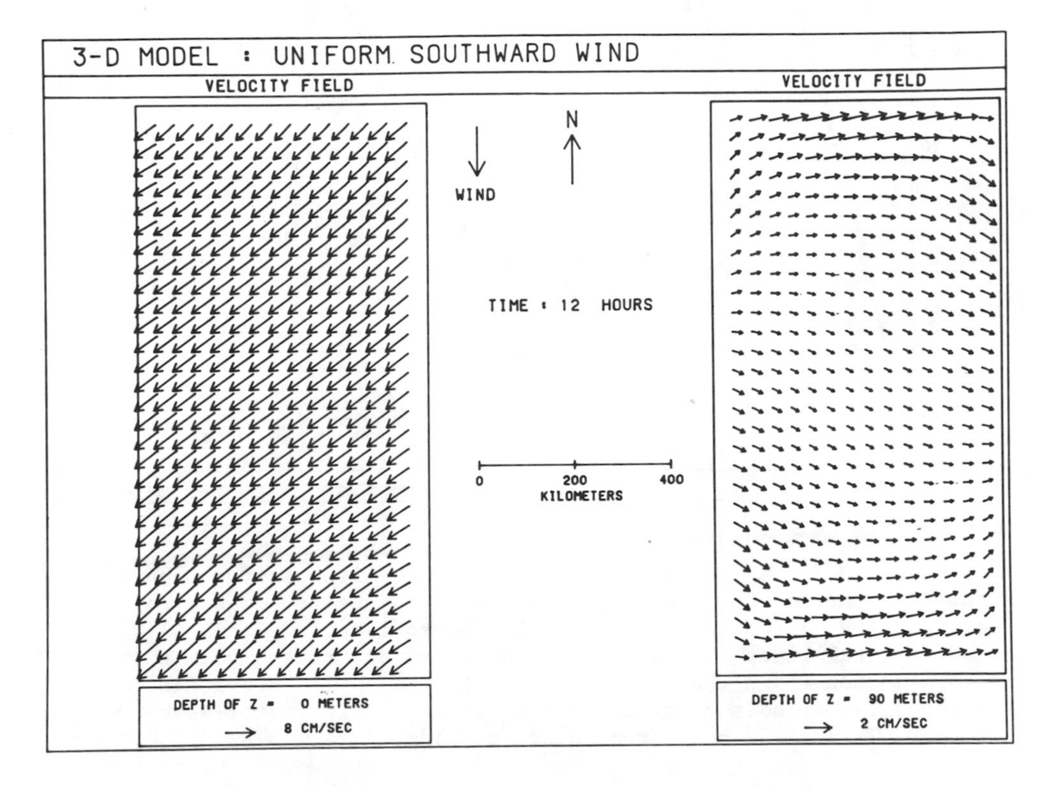

Figure 3. Horizontal velocity field, as calculated in the 3-D model, at the sea surface (left) and 10 meters above the bottom (right).

The period of the dominant oscillation is also somewhat larger in the 3-D than in the 2-D. This is clearly shown in Figure 4 where the time history of the surface elevation in the southwest corner of the basin is displayed for the 2-D and various 3-D calculations. Several possible explanations for that feature come to mind :

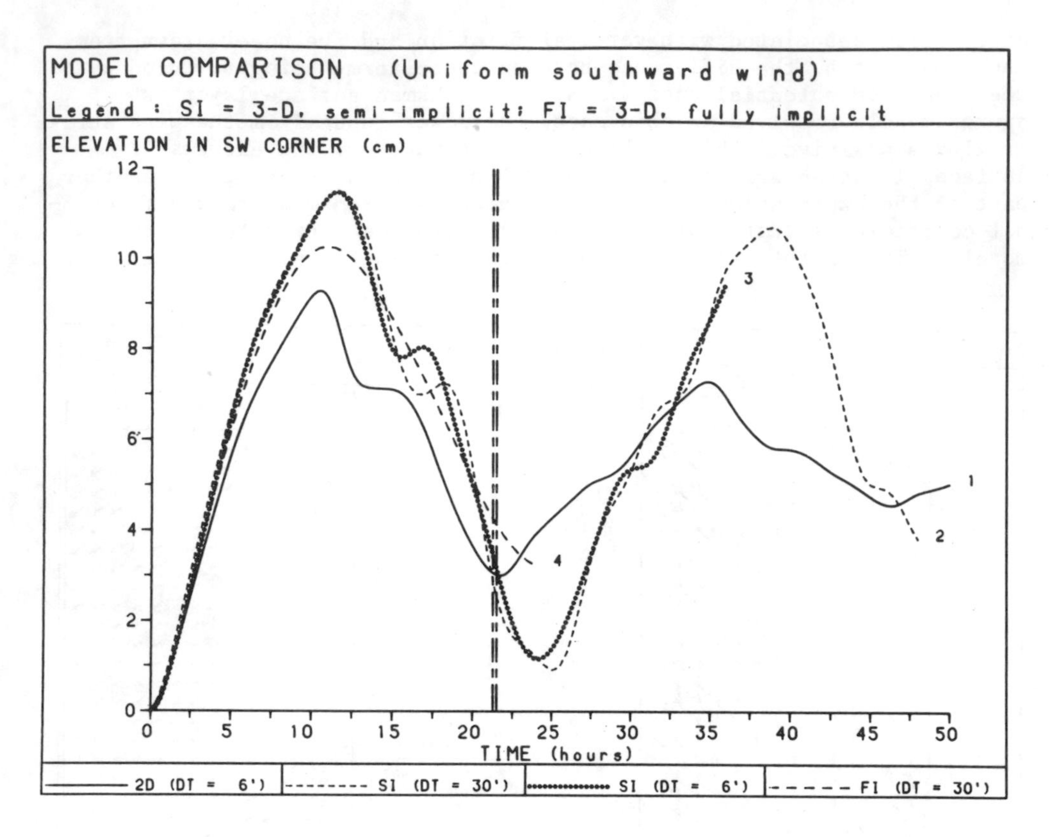

Figure 4. Time evolution of the elevation of the free surface in the southwest corner of the basin, as calculated in the 2-D model (solid line, labelled 1) and in the 3-D model (curves labelled 2, 3, 4). The dashed vertical lines correspond to the end of the first mode longitudinal oscillation i) in the absence of friction [T = 21.28 hrs] and, ii) with a linear bottom stress in the linear shallow water wave equations [T_1 = 21.52 hrs, from equation (9)] .

- The implicit or semi-implicit calculation in the 3-D model of the position of the free surface introduces some numerical damping and smoothing of the solution and can also modify the phase speed of long gravity waves. This effect can be judged by comparing in Figure 4 the solution obtained with a fully implicit and a semi-implicit scheme using the same timestep of 30 min. The importance of this error can be lessened by using a smaller timestep (see result obtained with Δt : 6 min). Other sources of numerical dissipation could also be invoked to explain the differences between models.
- The inevitable presence in the 3-D results of lateral boundary layers, even though such layers are not properly resolved with the

grid spacing used in this study, is also likely to affect the propagation of the gravity waves. However, it appears that the total amount of power dissipated by the horizontal diffusion terms is negligible in the energy balance. Similarly, we found that the linearization of the kinematic boundary condition, which has the effect of introducing two unconventional terms in the energy equation, leads to insignificant contributions to the integrated balance.

- In view of the no-slip boundary condition specified at the bottom, one would expect that the "effective" water depth, i.e., the depth that determines the speed of propagation of long gravity waves, might be somewhat smaller than the actual depth of the basin. This effect is not physically unrealistic (there *is* a bottom boundary layer), but its magnitude is likely to depend on the vertical structure of the current.

We believe that a combination of the factors just mentioned is responsible for the observed phase discrepancies between the results of the 2-D and 3-D models. We also believe that the latter point, i.e., the effect of the bottom stress on the speed of propagation of free gravity waves, is the dominant factor in the experiments reported here. Therefore, we have investigated this possibility in further detail.

4. BOTTOM STRESS EFFECT ON SEICHING MOTION

If the argument outlined at the end of the previous section is correct, it should be possible to evidence the gist of the mechanism on a simpler problem than the wind-driven circulation in a rotating basin. In this section we neglect rotation and wind forcing and we consider the motion generated by an initial, one-dimensional, first mode perturbation of the free surface. The length of the domain is taken as 1200 km, its depth as 100 m. Neglecting friction and advection, the periods of the first two modes are 21.28 and 10.64 hours, respectively.

With these assumptions, the response of the 2-D model is one-dimensional. The response of the 3-D model can also be made "horizontally one-dimensional", provided that the frictional lateral boundary layers are eliminated. This can easily be done by modifying the boundary conditions along the edges of the basin. Alternatively, the 2.5-D model described earlier can be used with all velocities normal to the vertical plane equal to zero : the model is then a two-dimensional, "x-z" model. We have verified the equivalence of the two codes running under such conditions. In the following, we shall therefore contrast "2-D" and "2.5-D" results.

4.1. 2-D Model

In a first set of experiments, we have verified the ability of the 2-D model to reproduce known analytical solutions to the seiche problem. In the absence of friction, and with linearized equations, the first mode oscillation is very well simulated by the model. The total "numerical dissipation", calculated on the basis of explicit formulae derived in setting up a discrete power balance equation, reaches a maximum value equal to 1.5 % of the maximum time rate of change of the potential or kinetic energy. That "numerical dissipation" oscillates in time around a zero mean value, so that the calculated oscillation is not damped. When the advective term of equation (1) is included in the computation, the first harmonic of the basic mode is excited, as expected, but a quantitative verification of this nonlinear effect escapes us. We also observe, in this case, a very slight long-term numerical dissipation.

When a bottom stress term linear in the depth-mean current is introduced in the 2-D model, the analytical solution for the elevation (neglecting advective terms again) is given by

$$\eta(x,t) = A_0 \cos \frac{\pi x}{L} \left(\cos \omega_1 t + \frac{k_2}{2D\omega_1} \sin \omega_1 t\right) \exp\left(-\frac{k_2 t}{2D}\right) , \quad (8)$$

where A_0 is the amplitude of the initial perturbation and ω_1 is given by

$$\omega_1 = \left(\frac{\pi^2 gD}{L^2} - \frac{k_2^2}{4D^2}\right)^{1/2} = \frac{2\pi}{T_1} . \quad (9)$$

We have also verified that the 2-D model reproduces this solution. The important point, though, is that for the set of parameters used in these experiments, the period of the first mode seiche is rather unsensitive to the value of the friction coefficient. (This is also true in the case of a quadratic friction law). For example, with the unusually large value $k_2 = 2.4 \ 10^{-3}$ m/s of the previous section, the period given by (9) is equal to 21.52 hours as opposed to 21.28 hrs for the frictionless case. The dashed vertical lines of Figure 4 correspond to these two periods.

4.2. 2.5-D Model

A second set of experiments was performed, with the 2.5-D model, using values of A_v equal to 1, 100, and 650 cm^2/s and different numbers of vertical levels. The main conclusions of these experiments, to be discussed below, are :

1. when the vertical resolution of the model is sufficient for numerical errors to be small, the presence of a no-slip bottom boundary layer under the oscillating flow causes not only damping of the wave but it can also lead to a significant lengthening of the period of the free oscillation;
2. when the vertical spacing is insufficient to resolve properly the bottom boundary layer, the numerical scheme used in the 2.5-D model

can still yield correct elevations. In this case, however, the "true" bottom stress can be severely underestimated and its phase in error.

It is, obviously, not realistic to use a constant and uniform value for the vertical eddy viscosity. This assumption, however, simplifies the interpretation of the numerical results by allowing some degree of algebraic analysis. Moreover, we believe that the assumption of a constant A_v does not limit the applicability of our conclusions. This remains to be verified.

An example of the results obtained with the 2.5-D model, using a vertical spacing Δz of 1 meter, is shown in Figure 5 (time series of elevation at one end of the basin). It is clearly seen that the free oscillation is more and more dampened as A_v increases. The period of that oscillation also increases with A_v. The latter result is in contrast with the findings of Heaps [1971] for the case of a "slippery" bottom. We offer the following "semi-analytical" explanation for our results.

Considering only the dominant terms of the equations, and taking z positive upwards and null at the bottom, the problem is defined by

$$\frac{\partial \eta}{\partial t} + \frac{\partial}{\partial x} \int_0^D u \, dz = 0 \tag{10}$$

$$\frac{\partial u}{\partial t} = - g \frac{\partial \eta}{\partial x} + A_v \frac{\partial^2 u}{\partial z^2} \tag{11}$$

with

$$\eta = A_0 \cos \frac{\pi x}{L} , \quad u = 0 \quad \text{at } t = 0,$$

$$\frac{\partial u}{\partial z} = 0 \text{ for all } t \text{ and } x \quad \text{at } z = D,$$

$$u = 0 \text{ for all } t \qquad \text{at } z = 0 , \; x = 0 , \; L.$$

A wave equation can be formed from (10) and (11) :

$$\frac{\partial^2 \eta}{\partial t^2} - gD \frac{\partial^2 \eta}{\partial x^2} - \frac{\partial}{\partial x} \left(\frac{\tau_b}{\rho}\right) = 0 \tag{12}$$

where

$$\frac{\tau_b}{\rho} = \left(A_v \frac{\partial u}{\partial z}\right)_{z=0} \tag{13}$$

Neglecting, as a first approximation, the divergence of the bottom stress in (12) yields the classical seiche solution

$$\eta^{(1)} = A_0 \cos \frac{\pi x}{L} \cos \omega t \tag{14}$$

$$u^{(1)} = A_0 \left(\frac{g}{D}\right)^{1/2} \sin \frac{\pi x}{L} \sin \omega t \tag{15}$$

where

$$\omega = \frac{\pi}{L} (gD)^{1/2} \tag{16}$$

At any given point, the frictional effects can now be estimated by solving the so-called "Stokes second problem" [Schlichting, 1960, p. 75] upside down :

$$\frac{\partial u}{\partial t} = - \frac{1}{\rho} \frac{\partial p}{\partial x} + A_v \frac{\partial^2 u}{\partial z^2}$$

with $u = 0$ at $z = 0$, and $u = u^{(1)}$ given by equation (15) at $z = D$. For D much larger than the "depth of penetration" $\delta \simeq \left(\frac{A_v}{\omega}\right)^{1/2}$, the velocity profile is given by

$$u = A_0 \left(\frac{g}{D}\right)^{1/2} \sin \frac{\pi x}{L} [\sin \omega t - e^{-\alpha z} \sin (\omega t - \alpha z)] \tag{17}$$

where

$$\alpha = \left(\frac{\omega}{2A_v}\right)^{1/2} .$$

From (17), a first estimate of the bottom stress can be obtained. The divergence of that stress can be written as

$$\frac{\partial}{\partial x} \left(\frac{\tau_b}{\rho}\right) = - \frac{\pi}{L} \left(\frac{g}{D}\right)^{1/2} \left(\frac{A_v}{2\omega}\right)^{1/2} \frac{\partial \eta^{(1)}}{\partial t}$$
$$- \left(\frac{g}{D}\right)^{1/2} \frac{L}{\pi} \left(\frac{\omega A_v}{2}\right)^{1/2} \frac{\partial^2 \eta^{(1)}}{\partial x^2} . \tag{18}$$

The two components of $\frac{\partial}{\partial x} \left(\frac{\tau_b}{\rho}\right)$ have been purposely written in terms of derivatives of $\eta^{(1)}$ to illustrate their physical meaning in equation (12). The wave equation, upon substitution of (18), is of the form

$$\frac{\partial^2\eta}{\partial t^2} + k\,\frac{\partial\eta}{\partial t}^{(1)} - gD\,\frac{\partial^2\eta}{\partial x^2} + \beta\,\frac{\partial^2\eta}{\partial x^2}^{(1)} = 0 \qquad (19)$$

where

$$k = \frac{\pi}{L}\left(\frac{gA_v}{2D\omega}\right)^{1/2} \qquad (20)$$

and, setting $gD - \beta = gD'$,

$$D' = D - \frac{L}{\pi}\left(\frac{\omega A_v}{2gD}\right)^{1/2} . \qquad (21)$$

The first contribution of (18) is a damping term similar to the one resulting, in the 2-D model, from a bottom stress linear in the depth-mean current. The second contribution amounts to a reduction of the depth that is actually "seen" by the mean flow. A more general derivation of these results is in preparation.

An estimate of the importance of those two terms can be obtained by using (16) to calculate a first guess of ω. With A_v = 650 cm^2/s, for example, we find an "effective depth" D' equal to 80.1 meters (recall that D = 100 m), which corresponds, using equation (16) to a period T' = 23.78 hours for the first mode oscillation. This period is displayed on Figure 5 (right dashed vertical line, the left line indicating the period for the frictionless case) : it is in excellent agreement with the numerical solution. As for the estimated damping coefficient (equation 20), we obtain $k = 1.63\ 10^{-5}\ \text{sec}^{-1}$. This term modifies the new period by only a few minutes and it leads, after one period of oscillation, to an amplitude of 49.4 cm (the initial height is taken as 100 cm) at the η-point closest to the left boundary. This figure compares well with the value of 42.7 cm calculated by the 2.5-D model. If the effective rather than the nominal depth is used in (20), the amplitude after one period is reduced to 43.1 cm.

In summary, both the numerical model and the above analysis concur to demonstrate that the main effect of the bottom boundary layer is a reduction of the effective depth of the basin which results in the lengthening of the periods of the free modes. The bottom stress given by the solution to Stokes problem is found proportional to $(A_v)^{1/2}$ and, in a first approximation, it leads the "surface" current (we have assumed that $D \to \infty$) by 45 degrees. These effects on the seiching motion cannot be produced by either a linear or a quadratic stress law in a 2-D model.

Finally, a peculiar feature of the numerical scheme used in the 2.5-D model, which came to light when calculations were performed with a vertical spacing too large to resolve the bottom boundary layer, seems worth mentioning because of its implications. The surprising result we want to discuss briefly is that on several occasions the seiching motion of the free surface was "perfectly" calculated by the 2.5-D model even though the vertical resolution was quite coarse compared to $(A_v/\omega)^{1/2}$.

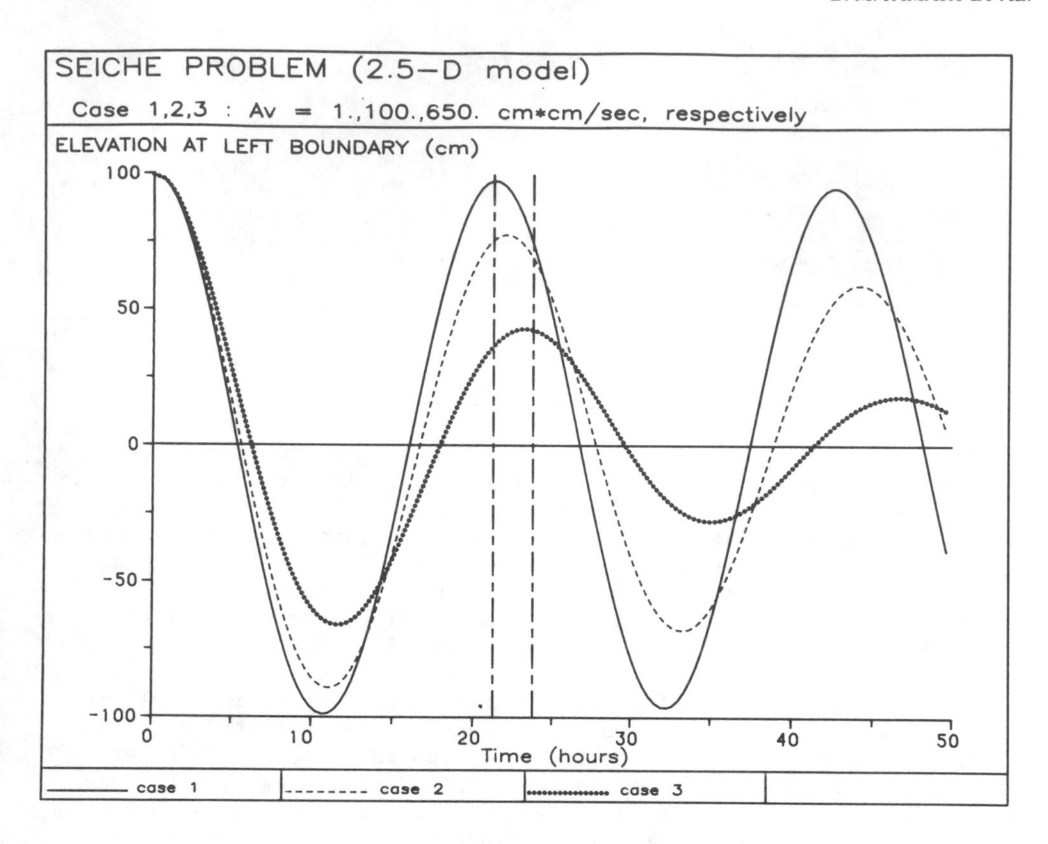

Figure 5. Time evolution of the elevation of the free surface at one end of a seiching basin, as calculated with the 2.5-D model for different values of the vertical eddy viscosity. The dashed vertical lines correspond to the end of the first mode oscillation i) in the absence of friction [T = 21.28 hrs] and ii) as predicted by the "semi-analytical" solution described in Section 4.2. for A_V = 650 cm^2/s [T' = 23.78 hrs, on the basis of equation (21)].

The explanation, as well as the implication, of this stroke of luck lies in the consideration of the bottom stress.

In a model that uses the no-slip condition at the bottom, the very concept of a bottom stress is not actually needed to set up the equations and their solution. The bottom stress is a quantity that can be computed, a posteriori, from the results of a simulation, provided a definition is agreed upon. The most obvious definition of the bottom stress, at least for a flat bottom, is that of equation (13). However, the numerical value obtained through this approach depends on the grid spacing and on the type of one-sided approximation used to evaluate the derivative.

An alternative way to calculate the bottom stress is to integrate numerically the discretized momentum equation and to define as "effective bottom stress" the sum of the terms that have no identifiable physical meaning. One of these terms is identical to (13); the others should be viewed as numerical errors inherent to the scheme. As pointed out by Jamart et al. [1982], such an exercise has to be done in a manner consistent with the discretization scheme and it requires that the momentum equation at the bottom itself be formally included in the integration (this equation is not used in the calculation of the solution but is useful to tidy up the numerically integrated momentum balance). In the limit $\Delta z \to 0$, these two definitions of the bottom stress converge towards the same value. Hence, we shall refer to definition (13) as the true bottom stress.

For the seiche problem under consideration, it can be shown that the "effective bottom stress" is made up of two main terms. The first term is the true bottom stress, and it cannot be correctly evaluated if Δz is too large : the reference velocity being too high above the bottom, the shear is underestimated and the stress has too small a phase lead with respect to the surface current. The second term, however, turns out to be proportional to $- \Delta z \frac{\partial \eta}{\partial x}$, and hence it acts as a depth-reducing factor and leads the surface current by 90 degrees. In several instances, even when the physically erroneous contribution was larger than the numerically wrong bottom stress, the sum of the two terms turned out almost identical to the true bottom stress calculated with a small Δz.

The implication of such a happening is that, using (13) as a definition, one could obtain a very poor estimate of the true bottom stress out of a model validated only on the basis of its ability to predict the motion of the free surface.

5. CONCLUSIONS

We have described three hydrodynamical models (2-D, 3-D, and 2.5-D) currently used in a model intercomparison experiment. Preliminary results for a simple test case (wind-driven circulation in an enclosed rectangular basin) show a similitude between the patterns of sea surface motion and mean currents calculated with the 2-D and 3-D models. Global energy balances in the 2-D and 3-D models yield differing pictures of the system response.

A difference is observed, between the 2-D and 3-D models, in the speed of propagation of free long gravity waves. A large fraction of that difference can be attributed to the choice of bottom boundary condition. In the 2-D model, the main effect of either a linear or quadratic stress law is to dampen the free oscillations without affecting their period to any significant extent. In the 3-D and 2.5-D models, on the other hand, if the bottom boundary layer created by the no-slip condition is resolved, the divergence of the bottom stress reduces the "effective depth" of the water column and thereby lengthen the periods of the free modes. A second contribution to that divergence attenuates the amplitude of the oscillations.

It is possible for some numerical models to yield basically correct free surface elevations while being grossly in error regarding the calculation of the true bottom stress.

6. AKNOWLEDGEMENTS

This work was supported by Det norske Veritas through the BSEX (Bottom Stress Experiment) agreement with the Belgian State. We thank Ms. A.F. Lucicki for her expert typing.

7. REFERENCES

Amsden, A.A., and F.H. Harlow, The SMAC method : A numerical technique for calculating incompressible fluid flows, *Report LA - 4370*, Los Alamos Scientific Laboratory, Los Alamos, N.M., 1970.

Heaps, N.S., On the numerical solution of the three-dimensional hydrodynamical equations for tides and storm surges, *Mémoires de la Société Royale des Sciences de Liège, Sixième Série, Tome II*, 143-180, 1971.

Jamart, B.M., R. Milliff, W. Lick, and J. Paul, Numerical studies of the wind-driven circulation in the Santa Barbara Channel, Final report to Exxon Production Research Company, 1982.

Paul, J.F., and W.J. Lick, A numerical model for three-dimensional variable density hydrodynamic flows, 150 pp., U.S. Environmental Protection Agency Report, 1981.

Renouard, D., Etude analytique et expérimentale des ondes internes engendrées par le vent dans un bassin rectangulaire tournant, Thèse de Doctorat, 145 pp., Institut National Polytechnique de Grenoble, 1980.

Ronday, F.C., Modèles hydrodynamiques, *Projet Mer, Rapport Final, vol. 3*, 270 pp., Services du Premier Ministre, Programmation de la Politique Scientifique, Bruxelles, Belgium, 1979.

Schlichting, H., *Boundary Layer Theory*, 647 pp., Mc Graw-Hill Book Company, New-York, 1960.

POSTFACE

ASI in Wonderland and Through the Drinking-Glass

One thing was certain; since June 83 the idea to have an ASI in Banyuls had ensnared three people. After NATO approbation in spring 84 and an inflation of the number of participants, the first thing to do was to assemble a triumvirate right in the place. This went on deciding important questions, as the attribution of the (in her girlhood) glamorous villa, and proceeding to an expertise of the local restaurants and wines. Once the course was chosen, the machine was put into gear with assurance and confidence.

Finally the grand moment came, and with it the puzzle of distributing each to his lodging in cadence with the times of delivery. Fortunately Madame Clara was on the bridge. And it was indeed an impressive-looking party that assembled in the conference-room the first day. After that things went like clockwork. In contrast to the maxim of an old and sagacious lady qualifying Banyuls "a paradise between two catastrophes", no extreme conditions, neither Tramontane of 80 mph, nor torrential rain, arose. In all respects the atmosphere and the spirit appeared favorable and the ASI sailed free. Along the way, two records fell: the greatest ratio of physicists to biologists in a biological marine laboratory, and the shortest possible visit in a French hospital.

Of course the balance between the two Banyulenc mottoes -- the one official, "in mare via tua", the other informal, "in vino veritas" -- was respected by following a rigorous schedule alternating the two traditional activities. Thanks to our Director the bowsprit did not get mixed with the rudder, and at the end of the game and of the banquet, the students gratefully and confidently put in his hands a splendid fishing rod.

Whether this ASI succeeded or vanished into the air, there is no way of guessing. However, if one makes a survey of the present volume for a minute or two, it can be not denied that the baby did not turn into a pig and that it is not absurd to carry it further.

B. Saint-Guily

J. J. O'Brien (ed.), Advanced Physical Oceanographic Numerical Modelling, 599.

SUBJECT INDEX